Chromatographic Methods

Fifth Edition

A. BRAITHWAITE
Department of Physical Sciences
Nottingham Trent University

and

F. J. SMITH
Department of Chemistry and Chemical Engineering
University of Paisley

KLUWER ACADEMIC PUBLISHERS
DORDRECHT / BOSTON / LONDON

Published by Kluwer Academic Publishers,
P.O. Box 17, 3300 AA Dordrecht, The Netherlands.

Sold and distributed in North, Central and South America
by Kluwer Academic Publishers,
101 Philip Drive, Norwell, MA 02061, U.S.A.

In all other countries, sold and distributed
by Kluwer Academic Publishers,
P.O. Box 322, 3300 AH Dordrecht, The Netherlands.

First edition 1963
Second edition 1967
Third edition 1974
Reprinted 1977
Fourth edition 1985
Fifth edition 1996
Reprinted 1999
Reprinted 2001

ISBN 0 7514 0158 7

A catalogue record for this book is available from the British Library

Library of Congress Catalog Card Number: 95-80922

Printed on acid-free paper

Preface

Since the inception of chromatography in 1903, the principal landmarks in its progress have been the virtual rediscovery of the technique in the 1930s, invention of synthetic resins in 1935, introduction of paper chromatography in the 1940s followed by that of thin layer, gas–solid and gas–liquid chromatography in the early 1950s. Whilst the theoretical aspects of HPLC were developed in the 1960s, it was the late 1970s before commercial instruments appeared. Developments through the 1980s in microelectronics and microprocessor technology afforded enhanced control, data acquisition and processing capabilities, and improved technologies for the manufacture of instrumentation. Developments in chromatographic media and packings and rapid scanning spectroscopic instruments have enabled combination techniques such as GC–MS, GC–IR, HPLC–MS and HPLC–IR to reach maturity and become standard routine techniques for the analyst. Further considerable research activity in the 1980s and early 1990s has led to the development of supercritical fluid chromatography (SFC), and high performance capillary electrophoresis is a technique that has proved invaluable in the genome project and the separation and typing of DNA fragments. Applications in environmental, health and safety, foods analysis and medical studies have contributed significantly to the advancement of these techniques. All of the instrumental chromatographic techniques are now used routinely by academic and industrial analysts. An understanding and experience of such techniques is fundamental to the training of today's science undergraduates studying a range of disciplines reflecting the application areas mentioned above.

A number of specialist books have been published over the years though many are concerned only with particular aspects of the subject whilst others are essentially literature surveys which, though comprehensive, are somewhat uncritical, lack background information and are rather formidable to someone seeking an introduction to chromatography. Thus the aim of the first edition published in 1963 'to present a unified account of the techniques in current use' is in many ways just as relevant today as it was then, although clearly a more extensive text is required to reflect the changes and advances described above.

The 5th edition of *Chromatographic Methods* reflects these many changes across all fields of chromatography. The principles of chromatography, however, remain the same although an understanding of the theoretical aspects and principles has become more important in order to appreciate the influence various parameters have on the separation process, particularly in high performance high resolution techniques such as capillary column GC, HPLC, HPIC and HPTLC. The 5th edition therefore includes an extended introduction to chromatographic techniques and a more in-depth discussion of the theory. Each subsequent chapter is updated to reflect the current status of the technique, particularly the instrumentation, detectors and column technology in for example, GC and HPLC. Hybrid techniques for GC and HPLC are now commonplace with bench-top instruments available. The chapter on spectroscopic techniques and chromatography looks at the principles and practice of these trends and discusses typical applications. Data systems have been a feature of the chromatography laboratory for many years; however, the power of modern PCs and workstations have been programmed for use as sophisticated integrators and to provide data processing and reporting and a gateway onto computer networks. Integrated laboratory data stations and information management systems are a familiar feature of QA and GLP laboratories. These and the many other aspects of the modern GC/HPLC laboratory are included in a much expanded chapter on data processing.

The 5th edition also includes an updated and expanded set of model experiments which reflect the current practice of chromatography. These are supported by a new chapter of a comprehensive set of problems with answers. A further new feature is a glossary of chromatographic terms and a list of symbols as an aide-mémoire and to support the preceding chapters.

The individual chapters have been written to be self contained so that readers may peruse particular topics but can pursue the background in more depth if they so wish. The comprehensive contents of *Chromatographic Methods* (5th edition) thus reflects the increased range of knowledge and expertise required of today's students and chromatographers.

Finally we would like to acknowledge the patience and support from our families and in particular our wives during the preparation of this text.

A. Braithwaite
F.J. Smith
November 1995

Contents

1 Introduction

1.1 Introduction to chromatography

Techniques related to chromatography have been used for centuries to separate materials such as dyes extracted from plants. However, it was the Russian chemist and botanist Michael Tswett (Mikhail Semenovich Tsvett, 1872–1919) who in 1906 first used the term *chromatography* (colour-writing, derived from the Greek for colour—chroma, and write—graphein) to describe his work on the separation of coloured plant pigments into bands on a column of chalk and other materials such as the polysaccharides, sucrose and indulin [1–4]. He had previously, on 21 March 1903, presented a related paper at the Warsaw Society of Natural Sciences on 'A New Category of Adsorption Phenomena and Their Application to Biochemical Analysis'. During his work, Tswett observed the effects on the separations of using different materials in the column and material of different particle size. The significance of his work was not realised until the 1930s when Lederer and others described their work on separation of plant pigments, including carotenoids and xanthophylls, and Zechmeister published the first book on chromatography in 1937 [5, 6]. Paper and thin layer chromatography developed rapidly during this period with many applications being published. The 1940s saw the development of ion exchange partition and column chromatography and the initial studies on gas adsorption chromatography. This resulted in gas adsorption and gas–liquid partition chromatography being developed in the 1950s. The 1960s saw the rapid rise in the use of chromatography until it is now an accepted routine technique, particularly in chemistry, biology, medical, pharmaceutical and environmental studies and in quality control. It is also used on a preparative scale in production processes and yet is sensitive enough for trace analyses particularly in monitoring environmental pollutants and metabolic studies. During the development years three chromatographers received Nobel Prizes, Tiselius (Sweden) for his work on 'Electrophoresis and Adsorption Analysis' and Martin and Synge (UK) for the 'Invention of Partition Chromatography'. Developments in engineering techniques, microelectronics and microcomputers and new materials, particularly over the last decade, have enabled manufacturers to produce reliable complex automated instruments that achieve reproducible chromatography. Typically, these incorporate sophisticated programmable features to set-up and control the instruments and collect and process data, producing chromatograms and analytical reports.

1.1.1 Definition of chromatography

In his early papers Tswett (1906) stated that "Chromatography is a method in which the components of a mixture are separated on an adsorbent column in a flowing system". Chromatography has progressed considerably from Tswett's time and now includes a number of variations on the basic separation process. Chromatography therefore encompasses a wide range of techniques many of which have their own terminology. Standards organisations in some countries such as the British Standards Institute (BSI) in the UK and the American Society for Testing Materials (ASTM) have produced their own definitions. In order to bring some order to the language of chromatography IUPAC, the International Union of Pure and Applied Chemistry, published in 1993 their updated nomenclature and definition for chromatography, 'Unified Nomenclature for Chromatography' [7]. The document is a result of 10 years' deliberations to produce a set of terms that were relevant to all branches of chromatography and yet were not too broad as to be almost meaningless and could be applicable, for example, to a 'moving phase' such as the wind in a tunnel or water in a river.

1.1.2 IUPAC definition of chromatography (1993)

> Chromatography is a physical method of separation in which the components to be separated are distributed between two phases, one of which is stationary while the other moves in a definite direction.

A glossary of chromatographic terms and a table of chromatographic symbols are included in the appendices at the end of this book. Further information on the terminology of chromatography is available in the literature [8].

1.2 History of chromatography

Chromatographic-like separations are literally as old as the hills. Naturally occurring phenomena such as the migration of gases through the earth's crust and soils, and percolation of water through rocks, clays and soils result in more rapid movement (separation) and concentration of some substances more than others. Through the ages man has copied these processes to help separate dyes from roots and leaves, process foods and extract metals.

However, the scientific value of these separations was not recognised until the industrial revolution of the nineteenth century. In 1850, Runge [9], working on the production of dyes and bleaches from coal tars, developed a crude form of paper chromatography for examining the mixtures of dyes. He spotted the mixtures onto a special paper and obtained colour separations which enabled him to characterise the dye mixtures. In 1861, Groppelsroeder

[10] used a form of paper chromatography, which he called 'capillary analysis', to separate coloured pigments. He achieved this using strips of paper with one end dipping into an aqueous solution and although he noted that capillary action carried the components up the paper he was unable to explain why separation occurred. It was a further 75 years before column, paper and thin layer chromatography were further refined and recognised by chemists and biochemists as a valuable method for obtaining information on complex mixtures [11]. Tswett's early papers are regarded as the first to describe the separation process. He accumulated information from over a hundred experiments using different adsorption media (stationary phase) in his columns to separate plant extracts and ligroin solutions. A comprehensive account of this work is given in Tswett's own book which was unfortunately only published in Russian [4].

The next major developments occurred in 1931 when Lederer and co-workers [5, 12, 13] separated lutein and zeaxanthine in carbon disulphide and the xanthophylls from egg yolk on a column of calcium carbonate powder 7 cm in diameter. The technique rapidly gained interest and Khun, Karrier and Ruzicka were each awarded the Nobel Prize (1937, 1938, 1939, respectively) for their work on chromatography. 'Flow through chromatography' rapidly gained acceptance and by the 1940s liquid adsorption column chromatography was an established laboratory separation technique on both analytical and preparative scales.

Classification of chromatography into three groups was proposed by Tiselius in 1940 [14] and Claesson in 1946 [15]. They had studied the properties of solutions during the chromatographic process and concluded that three forms of chromatography were occurring, differing only in the principle of the separation, namely frontal analysis, displacement chromatography and elution chromatography. Gradient elution, where the composition of the liquid moving phase (mobile phase) is varied during the separation process was introduced later in the early 1950s. Tiselius was awarded the Nobel Prize in 1948 for his contribution to chromatography.

The next important milestone in chromatography was made at about the same time by biochemists Martin and Synge who were developing a separation procedure for the isolation of acylated amino acids from protein hydrolysates by extraction of the aqueous phase with a chloroform organic phase [16, 17]. A series of 40 extraction funnels were used to effect a separation, the acetylated amino acids being separated in the sequence of funnels according to their distribution ratio and partition coefficients in the water–chloroform solvent mixture. The process was a tedious step-wise separation and illustrates the counter-current approach to explaining the theory of chromatographic separations, the Craig counter-current distribution theory [18, 19]. They soon replaced this with a chromatographic column containing silica gel particles with water retained on the silica gel, and the chloroform flowing through the column. This system successfully separated the acetylated amino acids according to their partition coefficients and marked the beginning of partition chromatography. The silica gel was soon replaced by cellulose, removing the need to derivatise the amino acids. Martin and Synge were awarded the Nobel Prize in 1952 for this work.

Evidence of the importance of the technique was the method development work sponsored by the American Petroleum Institute (API) for the analysis of constituents in petroleum products; for example, ASTM D-1319 [20] details the procedure for the determination of saturates, alkenes and aromatics using a fluorescent indicator on silica gel adsorbent with isopropanol as eluant.

During the 1930s and 1940s chromatography progressed rapidly with several parallel developments of the earlier work which have resulted in the various chromatographic techniques we use today. The text *Chromatography* (5th edn) [21] provides a comprehensive account of the techniques and applications of chromatography today. A brief note on the historical developments of the main techniques is presented below.

1.2.1 Paper chromatography

Paper chromatography (PC) was one of the earliest recognised forms of chromatography and although used by Runge (1850) and Groppelsroeder (1861) it was the developments due to Martin and co-workers [22] that resulted in the form of the technique we know today. Their work on partition column chromatography (vide supra) required an adsorbent that would retain water more efficiently than silica gel. This led to the use of cellulose, and hence filter paper as the planar 'column'. They were able to separate successfully over 20 amino acids by a two-dimensional technique using ninhydrin to locate the spots. The simplicity of PC ensured its rapid acceptance and reference texts were soon produced detailing organic and inorganic applications, thus illustrating the importance of the new analytical technique [23–26]. Although still used as a screening technique it has been overtaken by thin layer chromatography with its higher separating efficiencies.

1.2.2 Thin layer chromatography

The earliest thin layer chromatography (TLC) separations were reported by the Dutch biologist Beyerinck in 1889 [27] and Wijsman in 1898 [28], who used gelatine layers to separate strong acids and enzymes in malt extract respectively. Wijsman incorporated fluorescent bacteria and starch into the gelatine layer, fluorescence only occurring where the enzymes reacted with the starch. TLC as we know it today originated from the work of Izmailov and Shraiber in 1938, who analysed pharmaceutical tinctures, including extracts of cinnamon, belladonna, mint, foxglove and lily of the valley, by spotting samples on to a thin layer of alumina adsorbent on a glass plate and applying spots of solvent to give circular chromatograms [29–31]. Later, in 1949, Meinhard and Hall [32] used a starch binder with a mixture of celite and alumina on microscope slides, still obtaining circular chromatograms. In 1951, Kirchner *et al.* used an ascending development method analogous to PC and similar to present day TLC techniques [33]. Kirchner's book on TLC is regarded as one of the main TLC texts [34]. It was, however, the work of Stahl [35], and the development of standardised commercially available adsorbents that provided the impetus for the widespread use of TLC, illustrated in his book, a reference text on TLC [36].

1.2.3 Ion exchange chromatography

Ion exchange chromatography (IEC) came into prominence as a distinct chromatographic technique during The Second World War (1939–1945) as a separation procedure for the rare earth and transuranium elements. The use of IEC was first reported by Taylor and Urey in 1938 [37] to separate lithium and potassium isotopes using zeolite resins, and by Samuelson in 1939 [38] who demonstrated the potential of synthetic resins as a separation medium. Rapid development of the technique took place during The Second World War in the rush to discover the transuranic elements. Seaborg, McMillan, Ableson and Kennedy were involved with the discovery and separation of neptunium and plutonium isotopes. A summary of these developments and work carried out with Thompson is described in Seaborg's book [39]. Other texts describe the steady development of IEC [40]. The application of high performance liquid chromatography (HPLC) techniques to IEC in the 1980s led to the development of high performance ion chromatography, IC or HPIC [41–43]. High efficiency ion exchange columns and sensitive conductivity detectors enable samples containing ppm levels of anions or cations to be separated in minutes. Much of the development of HPIC is synonymous with Dionex. The company developed improved micropellicular column packings (5–20 μm beads) which had a 2 μm surface coating of an ion exchange resin; the addition of a suppressor column between the analytical column and detector considerably enhanced the sensitivity of the technique. HPIC is routinely used for the analysis of, for example, anions and cations in potable, river and ground water.

1.2.4 Gel permeation chromatography (size exclusion)

Gel permeation chromatography (GPC), sometimes called gel filtration or size exclusion, uses material with a controlled pore size stationary phase. The discovery by Flodin and Porath in 1958 of a suitable cross-linked gel formed by the reaction of dextran with epichlorohydrin provided the breakthrough [44, 45]. Subsequently, the commercial development of dextran and similar hydrophilic gels (e.g. agar), ensured rapid acceptance and application of GPC. Development of polystyrene and similar soft polymeric hydrophobic gels, such as Sephadex, with their semirigid structure and wide range of pore sizes permitted organic solvents to be used [46, 47]. Analysis by GPC of polymeric materials has revolutionised molecular weight analysis and preparative separation of high-molecular-weight synthetic polymers.

1.2.5 Affinity chromatography

Affinity chromatography (AC) is a relatively recent development attributed to Porath *et al.* in 1967 [48], specifically for the analysis of biological samples. The stationary phase is a peptide or protein which has a specific binding affinity for a particular analyte. It is covalently bonded to a ligand such as a nucleic acid or an enzyme on an inert open matrix of cellulose or agar. Only analytes with specific affinity for the ligand will be retarded and separated. AC is used for a range of biochemical analyses, including the separation of protein molecules [49–51].

1.2.6 Gas chromatography

Gas chromatography (GC) is one of the most important analytical techniques and because it became commercially available almost a decade before HPLC it may be regarded as the forerunner of modern instrumental analysis and routine analysis. GC evolved from earlier work on the adsorption of gases on various materials which had been observed for many years, and the pioneering work of Martin on partition chromatography. Martin, with co-worker James, developed and refined this earlier work to develop gas–liquid chromatography (GLC), a technique that has revolutionised analytical chemistry. They used a gas (nitrogen) instead of a liquid mobile phase and a stearic acid stationary phase on a celite support to separate C2–C4 fatty acids [52]. Steady development of GC theory, columns and detectors followed until the introduction of capillary columns as an alternative to packed columns. Golay had proposed in 1958 that column efficiency could be improved by eliminating the column packing, that is, using open tubular columns which could also be considerably longer [53]. Much interest and rapid developments of instrumentation and applications followed, initially using metal and glass columns with a stationary phase coating on the inner wall [54, 55]. More robust and higher temperature columns were developed in the 1980s with the introduction of pure silica columns and stationary phases bonded to the inner wall of the column. Modified and new stationary phases are continually being introduced, however, developments over the past decade have been in refinement of the engineering of the instruments, use of microprocessor technology and the linking of GC to spectroscopic techniques, particularly mass spectrometry to assist in identification of analytes.

1.2.7 *Supercritical fluid chromatography*

Supercritical properties of gases have been studied for many years and used in industrial processes from the 1930s. Use as a mobile phase was reported in the 1960s, however, it was over 10 years before reliable systems and chromatographic applications were developed [56, 57]. Supercritical fluid chromatography (SFC) has features that are common to both GC and HPLC. A supercritical fluid such as carbon dioxide with its low viscosity and low diffusion coefficients is used as the mobile phase with bonded phase HPLC or capillary GC columns modified to take the high pressures and temperatures. A GC flame ionisation detector is used. Recent developments have focused on refining the instrumentation and developing column technology so that trace analyses can be carried out [58].

1.2.8 *High performance liquid chromatography*

Liquid column chromatography has been used extensively for low resolution separation of sample mixtures. van Deemter and colleagues had described column efficiency in terms of velocity of the mobile phase and diffusion properties of the mobile and stationary phases [59]. It was not until 1963 when Giddings pointed out that if high efficiencies achieved in GC were to be obtained in liquid chromatography then the particle size would need to be reduced by a factor of 100 to 2–20 μm in diameter [60, 61]. However, manufacturing particles of this size and overcoming the pressure drop across the columns were not solved for a number of years. Kirkland developed 40 μm pellicular beads with a 2 μm surface coating of stationary phase and although impressive separations for that time were obtained, the size of the solid core limited column efficiencies [62]. Efficient high pressure pumps and injection valves were developed, supported by ultraviolet detectors capable of operating with 5–10 μl sample cells. Column packing technology developed too and when production of silica gel particles of 10 μm was solved it was realised that stable reverse phase packings could be produced by bonding alkyl chains to the particles through siloxane bonds [63]. Development over the past years has focused on improvements to instrumentation by refinements to the engineering and control of pumps and valves for gradient elution. Interfacing to spectroscopic techniques such as mass spectrometry to aid identification of separated components has added to the scope of an already important technique. Recent developments in microbore HPLC using columns 2.1 mm i.d. and 3 μm packing have produced column efficiencies in the order of 5–10 000 plates in a 100 mm column [64].

1.2.9 *Capillary (zone) electrophoresis*

Capillary zone electrophoresis (CZE) is a relatively recent separation technique based on the differential migration rates of ionic species in an electrical field applied across a capillary column. Moving boundary electrophoresis was pioneered by Tiselius in 1937, subsequently the technique has been used on a laboratory scale for the separation of proteins, nucleotides and similar applications [65]. Separation in capillary electrophoresis (CE) depends on the electrical properties of solutes under the influence of an electrical field, not distribution between a mobile and stationary phase. However, CE employs GC-like silica capillary columns and an HPLC ultraviolet detector. Initial separations in open tubes were carried out by Virtanen in 1974 [66] using 500 μm i.d. glass tubes and Mikkers in 1979 [67] using 200 μm i.d. PTFE tubes who obtained plate heights of <10 μm. Microcapillary columns were introduced by Jorgenson in 1983 for the separation of peptides and proteins [68]. The introduction of commercial CE systems in 1989 enabled many laboratories to use the technique for a wider range of applications [69]. Development of the technique reflects the use of improved engineering methods and introduction of microelectronics so that current instruments utilise a potential of >10 000 V across a 10 cm, 50 μm i.d. column, achieving an efficiency of over 500 000 plates.

The theoretical aspects of chromatography were first studied by Wilson in 1940 [70], who discussed the quantitative aspects in terms of diffusion, rate of adsorption and isotherm non-linearity. The first comprehensive mathematical treatment describing column performance (using the height equivalent to a theoretical plate, HETP) in terms of stationary phase particle size and diffusion was presented by Glueckauf in 1949 [71]. However, it was van Deemter and co-workers in 1956 who developed the rate theory to describe the separation processes following on from earlier work (1952) of Lapidus and Admunson [59, 72]. Giddings first looked at the dynamic theory of chromatography in 1955 and from the 1960s onward has examined many aspects of GC and general chromatography theory [60, 61, 73]. It was from this basis that modern chromatography has developed. There has been continuous development in all branches of chromatography, particularly in materials and refinement of instrumentation which has resulted in the efficient, reliable and sensitive chromatographic methods in use today and which form the backbone of modern analytical procedures and routine laboratory analysis.

1.3 Classification of chromatographic methods

1.3.1 According to separation procedure

Chromatography encompasses a number of variations on the basic principle of the separation of components in a mixture achieved by a successive series of equilibrium stages. These equilibria depend on the partition or differential distribution of the individual components between two phases; a mobile phase

(MP), which moves over a stationary phase (SP). The SP is present as a film on the surface of small particles or the walls of capillary columns and therefore presents a large surface area to the mobile phase. The sample mixture is introduced into the MP and undergoes a series of partition or adsorption interactions at the MP–SP boundary as it moves through the chromatographic system. The differences in physical and chemical properties of the individual components determine their relative affinity for the SP and MP. Therefore, components will migrate through the system at differing rates depending on their retardation resulting from attraction onto the stationary phase. The least retarded component, having an equilibrium ratio which least favours the stationary phase, will be eluted first, i.e. moves fastest through the system. The most retarded component moves the slowest and is eluted last. A wide range of SP and MP can be used making it possible to separate components with only small differences in their physical and chemical properties.

The MP can be a liquid or a gas and the SP a liquid or a solid. Separation involving two immiscible liquid phases is referred to as partition or liquid–liquid chromatography (LLC), that is, a solute is partitioned between a liquid MP and liquid film SP. When physical surface forces control the retention properties of the component on a solid SP, liquid–solid (adsorption) chromatography (LSC) occurs, correspondingly when the mobile phase is a gas we have gas–liquid chromatography (GLC) and gas–solid chromatography (GSC), respectively. The classification of chromatographic methods is shown in Figure 1.1.

Chromatographic separations may be carried out with the SP located in a column or as a planar film. A simplified diagram of column and planar chromatography (TLC) is shown in Figure 1.2. This illustrates the separation of three components A, B, C with A being the least retained on the SP (A < B < C). A therefore moves fastest down the column and furthest up the TLC plate. Note that the sample is introduced as a narrow band or small spot and as the chromatography proceeds the bands and spots separate and become broader. The concentration profile of molecules distributed in a band or spot is represented by a Gaussian curve, the familiar peak displayed in a chromatogram.

1.3.2 According to development procedure

In 1940, Tiselius [14] classified chromatography according to the separation principle, namely, elution development, displacement development and frontal analysis. In practice only elution and to a lesser extent, displacement development are commonly used (Figure 1.3, p. 13).

1.3.2.1 Elution development. Elution development is the technique most widely used in the various methods of chromatography (GC, GLC, LLC, HPLC and LSC). Consider a small sample mixture introduced on to a

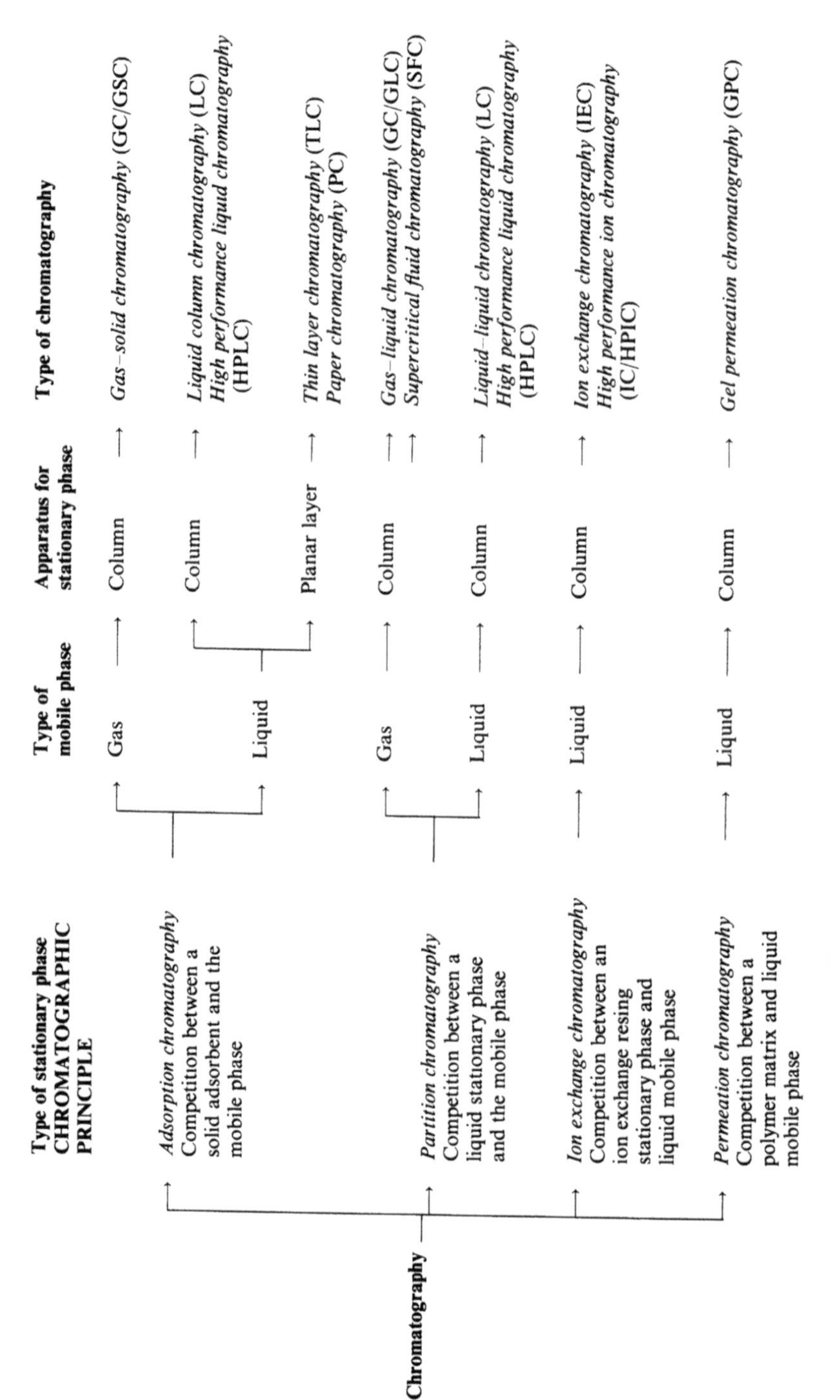

Figure 1.1 Classification of chromatographic methods.

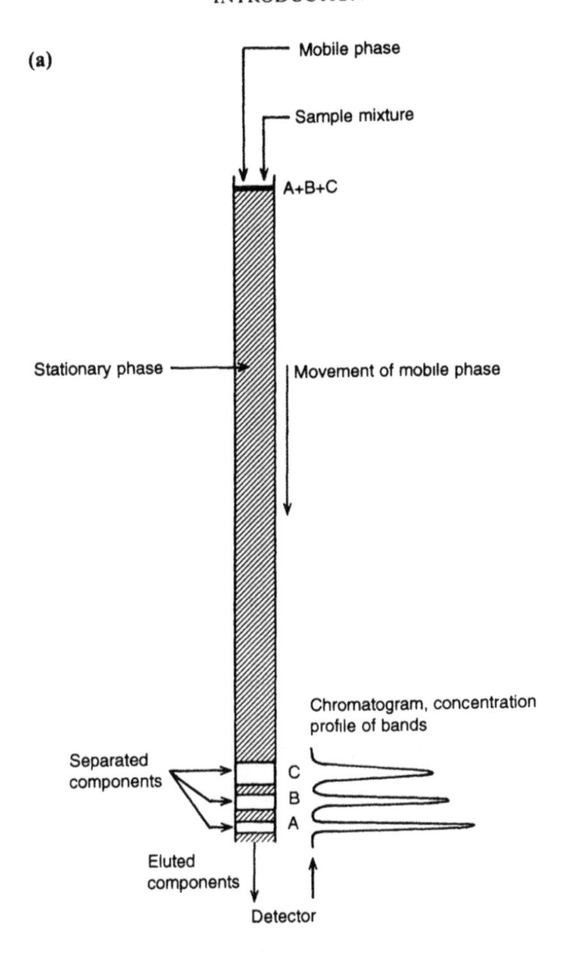

Figure 1.2 Separation using (a) column chromatography, and (b) planar chromatography (p. 12).

column which is eluted with an MP having a lower affinity for the SP than the sample components. The components therefore move along at a slower rate than the eluant, the rate being determined by the relative affinity of each component for the stationary phase with respect to the mobile phase, that is, the distribution ratio K:

$$K = \frac{C_{SP}}{C_{MP}}$$

where C_{SP} is the concentration of the component in the SP, and C_{MP} is the concentration of the component in the MP.

The components are eluted in order of their affinities but their relative rate of migration through the column is determined by the ternary interaction between components, SP and MP. Since the components can be completely

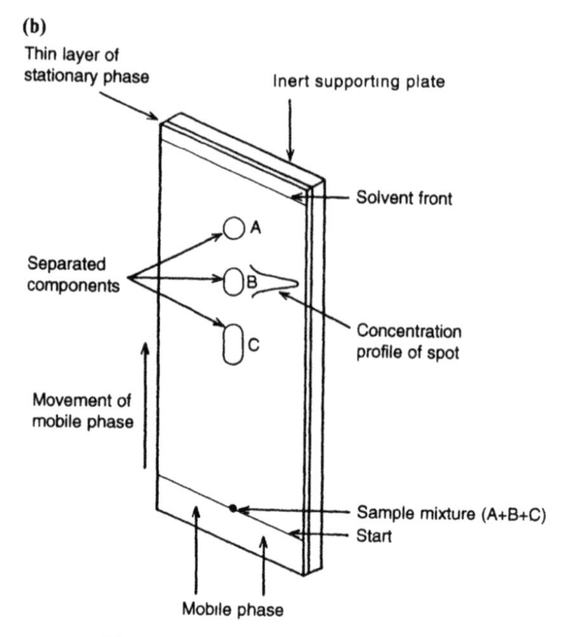

Figure 1.2 (b) Planar chromatography.

separated with a zone of MP between them, elution chromatography is used for analytical separations. In simple elution chromatography the column is eluted with the same MP solvent all the time. This is most suitable when the components have similar affinities for the SP and are therefore eluted rapidly, one after another. Similar principles apply for a gaseous MP except that the components are present in the vapour phase, retention being determined by their vapour pressure and attraction for the SP.

Stepwise elution is carried out by changing the eluant after a predetermined period of time. The eluants are chosen to have increasing eluting power, that is, increasing affinity of the MP for the components remaining on the column and therefore decreasing their affinity for the SP enabling them to move through the system faster. The value of K for each component decreases at each step.

Gradient elution uses a gradual change in composition of the eluting solvent to achieve separation of components of widely varying affinities for the SP. The ratio of two or more solvents is gradually changed to slowly increase the eluting power of the MP. Thus, the tailing part of a component band or peak emerging from a column is eluted by a solvent of slightly higher eluting power than the leading part. This eluant gradient narrows the bands or zones and reduces tailing. The value of K decreases for each component, therefore later eluting components move through the column at a progressively faster rate as the MP composition changes. The solvent composition

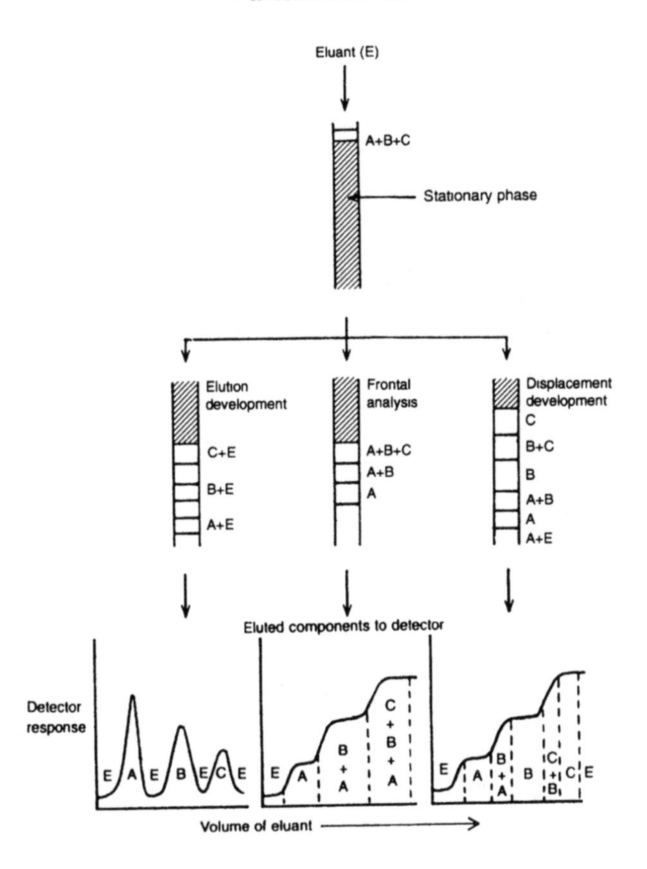

Figure 1.3 Classification of chromatographic methods according to development procedure, for components A, B, C and eluant E.

gradient may be linear, steadily increasing or decreasing, or logarithmic, and may be a concentration, pH, polarity or ionic strength gradient.

1.3.2.2 Displacement development. Displacement development consists of elution or development of the separation procedure by a solvent which has a greater affinity for the stationary phase than the sample components. The sample mixture is first introduced on to the end of the column and adheres to the SP. Elution occurs when a displacing solvent is passed through the column progressively displacing the components from the SP. The components separate due to their varying distribution ratios, K, and partition or adsorption properties, that is, their relative attraction for the SP with respect to the MP. Components with the least affinity for the stationary phase will be displaced first.

Generally, displacement development does not produce bands completely separated by eluant. Bands containing pure components are obtained, but between these bands are zones containing a mixture of the adjacent bands. Therefore, if preparative work is being carried out to obtain pure samples only the central parts of the bands are collected.

1.3.2.3 Frontal analysis. In frontal analysis the sample mixture is continuously added to the column. Initially, the component with the least affinity for the SP will pass along the column whilst a strongly adsorbed or attracted component builds up on the SP at the beginning of the column. However, there is a limit to the capacity of the SP and when this is exceeded this component also migrates along the column. Therefore, the first component is eluted from the column, initially in a pure form, then as a mixture with the next components to be eluted. This process can also occur with elution development if too much sample is placed on the column. Frontal analysis is clearly a preparative method primarily used for the separation of one readily eluted component from others with greater affinity for the SP.

References

[1] Tswett, M.S. *Ber. Deutsch Bot. Ges.*, **24** (1906) 316.
[2] Tswett, M.S. *Ber. Deutsch Bot. Ges.*, **24** (1906) 384.
[3] Sakodynskii, K.I. *J. Chromatogr.*, **73** (1972) 303.
[4] Tswett, M.S. *Khromophylii v Rastitelom I Zhivotom Mire* [Chromophylls in the Plant Kingdom]. Izd. Karbasnikov, Warsaw, 1910.
[5] Khun, R., Winterstein, A. and Lederer, E. *Z. Physiol. Chem.*, **197** (1931) 141, 158.
[6] L. Zechmeister and L. Cholnoty, *Die Chromatographische Adsorptions Method*, 2nd edn. Springer, Vienna, 1938; *Principles and Practice of Chromatography*, Wiley, New York, 1941 (translation).
[7] IUPAC (Analytical Chemistry Division Commission on Analytical Nomenclature). *Pure Appl. Chem.*, **64**(4) (1993) 819.
[8] Denney, R.C. *A Dictionary of Chromatography*. Wiley, New York, 1982.
[9] Runge, F.F. *Farbenchemie*, Vol. III. Berlin, 1850 in Cramer, E.J. *J High Res. Chrom. Comm.*, **2** (1979) 7.
[10] Goppelsroeder, F. *Verhandle. Naturforsch. Ges.* (Basel), **3** (1861) 3, 268.
[11] Liesegang, R.E. *Z. Anal. Chem.*, **126** (1943) 172.
[12] Ettre, L.S. and Zlatkis, A.J. *75 Years of Chromatography—A Historical Dialog*, J. Chromatogr. Lib. Vol. 17. Elsevier, Amsterdam, 1979.
[13] Ettre, L.S. *Anal. Chem.*, **43** (1971) 20A.
[14] Tiselius, A. *Arkiv. Kemi. Mineral Geol.*, **14B** (1940) 22.
[15] Claesson, S. *Kemi. Mineral Geol.*, **23A** (1946) 1.
[16] Martin, A.J.P. and Synge, R.L.M. *Biochem. J.*, **35** (1941) 91.
[17] Martin, A.J.P. and Synge, R.L.M. *Biochem. J.*, **37** (1943) proc. xiii.
[18] Craig, L.C. and Craig, D. *Techniques in Organic Chemistry*, Vol. III (Weissberger, A., ed.), p. 248. Interscience, New York, 1956.
[19] Craig, L.C. *J. Biol. Chem.*, **155** (1944) 519.
[20] ASTM *Hydrocarbon Types in Liquid Petroleum Products by Fluorescent Indicator Techniques*, ASTM D-1319-77. American Society for Testing Materials, Philadelphia, PA, 1954.
[21] Heftmann, E. *Chromatography*, 5th edn, Part A: *Fundamentals and Techniques*, Part B: *Applications*, J. Chromatogr. Lib. Vol. 51A and 51B. Elsevier, Amsterdam, 1992.
[22] Consden, R., Gordon, A.H. and Martin, A.J.P. *Biochem. J.*, **38** (1944) 38, 224.
[23] Lederer, M. *Anal. Chim. Acta*, **2** (1948) 261.

[24] Block, R.J., Strange, R. and Zweig, G. *Paper Chromatography—A Laboratory Manual.* Academic Press, New York, 1952.
[25] Pollard, F.H. and McOmie, J.F.W. *Chromatographic Methods for Inorganic Analysis with Special Reference to Paper Chromatography.* Butterworths, London, 1953.
[26] Sherma, J. and Zweig, G. *Paper Chromatography.* Academic Press, New York, 1971.
[27] Beyerinck, M.W. *Z. Phys. Chem.*, **3** (1889) 110.
[28] Wijsman, H.P., reported by van Klinkenberg, G.A. *Chem. Weekbl.*, **63** (1967) 66.
[29] Izmailov, N.A. and Shraiber, M.S. *Farm. Farmakol.*, **4** (1938) 8.
[30] Izmailov, N.A. and Shraiber, M.S. *Farmatsiya*, **3** (1938) 1; translated in *Advances in Chromatography*, Giddings, J.C. and Keller, R.A., p. 85, Dekker, New York, 1966.
[31] Ettre, L.S. and Zlatkis, A.J. *75 Years of Chromatography—A Historical Dialog*, J. Chromatogr. Lib. Vol. 17, p. 413. Elsevier, Amsterdam, 1979.
[32] Meinhard, J.E. and Hall, N.F. *Anal. Chem.*, **21** (1949) 185.
[33] Kirchner, J.G., Miller, J.M. and Keller, G.T. *Anal. Chem.*, **23** (1951) 420.
[34] Kirchner, J.G. *Thin Layer Chromatography.* Wiley, New York, 1978; reprinted by Sigma Chemical Company, St Louis, MO, 1990.
[35] Stahl, E. *Pharmazie*, **11** (1956) 633.
[36] Stahl, E. *Thin Layer Chromatography.* Academic Press, New York, 1962.
[37] Taylor, T.I. and Urey, H.C. *J. Chem. Phys.*, **6** (1938) 429.
[38] Samuelson, O. *Z. Anal. Chem.*, **116** (1939) 328.
[39] Seaborg, G.T. *Encyclopaedia of the Chemical Elements*, Reinhold, New York, 1968.
[40] Salmon, J.E. and Hale, D.K. *Ion Exchange—A Laboratory Manual.* Academic Press, New York, 1959.
[41] Gjerde, D.T. and Fritz, J.S. *Ion Chromatography.* Alfred Huthig, Heidelberg, 1987.
[42] Small, H., Stevens, T.S. and Bauman, W.C. *Anal. Chem.*, **47** (1975) 1801.
[43] Small, H. Modern Inorganic Chromatography, *Anal. Chem.*, **55** (1983) 235A.
[44] Flodin, P. and Porath, J. *Nature*, **183** (1959) 1657.
[45] Flodin, P. *Dextran Gels and their Application in Gel Filtration.* Pharmacia AB, Uppsala, 1962.
[46] Moore, J.C. *J. Polym. Sci.*, **835** (1964) A2.
[47] Bly, D.D., Kirkland, J.J. and Yau, W.W. *Modern Size Exclusion Liquid Chromatography, Gel Permeation and Gel Filtration Chromatography* Wiley, New York, 1979.
[48] Axen, R., Ernback, S. and Porath, J. *J. Nature*, **214** (1967) 1302.
[49] Frieburg, F. *Chromatogr. Rev.*, **14** (1971) 121.
[50] Turkova, J. *Affinity Chromatography*, J. Chromatogr. Lib. Vol. 12. John Wiley, New York, 1978.
[51] Mohr, P. and Pomerenig, K. *Affinity Chromatography—Practical and Theoretical Aspects*, Vol. 33. Dekker, New York, 1985.
[52] James, A.T. and Martin, A.J.P. *Biochem.*, **50** (1952) 679.
[53] Golay, M.J.E. *Theory and Practice of Gas–Liquid Partition Chromatography with Coated Capillaries, Gas Chromatography* (Coates, V.J. *et al.*, eds), p. 1. Academic Press, New York, 1958.
[54] Grob, K. *J. High Res. Chromatogr. Chromatogr. Comm.*, **2** (1979) 559.
[55] Jennings, W. *Gas Chromatography with Glass Capillary Columns.* Academic Press, New York, 1980.
[56] van Waisen, U., Swaid, I. and Scneider, G.M. *Angew. Chem.* (Int. Edn), **19** (1980) 575.
[57] Gere, D.G. *Science*, **222** (1983) 253.
[58] Smith, R.M. *Supercritical Fluid Chromatography.* Royal Society of Chemistry, London, 1988, reprinted 1993.
[59] van Deemter, J.J., Zuiderweg, F.J. and Klinkenberg, A. *Chem. Engng Sci.*, **5** (1956) 271.
[60] Giddings, J.C. *Anal. Chem.*, **35** (1963) 1999, 2215.
[61] Giddings, J.C. *Dynamics of Chromatography*, Part 1: *Principles and Theory.* Dekker, New York, 1965.
[62] Kirkland, J.J. *J. Chromatogr. Sci.*, **7** (1969) 7.
[63] Kirkland, J.J. *J. Chromatogr. Sci.*, **9** (1971) 206.
[64] Yang, F.J. *Chromatogr. Sci.*, **45** (1989) 1.
[65] Tiselius, A. *Trans. Faraday Soc.*, **33** (1937) 524.
[66] Virtanen, R. *Acta Polytech. Scand.* (1974) 123.

[67] Mikkers, F.E.P., Everaerts, F.M. and Verheggan, T.P. *J. Chromatogr.*, **169** (1979) 11.
[68] Jorgenson, J.W. and Lukacs, K.D. *Science*, **222** (1983) 266.
[69] Weinberger, R. *Practical Capillary Electrophoresis—An Integrated Approach.* Academic Press, 1993.
[70] Wilson, J.N. *J. Am. Chem. Soc.* (1940) 1583.
[71] Glueckauf, E. *Discuss. Faraday Soc.*, **7** (1949) 12.
[72] Lapidus, L. and Admunson, N. R. *J. Phys. Chem.*, **56** (1952) 984.
[73] Giddings, J.C. *Unified Separation Science.* Wiley-Interscience, New York, 1991.

2 Theoretical considerations

2.1 Introduction

The aim of this chapter is to present a detailed discussion of the principles of chromatographic separations and methods of assessing the quality of the chromatograms obtained. The reader is also referred to the Glossary at the end of this book for additional explanations of the terminology used in this chapter.

Chromatography is a separation technique where component molecules (solutes) in a sample mixture are transported by a mobile phase over a stationary phase. The mobile phase may be a gas or a liquid (solvent system) and the stationary phase may be a liquid film on the surface of an inert support material or a solid surface. The solute, mobile phase and stationary phase form a ternary system. Interaction occurs between the solute and stationary phase so that the solute is distributed between the stationary phase and mobile phase. Attraction of the solute for the stationary phase results in retardation of its movement through the chromatography system. Different components (solutes) will move at differing rates since each will have a slightly different affinity for the stationary phase with respect to the mobile phase. Each component or solute (A, B, C) is distributed between the two phases with an equilibrium established defined by the distribution ratio (previously known as the partition ratio); thus for component A

$$[A_S] \leftrightarrow [A_M]$$

where $[A_S]$ is the concentration of A in a unit volume of the stationary phase, and $[A_M]$ is the concentration of A in a unit volume of the mobile phase. The distribution ratio, K_A, for A, is therefore

$$K_A = \frac{[A_S]}{[A_M]}$$

Each component separated will have a different value for K, reflecting their relative affinities for the stationary phase; the generalised form of the distribution equation for each component is

$$K = \frac{C_S}{C_M}$$

where C_S is the concentration of a component in the stationary phase/unit volume, and C_M is the concentration of a component in the mobile phase/unit volume.

2.1.1 *Distribution ratio and separations*

A component in a sample mixture will be distributed between the mobile phase and stationary phase according to its distribution ratio, K, and will try to attain this ratio at all stages as it travels through the system. The larger the value of K the greater the affinity that component has for the stationary phase, that is, C_S is larger and C_M smaller.

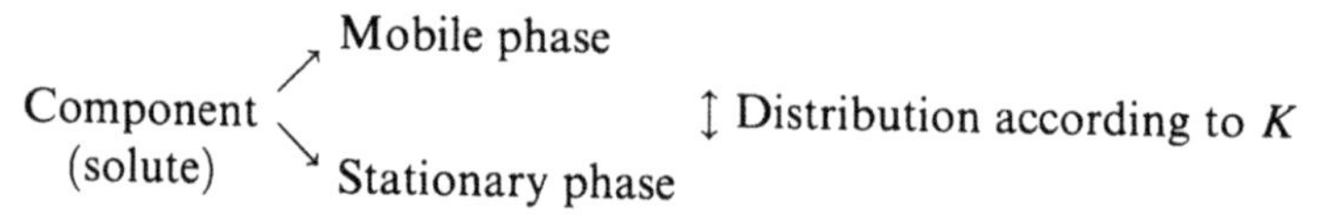

Although it is relatively straightforward to envisage the distribution of a solute between two phases in a simple extraction carried out in a separating funnel, the concept of a large number of sequential separation steps as in chromatography is more difficult to follow. Chromatography is a dynamic system with the separation process taking place continuously as the solute moves over the stationary phase trying to attain the equilibrium defined by the distribution ratio, K (Figure 2.1). The component can be considered to move through the column or plate in a series of theoretical separation steps with distribution between the two phases as

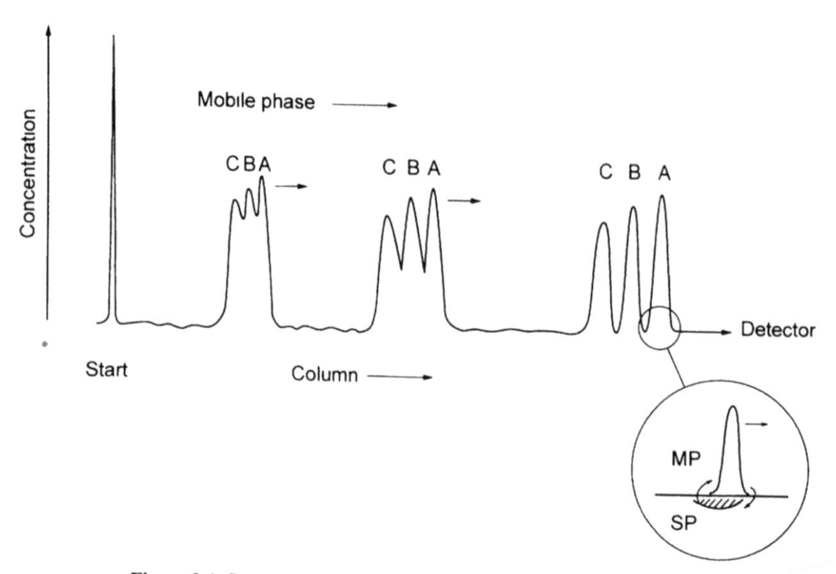

Figure 2.1 Separation of a three component mixture: A, B and C.

defined by K. The resulting time the solute molecules spend attracted to the stationary phase constitutes a delay with respect to the movement of the mobile phase. Consequently, a mixture of components would be delayed by varying amounts due to their differing retention by the stationary phase. They would be separated if sufficient equilibrium steps are available to magnify the small differences in their retention characteristics such that they arrive at the end of the column or plate one after another.

A solute is introduced onto the stationary phase as a narrow band (Figure 2.1). As this band moves through the chromatographic system, dispersion of the molecules occurs due to lateral diffusion and hence the band becomes broader. Too much dispersion can result in overlapping bands or peaks and incomplete separation. Therefore, in order to obtain efficient separations, the mobile phase, stationary phase and equilibrium conditions need to be carefully selected to achieve the desired separation as rapidly as is feasible but with minimum dispersion or band broadening. Figure 2.1 shows the various stages in the separation of a three component mixture A, B, C where A is the least retained component and C has the greatest affinity for the stationary phase

$$K_A < K_B < K_C$$

In planar chromatography the separation would be viewed as a series of spots (see Figure 1.2(b)). In instrumental methods such as gas chromatography (GC) and high performance liquid chromatography (HPLC) the composition of the mobile phase as it emerges from the system is continuously monitored by the detector, the signal produced being proportional to the amount of solute (component) present in the mobile phase, the eluant, and is plotted against time to form the chromatogram. Note that the concentration profile of an eluted band is a Gaussian peak representing the dispersion of the molecules during the separation. The longer a component takes to move through the system the greater the band broadening.

Figure 2.2 shows the chromatogram obtained for the separation of the three component mixture A, B, C and the symbol notation used to describe features of the peaks and chromatogram. The volume of mobile phase required to elute or transport a component through the system is the actual parameter to measure. However, this is not easily achieved so a constant mobile phase flow rate is used and the time a component takes to pass through the system, the retention time is measured. The magnitude of the detector signal reflects the amount of the component as it passes through the detector and the area under the peak is proportional to the total amount of that component in the eluted band. The detector signal is processed by an electronic integrator which plots the chromatogram and calculates retention time, peak area, peak height and peak width. A more detailed discussion of integration techniques and signal processing is presented in Chapter 8.

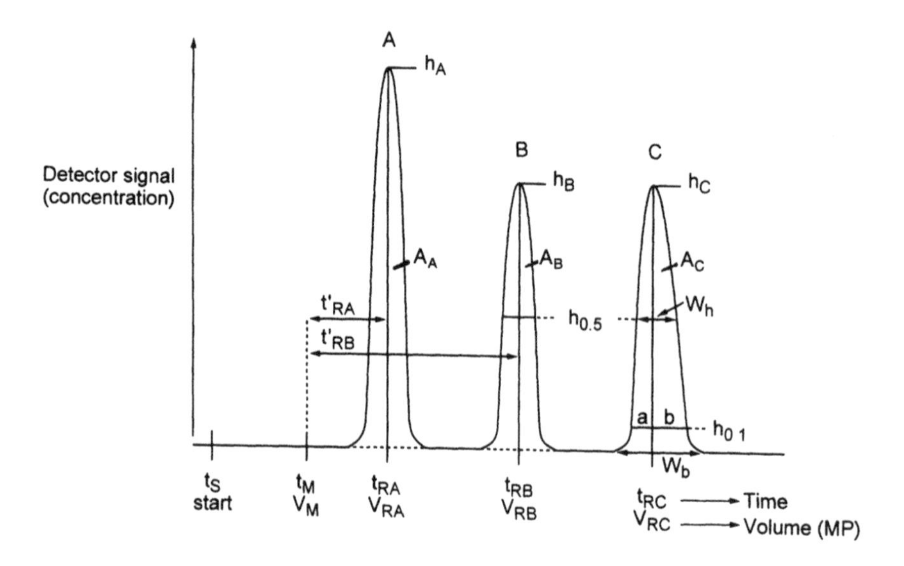

t_M dead time
V_M dead volume
t_{RA} retention time of A
t'_{RA} corrected retention time for A
V_A retention volume of A
V'_A corrected retention volume of A
t_{RB} retention time of B
t'_{RB} corrected retention time of B
V_B retention volume of B
V'_B corrected retention volume of B
t_{RC} retention time of C
t'_{RC} corrected retention time of C
V_C retention volume of C

A_A peak area for component A
A_B peak area for component B
A_C peak area for component C
h_A peak height for component A
h_B peak height for component B
h_C peak height for component C
$h_{0.5}$ peak half height
$h_{0.1}$ 10% peak height
w_b width at the base of a peak
w_h width of a peak at half height
F_C flow-rate of mobile phase
a, b the forward, rear part of a peak at 10% h

Figure 2.2 Data and symbols for chromatogram of A, B and C.

2.2 Factors influencing retention

The rate of migration of a component band through a column or thin layer chromatography (TLC) plate depends on the distribution of molecules between the mobile phase and stationary phase. The factors which influence the distribution and hence retention are

- composition and properties of the mobile phase;
- type and properties of the stationary phase;
- the intermolecular forces between the component(s) and stationary and mobile phases; and
- temperature.

The mobile phase in GC is often referred to as the carrier gas since its role is simply to transport the components through the column. The controlling factors in GC are choice of stationary phase and temperature of the column.

In liquid chromatography (LC, TLC and HPLC) the mobile phase and stationary phase both influence the distribution ratio and therefore LC separations can be influenced primarily by choice of stationary phase and modification of the mobile phase composition.

2.2.1 Coulomb's Law (like attracts like)

Intermolecular forces which influence retardation of components are based on Coulomb's Law—*like attracts like*—attraction occurs between molecules with similar electrostatic properties but molecules with dissimilar properties are repelled. Electrostatic interactions between molecules are of two main types:

- polar van der Waal's retention forces arising from interaction between molecules having a surface charge; and
- non-polar dispersion forces between neutral molecules or functional groups.

Ion chromatography separations are based on the strong attraction between ions of opposite charge and the exchange of ions between the analyte in the mobile phase and the stationary phase.

Hydrogen bonding between a component and the stationary phase (or mobile phase) is a relatively strong attractive force which in GC and HPLC can lead to slow equilibrium processes and tailing of peaks.

Components being separated by chromatography may have polar character, that is, the molecule contains polar functional groups or has non-polar alkane-like character. *Like attracts like*, therefore solutes with polar groups have a stronger interaction with a polar stationary phase than with a non-polar stationary phase and conversely non-polar groups are more attracted to a non-polar stationary phase than a polar stationary phase. It should be noted that many molecules can exhibit both polar and non-polar character, for example, *n*-butanol ($CH_3CH_2CH_2CH_2OH$) has a polar OH group and a non-polar C_4 alkane chain.

2.2.2 Polar retention forces

Polar van der Waal's retention forces are a consequence of dipole–dipole interactions and hydrogen bonding between molecules. Only components with dipoles similar to the 'solvent' (stationary phase) will disperse, producing solute–solvent pairs. Dipole induced dipole interactions arise from the charge on one molecule (component or stationary phase) disturbing the electrons in a second associated molecule, producing a shift in charge which then forms the induced dipole.

component molecule

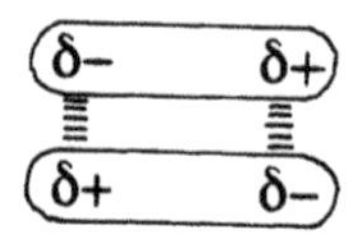

stationary phase

The average association energy (ϵ_D) is dependent on the respective dipoles (μ) of the two molecules (A ≡ analyte/component, S ≡ the stationary phase), the distance between centres (r) and temperature (T):

$$\epsilon_D = -\left(\frac{2\mu_A^2\mu_S^2}{r^6kT}\right)$$

The dipoles of a number of functional groups are shown in Table 2.1, giving an indication of the polar character of components (solutes) and stationary phases (solvents).

There is in practice about a 10-fold range in dipole moments which corresponds to a 10^4 range in dipole forces that gives rise to the selectivity of chromatographic separations. Non-polar groups will disperse in a non-polar solvent as described below. In addition to forming a single covalent bond hydrogen is able to form an associative or weaker bond with electron-rich molecules, hydrogen bonding, which may occur via inter- or intra-molecular association. The strength of the bond depends on stereochemistry, electronic effects of neighbouring atoms or groups and acid–base character. Hydrogen bonded solvents generally will attract polar solute molecules but will exhibit varying degrees of repulsion to non-polar molecules. Thus, in a chromatographic system the component or solute molecules will be attracted towards the phase of similar polarity.

2.2.3 Non-polar retention forces

Dispersion forces are the most universal intermolecular forces and therefore in non-polar solvents or stationary phase London's dispersion forces are

Table 2.1 Dipole moments of some organic groups, debye units

R–CH=CH_2	Alkene	0.4
R–O–Me	Methyl ether	1.3
R–NH	Amine	1.4
R–OH	Alkanol	1.7
R–COOH	Carboxylic acid	1.7
R–Cl	Chloride	1.8
R–COOMe	Methyl ester	1.9
R–CHO	Aldehyde	2.5
R–CO–R	Ketone	2.7
R–CN	Nitrile	3.6

the main interactions between molecules [1]. The association is formed as a result of induced dipoles formed by molecular electrons and nuclei interacting on the polarisable electronic systems of other molecules to induce coherent dipoles. Dispersion forces are relatively weak and therefore similar non-polar molecules are not repulsed, the net effect being mixing of molecules. The 'London forces' (ϵ_L) are dependent on the ionising potentials (I), polarisability (α) and distance between centres (r) for a pair of molecules (A $\equiv$ analyte/component, S $\equiv$ stationary phase):

$$\epsilon_L = \frac{3\alpha_A \alpha_P I_A I_S}{2r^6(I_A + I_S)}$$

Thus, dispersion forces decrease very rapidly with increase in distance from the interacting centres. When extended to large complex organic molecules encountered in chromatography the dispersion forces must occur mainly at the surfaces of the interacting molecules.

2.3 Retention and equilibrium in chromatography

An optimised chromatographic separation is achieved by varying the mobile and stationary phase properties and operating parameters to give the required retention of the components in a sample. The overall retention characteristics for each component are related to the kinetics and mass transfer processes, leading to retention forces.

2.3.1 *Retention time, retention volume and retardation*

The volume of mobile phase required to carry a band of component molecules through the system to the detector is termed the retention volume, V_R, and is measured from the start of the chromatography to the peak maxima (Figure 2.2). However, as stated earlier it is difficult to accurately measure volume flow rates of the mobile phase in column chromatography systems such as GC and HPLC. Therefore, a constant flow rate of mobile phase is maintained and the time taken by a component band to pass through the column is recorded as the retention time, t_R. If the flow rate of the mobile phase through the column is F_C then retention volume is calculated from

$$V_R = t_R F_C$$

F_C is calculated from the internal cross-section of a column, the average linear velocity of the mobile phase $\bar{u}$, and where appropriate, a term, ϵ_T, to account for the porosity of the particles in packed columns. ϵ_T is approximately 0.4 for solid particles, 0.8 for porous packings and 1.0 for columns without a packing such as GC capillary columns.

$$F_C = \epsilon_T \bar{u} \left(\frac{\pi d_C^2}{4} \right)$$

where d_C is the internal diameter of the column.

The average linear velocity of the mobile phase $\bar{u}$ is calculated from the length of the column, L, and the time the mobile phase takes to pass through the column, t_M:

$$\bar{u} = \frac{L}{t_M}$$

t_M is conveniently obtained by measuring the time an unretained component takes to go through the column. For example, in GC methane may be used and in HPLC a mobile phase of slightly different composition can be injected.

Retention volume and retention time are measured from the time the sample is introduced into the chromatograph to when the component(s) are eluted from the column; no allowance is made for the volume of mobile phase in the system nor the time the mobile phase takes to pass from the injector to the detector. A correction therefore has to be made to obtain a more accurate representation of the retention of a component by the stationary phase. V'_R, the corrected retention volume, accounts for the volume of mobile phase in the system, V_M, and t'_R, the corrected retention time, takes into account the time the mobile phase takes to pass through the system. V_M and t_M are often referred to as the dead volume and dead time, respectively (Figure 2.2).

$$V'_R = V_R - V_M$$

$$t'_R = t_R - t_M$$

In GC the retention volume must be corrected for the compressibility of the gaseous mobile phase due to the pressure differential along the column. A compression correction factor (j) is therefore used giving the net retention volume, V_N, and net retention time, t_N, for a given temperature and initial pressure:

$$V_N = jV'_R$$

$$t_N = \frac{V_N}{F_C}$$

In liquid chromatography the compressibility of the mobile phase is negligible so a correction factor is not required, therefore $V'_R = V_N$ and $t'_R = t_N$.

Retention volume and hence retention time are determined by the distribution ratio, K:

$$K = \frac{C_S}{C_M}$$

The greater the proportion of component or analyte in the mobile phase (C_M), the faster it will progress through the system; conversely the lower the proportion, the greater the retardation and so retention time will increase. Retardation, R_F, is therefore a function of the fraction of component in the mobile phase:

$$R_F = \frac{\text{amount of component/unit volume in the mobile phase}}{\text{total amount of component in the mobile phase} + \text{stationary phase}}$$

$$R_F = \frac{V_M C_M}{V_M C_M + V_S C_S}$$

$$\text{but} \quad K = \frac{C_S}{C_M} \qquad C_S = KC_M$$

$$\text{therefore} \quad R_F = \frac{V_M C_M}{V_M C_M + KV_S C_M}$$

$$R_F = \frac{V_M}{V_M + KV_S} \quad \text{for unit volume of the system}$$

R_F is <1, if $R_F = 1$ the component is not retained by the stationary phase and if $R_F = 0$ it is completely retained and does not migrate along the column or plate. The actual volume, V_R, required to elute the component, is a function of the retardation and dead volume of the system, V_M, therefore

$$V_R = \frac{V_M}{R_F} = \frac{V_M}{(V_M/(V_M + KV_S))}$$

$$V_R = V_M + KV_S$$

$$\text{and} \quad V'_R = V_R - V_M = KV_S$$

In adsorption chromatography V_S should be replaced by the surface area of the adsorbent. V_M may be obtained by measuring the dead time from the chromatogram and noting the flow rate of the mobile phase as discussed above, $V_M = t_M F_C$.

2.3.2 *Retention factor*

The retention factor (capacity factor) k is a more practical quantity than K for describing the retention characteristics of components and is the ratio of the number of molecules in the stationary to the number in the mobile phase/unit volume.

$$k = \frac{V_S C_S}{V_M C_M} \quad \text{but} \quad K = \frac{C_S}{C_M} \quad \text{therefore} \quad k = K\frac{V_S}{V_M}$$

$$\text{and } k = \frac{K}{\beta} \text{ for wall coated open tubular (WCOT) GC columns}$$

The ratio V_M/V_S is referred to as the phase ratio β in open tubular column GC and describes the retention characteristics of the column where $\beta = V_G/V_S$ (V_G = volume of carrier gas).

k describes the ability of a stationary phase to retain components and is a measure of the actual retention properties, the adjusted retention time with respect to the time the mobile phase takes to pass through the system:

$$k = \frac{t'_R}{t_M} = \frac{t_R - t_M}{t_M}$$

Retardation, R_F, can also be described in terms of the retention factor:

$$R_F = \frac{V_M}{V_M + KV_S} \quad \text{and from above} \quad K = k\frac{V_M}{V_S}$$

$$\text{therefore} \quad R_F = \frac{V_M}{V_M + V_S\left(k\dfrac{V_M}{V_S}\right)} \quad \text{and} \quad R_F = \frac{1}{1+k}$$

2.3.3 *Separation factor*

The ability of a stationary phase to separate two components A and B, where B is the more strongly retained component, is determined by their relative partition or distribution ratios and hence their retention factors for a given stationary phase. The separation factor (α) is therefore a function of the relative retention of each component by a stationary phase.

$$\alpha = \frac{k_B}{k_A}$$

Therefore

$$\alpha = \frac{t'_{RB}}{t'_{RA}} = \frac{V'_{RB}}{V'_{RA}}$$

The separation factor for an adjacent pair of bands is a function of the type stationary phase used, the mobile phase and column temperature, and should be optimised for the most difficult to separate early eluting pair of compounds. For separation to occur α should be greater than 1.0, however in high performance chromatography such as capillary column GC and HPLC, α is almost 1.0 for closely eluting compounds having 'narrow' peaks.

2.4 Separating efficiency of a column

Figure 2.1 shows the progress of a chromatographic separation. If the chromatography were ideal little dispersion would occur, the three components would separate and elute from the column, each having a similar narrow peak profile to the start of the column. However, some of the molecules of say component A elute before the main concentration at t_{RA} whilst others take longer. The resulting concentration profile of the band is approximately that of a bell-shaped Gaussian curve. Sufficient separation has to be achieved to resolve the total dispersion of each band into fully resolved peaks. Resolution is therefore a function of the number of separation steps in a column or TLC plate and the dispersion of the molecules during the separation process.

2.4.1 *Peak width*

The peak recorded in a chromatogram represents the distribution of molecules in a band as it elutes from the column, the overall broadness being conveniently measured in terms of the width of the peak. A number of independent factors such as sample-injector and detector characteristics, temperature and column retention processes, contribute to the dispersion of molecules in a band and band broadening. The cumulative effect of small variations in these factors is described in statistical terms as the variance, σ^2, in the elution process. Classical chromatography theory considers that the separation process takes place by a succession of equilibrium steps, the more steps in a column the greater the column efficiency with less band broadening (variance) occurring, therefore

$$\sigma^2 \propto \frac{1}{N}$$

$$\text{and} \quad \sigma \propto \frac{1}{\sqrt{N}} \quad \text{or} \quad \sqrt{N} \propto \frac{1}{\sigma}$$

where σ, the standard deviation of the Gaussian peak, describes the spread of the molecules in a band. Band broadening is also a function of time, the longer the band takes to elute the more time the molecules have to spread out, therefore

$$\sigma \propto t_R \times \frac{1}{\sqrt{N}} \quad \text{that is} \quad \sqrt{N} \propto \frac{t_R}{\sigma}$$

In practice the proportionality constant is 1, therefore

$$N = \left(\frac{t_R}{\sigma}\right)^2$$

95.5% of molecules in a band are contained within $\pm 2\sigma$ of the mean, t_R. This corresponds to the base width of the peak, w_b, measured at the intersection of the lines drawn at a tangent to the infection points (at $\pm\sigma$) of the Gaussian curve and the baseline (Figure 2.3).

$$w_b = 4\sigma \quad \text{and} \quad \sigma = \frac{w_b}{4}$$

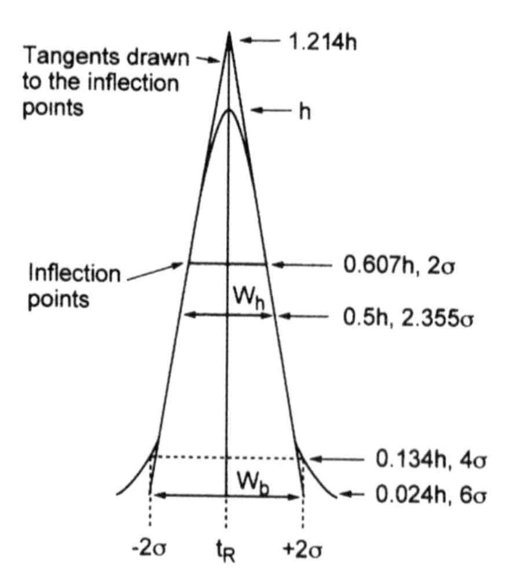

Figure 2.3 Peak heights of a Gaussian peak and width as a function of standard deviation.

2.4.2 *Column efficiency*

N, the theoretical number of separation steps (plates) in a column determines its separating capabilities; N is therefore an indication of column efficiency.

$$N = \left(\frac{t_R}{\sigma}\right)^2 \quad \text{therefore} \quad N = \left(\frac{t_R}{w_b/4}\right)^2$$

$$N = 16\left(\frac{t_R}{w_b}\right)^2$$

It is often more practicable to measure the peak width at half height, w_h; this corresponds to 2.35σ, therefore

$$N = \left(\frac{t_R}{w_h/2.35}\right)^2$$

$$N = 5.54\left(\frac{t_R}{w_h}\right)^2$$

N is the column efficiency observed in practice, however, the adjusted retention time, t'_R, reflects the effective retention and separation characteristics of the column. The effective column efficiency, N_{eff}, is therefore

$$N_{eff} = 16\left(\frac{t'_R}{w_b}\right)^2 = 5.54\left(\frac{t'_R}{w_h}\right)^2$$

Clearly, the number of theoretical plates in a column is directly proportional to the length of the column. If there are N equilibrium steps in a column of length L then the dimensions of a step, the height equivalent to a theoretical step or plate, H, is

$$H = \frac{L}{N} \quad \text{and} \quad \sqrt{N} \propto \frac{1}{\sigma^2}$$

$$\text{therefore} \quad H = \frac{\sigma^2}{L}$$

In 1952, Martin and Synge were the first to describe column efficiency in terms of H, an equilibrium step or HETP (height equivalent to a theoretical plate) and peak broadening or variance in the elution process [2, 3]. However, in 1956, van Deemter *et al.* realised that most band broadening occurs during the elution process in the column [4]. Column band broadening processes are the subject of the next section.

2.4.3 *Peak asymmetry*

The normal dispersion of component molecules as they move through the chromatographic system is represented by a bell-shaped Gaussian peak (Figure 2.3). However, if some molecules in the band are more strongly retained on the stationary phase due to high intermolecular forces, for example, hydrogen bonding, than predicted by the distribution ratio, K, then these molecules will lag behind the main band and will form a 'tail' on the main peak, peak tailing (Figure 2.4). Fronting occurs when some of the molecules move ahead of the main band due to less than expected retention by the stationary phase. This may be due to too large a sample being introduced onto the stationary phase such that the retention capacity is

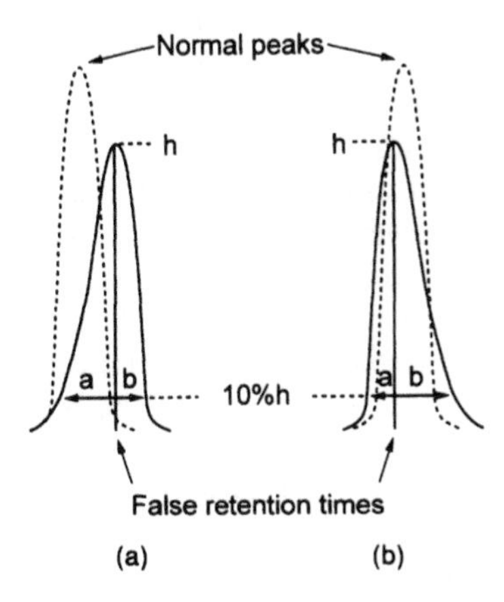

Figure 2.4 Peak asymmetry. (a) Fronting and (b) tailing.

exceeded. Ideally, a sample should be introduced onto a column such that it is distributed over the first one or two equilibrium steps. Sample capacity is defined as that amount (mg) of solute retained per g of stationary phase and will depend on the surface area and available volume of stationary phase. Overloading leads to broader peaks, a decrease in resolution and inaccurate retention times. Peak asymmetry (A_S) is measured at 10% peak height as the ratio of the forward part of a peak, a, to the rear part of the peak b and should be between 0.9 and 1.2 for acceptable chromatography (Figure 2.4). Ten per cent peak height corresponds approximately to $\pm 2\sigma$, at 13.4% $h(h_{0.134})$ (Figure 2.3).

$$A_S = \frac{b}{a} \quad \text{at } 10\%\ h$$

2.5 Band broadening processes

Band broadening during a chromatographic separation is a result of a number of random molecular processes. These may be broadly grouped into non-column and column effects.

Non-column band broadening is due to the dispersion of solute (component) molecules in the dead volume of the chromatographic system, that is, in the mobile phase in the injector and connections before the column and between the column and the active part of the detector. Dispersion results from mixing, caused by eddy or whirlpool effects and streaming. The effect is most significant in open tubular GC columns and HPLC columns with their low flow rates. Therefore, good design of injectors, detectors, tubing and fittings to minimise dead volume is essential for good resolution.

Column band broadening is a result of three molecular diffusion processes, described by the van Deemter terms, A, B, C, random movement through the stationary phase particles diffusion in the mobile phase, and interaction with the stationary phase. Giddings and co-workers proposed a generalised non-equilibrium theory for the band broadening process [5–7]. The theory, referred to as the 'random walk theory', assumes the progress of molecules through a column as a succession of steps with random dispersion processes occurring at each step. However, the mobile phase is continuously transporting the component molecules through the system. Thus, at a given equilibrium step the molecules try to attain the equilibrium defined by the distribution ratio, K, by association with the stationary phase, but in reality the true equilibrium will not be achieved. Molecules attracted onto the stationary phase will lag behind the central point of the moving band and the molecules remaining in the mobile phase will move more rapidly than the mean (Figure 2.5). Dispersion and hence band broadening increases with the number of transfer steps and decreases as the flow rate of the mobile phase decreases.

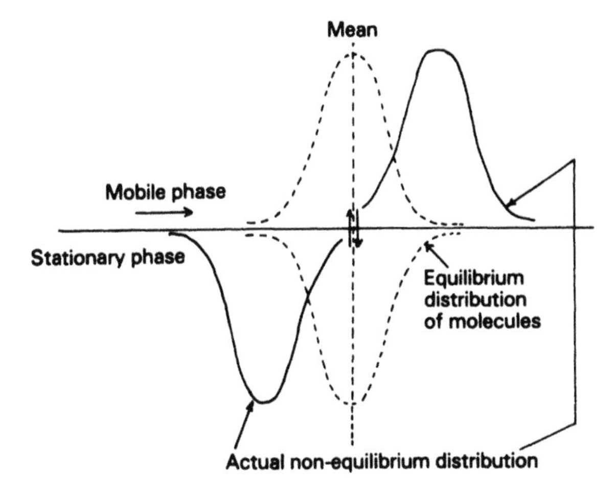

Figure 2.5 Equilibrium process during separation.

2.5.1 *Van Deemter model of band broadening*

Van Deemter and colleagues realised that there is an optimum velocity of the mobile phase at which band broadening would be a minimum and the separating capabilities of a column would be highest [4]. They identified three parameters, designated A, B, C, that contribute to band broadening based on the kinetic and thermodynamic processes that occur during elution. Section 2.4.1 explains that the separating capabilities of a column depends on the number of theoretical steps or plates, N, in a column. The maximum number of steps in a column will be achieved when the chromatographic parameters such as temperature, type of mobile phase and stationary phase and mobile phase velocity are optimised to give the minimum step or plate 'height'. Step height, H, is therefore used to express in simple terms the net effect of the band broadcasting processes in terms of the average velocity of the mobile phase, $\bar{u}$. The van Deemter equation, relating plate height to the average linear velocity and the band broadening parameters is

$$H = A + \frac{B}{\bar{u}} + C_S\bar{u} + C_M\bar{u}$$

Average linear velocity of the mobile phase is used rather than flow-rate since it is directly related to the speed of analysis, whereas flow-rate depends on the internal volume of the column. $\bar{u}$ is obtained from the column length, L, and the dead time, t_M:

$$\bar{u} = \frac{L}{t_M}$$

(a)

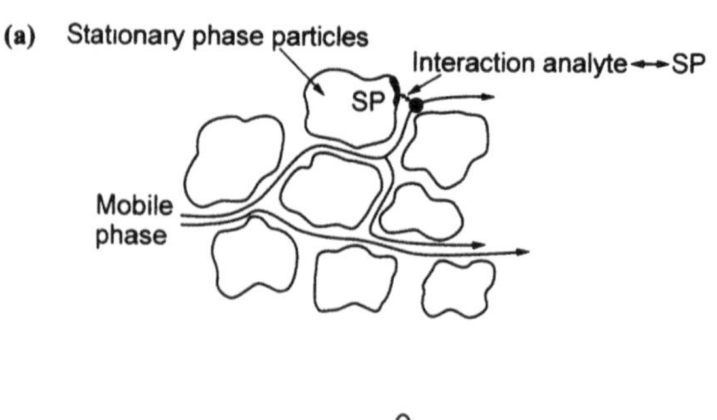

(b)

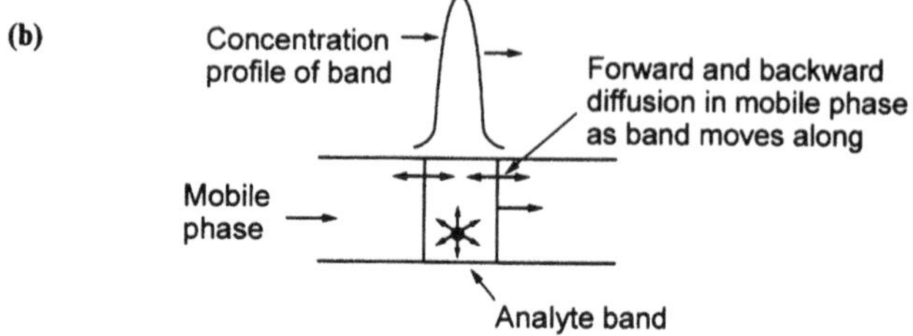

(c)

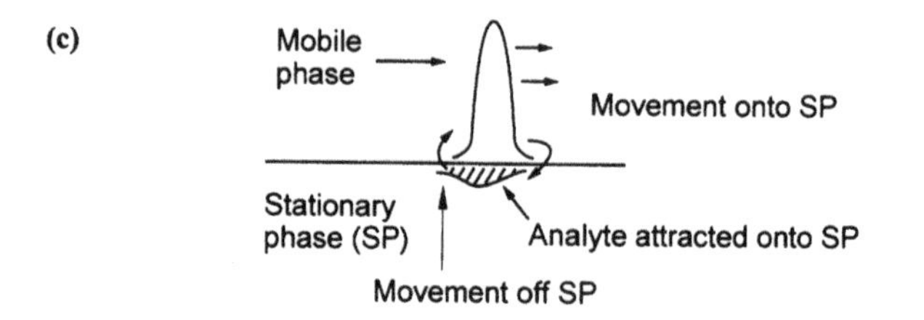

Figure 2.6 Terms of van Deemter equation. (a) unequal pathways; (b) molecular diffusion; (c) mass transfer.

2.5.2 A term ('eddy' diffusion and unequal pathways)

The A term describes eddy diffusion of component molecules and the variable pathways the mobile phase may follow through the stationary phase packing in the column. The component molecules may therefore travel different distances as they pass through a unit length of column (Figure 2.6(a)). The particles also cause eddys or turbulence in the mobile phase resulting in mixing and hence dispersion of the molecules. The A term is independent of mobile phase velocity, but is a function of the size of stationary phase particles and the way they are packed in the column or coated on a TLC plate. λ is a constant which takes into account the particle size range, packing uniformity and column dimensions and geometry.

$$A = \lambda d_P$$

where d_P is the mean diameter of the stationary phase particles.

Molecules finding relatively easy pathways including the more rapidly moving stream near the walls of the column, will move ahead of the main part of the band and elute first, whilst those following longer more erratic paths will take longer to move along the column or TLC plate and therefore lag behind.

Good separations and minimum band broadening will be achieved using small particles with a narrow size range that are packed uniformly into the column. Particles that are too small will lead to a high pressure drop across the column. In packed column GC optimum particle size is 80–100# mesh size (approximately 0.15 mm) and the flow-rate 30–50 ml min^{-1}. The A term is almost zero for WCOT GC columns. HPLC uses 5–10 μm particles and a high pressure drop across the column. Flow-rates in HPLC and WCOT columns are low (1–2 ml min^{-1}) and therefore the molecules in a band will have time to diffuse between mobile phase streams, thus reducing the band spreading effects. In 1983, Hawkes proposed a modified equation for band broadening which discarded the A term, diffusion being accounted for by the B term [8].

2.5.3 B *term (longitudinal diffusion)*

Molecules dispersed in a liquid or gaseous mobile phase will be in random motion and will diffuse in all directions independently of the direction of flow. Longitudinal diffusion is used to describe diffusion that takes place along the axis of the column and parallel to the movement of the mobile phase (Figure 2.6(b)). This forward and backward diffusion will therefore contribute to molecules moving ahead or lagging behind the main part of the band resulting in band broadening. Longitudinal diffusion is time dependent since the longer a band takes to elute the more time there will be for diffusion to take place; B is therefore inversely proportional to the velocity of the mobile phase, $\bar{u}$. The extent of diffusion will be hindered by the particles of column packing and the coefficient of diffusion of the component in the mobile phase, D_M

$$B = 2\gamma D_M$$

where γ is the hindrance factor dependent on the characteristics of the packing, γ is about 0.7 for packed columns and 1.0 for open tubular columns such as capillary GC columns.

The contribution of the B term to the van Deemter equation is most important at low mobile phase velocities. The B term is more important in GC than LC since diffusion coefficients of component molecules in gases are at least 10^4 greater than in liquids.

2.5.4 C terms (resistance to mass transfer)

Transfer of component molecules between the mobile and stationary phase is taking place continually during the elution process as the component molecules try to attain the equilibrium defined by the distribution ratio, K. At the leading edge of the band the component will be moving over 'fresh' stationary phase and will therefore move into the stationary phase (Figure 2.6(c)). At the trailing edge the converse will happen, the band will move on leaving some component on the stationary phase which subsequently moves off into the mobile phase. The C terms account for the finite time taken for the mass transfer processes to occur and are the most important band broadening process for both GC and HPLC. The C_S term relates to diffusion in the stationary phase and is directly related to the stationary phase film thickness, d_f. Thinner stationary phase films produce higher mass transfer rates, but there is an accompanying decrease in capacity of the column to retain a component. Although thicker films may have high retention characteristics and good separation factors this has to be equated against potentially greater band broadening.

$$C_S = \frac{d_f^2}{D_S}$$

where D_S is the diffusion of the solute in the stationary phase.

The C_M term describes diffusion of component molecules in the mobile phase as they move to the stationary phase (Figure 2.6(c)). Particles of column packing material offer a resistance to the mass transfer process. C_M is therefore a function of the diameter of the particles, d_P, and the diffusion coefficient of the component in the mobile phase, D_M:

$$C_M = \frac{d_P^2}{D_M}$$

In open tubular columns d_P is replaced by the internal diameter of the column, d_C. Diffusion rates in a gas are much higher than in a liquid, therefore in GC the mobile phase effects, C_M, are much smaller than the corresponding effects in the stationary phase, C_S. In HPLC, C_M and C_S are of comparable significance.

2.5.5 Optimum mobile phase velocity

A chromatography column is judged by its ability to separate complex mixtures using a given mobile phase. Highest column efficiencies, N, will be obtained when the equilibrium step or plate height is at a minimum, H_{MIN}, since $N = L/H$. This will be achieved when the velocity of the mobile

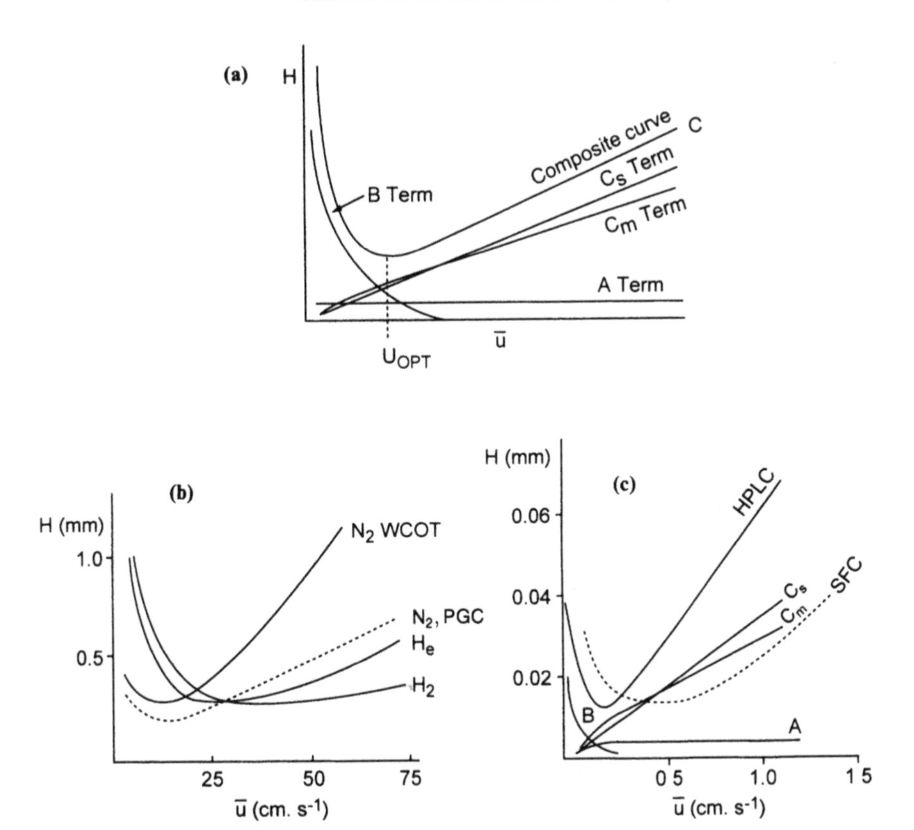

Figure 2.7 Van Deemter plots of plate height, H, against average linear velocity, $\bar{u}$, of the mobile phase: (a) the contribution of each term to the composite curve; (b) plots for W_{COT} GC columns using N_2, He, H_2 carrier gas and N_2 for packed column GC, PGC; (c) plots for HPLC and SFC composite curve.

phase is at an optimum, $\bar{u}_{OPT}$, and the band broadening processes described in the van Deemter equation are at a minimum (Figure 2.7). The A term is independent of $\bar{u}$, B is inversely proportional to $\bar{u}$ and C directly proportional to $\bar{u}$. Figure 2.7 shows the contributions of each term to the plate height at varying mobile phase velocities for GC and HPLC with plots for N_2, He, and H_2 carrier gases for capillary columns and packed columns. The advantages of helium over nitrogen as a GC carrier gas are also shown, the flat curve for He minimises the effects from slight variations in carrier gas flow rate. In practice a slightly higher velocity $\bar{u}_{PRAC}$ is used to ensure any slight variation does not drastically affect H and the column efficiency. A, B and C can be obtained by calculating H for three different settings of $\bar{u}$. H and the corresponding value of $\bar{u}$ are substituted in the van Deemter equation to give three simultaneous equations that can be solved for A, B and C.

$\bar{u}_{OPT}$ occurs at H_{MIN} and therefore $dH/d\bar{u} = 0$.

$$H = A + \frac{B}{\bar{u}} + C\bar{u} \quad \text{and} \quad \frac{dH}{d\bar{u}} = -\frac{B}{\bar{u}^2_{OPT}} + C = 0 \quad \text{therefore} \quad \bar{u}_{OPT} = \sqrt{\frac{B}{C}}$$

$$\text{substituting } \bar{u}_{OPT} \quad H_{MIN} = A + \frac{B}{\sqrt{B/C}} + C\sqrt{B/C}$$

$$H_{MIN} = A + 2\sqrt{BC}$$

GC capillary columns have a uniform geometry and in the absence of a stationary phase H_{MIN} is directly proportional to the column internal diameter d_C and retention characteristics, k, of the stationary phase film including polar character and film thickness [9–11]. Column design may be evaluated for efficiency and optimum mobile phase velocity using the following equation for H_{MIN}, proposed by Golay when he developed the theory for open tubular (WCOT) columns [9]:

$$H_{MIN} = \frac{d_C}{2}\sqrt{\frac{1 + 6k + 11k^2}{3(1 + k)^2}}$$

The equation is used to calculate H_{MIN} and coating efficiency, CE. The latter is used to indicate the separating efficiency of a column compared to the theoretical efficiency:

$$\text{CE\%} = \frac{H_{MIN}}{H_{PRACT}} \times 100$$

2.5.6 *Alternative models for band broadening*

In 1958, Golay first proposed GC separations using open tubular columns as a means of obtaining high column efficiencies [9]. His basic band broadening equation contained two terms: one to account for dispersion of the component band in the mobile phase resulting in longitudinal dispersion and the other for mass transfer processes. No unequal pathways term is required. The equation is equally applicable to HPLC. Although packed columns are used any unequal pathways effects are minimised due to the low flow rates and migration of the solute molecules between mobile phase streams. The reduced Golay equation is similar to the van Deemter equation:

$$H = \frac{B}{\bar{u}} + C\bar{u}$$

In an attempt to compare the band broadening characteristics and hence efficiencies of columns containing different packings of varying particle sizes, Giddings introduced the concept of reduced plate or step height, h_R, and reduced velocity, v, of the mobile phase [12]. Reduced plate height is the ratio of plate height to particle size and reduced velocity the ratio of mean velocity of the mobile phase with the velocity of the solute through the particle pores.

$$h_R = \frac{H}{d_P} \quad \text{and} \quad v = \frac{\bar{u}d_P}{D_M}$$

In the early 1970s, Knox and colleagues followed up Giddings work by studying the diffusion characteristics of a number of different packings and concluded that the A term is not independent of the mobile phase velocity [10, 11]. The Knox equation employing the reduced terms is

$$h_R = AV^{1/3} + \frac{B}{v} + Cv$$

2.6 Resolution

A good chromatographic separation is judged by the resolution between peaks, that is, adjacent peaks are resolved sufficiently so that accurate measurement of the peak areas can be obtained. There should therefore be a baseline separation between the peaks and no coelution or overlap of the tail of one peak with the leading edge of the next peak. Chromatographic peaks approximate to a Gaussian distribution so it is difficult to determine the minimum separation, in terms of retention times, achieving baseline resolution. Ideally, adjacent peaks (A and B) should be separated by 6σ ($3\sigma_A + 3\sigma_B$), that is, there would be $<0.3\%$ overlap since $\pm 3\sigma$ includes 99.7% of a Gaussian distribution. In practice, resolution (R_s) is assessed by comparing the difference in retention time for A and B with the half widths of the peaks since 95.5% of a Gaussian distribution is contained within $\pm 2\sigma$, therefore for a 2.3% overlap separation should be at least $2\sigma_A + 2\sigma_B$ or $\approx 4\sigma_B$, peak width of the broader peak (Figure 2.8).

$$R_s = \frac{t_{RB} - t_{RA}}{2\sigma_B + 2\sigma_A} \approx \frac{\Delta t}{4\sigma_B}$$

However, $2\sigma = \frac{1}{2}w_b = w_h$, therefore

$$R_s = \frac{\Delta t}{\frac{1}{2}(w_{bB} + w_{bA})} = \frac{2\Delta t}{(w_{bB} + w_{bA})} \approx \frac{2\Delta t}{2w_{bB}} = \frac{\Delta t}{w_{bB}}$$

$$\text{also} \quad R_s = \frac{\Delta t}{w_{hB} + w_{hA}} \approx \frac{\Delta t}{2w_{hB}}$$

Ideally, R_s should be between 1.2 and 1.5, corresponding to peak overlaps of approximately 1% ($\pm 2.5\sigma$) and 0.2% ($\pm 3\sigma$) (Figure 2.8). $R_s = 1.0$ corresponds to a peak overlap of $\sim$2.3% ($\pm 2\sigma$) and $R_s = 0.75$ to an overlap of 6.5% ($\pm 1.5\sigma$). R_s values of 1.8 or greater indicate too much separation which can lead to long analysis times and band broadening of the later eluting peaks.

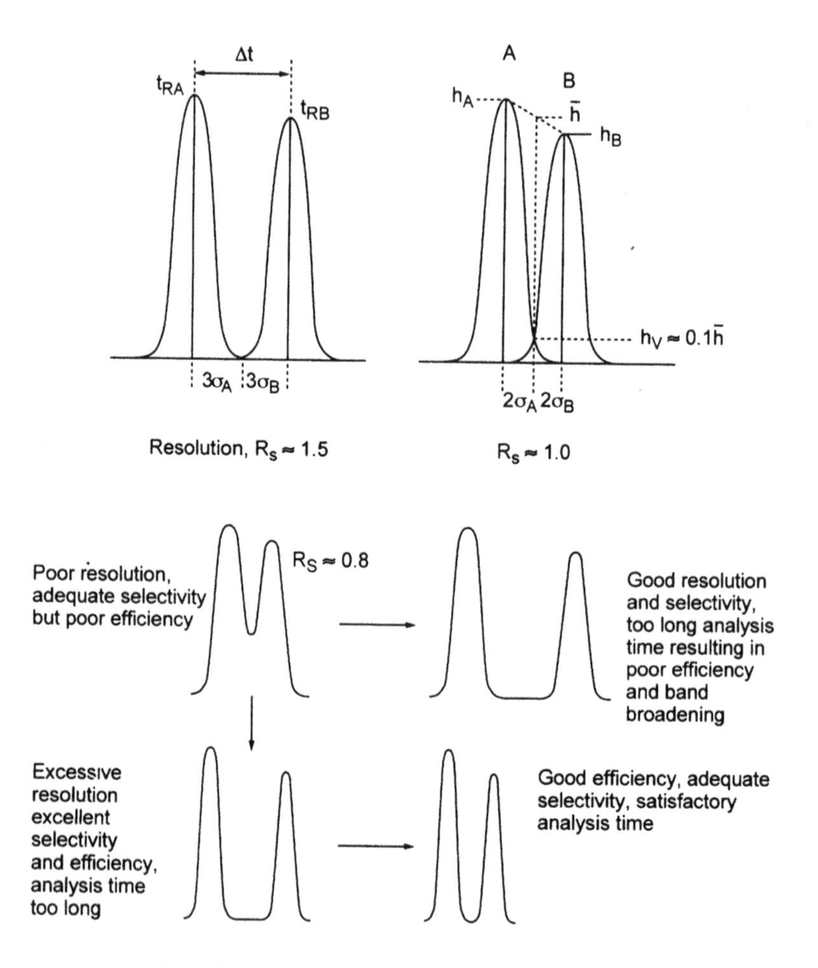

Figure 2.8 Resolution, selectivity and column efficiency.

2.6.1 10% valley resolution

The resolution between peaks can also be measured in terms of the height of the valley between the peaks above the base line (Figure 2.8). At a peak width of 4σ corresponding peak height is $0.134h$ (Figure 2.4), therefore, for a resolution giving <2.3% overlap, $R_s \approx 1.0$, the valley height should be $\leq 0.134h$ or $\sim\leq 0.1h$. Since adjacent peaks are unlikely to be of the same height, the mean height, $\bar{h}$, is used.

$$R_s \approx 0.1\left(\frac{\bar{h}}{h_V}\right)$$

$$\text{where} \quad h_V \leq 0.1\left(\frac{h_A + h_B}{2}\right) \leq 0.1\bar{h} \quad \text{for } \approx 2.3\% \text{ peak overlap}$$

2.6.2 Separation number

Separation number (SN) or Trennzahl number (TZ) is an alternative to separation factor as a means of describing the separating capabilities of high efficiency chromatographic systems such as HPLC and capillary column GC. It is a practical method of assessing the ability of a column to separate a sample mixture into components with a resolution of 1.0 between adjacent peaks. SN is obtained from the retention times and peak widths of two consecutive homologous *n*-alkanes (Z and $Z+1$) and indicates the number of component peaks that could be resolved, with an $R_s \approx 1.0$, between the two alkanes in a given part of the chromatogram.

$$\mathrm{SN} = \frac{(t_{R(Z+1)} - t_{RZ})}{(w_{h(Z+1)} + w_{hZ})} - 1$$

The value of SN depends on the alkanes used for the calculation, these are therefore specified with any value of SN quoted.

2.6.3 Resolution and selectivity

Resolution between peaks is dependent on the retention characteristics of each component (k), the ability of the stationary phase to selectively retain the components (α) and the overall efficiency of the column (N). The equation derived below was first developed by Purnell in 1959 to accurately relate the degree of separation attainable in a GC column in terms of the separating capability and efficiency [13–15].

$$\text{Using} \quad N = 16\left(\frac{t_{\mathrm{RB}}}{w_{\mathrm{bB}}}\right)^2 \quad \alpha = \frac{k_{\mathrm{B}}}{k_{\mathrm{A}}} \quad \text{and} \quad k = \frac{t_{\mathrm{R}} - t_{\mathrm{M}}}{t_{\mathrm{M}}}$$

$$R_s = \frac{t_{\mathrm{RB}} - t_{\mathrm{RA}}}{w_{\mathrm{bB}}} \quad \text{and} \quad w_{\mathrm{bB}} = \frac{4t_{\mathrm{RB}}}{\sqrt{N}} \quad \text{therefore} \quad R_s = \frac{\sqrt{N}}{4}\left(\frac{t_{\mathrm{RB}} - t_{\mathrm{RA}}}{t_{\mathrm{RB}}}\right)$$

$$\text{Substituting for} \quad t_{\mathrm{RA}} = t_{\mathrm{M}}(k_{\mathrm{A}} + 1) \quad \text{and} \quad t_{\mathrm{RB}} = t_{\mathrm{M}}(k_{\mathrm{B}} + 1)$$

$$R_s = \frac{\sqrt{N}}{4}\left(\frac{k_{\mathrm{B}} - k_{\mathrm{A}}}{1 + k_{\mathrm{B}}}\right) \quad \text{but} \quad k_{\mathrm{A}} = \frac{k_{\mathrm{B}}}{\alpha}$$

$$R_s = \frac{\sqrt{N}}{4}\left(\frac{k_{\mathrm{B}} - k_{\mathrm{B}}/\alpha}{1 + k_{\mathrm{B}}}\right)$$

rearranging and using $k \equiv k_{\mathrm{B}}$ or the mean of k_{B} and k_{A}

$$R_{\mathrm{R}} = \frac{\sqrt{N}}{4}\left(\frac{k}{k+1}\right)\left(\frac{\alpha - 1}{\alpha}\right)$$

The equation is important in column design, for example, to achieve a given separation in the minimum column length and analysis time. Resolution is a function of the square root of N so large changes in N are required to have a marked effect. Efficiency may be improved by increasing column length but for a given column geometry it is preferable to optimise k and α first. The mobile phase, stationary phase system is chosen so that a critical pair of peaks are not eluted at low k values, that is, close to t_M. k is the average retention factor for the two peaks and severely limits resolution at low values, larger values will increase relative retention times improving the separation. For closely eluting peaks α is close to 1.0 so small changes in α have large effects on resolution. Modifying the stationary phase/mobile phase system is used to optimise k and α. However, optimising retention times for a given sample mixture whilst maintaining adequate resolution is also important for routine analysis. This is to avoid excessively long analytical procedures particularly when automated chromatographic systems are used.

2.7 Quantification in chromatography

Data from chromatograms may be used to obtain the relative concentrations of components in a mixture, providing good resolution is achieved. Peak area, from integration of the detector signal during elution of a component, is proportional to the amount of that component in the sample. However, the response of a detector varies from one compound to another; for example, the HPLC ultraviolet detector depends on absorptivity, a GC flame ionisation detector on the formation of ions and a GC electron capture detector depends on electron affinities. Thus, a set of detector response factors needs to be determined for a particular analysis. Although many integrators include area% in the printout this is not the true ratio of the components. Area% is simply the area of an individual peak calculated as a percentage of the total areas recorded for all peaks in the chromatogram. It can be useful for a quick check of replicate analyses. There are four principal methods for obtaining quantitative information: normalising peak areas, internal standards, external standards and standard addition.

2.7.1 Normalising peak areas

The area of each peak is obtained from a series of replicate (5+) injections of a mixture containing equal (or known) amounts of all the components. Acceptable precision is essential to obtain satisfactory data. One component is chosen as the reference and the relative responses of the other components are determined by dividing the peak areas by that of the reference component. The detector response factors (D_{RF}) may then be used to calculate corrected peak areas ($A_{correct}$) for other analyses involving these components and hence their percentage ratios in the mixture may be determined.

$$\text{For component } x, \quad D_{RF_x} = \frac{\text{Area for component}}{\text{Area for reference}} = \frac{A_x}{A_{REF}}$$

Corrected area for a component in a sample is $A_{correct} = D_{RF} \times A_{chrom}$

% of component x in the sample is calculated as $\text{area\%}_x = \dfrac{A_{correct_x}}{\sum A_{correct}} \times 100$

where A_{chrom} is the area from the chromatogram and $\sum A_{correct}$ is the sum of the corrected peak areas for all peaks in the chromatogram.

2.7.2 *Internal standard*

The internal standard method is a variation on the above, and is recommended for accurate quantitation. It eliminates the need for accurate injections since a reference standard is included in each sample analysed. An internal standard is selected which has a retention time such that it is eluted in a suitable 'gap' in the chromatogram. The procedure involves analysing a test sample containing known amounts of each component plus a predetermined amount of the internal standard. Since peak area is proportional to the amount of an eluted component and the detector response factor (D_{RF})

for an individual component x: $A_x = D_{RF_x} \times C_x$

and for the internal standard: $A_{IS} = D_{RF_{IS}} \times C_{IS}$

where C is the amount of component x or internal standard, IS. The relative response of a component ($D_{RF'_x}$) to the internal standard is therefore

$$D_{RF'_x} = \frac{D_{RF_x}}{D_{RF_{IS}}} = \frac{A_x/C_x}{A_{IS}/C_{IS}} = \frac{A_x}{A_{IS}} \times \frac{C_{IS}}{C_x}$$

Response factors for all components are calculated in the same way. Analysis of an unknown mixture is achieved by adding an accurately known amount of internal standard and then carrying out the chromatography. The concentration of each component calculated using the equation above rearranged to give

$$C_x = \frac{A_x}{A_{IS}} \times \frac{C_{IS}}{D_{RF_x}}$$

The precision of the analysis is not dependent on injection of an accurately known amount of sample, but does depend on accurate measurement of peak areas. This is not a problem with electronic integrators and an overall precision or covariance of <4% should be readily obtained.

2.7.3 *External standards*

Automated sample injection systems and multiport injection valves (HPLC) have good reproducibility so that a series of injections can be made with a variation in sample volume of <1%. A set of standard mixtures containing known concentrations of the analytes is analysed and their peak areas recorded. A calibration graph of area versus concentration can be drawn for each analyte to confirm a linear detector response and from which the amount of the analytes in a mixture can be determined. Alternatively for an established method a replicate series of one standard mixture is injected and the area/unit amount of analyte calculated.

$$A_{\mathrm{STANDARD}} \equiv x \text{ mg litre}^{-1}$$

The mixture is then analysed and the amount of the components in the sample calculated using the peak area data for the standard mixture. Therefore, if the recorded peak area for the component in a sample mixture is A_{MIX} then the amount of component x is

$$\mathrm{Amount}_x = \frac{x \times A_{\mathrm{MIX}}}{A_{\mathrm{STANDARD}}} \text{ mg litre}^{-1}$$

2.7.4 Standard addition

Standard addition is used in many techniques in analytical chemistry. It is of limited use in chromatography because of the difficulty of injecting accurately known amounts of sample. A sample mixture is analysed for the analyte of interest by adding a specified amount of this analyte to the sample, thus increasing its concentration. The analysis is then repeated and the resulting increase in peak area due to addition of the standard amount is noted. Hence, the concentration of the analyte in the original sample may be calculated. If the peak area for the first analysis is A_1 and with the standard addition of x mg is A_2, then the peak area corresponding to x mg (or x mg litre^{-1}) is $(A_2 - A_1)$. Thus, the original amount of the analyte x in the sample corresponding to A_1, is given by

$$\mathrm{Amount}_x = \frac{x \times A_1}{(A_2 - A_1)} \text{ mg litre}^{-1}$$

Allowance for any dilution due to addition of the standard amount has to be made. The main difficulty with this method concerns the reproducibility of the sample injection. A precision of better than ±1% should be achieved if valid quantitative results are to be obtained.

An alternative approach is first to analyse the sample, noting the area, A_1, for the analyte. Successive standard amounts of the analyte are then added, each sample + standard mixture being analysed and the areas recorded. A graph of peak area versus concentration is drawn and the amount of analyte in the sample obtained by projection of the calibration line to intersect the abscissa as shown in Figure 2.9. This approach is conveniently illustrated by the determination of water in methanol by GC using thermal conductivity detection experiment 24, Chapter 9.

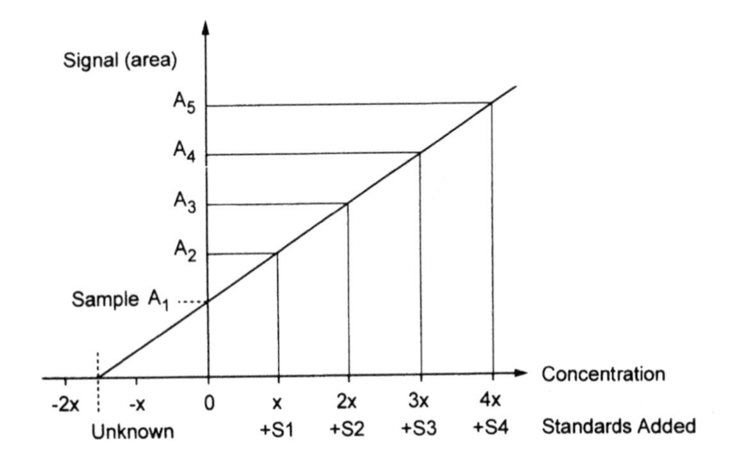

Figure 2.9 Standard addition.

References

[1] Littlewood, A.B. *Gas Chromatography*, p. 71. Academic Press, New York, 1970.
[2] Martin, A.J.P. and Synge, R.L.M. *Analyst*, **77** (1952) 915.
[3] Desty, D.H. *Committee Recommendations in Vapour Phase Chromatography* (Proceedings of the First Symposium), p. 30. Academic Press, New York, 1957.
[4] van Deemter, J.J., Zuiderweg, F.J. and Klinkenberg, A. *Chem. Engng Sci.*, **5** (1956) 271.
[5] Giddings, J.C. *J. Chem. Phys.*, **35** (1958) 588.
[6] Giddings, J.C. *Dynamics of Chromatography*, Part I: *Principles and Theory*. Marcel Dekker, New York, 1965.
[7] Giddings, J.C. *Unified Separation Science*, p. 86. Wiley Interscience, New York, 1993.
[8] Hawkes, S.J. *J. Chem. Educ.*, **60** (1983) 393.
[9] Golay, M.J.G. *Gas Chromatography* (Desty, D.H., ed.), p. 36. Butterworths, London, 1958.
[10] Knox, J.H., Kennedy, G.J. and Done, J. *J. Chromatogr. Sci.*, **10** (1972) 606.
[11] Perry, S.G. *Gas Chromatography 1972*, p. 145. Applied Science Publishers, London, 1973.
[12] Giddings, J.C. *J. Chromatogr.*, **5** (1961) 46.
[13] Purnell, J.H. *Nature*, **184** (1959) 2004.
[14] Purnell, J.H. *J. Chem. Soc.* (1960) 1268.
[15] Purnell, J.H. and Quinn, C.P. *Gas Chromatography 1960* (Scott, R.P.W., ed.), p. 184. Academic Press, New York, 1960.

3 Planar chromatography

3.1 Thin layer chromatography

Thin layer chromatography (TLC), or planar chromatography, began as a method of separating mixtures by eluting them through a planar chromatographic bed then visualising the separated components by staining or charring. In its simplest form TLC plates can be prepared in the laboratory, the plate placed in any suitably sized container and the resultant chromatogram scanned visually. At its most sophisticated there is a large variety of plates, sample application aids including automated applicators, developing chambers, visualisation aids, scanning devices and adsorbents commercially available. TLC can be an excellent qualitative and quantitative method and more detailed descriptions of the techniques employed will follow.

The first recorded works on TLC are those of Beyerinck in 1889 [1] and Wijsmann [2] in 1896; however it is generally accepted that it was Izmailov and Shraiber in 1938 [3] who enunciated the ideas and fundamental principles of using a chromatographic adsorbent in the form of a thin layer fixed on an inert rigid support. Meinhard and Hall [4] in 1949 developed this notion of an 'open column', and in 1951 Kirchner *et al.* [5] reported the separation of terpenes on a 'chromatostrip', prepared by coating a small glass strip with an adsorbent mixed with starch or plaster of Paris, which acted as a binder.

Several years passed before the method became widely used, probably because at that time much effort was being devoted to the development of paper and gas chromatography. In the late 1950s, however, Stahl [6, 7] devised convenient methods of preparing plates, and showed that thin layer chromatography could be applied to a wide variety of separations. He introduced a measure of standardisation, and since the publication of his work and the appearance of commercial apparatus based on his designs, TLC has been accepted as a reproducible analytical technique. As usually happens, once the initial stimulus had been given, many variations of the original procedures were proposed and developed.

Over the past decade there have been a number of further significant improvements in the technique. For instance, there are now a wide range of sorbents available in the form of pre-coated plates and the applicability of the technique has also been extended with the increasing range of bonded phase sorbents, e.g. reverse phases (C_8 and C_{18}), those of medium-polarity (amino and cyano) and other specialised layers featuring chiral and mixed stationary phases.

As well as being convenient these commercially produced plates give improved performance due to the narrower particle size range and reproducible physical properties such as surface area and pore size. For standard TLC, using silica gel, the particle size is between 5 and 17 μm, while typical pore size and layer thickness are 60 Å and 0.25 mm, respectively. The use of a refined silica gel with a mean particle size of 5 μm and particle size range of 2–10 μm (*cf.* HPLC) has led to the development of high performance TLC (HPTLC) which uses thinner layers (100 μm) and requires smaller samples.

Further advances have resulted from developments in instrumentation, particularly in the areas of scanning densitometry and automated sample application, which have now made fully instrumental quantitative TLC a reality far removed from the basic practice. TLC is now regarded as an indispensable tool in both quality control and research laboratories. The technique is easy to learn and is fast and versatile and in many instances may be preferred to the techniques of gas chromatography and high performance liquid chromatography.

3.1.1 *Theory and principles*

Major features controlling the resolution capability of the TLC plate are 'spot' size and the physical dimension of the plate; with a spot diameter of 0.5 cm and a plate length of typically 10 cm, then optimally only 20 analytes could be completely resolved. However, with normal capillary delivery of eluant the maximum number of theoretical plates is <5000 and a given analyte *only* experiences those plates through which it travels. Furthermore, as the mobile phase velocity varies throughout the length of the plate, decreasing as it moves further from the origin, it is unlikely ever to be at the optimum, the number of analyte spots which can be fully separated in practice will be no more than six to eight.

In chromatography the distribution coefficient (K) of a component is defined by the relation

$$K = \frac{C_{SP}}{C_{MP}}$$

where C_{SP} is the concentration of the component in unit volume of the stationary phase, and C_{MP} is the concentration of the component in unit volume of the mobile phase.

K describes the relative affinity of a component for the two phases and hence relates to the distance and speed with which it moves through the plate on elution. Distribution coefficients and rates of migration, however, are difficult to evaluate, certainly in routine analysis, and a more practical evaluation of the chromatogram is required. In TLC and paper chromatography the results obtained are described by quoting the

R_f values which refers to migration '*r*elative to the solvent *f*ront' and was shown by Martin and Synge [8] to be related to the distribution coefficient of the component.

$$R_f = \frac{\text{Distance travelled by solute}}{\text{Distance travelled by solvent front}}$$

The R_f value, though closely related to the retardation factor (R), is not exactly equal as it is a ratio of distances moved compared to a ratio of velocities. In planar chromatography the solvent front migrates slightly ahead of the bulk of the mobile phase and hence the retardation factor $R \sim 1.15\, R_f$.

While the R_f value is not an absolute physical value for a component, with careful control of conditions, the value can aid identification. Due to the large number of variables which can influence the R_f value—for instance, minute differences in solvent composition, temperature, size of tank, the sorbent layer and the nature of the mixture—coincidence of R_f values, even in more than one solvent system, should never be taken as unequivocal proof of the presence of a component.

An alternative approach which minimises differences in R_f values due to the influence of the variables enumerated above, is to quote the R_f value relative to a carefully chosen standard which is run on the same plate. This relative value R_{st}, should be constant since under any given conditions the relative R_f values remain the same.

$$R_{st} = \frac{\text{Distance to centre of component spot}}{\text{Distance to centre of standard spot}}$$

Coincidence of R_{st} values should not be taken as absolute proof of identification and for complete structural characterisation the component should be eluted from the sorbent layer and spectroanalytical studies, such as infrared–ultraviolet nuclear magnetic resonance (^{13}C and ^{1}H) and mass spectroscopy carried out to aid identification.

3.1.1.1 Plate theory. The effectiveness of plane chromatography can be extended by using stepwise elution with different eluants and by using layers comprising mixed stationary phases. However, in order to achieve optimum results, effects which lead to spot broadening must be minimised. In the context of plane chromatography the number of theoretical plates is given by

$$N = 16\left(\frac{d_A}{d_W}\right)^2$$

where d_A is the distance to the centre of the spot and d_W is the width of the spot (Figure 3.1).

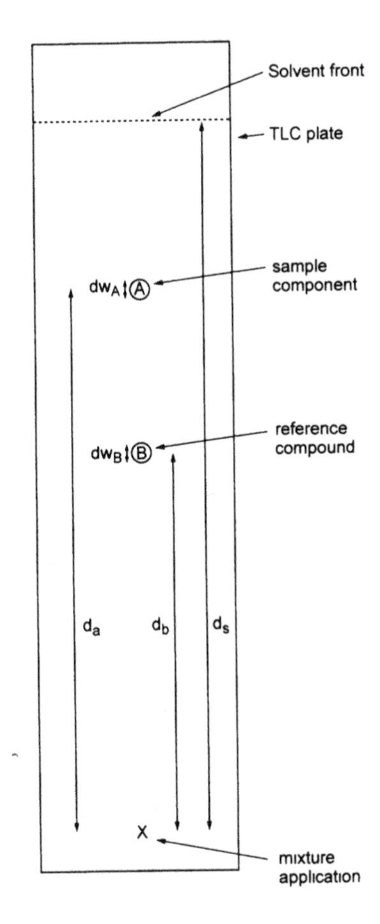

Figure 3.1 Determination of R_f and R_{st} from TLC chromatoplate.

The retention ratio, i.e. capacity factor k, is a measure of the retention of the analyte in the sorbent bed compared to its retention in the mobile phase.

$$k = \frac{t_R}{t_M}$$

In thin layer or paper chromatography the time spent in the sorbent bed or stationary phase is proportional to the (distance to the solvent front − the distance to the sample spot) whereas the time spent in the mobile phase is proportional to the distance to the centre of the spot, hence the capacity factor can be expressed as

$$k = \frac{d_{\text{solvent}} - d_A}{d_A} = \frac{1 - R_f}{R_f}$$

The resolution (R_S) between an adjacent pair of spots and separation factor (α) can also be expressed in terms of distances travelled by solute and solvent:

$$R_S = 2\frac{(d_B - d_A)}{(d_{W_A} + d_{W_B})}$$

$$\alpha = \frac{k_B}{k_A} = \frac{d_A(d_{solvent} - d_B)}{d_B(d_{solvent} - d_A)}$$

3.1.1.2 Rate theory of band broadening. The van Deemter equation in planar chromatography can be reduced to

$$H = A + \frac{B}{\mu} + (C_S + C_M)\mu$$

The multipath term A arises, as in other forms of chromatography, from the different pathways the analyte molecule can follow in permeating the absorbent layer.

It is important in minimising longitudinal diffusion, the second term to ensure that the tank is saturated with eluant vapour and that the developed plate is removed before the eluant reaches the top of the plate. Failure to follow these procedures will lead to spot broadening due to diffusion of the analyte concentration zone. These effects are proportional to the time spent in the mobile phase and hence with HPTLC, where solvent flow is low due to small particle size, shorter run times are employed, less spot broadening occurs and hence resolution is higher.

The third term encompasses slow equilibrium or mass transfer effects; if a choice of solvent is available then that with the lower viscosity should be chosen to minimise this effect. Smaller particle size will also give improvements though there is a limit due to the increasing resistance to flow.

3.1.1.3 Mobile phase flow rate. Planar chromatography is atypical of chromatographic methods in that it is difficult to control the mobile phase flow rate and furthermore it is not uniform throughout the sorbent bed. The flow rate is dependent upon a number of features; for example, the viscosity and surface tension of the solvent and the particle size and permeability of the packing. In ascending development where the eluant is drawn through the sorbent bed by capillary forces it can be shown that the mobile phase velocity is inversely proportional to the distance the solvent front has moved. At the limit the velocity is zero at which point lateral diffusion starts and the spots begin to broaden with the concomitant loss in resolution.

These limitations have led to the development of a number of systems which attempt to deliver eluant in a controlled manner, for example, centrifugal chromatography drives the eluant through the sorbent bed whilst in over pressured chromatography the plate is located in a sealed holder and eluant is pumped through. These techniques are discussed in further detail within the chapter.

A fuller treatment of the theoretical aspects of planar chromatography is presented by Geiss [9] and a general account is given in Chapter 2.

3.1.2 Outline of the method

In thin layer chromatography a solution of the sample in a volatile solvent is applied via glass capillary tubes approximately 1–2 cm from the bottom of a uniform layer of inert adsorbent, such as silica gel or alumina, which has been uniformly spread over a suitable supporting plate such as glass or plastic and dried under standard conditions. When the spot has dried the plate is placed vertically in a suitable tank with its lower edge immersed in the selected mobile phase. The solvent rises by capillary action and an ascending chromatographic separation is thus obtained, resolving the sample mixture into discrete spots.

At the end of the run the solvent is allowed to evaporate from the plate and the separated spots are located and identified either by physical methods such as visual inspection, fluorescence or radiation monitoring or by chemically reacting with a developing reagent; the method chosen is dictated by the composition of the starting mixtures.

In thin layer chromatography only small amounts of adsorbent and minute samples are needed. The separated spots are located on the plate using a variety of visualisation techniques in common with paper chromatography, so that normally no collection of fractions is necessary. There is, however, no difficulty about preparative separations (preparative layer chromatography), which are achieved by increasing the thickness of the layer and using a higher loading of sample. After separation, it is easy to recover an individual substance by scraping off and collecting the part of the layer on which the spot is adsorbed. The substance can then be extracted with a suitable solvent.

3.1.3 Surface adsorption processes and spot shape

The possible interactions which lead to solute molecules being adsorbed are

- London dispersion or van der Waals forces;
- dipole–dipole interactions;
- intermolecular hydrogen bonding; and
- chemisorption.

Where London dispersion forces are the principal interaction between solute and sorbent, then because equilibrium is readily established the Langmuir isotherm tends to have a wide linear range and the spots tend to be symmetrical. An S-shaped isotherm indicates that the solute itself is acting as an adsorbent giving solute double layers, which results in teardrop-shaped spots comparable to tailing of peaks in gas chromatography. In the situation where all the available adsorption sites are occupied then adsorption of more solute molecules is hindered as monolayer coverage is approached and the surface is said to be saturated. In this instance rocket shaped spots result akin to fronting of peaks in gas chromatography. Many problems associated with spot shape can be overcome by reducing sample loading thus ensuring operation in the linear region at the bottom of the isotherm curves. A fuller treatment of isotherms and isotherm curves is presented in Chapter 2.

3.1.4 Comparison of thin layer with other forms of chromatography

Traditional open column liquid chromatography (LC) suffers poorly in comparison with the thin layer method in that it is a fairly slow process which requires large amounts of packing material, eluant and sample. The major disadvantages of LC methods are with speed, scale, resolution and characterisation.

The newer LC techniques of flash column chromatography and short path or medium pressure column chromatography though offering resolution approaching that of thin layer are still comparatively slow and expensive in terms of packing and solvents. The principal advantage of these techniques is in their preparative capability.

In recent years HPLC has arguably been pre-eminent amongst chromatographic techniques and has tended to be the method of choice for the rapid analysis, both qualitative and quantitative, of complex samples. However, despite this trend, TLC has always offered a number of advantages:

- TLC samples require minimum sample preparation;
- simultaneous analysis of multiple standards and samples can be carried out under identical conditions in a time comparable to HPLC;
- strongly retained compounds in comparison to HPLC from the most compact chromatographic zones and therefore can be detected with the highest sensitivity; in addition the bands can be removed and purified;
- all components are locatable, *cf.* HPLC, where highly polar materials may be overlooked as the peaks are very broad and difficult to discern;
- as the solvent can be removed there are far greater post-chromatographic detection possibilities; and
- the analyte need not possess a chromophoric group as post-chromatographic reactions can be used for location.

In comparison with paper chromatography, the thin layer method has the main advantages of greater speed, and in most cases, better resolution. The average time for a 10 cm run in TLC on silica gel is 20–30 min (depending on the nature of the mobile phase), whereas the same separation on a fast paper might take 2 h.

Rough qualitative separations on small plates may take as little as 5 min. The better resolution arises from the fact that the adsorbent in thin layer chromatography has a higher capacity than the paper in paper chromatography and the particle size of the adsorbent layer materials are very small compared to the large cellulose fibres from which the paper matrix is formed. The separated spots therefore retain fairly closely the shape and size of the original applied spot, without the spreading associated with partition chromatography on paper. This advantage is largely lost when a partition system is used on a thin layer. A further, and very important, advantage of the adsorption system is that it can be used to separate hydrophobic substances, such as lipids and hydrocarbons, which are difficult to deal with on paper, even with a reversed phase system. Thin layer separations have been applied, however, in most fields of organic, and some of inorganic chemistry.

Location of separated substances on thin layers is done in the same way as it is on paper, but more reactive reagents, for example, concentrated sulphuric acid, can be applied on thin layers, provided that the thin layer material is an inert substance such as silica gel or alumina.

3.1.5 Adsorbents

The general properties of adsorbents for thin layer chromatography should be similar to those described for adsorbents used in columns, and the same arguments about 'activity' apply. Two important properties of the adsorbent are its particle size and its homogeneity, because adhesion to the support plate largely depends on them. A particle size of 10–25 μm is usually recommended. A coarse-grained material will not produce a satisfactory thin layer, and one of the reasons for the greatly enhanced resolution of thin layer chromatography is this use of a fine-grained adsorbent. In a column a very fine material will give an unacceptably slow flowrate, whilst on a thin layer the fine grain gives a faster and more even solvent flow. Some examples of adsorbents which have been used for representative separations by thin layer chromatography are given in Table 3.1.

The methods of preparation of the chromatographic layer are most important if optimum and reproducible results are to be obtained. Small traces of impurities greatly influence the behaviour by occupying adsorption sites, and one of the most common impurities is water. The sorbent can be activated by simply heating to drive off the moisture giving a sorbent layer which gives sharper and better separations.

3.1.5.1 Additives. A wide variety of adsorbents are available with or without binders and/or fluorescent indicators; however, these are usually

Table 3.1 Adsorbents for thin layer chromatography

Solid	Used to separate
Silica gel	Amino acids [6, 7], alkaloids [10], sugars [11], fatty acids [12, 13], lipids [14], essential oils [15], inorganic anions and cations [16], steroids [17, 18], terpenoids [19]
Alumina	Alkaloids [10], food dyes [20], phenols [21], steroids [18, 22], vitamins [23], carotenes [24], amino acids [25]
Kiselguhr	Sugars [11], oligosaccharides [26], dibasic acids [27], fatty acids [28], triglycerides [29], amino acids [30], steroids [31, 32]
Celite	Steroids [33], inorganic cations [16]
Cellulose powder	Amino acids [34, 35], food dyes [36], alkaloids [37], nucleotides [38]
Ion exchange cellulose	Nucleotides [39], halide ions [40]
Starch	Amino acids [41]
Polyamide powder	Anthocyanins [42], aromatic acids [42], antioxidants [43], flavanoids [44], proteins [45]
Sephadex	Nucleotides [46], proteins [47, 48], metal complexes [49]

incorporated in the sorbent for improved performance and convenience. Binders are added to sorbents to improve the adhesion of the chromatographic layer to the substrate, i.e. the glass or plastic sheet and also the layer's durability and abrasion resistance. However, there are certain constraints with such sorbent layers; for example, where the binder is starch or some other polymeric material then these plates are unsuitable when corrosive reagents are to be employed for the visualisation of spots. Calcium sulphate bound layers, on the other hand are not compatible with aqueous based eluants due to the solubility, albeit slight, of the binder.

The use of adsorbents with added fluorescent indicator provides an extremely versatile non-destructive means of locating analyte spots. The fluorescent reagent, a manganese activated zinc silicate of a particle size comparable to that of the adsorbent absorbs at 254 nm emitting a light-green fluorescence. Analytes appear as dark spots against this background due to quenching of the indicator fluorescence.

3.1.5.2 Silica gel. Silica gel (Figure 3.2) is the most commonly used adsorbent in TLC studies. It is prepared [50] by the hydrolysis of sodium silicate to polysilicic acid which on further condensation and polymerisation yields silica gel material.

The synthesis can be controlled so as to yield silicas of high purity and of the required specification with regards to pore volume, pore size and specific surface area. Binders as well as enhancing adhesion to the backing plate are commonly added to the silica gel to confer greater mechanical strength to the layer. The suffix 'G' is used universally to denote silica gel with a gypsum binder, namely calcium sulphate hemi hydrate ($CaSO_4$ ($0.5H_2O$)). The presence of calcium ions does not affect most separations; however, there are available silica gels which adhere sufficiently without the use of a

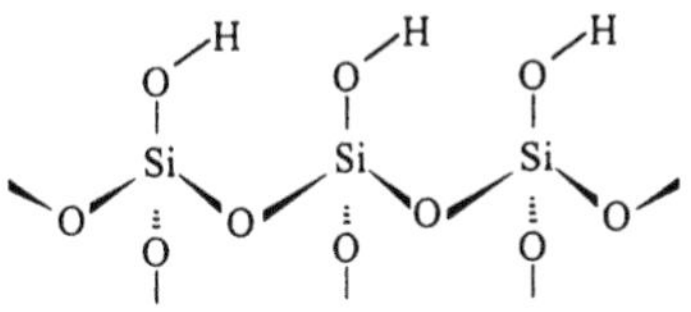

Figure 3.2 Silica gel matrix structure.

binder. Another binder which has found a limited use is starch, though it places restrictions on the use of corrosive location agents. The binder is typically present at about 10% (w/w).

The resolution and separating efficiency achieved are, as in other forms of chromatography, dependent upon particle size and particle size distribution. The resolution improves as particle size becomes smaller and particle size distribution narrower. The trend towards smaller and more uniform particles continues and the silica gel now commonly in use for TLC studies has a mean particle size of 12 μm with a particle size range of 5–25 μm and a pore diameter of 6 nm which gives partial exclusion for compounds between 500 and 1000 Da. Commercially available precoated plates have a mean particle size of 10 μm with a correspondingly narrower particle size range, with thinner layers of 250 μm for increased speed and resolution. The newer technique of HPTLC uses silica gel with a particle size of 5–6 μm.

Plates are available with a fluorescent indicator, a manganese activated zinc silicate or a phosphor, which emits a green fluorescence when irradiated with UV light (Hg lamp) of 254 nm. The absorbing substances appear as dark spots against the green fluorescent background due to quenching of the fluorescence.

It is also possible to modify the adsorption properties of the silica gel by incorporating substances such as bases or buffers enabling coatings with accurately defined pH to be prepared. In a like manner silver nitrate can be added; this admixture changes the adsorptive properties to permit increased discrimination and separation of unsaturated compounds, especially alkenes. The technique is commonly known as ‘Argentation TLC’.

3.1.5.3 Kieselguhr (Celite). Kieselguhr and Celite are diatomaceous earths composed of the silica-rich fossilised skeletal remains of microscopic sea organisms called diatoms. The material has high porosity and large surface area and exhibits very little adsorptive property; for these reasons it is used primarily as a support for the stationary phase in partition chromatography. For TLC applications calcium sulphate is normally added as a binder at about 15% (w/w). The adsorptive capacity can be further reduced by treatment with acids or alkali or by silanising.

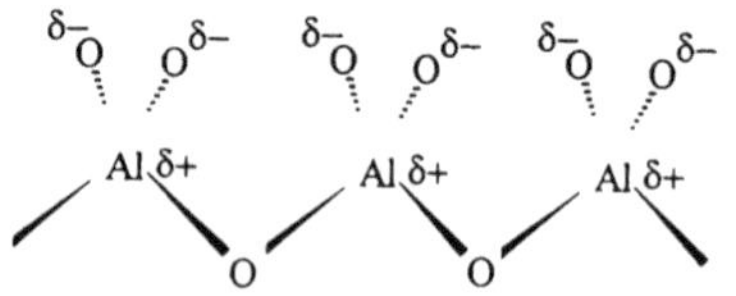

Figure 3.3 Partial structure of alumina.

Celite when mixed with clay forms firebrick which can subsequently be crushed and sieve graded. The firebrick has good flow and packing characteristics with an increased adsorptive capacity.

3.1.5.4 Alumina. Alumina (aluminium oxide) (Figure 3.3) can be synthesised to the same degree of purity and specification as silica gel by a series of non-uniform dehydration processes of various crystalline modifications of aluminium hydroxide [51]. Furthermore, the reaction conditions can be adjusted to produce aluminium oxide with either an acidic, basic or neutral surface; it can be used with or without a binder and the use in the latter form is more common than with silica.

In order to obtain reproducible results and optimum separations it is necessary to activate the alumina to control the amount of adsorbed water which can block the adsorbing sites on the alumina. Typically, alumina sorbent will be activated by heating at a specified temperature (125–150°C) for a defined time.

Alumina is a strong adsorbent and can function as an amphoteric ion exchanger, depending upon the nature of the surface and the solvent; for instance, basic aluminium oxide when used with organic eluants will adsorb aromatic and unsaturated hydrocarbons, carotenoids, steroids, alkaloids, and other natural products. In aqueous or aqueous alcoholic solutions its exchanger properties become more pronounced and it can adsorb basic dyes, basic amino acids, as well as inorganic cations. Neutral aluminium oxide is principally employed with organic eluants and is suitable for use with substances that are either labile or bound to strong alkalies. Acidic aluminium oxide is used for the separation of neutral or acid materials that are not acid labile. In aqueous and alcoholic media it serves as an anion exchanger.

Alumina can catalyse both inter- and intramolecular reactions, especially with compounds which are base sensitive due to its alkaline nature, e.g. dehydration, double bond migrations and ring expansions have all been observed. This disadvantage accounts largely for the preferred use of silica gel.

3.1.5.5 Cellulose powders. It may at first seem unnecessary to go to the trouble of preparing cellulose powder plates when paper could be used more conveniently, but there are important advantages in the thin layer method. In paper the cellulose is fibrous, and the fibres, however closely they are matted together, inevitably form a network with large gaps. The solvents flow along the surface of the fibres, and the gaps become filled with liquid, with the result that excessive diffusion of solutes takes place, and the separated zones tend to be larger than the original spot. If the fibres are too tightly compressed the flow-rate becomes unacceptably slow.

Layers of cellulose powder, on the other hand, are aggregations of very small particles, all of much the same size. The interstices are therefore much smaller and more regular, and the adsorbent surfaces are more evenly distributed. In consequence, there is a much more even flow of mobile phase, with less diffusion of the dissolved substances. The flow is also much faster.

Cellulose contains adsorbed water which is held in the glucopyranose structure by hydrogen bonding, hence the separation proceeds via a partition mechanism. Cellulose materials are used almost exclusively for separating hydrophilic substances, for instance, amino acids and sugars in contrast to silica gel and alumina which are used for the separation of lipophilic compounds. Similar eluants, as for the PC application, can be selected. The partial structure of the cellulose molecule is shown in Figure 3.4.

There are a variety of chemically modified celluloses each of which behaves as an ion exchange medium.

- *Diethylaminoethyl (DEAE) cellulose* is formed by the reaction between cellulose and 2-chloro-1-diethylaminoethyl hydrochloride:

$$\text{Cel–OH} \xrightarrow{\text{ClCH}_2\text{COONa}} \text{Cel–OCH}_2\text{COONa}$$

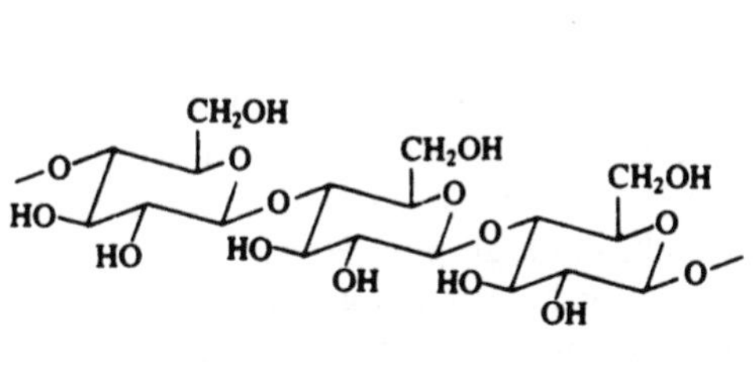

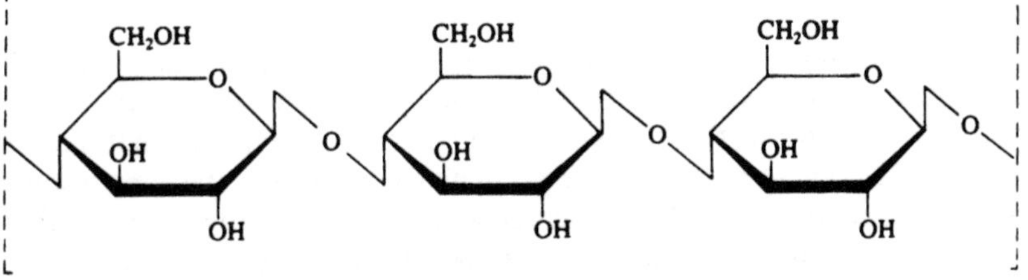

CH₂OH O HO HO O CH₂OH O HO OH O CH₂OH O HO OH O

CH₂OH O OH OH O CH₂OH O OH OH O CH₂OH O OH OH O

Figure 3.4 Partial structure of cellulose molecule.

DEAE functions as a strong anion exchanger and carries positive charges at neutral and acidic pH. It is commonly employed for separation of negatively charged molecules by ion exchange chromatography. Due to the hydrophilic nature of the cellulose substrate, DEAE cellulose is particularly well suited to ion exchange separations of delicate biomolecules such as proteins and nucleic acids.

- *Carboxymethyl (CM) cellulose* is synthesised from monochloroacetic acid and cellulose and has weak cation exchange capability:

$$\text{Cel-OH} \xrightarrow{\text{ClCH}_2\text{CH}_2\text{N(C}_2\text{H}_5)_2} \text{Cel-OCH}_2\text{CH}_2\text{N(C}_2\text{H}_5)_2$$

- *Eteola cellulose* is a weak basic anion exchanger and is the product of reaction between cellulose, triethanolamine and epichlorohydrin.
- *Polyethyleneimine (PEI) cellulose* is formed by impregnating the 'support', microcrystalline cellulose, with PEI. The latter is prepared by the copolymerisation of aziridine in the presence of an acid catalyst and the resultant product is highly branched and has primary, secondary and tertiary amino groups. The PEI cellulose behaves as a strong anion exchanger with high capacity and is of considerable utility in the analysis of nucleotides, nucleosides, nucleo-bases and sugar phosphates. The layers can be obtained with the common fluorescent indicators and should be stored at 0–5°C to prevent (minimise) deterioration.

These modified cellulose powders can be used to obtain ion exchange separations on thin layers, with similar advantages over column or paper sheet methods. Both normal and modified cellulose powders can be used with or without a binder and fluorescent indicator.

3.1.5.6 Molecular sieve layers. Thin layers on glass plates can be made from Sephadex gel chromatography media, with a superfine grade having a particle size 10–40 μm. The layers range in thickness from 0.4 to 1.00 mm so chromatography is rather slower and more troublesome than other forms of thin layer chromatography. However, it is marginally faster and more convenient than gel chromatography in a column. Separation is achieved by partition, augmented by size exclusion in the solvent filled pores of the swollen gel.

3.1.5.7 Polyamide layers. A number of polyamides, for instance, polycaprolactam and Nylon 6.6 (polyhexamethylenediaminoadipate), can be coated onto plates in the conventional manner to give a chromatographic material [52]. These materials are particularly useful for the separation of closely related phenols which interact with the chromatographic layer by hydrogen bonding. These bonds are weak and reversible and the adsorbed analytes can be displaced by elution with solvents capable of H bonding.

The desorptive power of commonly used solvents is of the order:

dimethylformamide > formamide > acetone > methanol > water

All the adsorbents so far mentioned can be used for adsorption chromatography, but some such as silica gel, celite, kieselguhr, or cellulose, can be used for thin layer partition chromatography (as they can be used for partition chromatography in columns), if an appropriate mobile phase is selected. If the plate is dried so that it retains very little adsorbed water, and the mobile phase is a non-polar mixture, separation will be by adsorption chromatography. If, however, the plate retains an appreciable amount of water, or the solvent mixture contains a highly polar constituent, separation will be largely by partition chromatography.

3.1.5.8 Reversed phase packings. The sorption mechanism predominant on silica gel is adsorption and the plates with suitable choice of eluant can be used to separate neutral, basic and acidic hydrophilic substances. This mode of separation is referred to as normal phase. In contrast when a non-polar stationary phase is eluted with a more polar eluant the order of elution of analytes is reversed and this is referred to as reverse phase chromatography.

At first reverse phase sorbents were prepared by impregnating the support with long-chain hydrocarbons such as paraffins and silicone oils to give packings of defined composition with the stationary phase held to the support by purely physical forces of attraction [53]. These materials though of use for the analysis of lipophilic substances, fats and waxes, steroids and fat soluble vitamins and dyes were unsatisfactory as the stationary phase can be washed off by the eluant which consequently lowers the capacity and effectiveness of the partition separation.

As in HPLC (see Chapter 6) these failings have been largely overcome by chemically bonding the stationary phase to the support. The most suitable base material has been found to be silica gel with a pore width of 6 nm. Special alkylsilanes are used to react with the accessible silanol groups on the surface of the silica particles. For instance reaction with dimethyldichlorosilane transforms the silanol groups to dimethyl silyl ethers which comparatively have little adsorptive capacity. Reaction with other silanising reagents such as diphenyl-, ethyl-, octyl-, octadecyl-, 3-aminopropyl- and 3-cyanopropylsilyl reagents as well as spacer bonded polar moieties, e.g. the propanediol group. As in HPLC the most popular alkyl chain length is C_{18}. The preparation of silanised silica gel is shown in Figure 3.5.

A high degree of modification of the silanol groups would result in a surface with pronounced hydrophobic character reducing the wetability of the plates in aqueous-based eluants thus severely limiting the composition and type of eluants which could be used. This problem has been overcome by preparing stationary phases with clearly specified lower degrees of

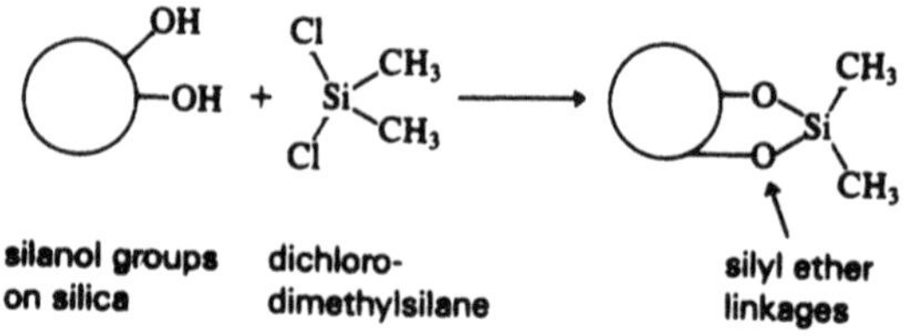

Figure 3.5 Preparation of silanised silica gel.

modification or using greater particle size. Such layers can be effectively wetted with eluants containing up to 50% water. A related difficulty, encountered in the application of samples from polar or aqueous media, has been overcome by using plates with pre-adsorbent layers. This approach also avoids the occurrence of 'ringing' where the analytes separate out as concentric rings within the sample spot when spotting from an organic solvent.

The expectation that such plates could be used to mimic or scout HPLC methods and vice versa has not been fulfilled due in part to differences between TLC sorbents and HPLC packings such as carbon loading, the degree of end-capping and particle and pore size of the base silica. In addition, in HPLC all the chromatographic media is in equilibrium with the eluant, whilst in TLC the solvent front is continually encountering dry sorbent.

Nonetheless, RPTLC is the method of choice in a wide range of application areas and has a number of advantages compared with normal phase TLC:

- superior selectivity;
- good for separation of non-polar compounds;
- better recovery in preparative mode;
- simplified sample clean-ups; and
- activation unnecessary.

3.1.5.9 Chiral separations. It has become increasingly important to be able to resolve racemates, as pharmacological activity is principally if not exclusively the property of one or other enantiomer. TLC has become important as a rapid means of monitoring optical purity [54]. Separation of mixtures of optically active compounds can be achieved by either separating on a chiral stationary phase, by incorporating a chiral additive in the eluant which induces temporary diastereoisomerism, thus affording separation, or by derivitising the analytes with a chiral reagent then separating the resulting diastereoisomers on an achiral stationary phase.

A disadvantage of this later approach is that the diastereoisomers may have different detector responses rendering quantitation difficult or they may not be formed at the same rate. As in HPLC β-cyclodextrins have proved particularly useful in the separation of mixtures of chiral compounds. These chiral-cavity type media act by allowing selective occlusion or intercalation of one enantiomer into chiral cavities in the matrix of the phase. In principle 'Pirkle' or 'brush type' chiral stationary phases also find application in TLC. Separation in this instance is dependent on at least three points of interaction between the enantiomers and stationary phase, so forming transient diastereoisomers of differing stability.

A complementary approach is to use reverse phase silica gel impregnated with a chiral selector [55, 56]. Chiral plates comprising reverse phase C_{18} impregnated with copper *N*,*N*-dialkyl-α-amino acids, e.g. proline have been used successfully for the separation of a variety of enantiomeric mixtures, e.g. D- and L-dopa. The resolution is based on ligand exchange, the enantiomers to be separated forming chelate complexes with the copper ion of differing stabilities.

3.1.5.10 Dual phase TLC. Dual phase TLC first described by Kirchner in 1951 [57] is an extension of two-dimensional thin layer chromatography and uses two sorbents laid side by side. Typically, a normal and reverse phase media would be used though other combinations dictated by sample characteristics could be employed. A sample is applied at one corner of the plate and chromatographed on sorbent 1; the plate is then removed from the tank, dried rotated through 90° and then chromatographed on sorbent 2 with the appropriate eluant. Thus separation of complex mixtures can be achieved of complex mixtures where either the analytes are too similar or too different to resolve effectively using a single phase. The use of such plates can also be used (*cf.* two-column GC analysis) to confirm analyte identity as it is extremely unlikely that two analyte species would have the same retention characteristics on two different phases.

3.1.6 Preparation of plates

The adsorbent is spread on a suitable firm support, which may be quite rigid, or flexible. Glass plates were the original support; however, flexible plates have become increasingly popular. The size used depends on the type of separation to be carried out and on the type of chromatographic tank and spreading apparatus available.

Most of the commercial apparatus is designed for plates of 20 × 5 or 20 × 20 cm, and those are now regarded as 'standard'. It is important that the surface of the plate shall be flat, and without irregularities or blemishes. Glass plates are cleaned thoroughly before use, washed with water and a detergent, drained and dried. A final wash with acetone may be included, but it is not essential. It is important not to touch the surface of the cleaned plates with the fingers. The first step is to make the adsorbent into a slurry with water, usually in the proportion x g of adsorbent and $2x\,\text{cm}^3$ of water. The slurry is thoroughly stirred, and spread on the plate by one of the methods described below.

If a binder is used, the time available from mixing the slurry to completion of spreading is about 4 min (after which setting will have begun). The properties of the adsorbent can be modified by using buffer solutions instead of water, to give a layer of desired acidity, or to modify the water-retaining properties of the layer. Similarly, complexing agents or fluorescent indicators can be incorporated in the layer by mixing the slurry with solutions of the appropriate substances. An important example is the use of silver nitrate in the separation of lipids and related materials.

The film thickness is a most important factor in thin layer chromatography. The 'standard' thickness is 250 μm, and there is little to be gained by departing very much from that in analytical separations (thinner layers may give rise to erratic R_f values). Thicker layers (0.5–2.0 mm) are used for preparative separations, with a loading of up to 250 mg on a 20 × 20 cm plate.

There are, in principle, four ways of applying the thin layer to its support: spreading, pouring, spraying and dipping.

3.1.6.1 Spreading. It is possible to spread the adsorbent in a number of ways, but the main objective is to produce an absolutely even layer with no lumps or gaps, which adheres evenly and securely to the support. Commercial spreaders are in general of two types, which might be called the 'moving spreader' and 'moving plate' types. The difference is shown diagramatically in Figure 3.6. Both types are made in varying degrees of sophistication (and cost), and since each maker supplies detailed instructions for the use of his equipment, only the general principles are mentioned here.

In the 'moving spreader' type, the glass plates (usually five 20 × 20 cm, or an equivalent number of smaller size) are held in a flat frame, and a rectangular hopper containing the slurry is passed over them. The hopper has no bottom (or the bottom can be opened when required), and its trailing face has an accurately machined lower edge to give an even layer of the required thickness. In some appliances the thickness is adjustable. The lower edge of the leading face of the hopper rests on the glass plates. In the moving plate apparatus the hopper containing the slurry is fixed and the plates are pushed through under the hopper as indicated.

Makers have produced spreading 'beds' in which the long sides have an overhanging lip (Figure 3.7). The plates are pushed from below against the lip, by means of spring strips, or by an inflatable air-bag, so that the upper

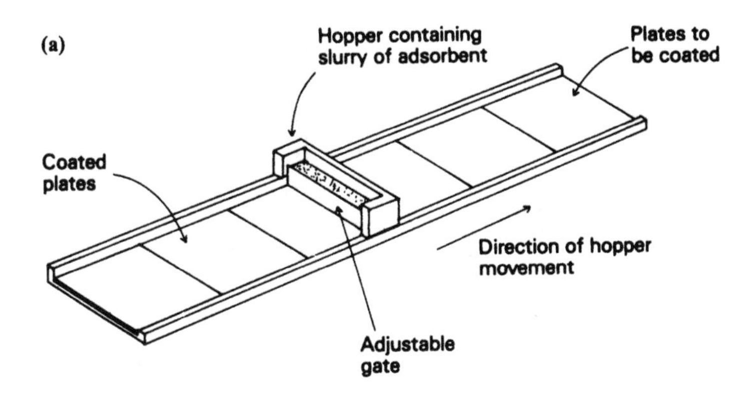

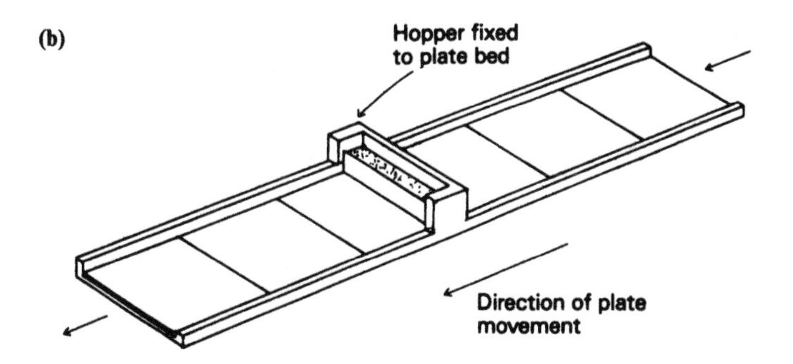

Figure 3.6 Commercial plate spreaders: (a) moving spreader apparatus; (b) moving plate type.

surfaces of the plates are at the same level, regardless of their thickness. A foam-rubber base on which the plates are laid has also been suggested as another way of getting an even spreading surface. With a little practice the 'moving spreader' method is a very satisfactory way of making thin layer plates.

3.1.6.2 Pouring. Many workers prefer not to use mechanical spreading methods at all. If the adsorbent is very finely divided and of homogeneous particle size, and if no binder is used, a slurry can be poured on a plate and allowed to flow over it so that it is evenly covered. Some manual dexterity is required to do this properly. Preparation of plates by pouring is particularly easy with certain types of alumina, but water alone is not usually suitable for making the slurry; a volatile liquid such as ethanol (or an ethanol–water mixture) or ethyl acetate is preferable. The appropriate amounts of liquid and solid adsorbent needed to cover a plate have to be found by trial and error, and exactly those quantities should

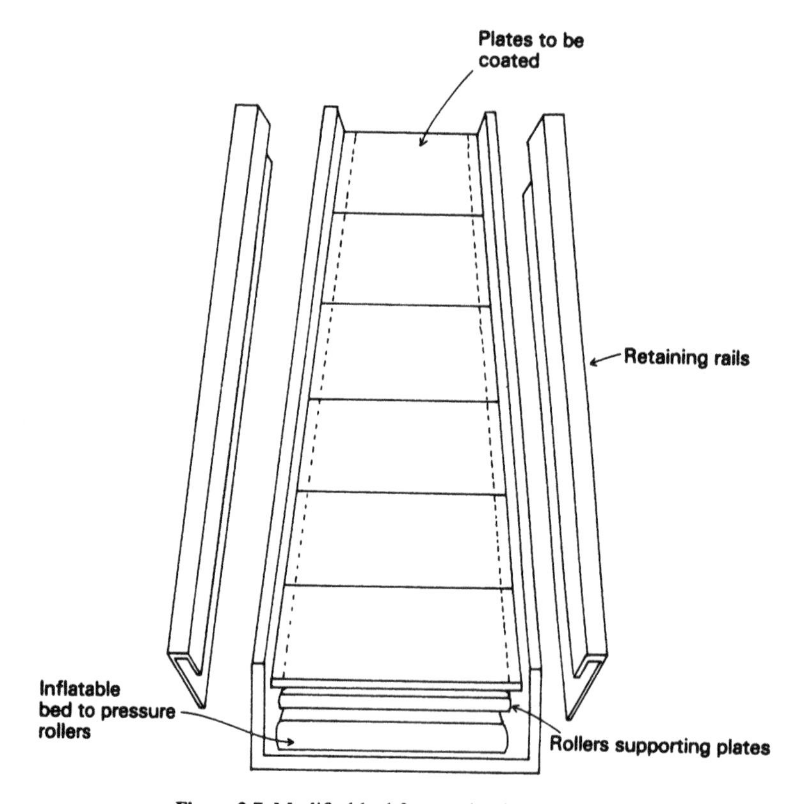

Figure 3.7 Modified bed for moving bed apparatus.

be used to ensure that the thickness of the layer is reproducible. Good, even plates can be made by this method, but the thickness of the layer is not known.

3.1.6.3 Spraying. Spraying methods do not seem to have any particular advantage over spreading, and are open to the objections usually associated with spraying, among which are the difficulty of getting even coverage, and the fact that there is no easy way of ensuring reproducible thickness from batch to batch.

3.1.6.4 Dipping. Small plates, such as microscope slides, can be spread by dipping in a slurry of the adsorbent in chloroform, or other volatile liquid. Again, the exact thickness of the layer is not known and the evenness of the layer may not be very good, but this is a most convenient method for making a number of plates for rapid qualitative separations. After spreading, the plate is allowed to dry for 5–10 min, and, if it has been made with an aqueous slurry, it is further dried and 'activated' by heating at about

100°C for 30 min. Plates made with volatile organic liquids may not need this further drying. It is important to standardise this part of the preparation of plates, because the activity of the adsorbent may depend rather critically on it.

Plates may be kept for short periods in a desiccator, but long storage is not recommended. When inspected for imperfections before use, they should appear uniform in density when viewed by transmitted or reflected light, and there should be no visible large particles. Gentle stroking with the finger should not remove the layer, and there should be no loose particles on the surface.

3.1.6.5 Activation. The activity of a sorbent, i.e. its ability to retain analyte species, is determined by its chemical structure and properties. In silica gel, for example, there are several different types of active sites though the OH groups, the surface silanol groups, are mainly responsible for its adsorptive properties. Substances are adsorbed onto the surface of the silica via hydrogen bonding, with the surface silanol groups serving as hydrogen donors. Silica gel is heated typically at 120–150°C to drive off surface adsorbed water otherwise the active sites are masked/blocked, this effect can simply be reversed by exposing the silica to water or water vapour. However, if higher temperatures of the order of 400–500°C are used then deactivation of the silica gel can result due to elimination of water from adjacent silanol groups—this effect is permanent and irreversible.

The exact nature of the active sites on alumina is not well understood and may be due to exposed aluminium or other cationic species and perhaps strained Al–O bonds; however, in contrast to silica gel it is found that the activity of alumina increases steadily even on heating up to 1000°C. The optimum activation temperature for alumina is not certain but fully activated alumina can be produced by heating for 12–16 h at 400–450°C. Alumina of suitable activity for TLC can be produced by heating at 150–200°C for a few hours.

Its activity can then be lowered by adding precise amounts of water. Indeed the activity of alumina may be defined according to the Brockmann scale where alumina with 3% water added is classified as grade II and with 6%, 10% and 15% grades III, IV and V, respectively. Test dyes, namely, azobenzene, *p*-methoxyazobenzene, Sudan yellow, Sudan red and *p*-amino-azobenzene can also be used to evaluate the activity of sorbent layers, the higher the R_f the lower the activity and Brockmann number. Silica plates can also be assigned a Brockmann number.

Both the above sorbents can adsorb significant quantities of moisture on standing and hence plates must be stored over desiccant and prepared in a controlled and reproducible manner in order to ensure the same level of activity.

3.1.7 *Application of samples*

Samples must be applied to TLC plates with extreme care and minimal disturbance of the adsorbent layer. Normally samples are manually applied via a capillary tube, a micropipette or a calibrated glass microsyringe such that the emerging drop just touches the surface of the plate with the appliance tip remaining just above the sorbent layer. A hole in the sorbent causes an obstruction to uniform solvent flow. The resultant channelling of solvent causes distortion of the moving spots and culminates in a loss of resolution.

Spots should not be closer than 1 cm centre to centre, they should be 2–5 mm in diameter and should not be nearer to the edge of the plate than 1.5 cm on a 20 × 20 cm plate. A volume of 0.1–0.5 mm^3 should be applied, and if a larger volume of the sample solution is needed to give the required loading, it should be applied in stages. TLC is a microanalytical method and the loading should not be more than about 12 μg per spot on a layer 250 μm thick, 10 μg being the optimum amount for most substances. Normally, 1–5 $\mu g\,\mu l^{-1}$ solution concentrations are practicable.

In semipreparative work the sample is often applied as a streak along the start line; up to about 4 mg may thus be loaded on a 20 × 20 cm plate, 250 μm thick. Mechanical devices can be obtained for use in conjunction with a syringe to give rapid and even streaking. A stop line created near the top of the plate with a sharp pencil removes a fine line of adsorbent down to the glass. The solvent flow is forced to stop when the front reaches the line, a useful feature which enables one to standardise conditions very easily. The plate should not, however, be left standing in the tank for a long time after the front has reached the stop line, as diffusion and evaporation may cause spreading of the separated spots. The edges of the plate should be rubbed clear with a finger before spotting to a width of about 0.5 cm, to give a sharper edge to the adsorbent layer.

The solvent in which the sample is dissolved for spotting should be as volatile as possible, and also have as low a polarity as possible. If the spotting solvent is strongly adsorbed by the layer, marked irregularities may be observed as the mobile phase passes the position of the spots, and the separated spots may be seriously distorted.

3.1.7.1 Autospotters. There are a wide variety of semiautomated and fully automated devices commercially available for spotting samples. These devices have improved the reproducibility of the technique as they can deliver exact volumes as precisely defined spots on the plate, thus removing the greatest source of error in quantitative work. In their simplest form a sample vial is presented by hand; the sample is drawn into the syringe which is driven by a stepper motor and is then applied to the plate. More sophisticated, fully automated systems are available where the ‘spotter’ is microprocessor controlled and may be programmed to apply precise and varying amounts of up to 30 samples/standards. The sample/standard vials are delivered to the syringe module by an HPLC style autosampler unit. In the extreme the 'spotter' can be used in conjunction with a scanning densitometer with a single PC controlling both instruments.

3.1.8 *Documentation*

It is obvious from the above discussion that in any description of a chromatographic procedure the fullest possible details should be presented. Not only are R_f values important, but the shape and colour of spots give valuable information also. A minimum record of the chromatogram would be to sketch the chromatoplate and annotate the diagram fully, for instance, colour of spots, locating reagent, solvent used, etc. Alternatively, the pre-prepared plates may themselves be retained and detail entered on the plate with pencil or a suitable marker. A popular method is to photocopy or photograph the plate—commercial photographic apparatus is available. More recently video documentation systems have been developed. These systems are available in monochrome or colour. A powerful CCD camera is the keypiece of the image gathering module and typically has a resolution of 500 lines. The high light sensitivity makes it possible to record even weak fluorescence chromatograms. These systems can be used with incident or transmitted fluorescence and the digitised data stored either on hard or floppy disk.

3.1.9 *Development*

Once the chromatographic plate has been prepared and the samples have been applied to it, it is placed in a suitable chamber with the lower edge immersed in the eluant to a depth of 0.5–1.0 cm. A suitable solvent should give R_f values of between 0.3 and 0.7 for the sample components of analytical interest.

The purpose of allowing equilibration time is to allow homogeneity of the atmosphere in the tank, thus minimising evaporation of the solvent from the TLC plate during development. It has been shown that equilibration is more important in thin layer than in paper chromatography, and therefore it is important to keep the atmosphere in the tank saturated with solvent vapour. Saturation can be assisted by lining the walls with filter paper soaked in solvent. With these precautions, however, it is not usually necessary to allow any 'equilibration time' before the start of the chromatographic run. The size of the tank and the actual volume of a mixed solvent used have an effect of R_f values, because between them these two factors control the composition of the vapour in the tank. The smallest tank possible should be used, so that the enclosed atmosphere has the smallest possible volume. This, in turn, controls the rates of evaporation of solvents from the plate during the run. Standardisation of the geometrical parameters of the plate is desirable; for instance, length of run, distance of the start line from the solvent surface (or position of solvent feed), and positions of initial spots, should be kept constant. An indication that the atmosphere in the tank is not saturated with vapour, when a mixed solvent is being used, is the development of a concave solvent front, the mobile phase advancing faster at the edges than in the middle. This condition should be avoided.

It is not essential to use a tank at all. A sandwich can be made consisting of a second glass plate clamped firmly over the sorbent surface (Figure 3.8). This sandwich can then be placed into a normal development chamber or into a special trough designed to prevent solvent loss.

Regardless of the procedure used it must be documented in its entirety. A few important parameters are solvent composition, tank volume, position of application, and migration distance.

A number of methods of development have been investigated in TLC studies. Those still commonly employed are detailed below.

3.1.9.1 Ascending. In the ascending technique, the TLC plate is positioned in the development tank after it has come to equilibrium with the solvent; the application of sample spots should be above the solvent

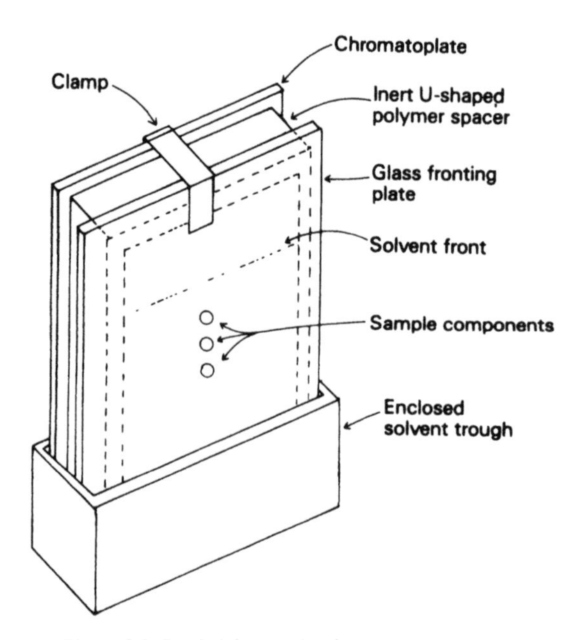

Figure 3.8 Sandwich type development apparatus.

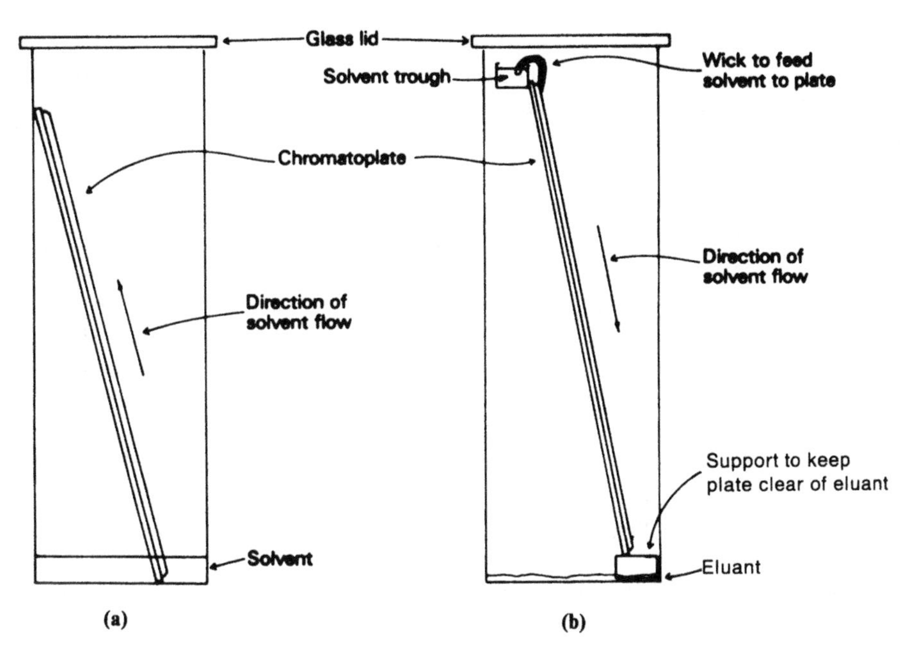

Figure 3.9 TLC plate development: (a) ascending; (b) descending.

level. The solvent percolates through the sorbent material by capillary action moving the components to differing extents, determined by their distribution coefficient, in the direction of flow of the eluant. This is the simplest technique and remains the most popular. Ascending and descending TLC plate development are shown in Figure 3.9.

3.1.9.2 Descending. The descending technique, though more common in paper chromatography, can be used in TLC. The top of the plate, where the spots are located, has solvent from a trough fed onto it via a wick; some solvent of the same composition is placed in the bottom of the tank but the plate is supported above the solvent level. In order to improve the resolution for particularly difficult separations a number of modifications to the above technique have been developed.

3.1.9.3 Continuous development. It is not necessary in either ascending or descending techniques to remove the plate once the solvent front has traversed the sorbent layer. In ascending chromatography the eluant is allowed to wash off the top of the plate and a continuous flow of solvent is obtained. Descending chromatography readily lends itself to this modification, the eluant simply being allowed to wash off the plate into the bottom of the tank. Descending development has the advantage of quicker solvent flow, due to the action of capillary and gravity forces on the eluant, though it has the disadvantage of requiring additional equipment and additional expertise to set up. Continuous development can be used to good effect for the separation/resolution of compounds using low polarity solvents.

3.1.9.4 Multiple and stepwise development. There is no reason why the development process cannot be interrupted, the plate removed from the tank, solvent allowed to evaporate off and the whole sequence repeated. If the same eluant is employed and the elution distance kept constant from one run to another, this is termed multiple development. Alternatively, the subsequent elution procedure may be modified: for instance, different solvent systems can be utilised and the solvent allowed to migrate to different extents. The sequential use of a series of eluants of differing elutropic strength can be used for the separation of mixtures of wide-ranging polarity. Depending upon the nature of the mixture either an increasing or decreasing series of solvent strengths may be used. This latter approach is termed stepwise development.

Both these development procedures are complementary to gradient techniques [58, 59] and offer the following advantages:

- enhanced efficiency due to the refocusing/reconcentration of the analyte spots on each run;
- improved detection due to smaller separated zones; and
- separation of wider polarity range of analytes.

3.1.9.5 Two-dimensional development. Sometimes, particularly in the case of large groups of compounds of similar chemical structure and properties, such as amino acids, the R_f values are too close together to give a good separation using one-dimensional linear development techniques. In these instances, improved resolution can be obtained with two-dimensional development, a technique developed by Martin, which employs a second eluant system run at right angles to the first. The sample is spotted in the normal manner and developed with the edge (AB) of the plate in contact with the first solvent system. The plate is then removed and allowed to dry. It is then developed with a second solvent at right angles to the first, edge AC of the plate in the solvent. The diagram shows the form of chromatogram obtained and the use of standards and identification of the components of the unknown by instructing tie-lines as illustrated (Figure 3.10). Where the pure components of a mixture are not available or are unknown then the chromatogram obtained may serve as a fingerprint/map in identifying and characterising the sample.

3.1.9.6 Radial development. The technique of radial development (Figure 3.11), sometimes referred to as horizontal chromatography, involves a

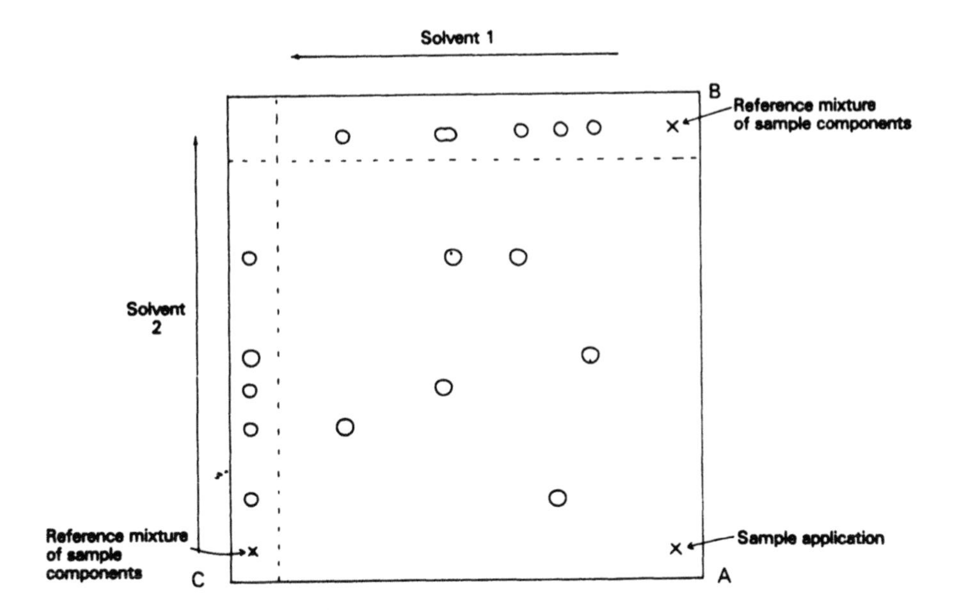

Figure 3.10 Two-dimensional development in TLC.

slightly different principle from those so far described. In this method the sample spot is applied to the centre of the plate (a disc) and the solvent is supplied through a hole in the plate via a wick which dips into a solvent reservoir. Alternatively the plate is placed with the sorbent layer facing downward with its centre in contact with a porous 'spring loaded' wick standing in the solvent. As development proceeds the components move out radially forming circles of increasing diameter. There is an inherent advantage here, since there is a concentration effect, as the annular zones are formed, due to the solvent moving the trailing edge faster than the leading edge. Due to this concentration effect the resolution of components of low R_f with linear development is much improved. The relationship between linear R_f values and circular RR_f is

$$R_f = (RR_f)^2$$

A further feature in circular development is suppression of trailing of analyte bands as the mobile phase feed is always faster at the trailing edge which together with the progressive decrease in sample loading tends to compress the zones. Specially developed systems are now commercially available where the plate is placed face down in an enclosed chamber and the eluant and vapour phase are delivered by motor-driven syringes in a controlled manner. As the eluant and vapour phase are held in containers external to the chromatoplate compositions remain constant throughout development.

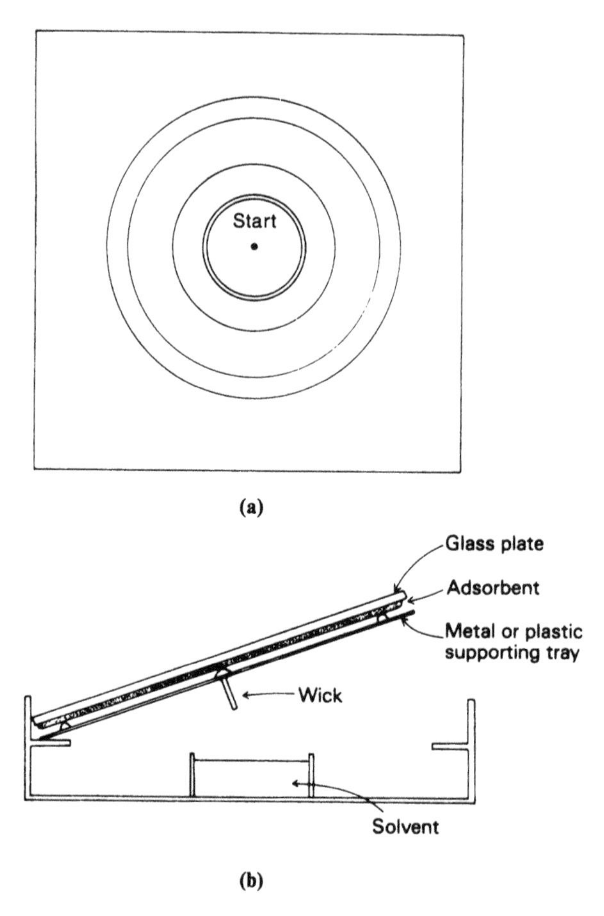

Figure 3.11 Radial development system for TLC: (a) Radial chromatogram; (b) radial development system for TLC.

All of the development techniques, except two-dimensional, are equally applicable to qualitative and quantitative studies.

3.1.9.7 Gradient elution. Where samples contain analytes having a wide range of polarities, then in order to separate out components at the sample origin or solvent front, gradient techniques can be used rather than the stepwise or multi development approaches described above.

One approach to generating a gradient is to stand the plate as usual in one solvent while a second is added slowly, with stirring. A constant volume device is employed. Recent advances in instrumentation, e.g. OPLC now allow for fully automated development procedures including gradient elution [60]. One approach is to repeatedly develop the chromatogram over increasing solvent migration distances but with decreasing solvent strength. The

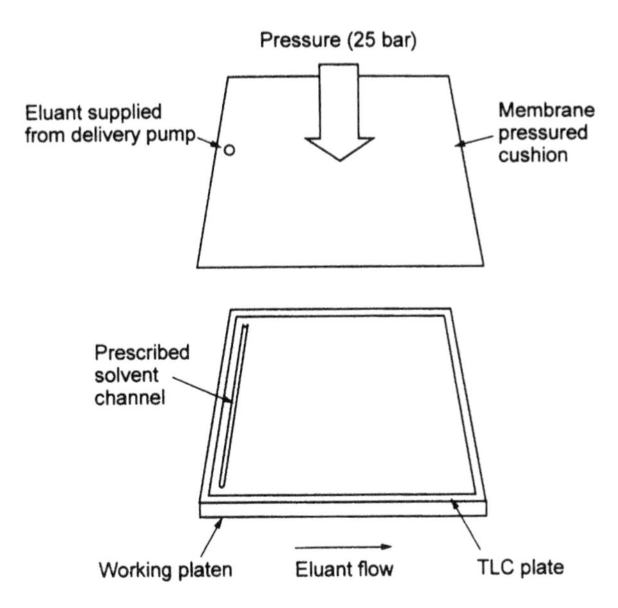

Figure 3.12 Schematic OPTLC apparatus.

whole process may involve between 10 and 30 cycles and can be totally automated. The principal function of the first cycle, using eluant of strongest strength and the shortest elution distance, is to refocus the sample spot. Eventually the mobile phase strength will decline to such an extent that a particular analyte shall migrate no further after a certain cycle. It is also possible to employ gradient techniques in the stationary phase using defined pH layers. These are discussed in more detail by Stahl and Dumont [61].

3.1.10 Overpressured thin layer chromatography

In standard TLC the eluant is drawn through the adsorbent layer by capillary forces (Figure 3.12). Under these conditions there is no control over the flow-rate, which is not uniform throughout the bed and furthermore it cannot be optimised.

These limitations have led to the development of forced flow development systems and to the technique of overpressured thin layer chromatography. The special feature of this method is that the adsorbent layer is in a completely sealed unit and the solvent is delivered under pressure at a controlled uniform flow-rate by a pump module as in HPLC. Thus, overpressured TLC (OPTLC) takes place in the absence of a vapour pressure and the migration of the solvent front is free from both evaporation and adsorption effects. As the eluant is delivered under controlled conditions it is possible to optimise the separation conditions by adjusting the flow-rate of the eluant and also to undertake continuous development procedures.

This technique has other advantages compared with TLC and HPTLC, e.g. high-speed sample development. low consumption of eluant and reduced diffusion of analyte spots. Newman–Howells Associates and Supelco market horizontal sandwich chambers which are overpressurised; in the latter system the chromatoplates can be eluted simultaneously from both ends, thus doubling sample throughput.

3.1.11 Solvents

Separations in TLC are controlled by the distribution ratios of the components for a given system of sorbent and eluant. The separation profile may be modified by altering the distribution ratios which in itself is readily achieved by changing solvent composition with regards to strength and polarity. The choice of eluant as in column chromatography is determined by the sorption process employed and by the nature of the sample components. The polarity of solvents is typically expressed in an elutropic series in which they are arranged in order of increasing polarity as indicated by their dielectric constant (Table 3.2).

Generally, a solvent or solvent mixture of lowest polarity consistent with a good separation should be employed. Suitable mixing gives mobile phases of

Table 3.2 Elutropic series of solvents

Solvent	Dielectric constant (at 25°C)
Hexane	1.89
Cyclohexane	2.02
1,4-Dioxan	2.21
Carbon tetrachloride	2.24
Benzene	2.28
Toluene	2.38
Acetonitrile	3.88
Diethyl ether	4.34
Chloroform	4.87
Formic acid	5.0
2-Methylbutan-2-ol	5.82
Ethyl acetate	6.02
Acetic acid, glacial	6.15
Tetrahydrofuran	7.58
Dichloromethane	9.14
2-Methylpropan-2-ol	10.9
Pyridine	12.3
Butan-2-ol	15.8
2-Methylpropan-1-ol	17.7
Butan-1-ol	17.8
Propan-2-ol	18.3
Propan-1-ol	20.1
Acetone	20.7
Ethanol	24.3
Methanol	33.6
Water	78.3

intermediate eluting power, but it is best to avoid mixtures of more than two components as much as possible, chiefly because more complex mixtures readily undergo phase changes with changes in temperature. When mixtures are used, greater care is necessary over equilibration. The purity of the solvents is of much greater importance in thin layer than in most other forms of chromatography, because of the small amounts of material involved.

3.1.11.1 Principal requirements. The solvents used should be reasonably cheap, since large amounts are often used. They must be obtainable in a high level of purity. It is now possible to buy pure solvents for chromatography which for most purposes need no further treatment. Mixtures of isomers, such as xylene, pyridine homologues, or petrol tend to be of variable composition, and are best avoided. Traces of metals are undesirable, even for organic separations. The solvent should not be too volatile, because of the necessity for more meticulous equilibration. On the other hand, high volatility makes for easy removal of the solvent from the sheet after the run. Its rate of flow should not be greatly affected by changes in temperature.

The choice of solvent for a particular purpose is still to a large extent empirical. So many separations have been reported, however, that it is not usually difficult to find a suitable one to use as a starting point for new work, but a knowledge of the chemical properties of the substances to be separated is clearly desirable.

Some examples of the types of solvent mixture which have been used in various representative separations are shown in Table 3.3. They are given simply as examples, and are not necessarily the best for the separations quoted, although they are all perfectly satisfactory; it is not really possible to recommend a 'best' solvent for any particular purpose, as the views and requirements of authors vary considerably, even when they are working in closely related fields.

3.1.12 Location of separated substances

The success of a chromatographic separation depends ultimately on the location process. Coloured substances are, of course, visible as separate spots at the end of the run. Colourless ones require chemical or physical detection.

3.1.12.1 Chemical methods. Chemical methods of detection involve the application of a derivatising agent, commonly referred to as a locating reagent, or chromogenic reagent, to the TLC plate. The reagent in a suitable solvent is applied as a spray to the plate, following which a coloured derivative is formed *in situ*. The reagents may be classified as non-specific if they produce coloured spots with a wide range of compounds, classes—for instance, iodine, sulphuric acid, Rhodamine B and fluorescein—or they may be specific and only react with compounds containing a particular

Table 3.3 Solvent mixtures employed in TLC for various representative separations

Coating material	Dominant sorption process	Substances separated	Solvent system	Method of location	Remarks
Silica	Adsorption	Amino acids	1. $BuOH/AcOH/H_2O$ (4:1:1) 2. $PhOH/H_2O$ (3:1)	Ninhydrin or densitometer	Two-way development
		Fatty acids	1. $MeCN/Me_2CO/C_{12}H_{26}$ 2. $Pr_2O/n\text{-}C_6H_{14}$	pH indicator or 50% H_2SO_4	Two-way development, adsorbent with binder and $AgNO_3$
		Unsaturated fatty acids	Pet. ether (40–60°)/Et_2O	I_2 or diphenylcarbazone	
		Lipids	Pet. ether/Et_2O/AcOH (80:20:1)	I_2 or 50% H_2SO_4	Suitable for neutral lipids
		Lipids	$CHCl_3/MeOH/H_2O$ (65:25:4)	I_2 or 50% H_2SO_4	Suitable for phospho and neutral lipids
			Hexane/Et_2O (4:1)		Two-way development in same direction
		Hydrocarbon oils Synthetic ester oils	$CHCl_3/C_6H_6$	Fluorescence or conc H_2SO_4	
		Sterols	$CHCl_3$/MeCO (95:5)	50% H_2SO_4	Adsorbent with binder + $AgNO_3$
		Sugars	EtOAc/AcOH/MeOH/H_2O (60:15:15:10)	α-Naphthol–sulphuric acid	
Alumina	Adsorption	Amino acids	$BuOH/EtOH/H_2O$	Ninhydrin	
		Vitamins	Hexane–acetone	Antimony chloride in AcOH	Adsorbents with $AgNO_3$
		Sugars	$PrOH/H_2O/CHCl_3$ (6:2:1)	α-Naphthol–sulphuric acid	
Kieselguhr	Adsorption or partition	Disaccharides	PrOH/EtOAc (65:35)	α-Naphthol–sulphuric acid	
		Carotenoids	PrOH/Pet. ether/Et_2O	Visual	
Cellulose	Partition	Amino acids	$BuOH/AcOH/H_2O$ (60:15:25)	Ninhydrin	
		Carbohydrates	BuOH/Pyridine/H_2O (6:4:3)	*p*-Anisidine phthalate	
PEI-Cellulose	Ion exchange	Nucleic acid components	Dil. HCl (0.01–0.1 M)	Autoradiography	Also two-way development
DEAE-Cellulose	Ion exchange	Amino acids	0.01–1 M aq. NaCl	Ninhydrin	
Reverse phase silica	Partition	Nucleo bases	Acetonitrile/H_2O (20:80)	UV 254 nm	C_8 alkyl-bonded phase
	Ion pair	Alkaloids	$C_4H_9SO_3Na$ in $(CH_3)_2CO/H_2O$ (40:60)	UV 254 nm	C_{18} alkyl-bonded phase
Sephadex gels	Exclusion	Serum proteins	Phosphate buffers	UV absorbance	

Table 3.4 Locating reagents for TLC

Reagent detected	Colour of spots	Component
Iodine vapour	Brown	General organic Unsaturated compounds
2,7-Fluorescein	Yellow–green	Most organic compounds
Ninhydrin	Pink–purple	Amino acids and amines
2,4 DNP	Orange–red	Ketones and aldehydes
Antimony chlorides	Various Characteristic	Steroids, alicyclic Vitamins and carotenoids
Bromophenol blue or bromocresol green	Yellow	Carboxylic spots
Diphenylcarbazide	Various Characteristic	Metals

functional group—e.g. dinitrophenylhydrazine for carbonyl compounds. Thus, specific chromogenic reagents could be applied successively and judicious choice would not only make the spots visible but would also aid in the component identification (Table 3.4).

It is frequently necessary to heat the plate after spraying to accelerate the chemical reaction between reagent and components and this requires specialised heating chambers to provide uniform conditions for even spot development. An alternative to spraying is to dip the TLC plates into a solution of the reagent; however, though this may give a more uniform application of reagent to the plate, samples can be lost from the plate and spreading of the spots may occur which leads to a loss in resolution and sensitivity.

As a rule these methods of location are 10–100 times more sensitive on TLC than paper with the added advantage that more corrosive agents can be employed. Comprehensive lists of spray reagents for TLC and PC are available from manufacturers and in the chemical literature [62].

3.1.12.2 Physical methods

3.1.12.2.1 Ultraviolet detection. The most common method of location uses an adsorbent layer containing a fluorescent indicator. Plates commercially available use an indicator (e.g. Mn activated Zn silicate) which absorbs light at 254 nm and re-emits or fluoresces light at the green end of the spectrum, thus the plate when irradiated at 254 nm takes on a striking green colour. If a spot of compound is present which itself absorbs at 254 nm, this will quench the fluorescence and the component will show up as a dark spot against the green background. While this is not a specific identification the technique has the advantage, as the indicator is insoluble in common solvents and the location is non-destructive, of allowing isolation of the component for subsequent spectroanalytical analysis. This is achieved by scraping the sorbent where the spot is positioned from the plate and extracted with a suitable solvent. Removal of solvent under reduced pressure leaves the pure components. Other reagents used fluoresce at 370 nm.

3.1.12.2.2 Densitometers. Spectrodensitometers or TLC scanners are available for the quantitative evaluation of thin layer chromatograms based on measuring transmittance, fluorescence or reflectance intensity. Derivative spectra can also be generated which can aid the identification of minor and/or masked components [63]. Measurements with these instruments are precise with standard deviations of 1% beinging achieved. For accurate quantitative measurements double beam instruments should be used to reduce background 'noise'. They all have the same principal components, namely, a light source, condensing and focusing systems and either gratings or filters for selecting the necessary wavelengths. The instruments can also be used for the scanning of separations on electrophoretic gels.

(1) *Absorption measurements*
Compounds which absorb ultraviolet or visible light can be quantified for absorption measurements. Absorption measurements can be obtained in either the reflectance or transmission modes. The instrumentation required for these techniques is shown schematically in Figure 3.13. The vast majority of compounds absorb in the ultraviolet or visible region of the spectrum and deuterium and tungsten lamp sources can be used with prisms or gratings, cheaper instrumentation uses band-pass filters for monochromation.

(2) *Transmission*
For transmission measurements the chromatoplate is scanned by a point light source. The intensity of the beam transmitted through the adsorbent is taken as I_0 while the intensity of the beam transmitted through adsorbent components spot is I_t. Quantitation using transmission

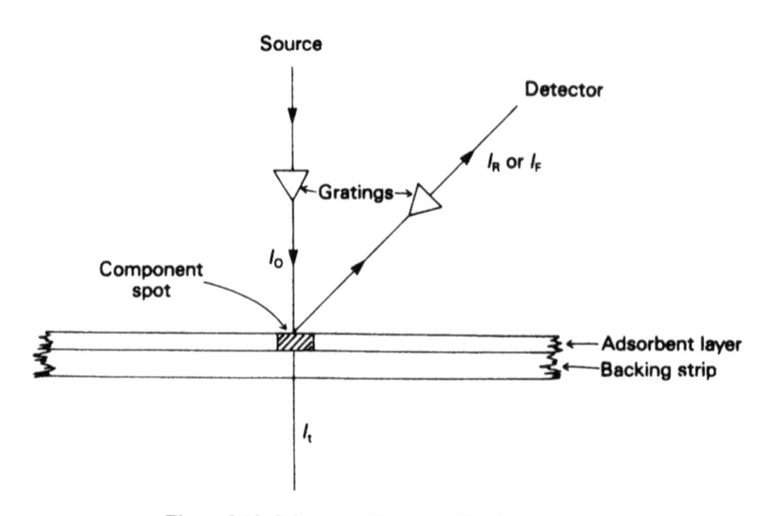

Figure 3.13 Schematic diagram of a densitometer.

values is not simply governed by the Beer–Lambert Law:

$$A \propto \epsilon c l$$

where A is $\log I_0/I_t$ (absorbance), ϵ is the absorptivity coefficient, c is the component concentration, and l is the pathlength (sorbent thickness). This deviation is due to (i) considerable scatter of the incident radiation, and (ii) variable thickness and irregularity of the surface.

Quantitative studies require calibration graphs to be constructed based on measurements of standard samples of known concentration.

(3) *Reflectance*

The instrument components required for reflectance measurements are similar to those for transmission. There are two standard geometric configurations for light source and detector, an incidence angle of 0° with the detection instrumentation at 45° or vice versa. Reflectance measurements are not so sensitive to variation in thickness and uniformity of surface layer, and give more precise and accurate measurements. A modified form of the Rubella–Munk function [64] relates intensity of reflectance beam to concentration, however, for routine work calibration curves would be used.

3.1.12.2.3 Fluorescence measurements. Compounds which emit light after irradiation are said to fluoresce. The intensity of the fluorescence is related to concentration and the measurement of this intensity is the basis of quantitation. As for absorption studies, fluorescence measurements can be carried out in the transmittance or reflectance modes, and the arrangement of the instrument components is similar to that depicted in Figure 3.13. The differences are minor. Mercury or xenon lamps are used due to their high-intensity and spectral range, the detection monochromator is adjusted for the appropriate fluorescence wavelength.

Where sources of this type are used then the beam of incident radiation is sufficiently intense to excite all the molecules and measurements in either the reflectance or transmittance mode are similar. As fluorescence measurements are absolute then this affords high signal amplification and fluorescence studies are 10 times more sensitive than absorbance. In practice, calibration graphs are used for quantitative measurements.

Measurements of fluorescence quenching relies on the uniform distribution of a fluorescent indicator throughout the matrix. Substances which absorb ultraviolet light, most organic compounds, thus appear as dark spots. The difference in the fluorescence intensity is measured and can be related to concentration. The advantage of the technique lies in the fact that it does not rely on the compounds themselves to fluoresce or being derivatised with fluorescent labels; however, though used routinely in qualitative work the lack of homogeneity of the fluorescent indicator restricts its use in quantitative analysis.

A variety of instrumentation ranging in complexity and sophistication is commercially available, including combined reflectance–transmission systems and double beam instruments. The systems are capable of giving quantitative accuracy ±2%.

3.1.12.2.4 Radiochemical detection. Radiolabelled compounds are widely used in radiotracer methods for following the course of chemical and biochemical reactions. For instance, the study of metabolic pathways of drugs involves adding a substrate, analysing the reaction mixture by taking aliquots at various times, separating the products by chromatography, then detecting the radiolabelled compounds by autoradiography, liquid scintillation counting or *in situ* measurement of radioactivity.

(1) *Autoradiography*
In this technique the thoroughly dried and developed chromatoplate is laid on top of a piece of film sensitive to X-rays. The film is exposed to the chromatogram for a precise period of time, determined by the type of film being used, and the type of radiation given off by the isotope. The film is then developed by normal photographic procedures. Spots appear as dark areas and can be quantified by densitometric procedures. The major disadvantage of this technique is that, depending on the activity of the isotope, suitable exposure times can vary from seconds to weeks.

(2) *Liquid scintillation counting*
This method is also referred to as scintillation autoradiography. In TLC, portions of the sorbent, where the components are located, are scraped off, placed in a vial and mixed with scintillation solution. The radiation emitted, for instance by ^{3}H nuclei, is converted into light, and can be detected either with photographic film or with a scintillation counter. The technique for paper chromatography is somewhat different in that the paper chromatogram is attached to a similarly sized piece of photographic film and dipped into a tank containing the scintillator.

(3) In situ *measurements*
In situ or direct measurements are carried out by scanning the chromatogram with a radiation sensitive device such as a thin end-window Geiger–Müller tube, with automated instrumentation which can be calibrated to give a quantitative evaluation of the spots. This produces a trace similar to a densitometer result. The result can be confirmed by elution of the substances and measurement of the active material in the eluate. In paper chromatography the area of a paper containing each spot can then be cut out, put on an ordinary planchette and counted accurately in a lead castle.

Perhaps the major advantage of radiotracer methods lies in their sensitivity. Due to the low concentration of metabolites present in biological samples, radiolabelling of the sample followed by thin layer radiochromatography (TLRC) is often the only reasonable method for following the course of biochemical reactions.

A more detailed discussion of isotope techniques and the detection of radioisotopes is outside the scope of this text. The interested reader is directed to the excellent exposes presented by Roberts [65] and references therein.

3.1.12.3 Flame ionisation detector and specific detector systems. TLC and HPTLC are fast and versatile analytical techniques and considerable time and energy has been expended to automate the various operational stages. In recent years there have been significant advances in the areas of sample application, solvent delivery, documentation and quantitation.

A further innovation is based on the use of silica sorbent fused to a thin rod which may be made of copper, steel or quartz. The support or Chromarod™ is in effect a short column typically 150 mm in length × 1 mm diameter, onto which sorbent is fused to a thickness of 75 μm. The sample (1–20 μg) is spotted on the rod. The rod is then located in a frame and the frame placed in a modified tank and eluted as per conventional TLC.

The solvent is allowed to evaporate, then the rod, still in its frame, is passed through the hydrogen flame of an FID [66], where the organic components (analyte bands) are combusted. The detector response is thus recorded against displacement distance along the rod. The frame can be moved in a controlled fashion and each rod is passed through the flame in sequence.

The rods are effectively regenerated and can be reused. If necessary, they can be decontaminated by dipping in chromic acid, washing thoroughly with water and then fully reactivated by passing the rod through the flame of the FID.

The Introscan system described above has been extended to use specific detectors, for example, for the selective determination of nitrogen and halogen compounds.

3.1.12.4 Miscellaneous detectors. TLC, as with other chromatographic methods, is a separation not an identification technique and thus for unambiguous identification of analytes the separated components must be examined by spectroanalytical techniques. Mass [67] and Fourier-transform infrared spectrometry [68] have both been used to good effect and considerable effort is currently being expended to develop robust methodologies and instrumentation in these areas. Instrumentation has recently been developed, for example, which elutes separated components directly onto a measured amount of potassium bromide which is then automatically pressed and introduced into an infrared spectrometer.

3.1.13 Preparative thin layer chromatography

TLC can be scaled-up and used for the isolation of large (10–100 mg) quantities of pure component. The practice of the technique is similar to that for analytical, qualitative scale work. The main difference lies in the plates used. Almost all preparative scale work is carried out, in the adsorption mode, principally on silica gel plates of varying thickness, 1–5 mm, and of 20 × 20 cm dimensions. The sample is applied as a streak, either by a pasteur pipette, syringe or a motorised 'streak applicator'. Advantage can be taken of multiple development techniques, which allow efficient separation of components of markedly different polarities. Bands incompletely resolved can be applied to a fresh plate and rechromatographed with a suitable solvent and development procedure. Once development is complete the 'bands' of component can be scraped off with a razor blade or spatula and the component washed off the adsorbent with a suitable solvent. Plates for preparative chromatography are available with added fluorescent indicator which facilitates non-destructive location of the components. The fluorescent indicator is irreversibly bound to the silica.

Preparative TLC is an ideal quantitative technique for radioactive and toxic substances and for feasibility studies of reactions of expensive pharmaceutical mixtures. A recent innovation has been the introduction of taper plates (Analtech) for use in preparative TLC work. The dimensions and features of the taper plate are illustrated in Figure 3.14. Sample concentration prior to separation occurs in the pre-adsorbent zone. The tapered adsorbent layer causes low R_f bands to separate further than on a preparative plate of constant thickness. A more uniform mobile phase flow pattern and reduced vertical band spreading, further enhance the performance.

3.1.14 High performance thin layer chromatography

As in other chromatographic techniques overall performance in TLC is inextricably linked to the physical characteristics of the sorbent, in particular its particle and pore size and the particle size distribution. Normal TLC plates, i.e. those spread with 5–40 μm particle size distribution are capable of giving 1000–2000 theoretical plates per 5 cm migration. High performance

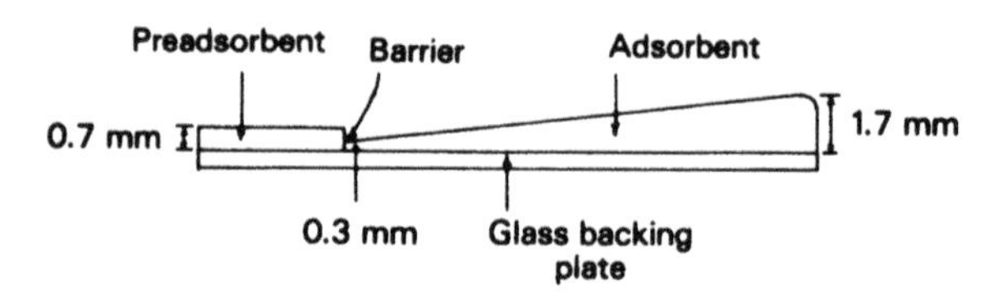

Figure 3.14 Cross-section of a taper plate with pre-adsorbent layer (Analtech).

thin layer chromatography (HPTLC) differs from ordinary TLC in that it uses a refined silica gel with a mean particle size of 5 μm, a very narrow particle size distribution and uses thinner layers (100 μm) which gives an extremely smooth homogenous surface. An advantage of using smaller particle size of sorbent is a reduction in the extent of lateral spreading during development thus affording improved resolution, up to 10 000 plates over 5 cm migration, with higher spot concentrations easing detection and giving improved sensitivity. Further advantages are in the marked—five- to 10-fold—reductions in analysis times and also the smaller the volume of eluant required for development. The latter feature opens up the possibility of using more specialised and exotic eluants. A consequence of the reduced particle size is that the sample capacity is also reduced this can be a disadvantage where lengthy clean-up involving dilution of samples is required which restricts the precision and accuracy in quantitative analysis.

However, due to the greater resistance to capillary flow eluant runs are only of the order of 8 cm and hence to capitalise on the benefits of HPTLC it is necessary to apply analytes to the plates in as small a spot diameter as possible. There are a range of semi automatic and automatic sample applicators available which use the technique of 'contact spotting' which allows as much as 1000 μl of analyte solution to be applied with a spot diameter of 0.1–0.5 mm. Pre-adsorbent layers as discussed earlier have also been used to advantage.

The full range of development procedures have been exploited with HPTLC and detection, whether by colour development, quenching of *in situ* fluorescent agents or by scanning densitometry techniques can be undertaken as for conventional layers.

In addition to the established reversed phase layers a wide range of chemically modified silica layers have been developed for HPTLC in many ways paralleling the developments in HPLC. Plates with amino and nitrile modification are available with and without fluorescent indicator; microcrystalline cellulose with a mean degree of polymerisation between 40 and 200 is also available in ready to use sheets.

3.2 Paper chromatography

Although published work in which paper chromatography (PC) is used continues to appear, use of the method has long since passed its peak. Developments in thin layer, gas and high performance column chromatography, with the increased speed and degree of instrumentation that they offer, have diverted attention from the cheaper, less sophisticated and poorer performing—inferior resolution and long analysis times (several hours)—paper methods. Nonetheless, PC is useful as a low cost approach to teaching the fundamentals and practice of chromatography.

The following is intended as no more than an overview and the interested reader is directed to the references cited herein and to the numerous reviews and bibliographies containing detailed information on the techniques of paper chromatography; a current listing of papers published appears in the bibliography section of the *Journal of Chromatography*.

3.2.1 Origin

Various types of simple separations on paper have been described as forerunners of paper chromatography, among them a method of Runge in 1850 for separating inorganic mixtures, and the process called 'capillary analysis'. However, chromatography and specifically PC, was not established until the late 1940s, arising from the pioneering work of Consden *et al.* [69]. Following their studies of the separation of amino acids using cellulose powder columns, they used paper in the expectation that the bound water in the paper would serve as a basis for partition chromatography. Using a two-way development technique they achieved the separation of eighteen amino acids and peptides in wool protein hydrolysates. The PC technique devised was rapidly and successfully applied to the analysis of amino acids and peptides in a wide variety of matrices and was quickly adapted and modified for the analysis of numerous classes of organic compound and inorganic ions—both cationic and anionic.

The use of paper as a chromatographic medium is usually regarded as a typical partition system, where the stationary phase is water, held by adsorption on cellulose molecules, which in turn are kept in a fixed position by the fibrous structure of the paper. It is now realised, however, that adsorption of components of the mobile phase and of solutes, and ion exchange effects, also play a part, and that the role of the paper is by no means merely that of an inert support. The technique devised by Consden *et al.* has not undergone any fundamental changes, but there has been considerable improvement in detail, resulting partly from the wide variety of commercial apparatus which is available. Elaborate or expensive equipment is not essential, however, and good results can be obtained with quite simple apparatus and materials.

3.2.2 Overview of the technique

PC may be described as the technique for the separation of substances using paper (a mat of cellulose fibres) as the chromatographic sorbent with a liquid mobile phase. The forces retarding the components and preventing them being located at the solvent front derive from the special physical/chemical or structural properties of the paper used as chromatographic medium, and are in part due to partition, adsorption and ion exchange sorption phenomena. The practice in terms of sample application and development and location techniques is in essence the same as for TLC. Paper chromatography is considered and used primarily as an analytical technique and though a range of preparative procedures have been developed their use has declined dramatically in favour or preference to thin layer preparative techniques.

3.2.3 Sample preparation

3.2.3.1 Preparation of specimens. The mixture to be separated is applied to the paper as a solution. The solution should be ~0.5% of which 10 μl are applied to the paper. The precise amount and concentration will depend on the complexity of the sample mixture and the sensitivity of the detection system. The nature of the solvent is immaterial, as long as it will evaporate completely without leaving a residue, and without attacking the paper. Solid samples, such as oils, or biological cell or tissue material, are macerated with the solvent, or submitted to some standard extraction procedure, such as Soxhlet. Many important samples, such as urine or other biological fluids, are already in an aqueous medium. In other cases water is used as the solvent when the substances are soluble in it.

3.2.3.2 Removal of matrix interferants. The extraction procedures used in sample preparation inevitably extract more than just the compounds to be tested for analytical interest, and in addition biological materials frequently contain substances not of analytical interest, which when present in large amounts may have a deleterious effect on the chromatography. For instance, aqueous biological extracts, urine, neutralised protein hydrolysates and other solutions which may have to be examined for amino acids and sugars always contain appreciable amounts of inorganic material which impairs the separation process. Removal of these is called 'desalting'; it is important, and should always be carried out if it can be done without affecting the organic compounds.

Organic interferants can be removed by extraction or ultrafiltration.

3.2.4 Types of paper

The majority of paper chromatography has been carried out on standard filter paper material, however there are still commercially available a range of chromatography papers:

- pure cellulose;
- silica gel loaded;
- ion exchange cellulose; or
- resin loaded.

These papers are manufactured to a high specification with controlled porosity, thickness and matting characteristics and are low in metal content.

Even so variations in these properties can be encountered with the same type of paper and thus procedures for the storage, pre-treatment and handling of the paper should be standardised in order to ensure reproducible chromatography. In order to appreciate more fully the processes occurring during development, it is necessary to look closely at the structure of paper itself. Paper is a random pile of cellulose fibres, each fibre comprises a number of chains of ~2000 anhydroglucose units. The structure of the cellulose chain is shown in Figure 3.4.

As indicated, there may be up to three hydroxyl groups per glucose monomer but during the manufacturing process many of these undergo oxidation to carbonyl and carboxyl functional groups. These polymeric chains can then interact via hydrogen bonding to give a film with highly ordered crystalline regions, with some non-interacting amorphous regions. Cellulose material exhibits a strong affinity for water and adsorbs between 5 and 20% by weight depending on the nature of the paper, the humidity and temperature. It is estimated that about ~7% of the water is strongly held via H bonding probably in the crystallite region, the remainder being present as more loosely bound surface water.

Thus in practice a range of sorption mechanisms—partition, ion exchange and adsorption—will all be at work during a separation though one or other may dominate depending on the nature of the sample and on whether a modified paper is being used.

3.2.4.1 Ion exchange papers. A combination of the specificity of ion exchange with the convenience of paper chromatography is afforded by ion exchange papers. There are two kinds. One consists of cellulose where acidic groups have been introduced by chemical modification of the –OH groups, e.g. conversion to carboxylic, phosphoric and sulphonic acids and to *N*,*N*-dialkylamines. These papers are suitable for the separation of cations, amines and amino acids.

Alternatively, ion exchange capability may be introduced by blending an ion exchange resin with cellulose, and making sheets with the mixture in the normal way. Paper so obtained contains about 45% of resin by weight. By using the appropriate ion exchange resin papers with strong and weak acid character and strong and weak basic character can be produced. These papers have been used successfully for the chromatography of inorganic anions and cations.

Other modified forms of paper have been utilised in which the paper has been impregnated with alumina, silica, ion exchange resins and organic stationary phases giving improved separations for a number of applications.

3.2.4.2 Reversed phase methods. If the substances being chromatographed are only very sparingly soluble in water, they merely move with the solvent front, and thus no separation results. In such a case it may be advantageous to impregnate the paper with a non-aqueous medium, to act as the stationary phase. In 'normal' chromatography, the stationary phase, being aqueous, is more polar than the mobile. The technique where the mobile phase is the more polar is called 'reversed phase' chromatography. The mobile phase is not necessarily water, although the mixtures used normally contain some water.

A number of substances have been employed as supports for the stationary phase, among them rubber latex, olive oil and silicon oils and similar materials. A number of papers impregnated with these materials are now available commercially. If the paper is treated with rubber latex or silicones it is rendered water-repellent, and it absorbs the organic component of the solvent mixture, which then becomes the stationary phase, in preference to the water. Where liquids such as olive oil are used they perform a similar function, but may also act as the stationary phase themselves. There must also be some modification of the adsorption and ion exchange effects of the paper. Reversed phase methods can be applied to all the various forms of paper chromatography. The general procedure is the same as for normal methods. The solvent for the locating reagent must be one in which the stationary phase support is not soluble.

3.2.4.3 Dual phase PC. The combination of ion-exchange separation in one direction with a conventional chromatographic separation in the other has been used to good effect in the separation of amino acids. The following conditions are suitable:

(a) cellulose phosphate paper: H+ form;
 solvent 1: 0.02 mol dm^{-3} sodium buffer, pH 4.7;
 solvent 2: *m*-cresol/1% ammonia;
(b) diethylaminoethyl (DEAE) cellulose paper: free base form;
 solvent 1: 0.02 mol dm^{-3} acetate buffer, pH 7.5;
 solvent 2: *m*-cresol/1% ammonia.

In the first solvent for each separation the amino acids were in the cationic form in the pH 4.7 buffer in cellulose phosphate paper and the anionic form in the pH 7.5 buffer on DEAE cellulose. In the second solvent they were in the anionic form in both cases, and thus on cellulose phosphate paper no ion exchange occurred, the compounds behaved as they do on ordinary paper. On the DEAE cellulose the partition solvent probably inhibits the ionisation of the exchange groups, giving a separation mainly of the ordinary chromatographic type. The separation obtained is illustrated in Figure 3.15, with a diagram of the same separation carried out by a normal two-way procedure for comparison. It will be observed that the resolution is much better on the strongly acidic cation-exchange paper, and that the acids are more evenly spread over the sheet.

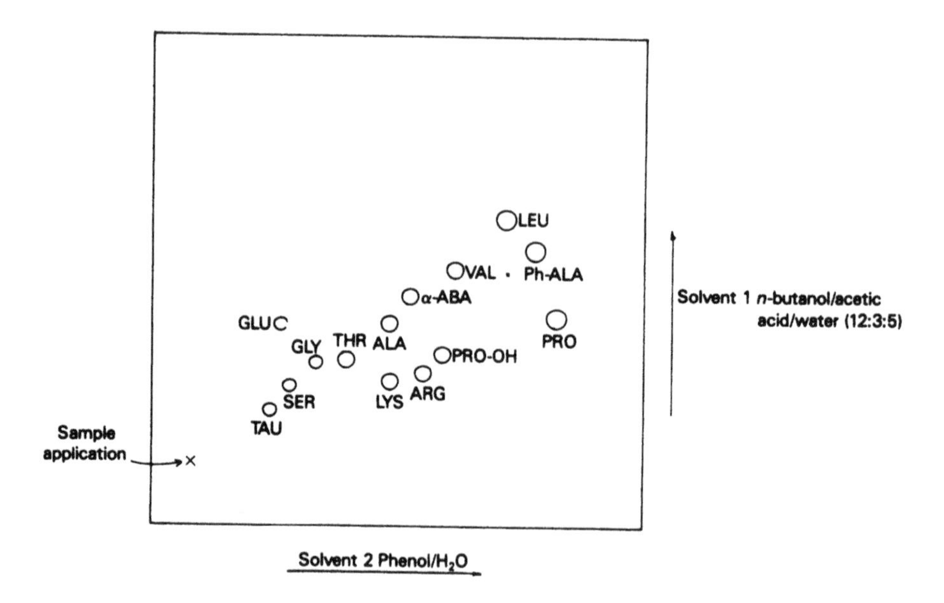

Figure 3.15 Separation of amino acids.

3.2.5 Solvents

The mobile solvent is normally a mixture consisting of one main organic component, water, and various additions, such as acids, bases or complexing agents, to improve the solubility of some substances or to depress that of others. An antioxidant may be included to stabilise the solvent.

The choice of eluant is largely empirical and is dictated by the complexity of the mixture and whether the stationary phase is hydrophilic or hydrophobic; however, a few general principles can be highlighted.

For polar organic substances more soluble in water than in organic liquids, there will be little movement if an anhydrous mobile phase is used; adding water to the solvent will cause those substances to migrate. Thus, butan-1-ol is not a suitable solvent for amino acids unless it is saturated with water; addition of acetic acid allows more water to be incorporated, and hence increases the solubility of amino acids, particularly basic ones; the addition of ammonia increases the solubility of acidic materials. *tert*-Butanol and water mixtures are the primary solvent for the separation of many polar anionic species, and many other polar substances with solubility characteristics similar to those of amino acids, such as indoles, guanidines and phenols, can be separated with this mixture. For hydrophobic stationary phases, various mixtures of benzene, cyclohexane and chloroform have been used to good effect as eluants.

Inorganic ions are usually separated as complex ions or chelates with some solubility in organic solvents; for instance, iron forms a complex chloride ion

Table 3.5 Solvent systems for PC applications

Compound class	Solvent	Proportions	Analyte
Hydrophilic compounds	Phenol/water	Sat. soln.	Amino acids
	Phenol/water/ammonia	200:1	
	Butanol/water/acetic	4:1:5	
	Butanol/water/pyridine	1:1:1	
	Isopropanol/water/ammonia	9:1:2	
	Butanol/ammonia	Sat. soln.	Fatty acids
Moderately hydrophilic substances	Pyridine/EtOAc/water	2:1:2–12:5:4	Sugars
	Formamide/chloroform	1:9–9:1	
	Formamide/$CHCl_3$/benzene[a]	1:9–9:1	
Inorganic	Acetone/water/conc. HCl	87:8:5	Co, Mn, Ni, Cu, Fe (chlorides)
	Pyridine/water	9:1	F, Cl, Br, I (Na salts)
	n-Butanol/HCl (3 mol litre^{-1})	Sat. soln.	Hg, Pb, Cd, Cu, Bi (chlorides)
	Pentan-2,4-dione (sat. soln. in water)/acetone/conc. HCl	149:1:50	As, Sb, Sc (chlorides)

[a] Paper impregnated with 60% ethanolic formamide prior to use.

which is very soluble in aqueous acetone, whereas nickel does not so readily form such an ion; iron and nickel can therefore be separated with this solvent, so long as hydrochloric acid is present to stabilise the complex ion.

Similar arguments can be applied to aid the choice of solvent system for any particular application. However, a suitable choice of eluant for new work can usually be made from the mass of applications literature. Whilst there have been a vast number of solvent systems developed many are simply modifications of the primary solvents listed in Table 3.5.

In general a solvent system with as low a polarity as possible consistent with an adequate separation should be used. Polar components of the eluant can be strongly adsorbed onto the cellulose matrix forming a stationary phase which may not give the desired properties.

3.2.6 *Equilibrium*

Study of the literature of paper chromatography reveals many views on the establishment of 'equilibrium' in the tank and some authors have given it special attention [70]. It is not possible to generalise because the degree of equilibration needed depends on the size of the apparatus, the solvent system and the nature and purpose of the separation. The object is to prevent evaporation of the solvent from the paper. For most applications it seems that it is not necessary to go to extreme lengths, and some of the meticulous and elaborate equilibration carried out by earlier workers was probably not really necessary.

Since the basis of the paper chromatography process is distribution between cellulose-bound water and a moving organic solvent, it has been

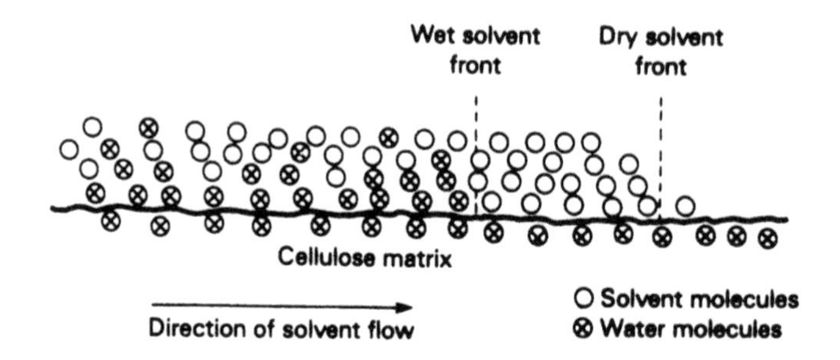

Figure 3.16 Schematic diagram of the solvent layers in a paper chromatogram.

usual to choose as the mobile phase an organic liquid which is only partly miscible with water, and to saturate it with water before use. It is probably best to consider that there is a layer of water molecules held on the cellulose by hydrogen bonding, and that this (the stationary phase) absorbs the solute molecules from the moving solution. There is thus no definite solvent–solvent interface, but there is a gradual change in composition from polar solvent (water) to the less polar organic solvent in the direction away from the surface of the cellulose molecules (Figure 3.13). This is only possible if the layers concerned are fairly thin, and explains why an excess of either phase ('flooding') prevents separation by causing extensive diffusion.

As the mobile phase flows through the fibres of the paper the cellulose will tend to adsorb more water from it. The result is that the moving solvent tends to become denuded of water as it advances, and its composition is not constant along the sheet (Figure 3.16). Sometimes there is a definite boundary where the solvent composition changes, for example, in some inorganic separations where acetone–water–hydrochloric acid mixtures are used as the solvent there is a 'dry' solvent front, and, some instance behind it, a 'wet' solvent front. The forward area consists of acetone from which the water has been removed, and the area behind the wet solvent front consists of aqueous acetone, and therefore contains all the acid.

This purely descriptive analysis reveals a situation of great complexity. If the composition of the solvent varies continuously along the paper, the composition of the vapour with which it is in equilibrium also varies. As it is not possible to arrange a corresponding concentration gradient in the atmosphere in the tank, there may tend to be further changes in solvent composition due to evaporation of the more volatile constituents.

3.2.7 Development

The range of development techniques available in paper chromatography is similar to those employed in TLC, namely, ascending, descending, radial,

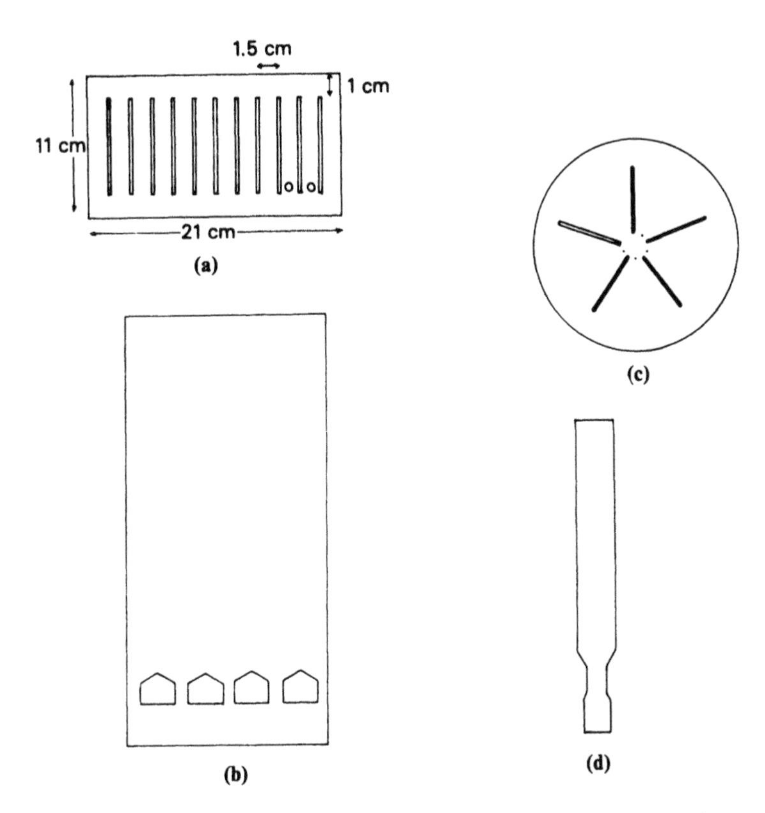

Figure 3.17 Various designs of chromatographic paper: (a) slotted paper for multiple developments; (b) multiwedge strip; (c) paper with radial slots for sector circular development; (d) constriction of paper to reduce the flow of solvent.

two-dimensional, multiple and stepwise (Figure 3.17). The principles of the techniques are identical the only variations arise from the difference in design of the various equipment, for instance, special frames are commercially available for supporting the papers and the variety of paper sheet designs available. A special paper for radial chromatography, developed by Kawerau has five radial slits, which help to homogenise the atmosphere in the container and keep the migrating zones separate from each other.

The modifications of vertical development techniques such as two-dimensional, stepwise and multiple are as for TLC and will not be discussed further. Typical apparatus and experimental set-up for each of the principal development procedures is shown in Figure 3.18.

3.2.7.1 Radial development techniques. The advantages and practice of radial development have previously been discussed in the context of TLC. One of the major restrictions to the wider application and advancement of

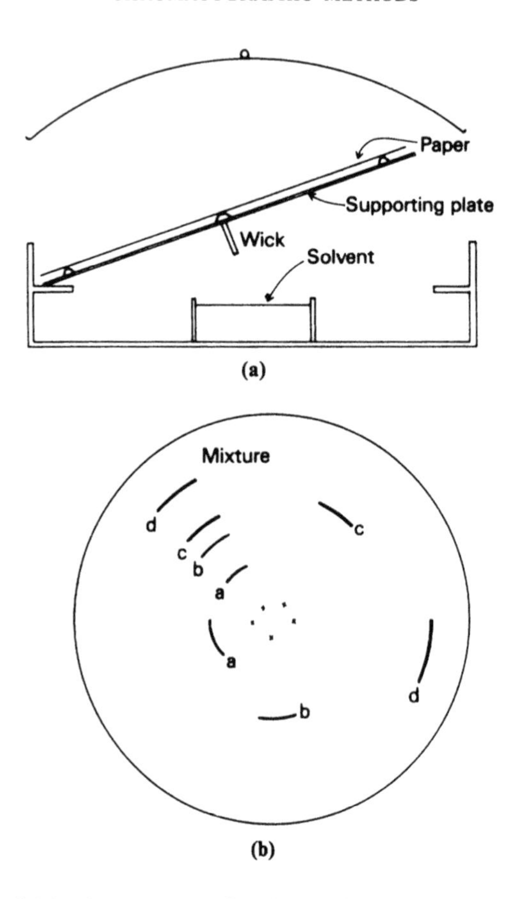

Figure 3.18 (a) Radial development system for PC; (b) radial chromatogram of a simple mixture.

paper chromatography as an analytical separation technique was the lengthy development times (~several hours). A development/modification of horizontal radial chromatography, known as centrifugally accelerated chromatography, uses circular paper; the sample(s) are applied at the centre and the solvent is allowed to drip onto the paper from above. The innovation was to spin the paper at 300–1500 rpm, thus reducing the development time for many applications to a few minutes [71], due to the centrifugal acceleration of the solvent flow. The centrifugal method has been found useful in the separation of compounds labelled with radioactive isotopes with short half-lives [72].

3.2.8 Sample application and detection

The majority of the techniques employed in TLC for the application and location of sample components can be used with the minimum of modification in paper chromatography. In the latter case for instance, chemical derivitisation (though the more corrosive agents such as sulphuric acid cannot be used on paper), physical methods, e.g. ultraviolet, fluorescence and densitometer techniques and radiochemical methods. Due to the more diffuse spots encountered in paper compared to TLC, detection sensitivities are ~100 times less sensitive on the former. Methods of quantitation similar to those used in TLC and based on the development techniques mentioned above, have been developed for paper chromatography.

3.2.9 Identification

As in TLC studies the accepted method of evaluating paper chromatograms is by determination of retardation factors. These can be obtained with equal facility for vertical and horizontal development techniques and the underlying principles and concepts are as for TLC. However, coincidence of R_f values should never be taken as absolute identification and for complete and unambiguous structural characterisation the components should be eluted from the sorbent layer, isolated and further spectroanalytical studies carried out.

3.2.10 Quantitative methods

Chromatography is a method of separating substances in a mixture. Quantitative use of the technique requires not only a quantitative separation, but also quantitative location and evaluation of the substances present. If a good chromatographic separation can be obtained, then the quantitative application depends solely on the last factor. A satisfactory qualitative separation is not necessarily useful quantitatively. The quantitative finish can be either by estimation of the amount of substance in the spot on the paper *in situ*, or by removal of the substance from the paper, and analysis of the separate fractions by conventional quantitative techniques. The techniques and approaches employed in TLC are for the most part applicable to PC.

3.2.11 Applications of paper chromatography

The literature on analytical methods and on the investigation of natural compounds shows that there are hardly any fields in which paper chromatography has not found some use, although it was most widely employed in separations of a biochemical nature.

Since the early 1970s paper chromatography has gradually given way to thin layer though it still offers a few advantages in comparison to TLC, for instance, in cost, effectiveness in separating polar and water-soluble compounds and without elaborate modification continuous development techniques lead to the separation of compounds of low R_f. Some of the principal uses, with examples, are summarised below. It is emphasised that the actual separations mentioned are given merely for illustration, and it is not suggested that the list is in any way comprehensive.

3.2.11.1 Clinical and biochemical. Separation of amino acids and peptides is regularly carried out to aid in the investigation of protein structures. Routine examination of urine and other body fluids for amino acids and sugars (this is most important, as it can be used for diagnosis of a number of pathological conditions, with the 'standard map' technique). Separation of purine bases and nucleotides in the examination of nucleic acids. Separation of steroids.

3.2.11.2 General analytical. General analytical applications include, analysis of polymers, detection and estimation of metals in soils and geological specimens, investigation of phenolic materials in plant extracts, separation of alkaloids and separation of radioisotopically labelled compounds.

3.3 Electrophoresis

The basis of electrophoresis rests in the differential rate of migration of ion molecules in an electrolyte solution when under the influence of an applied electric field. Although not in principle a chromatographic technique, electrophoresis used in conjunction with paper chromatography and gel materials, proves an extremely useful method for the separation of charged substances, ranging from small ions to large charged macromolecules, of biological and biochemical interest. Many of these compounds contain acidic and basic groups which are readily ionisable, the extent of ionisation being dependent upon the composition or pH of the matrix for instance amino acids have two distinct pK_a values and the actual form of any amino acid depends upon the pH of the solution. The ionic forms of glycine which predominate at various pH values are illustrated in Figure 3.19.

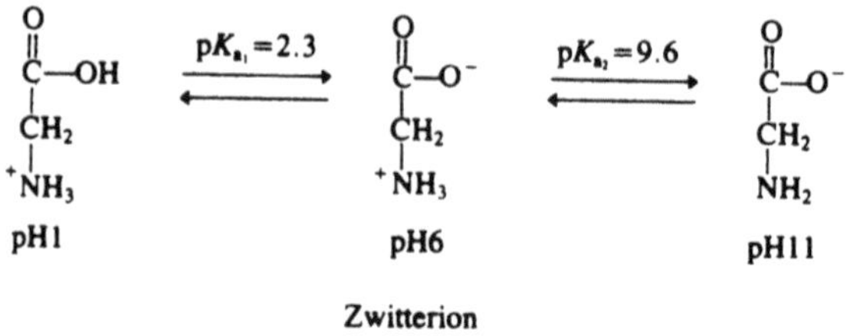

$$HOOC{-}CH_2{-}NH_3^+ \underset{}{\overset{pK_{a_1}=2.3}{\rightleftharpoons}} {}^-OOC{-}CH_2{-}NH_3^+ \overset{pK_{a_2}=9.6}{\rightleftharpoons} {}^-OOC{-}CH_2{-}NH_2$$

pH1 | pH6 (Zwitterion) | pH11

Figure 3.19 The different ionic forms of glycine and the pH at which the form predominates.

The dipolar form of the amino acid is known as the zwitterion, and the pH at which the net charge on the amino acid is zero, the isoelectric point (pI). At other pH values, amino acids and like species can exist in solution either as cations or anions, the charge carried depending upon the solution pH. When placed in an electric field these charged species will migrate to the appropriate electrode. The driving force of migration is the resultant of the electrostatic forces of attraction, between the electric field and the charged molecule, and the retarding forces due to friction and electrostatic repulsion from molecules of the transport medium. The electrostatic forces of attraction are proportional to the mass/charge ratio and thus molecules of different mass/charge ratio will migrate at different rates, when placed in an electric field. The technique can be extended to non-polar compounds such as carbohydrates and sugars by derivatisation as the borate and phosphate [73].

Migration usually takes place in a buffered medium to ensure constant pH as changes could result in alteration of the charges borne by the species under examination. Electrophoresis may be conducted in free solution, where the species are free to move as soon as the applied field is in force. Migration is rapid as there is minimal frictional resistance. This technique is known as moving boundary or frontal electrophoresis, and was pioneered by Tiselius and co-workers [74]. The method found initial application in the separation of proteins. However, it requires sophisticated scanning optics to detect changes in the refractive index at boundary interfaces, and is further limited by convective instabilities and by the development of 'false' salt boundaries. Diffusion begins after the voltage is removed, resulting in rapid remixing. Of greater utility is zone electrophoresis, in which the separation, according to electrophoretic mobility, is carried out on relatively inert supporting media, such as paper, agarose gels, cellulose acetates and acrylamides. The frictional resistances are considerably increased resulting in the components migrating and separating out as distinct zones or bands. The position of spots on the electrophoretogram can be readily located by procedures previously discussed for paper chromatography.

3.3.1 Procedure of zone electrophoresis

A paper strip, suitably supported, dips at its ends into electrode vessels containing a buffer solution, acting as the electrolyte. The paper is soaked in the buffer, and the sample is applied at some point on the strip as a thin transverse streak (by the same method as used in paper chromatography). The electrodes are connected to a d.c. source, and field applied for a predetermined time; see Section 3.3.4 for further details. The strip is removed from the apparatus, dried, and bands located—again as in paper and thin layer chromatography.

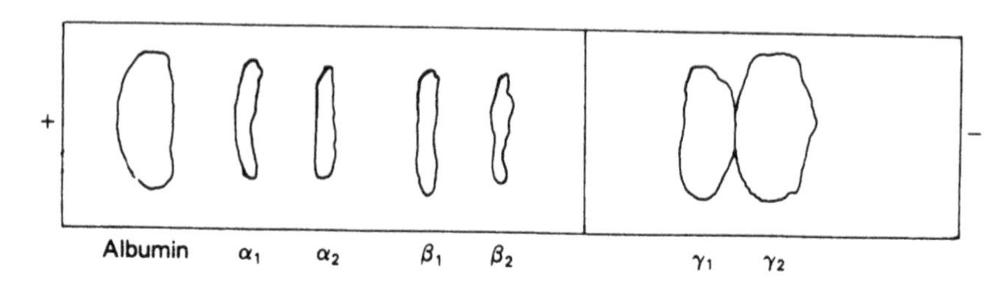

Figure 3.20 Electrophoretogram of plasma proteins on cellulose acetate at pH 8.6 (0.5 M barbitone).

The separated substances are then apparent as a series of bands, whose distance from the origin depends on the charge on the ion, its mobility in the field applied (and thus on the applied voltage and the current) and the pH of the buffer. Whereas in chromatography the separation is achieved by means of a flowing solvent which differentially moves the various solutes, in electrophoresis the electrolyte is stationary and the flow of ions occurs by virtue of their charge. The only movement of solvent is the electro-osmotic flow (produced by ionic charges induced on the supporting medium), which is only slight, but which may restrain the movement of some of the ions of lowest mobility.

This technique, with its similarity to paper chromatography, is called 'paper electrophoresis'. The paper acts as a support for the electrolyte, and also for the separated substances. It restrains their diffusion in the buffer solution, and holds them on drying so that the locating reagent can be applied (Figure 3.20).

The nature of the paper has very little bearing on the separation, and, since other supports are now available the terms 'zone electrophoresis' or 'electrochromatography' are commonly used. Electrophoresis can be used in conjunction with conventional chromatography. A paper sheet cut as shown in Figure 3.21 can be used, in suitable apparatus, to give useful two-way separations of substances such as amino acids and similar polar species.

3.3.2 Factors affecting migration rates

3.3.2.1 Electric field. The movement of ions on the paper is caused by the potential difference applied across the paper. The important parameter determining the extent of migration (d) is the voltage gradient, i.e. the applied voltage (V) divided by the distance (L) between the electrodes:

$$d \propto (V/L)$$

The extent of the migration increases with the increase in the voltage gradient. The speed of migration has been found to quickly attain a constant

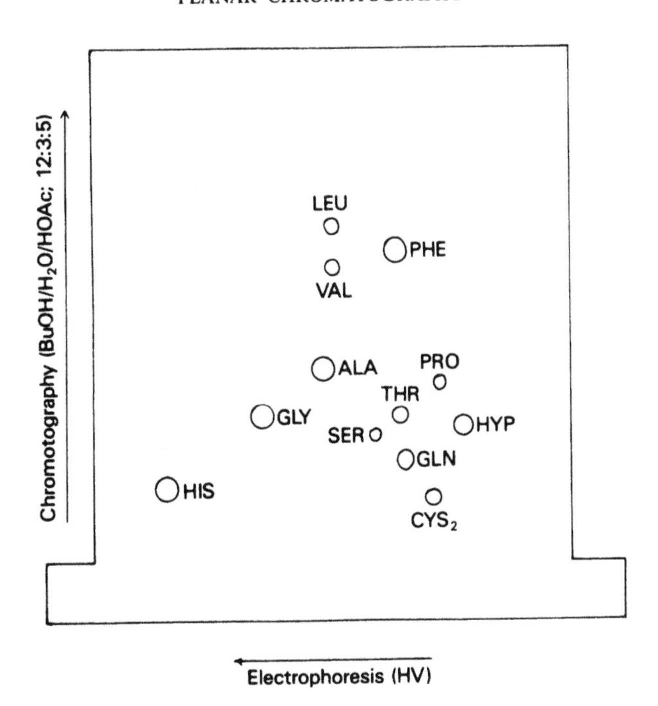

Figure 3.21 Arrangement of paper sheet for combined electrophoresis and chromatography.

value, the terminal velocity, when no net force acts on the charged ion, the frictional forces arising from movement through the medium being exactly balanced by the applied electrical force. For species of the same charge, the terminal velocity achieved is determined by their size and shape.

The larger the molecule, then the greater the frictional and retarding electrostatic interactions and hence the more slowly it will move. However, molecules of similar size can have different shapes; for instance, secondary and tertiary interactions in protein macromolecules, leads to fibrous and globular proteins, which due to the differential effect of frictional and electrostatic forces can result in different migration characteristics. The characteristics of each species, such as net charge, shape and size, overall charge distribution, electrostatic interactions, and their influence on the extent of electrophoretic migration are embodied in its electrophoretic mobility, μ. Thus, the distance moved, d, is related to the voltage gradient applied and the time, t, of migration by the compounds electrophoretic mobility constant (μ):

$$d = \mu t(V/L)$$

The separation of two species of different electrophoretic mobility, after time, t, is given by

$$(d_1 - d_2) = (\mu_1 - \mu_2)t(V/L)$$

During the course of electrophoresis current flows and as in electrolysis the products are oxygen and hydrogen.

$$H_2O + 2e \rightarrow 2OH^- + H_2\uparrow \qquad \text{Cathode}$$

$$2H^+ + 0.5\,O_2 + 2e \rightarrow H_2O \qquad \text{Anode}$$

(The applied voltage is normally removed before the ions of analytical interest reach the electrodes.) The current flowing generates heat in the chromatostrip which results in evaporation of the solvent. The effective increase in electrolyte concentration, results in a decrease in electrical resistance, in part, due to the increased mobility of ions due to reduced electrostatic repulsion from solvent molecules. Consequently, the current flowing will increase, resulting in greater heat generation and solvent evaporation. In paper electrophoresis, the heat generated is relatively small and easily dissipated at small voltages (~100–400 V). Evaporation is reduced to a minimum by enclosing the apparatus under an airtight cover. Other media, such as layers of cellulose acetate, are preferable for work under constant current condition. Likewise, for certain separations, e.g. of low molecular weight. A further procedure to minimise heat generation is to use organic buffers, for instance, aqueous pyridine and acetic acid solutions, as these ionic species are not nearly as inducting as their inorganic counterparts. High voltage electrophoretic studies require cooling plates to aid dissipation of heat.

3.3.2.2 Buffer. The importance of buffer pH on the mode of ionisation or organic compounds has previously been discussed. Amino acids (and ampholytes) at their isoelectric point exist as the zwitterion and will not migrate in an electric field; however, if the pH is made slightly more acidic the protonated form of the amino acid dominates and migration will occur to the cathode. The converse is true if the pH of the buffer solution is slightly increased, i.e. migration occurs to the anode. The extent and direction of migration of many compounds of biological interest is thus pH dependent.

A range of buffers is available both inorganic and organic, for instance, EDTA, phosphate, Tris, citrate, barbitone, acetate and pyridyl. In addition, the use of borates as the electrolyte has allowed the technique to be extended to carbohydrates and polyhydric alcohols. These compounds form ionised complexes with borate reagents (e.g. cetyltrimethylammoniumborate). Organic buffers as well as producing less heat during development of the electrophoretogram are also easily removed prior to location of the components on the chromatostrip.

The migration rate of the solutes is controlled not only by the pH of the electrolyte but also by its concentration, normally referred to as its ionic strength. As the ionic strength of the buffer solution increases, it carries a higher proportion of the current, with a concomitant reduction in the proportion carried by the sample. Although this gives compact bands, the time required to develop the electrophoretogram is excessive and the problems encountered with heat generation and solvent evaporation, prevent high concentrations being used. On the other hand, if too low a concentration is used, migration times are very short and this, with extensive diffusion of the solute bands, results in a loss of resolution. Therefore, as a compromise, buffer concentrations used, normally lie in the range 0.05–0.1 M.

3.3.3 Supporting media

Several different supports and numerous designs of apparatus can be used for electrophoresis. The supports can be classified broadly as strips, gels, and thin layers; they may be totally inert, or they may have a physical effect on the separation.

Materials such as filter paper, cellulose acetate, gels made from starch, agar or polyacrylamide and thin layers of silica and alumina have been used as the supporting medium for the electrolyte solution and sample.

3.3.3.1 Paper. Filter paper such as Whatman no. 1 and no. 3MM in strips 3 or 5 cm wide have been used to good effect. The latter has a slightly greater wet strength, which is an advantage when aqueous solutions are being used. Since the paper is thicker it will pass a higher current than no. 1 for a given voltage. Apart from the use of still thicker papers for preparative separations of proteins and oligonucleotides there is nothing to be gained by changes in the type of paper, except that ion exchange papers can be used to improve separations of some mixtures by the two-way electrophoresis chromatography technique [75].

While paper offers the advantages of economy and ease of use, certain classes of compound, such as proteins and other large hydrophilic molecules, cannot be adequately resolved due to the adsorptive and ionogenic properties of paper which results in tailing and distortion of component bands, e.g. biological molecules such as proteins. Another significant problem with paper is caused by the effect known as electro-osmosis which results in the paper taking on a negative charge when in contact with water due to migration of labile hydrogens into the buffer solution. Paper electrophoresis is typically restricted to low field strengths 20 V cm^{-1}; field strengths up to 200 V cm^{-1} have been used but require very efficient cooling systems.

3.3.3.2 Cellulose acetate. Due to the limitations of paper enumerated above, cellulose acetate has been developed as an alternative. Cellulose acetate contains two to three acetyl groups per glucose unit and its adsorptive capacity is substantially less than paper. Thus, for many purposes strips of cellulose acetate are preferred to paper, since they give sharper bands, and are more easily rendered transparent, by impregnation with mineral oil of similar refractive index, for photoelectric scanning.

The low solvent capacity of cellulose acetate chromatostrips enables higher voltages to be used further enhancing the resolution. In practice, no tailing of proteins or hydrophilic materials occurs. The presence of residual sulphonic and carboxylic acid residues within the cellulose acetate matrix is a disadvantage due to the induced electro-osmosis during electrophoresis. They are available in a range of particle size and layer thickness. Cellulose acetate though expensive gives fat run times and is used extensively for the analysis of clinical and biological protein samples.

3.3.3.3 Gels. Gels are three-dimensional semisolid colloids which are commonly prepared from starch, agar, or polyacrylamide for use as supports in electrophoresis. As discussed earlier the rate of migration of a species is determined by its electrophoretic mobility. The frictional resistance is dependent on the pore size of the gel relative to the radius of the molecules. In gels the pore size can be finely controlled by adjusting the amounts of monomer/polymer and cross-linking reagent. Hence the resolving power in electrophoretic studies using these materials is considerably enhanced due to the molecular filtration or sieve effect operating. The first successful gel electrophoretic study was carried out on a starch medium on human serum and showed ~15 bands. The paper electrophoretogram showed only five zones.

3.3.3.3.1 Starch. For the preparation of a starch gel, a suspension of granular starch should be boiled in a buffer to give a clear colloidal suspension. The suspension on cooling sets as a semisolid gel due to the intertwining of the branched chains of amylopectin. The porosity varies with the percentage of starch in the buffer solution; 15% solutions give low porosity gels, 2% solutions afford high porosity. Though some molecular sieving is achieved with starch gels, difficulty is encountered in achieving consistency of pore size from batch to batch. A further disadvantage is the presence of charged groups which can set-up an electro osmotic effect. Starch gel electrophoresis is carried out with the gel mounted horizontally.

3.3.3.3.2 Polyacrylamide gels. Polyacrylamide gels are synthesised by cross-linking with *N,N*-methylenebisacrylamide, the chain polymer of acrylamide ($CH_2=CHCONH_2$) (Figure 3.22). The extent of cross-linkage is oxygen dependent and thus enables gels of varied but precise porosity to be synthesised.

Cross-linkage may also be carried out using *N,N′*-bisacrylcystamine. The co-polymer contains disulphide cross-linkage units which may be cleaved after hydrolysis and the gel solubilised, thus facilitating recovery of the separated components. Other solubilisable polyacrylamide media are

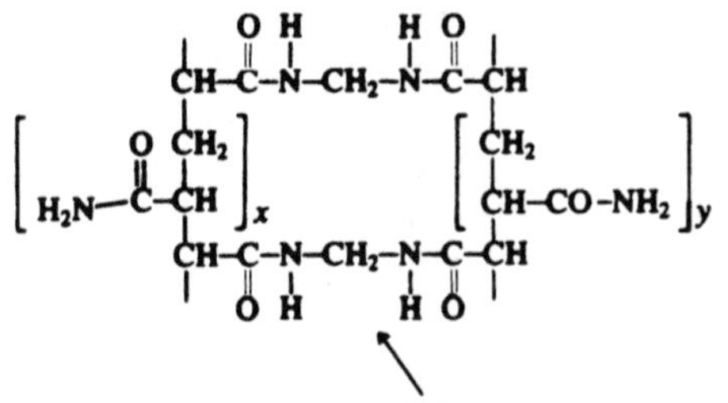

Cross-linkage unit

Figure 3.22 Partial structure of polyacrylamide gel.

commercially available which have suitable properties for isoelectric focusing, isotachophoresis and gradient work.

Though proteins are more stable in starch gels, acrylamide-based gels have a number of advantages. They are superior to starch gels with respect to resolution, versatility, R_f reproducibility, ease of handling and heat dissipation. They may be synthesised with a high degree of reproducibility and the extent of porosity can be varied within wide limits to enhance the separation of molecules of similar charge, but different size and shape, such that macromolecules can either be allowed access to the gel or excluded.

The applicability of acrylamide gels is wide and they have been used in a variety of techniques, for example, pore-gradient electrophoresis, molecular weight characterisation and isoelectric focusing. Arguably the most important application is in the molecular weight characterisation of proteins. The sample pre-treatment required, however, to produce similar mass/charge ratios is involved; detailed information of the subunit structure of the proteins is essential and furthermore standards composed of the proteins of analytical interest must be available for calibration studies. These gels can be used horizontally or vertically in the form of slabs or vertically in the form of tubes.

3.3.3.3.3 Agarose. Agarose is the name given to a mixture of polysaccharides, principally agarose and agaropectin, and commonly referred to as agar. The porosity is determined by the percentage of agarose in the gel. Agarose gel electrophoresis can be carried out in both the horizontal and vertical configurations. For the former, concentrations as low as 0.2% (w/v) can be used which allows molecules of 150×10^6 Da to migrate; the latter requires 0.8% (w/v) agarose which still allows passage of molecules of 50×10^6 Da. The methods of preparation of agar gels are principally concerned with removing those polysaccharides with charged and highly polar substituent groups, as these confer undesirable adsorptive effects and can denature the protein. Agarose gels are not covalently cross-linked but are stabilised by helical formation between individual polymer molecules, they have large pore size, low resistance to macromolecular measurement and hence exert little if any molecular sieving effect. The latter feature makes it especially suitable for the separation of high molecular weight proteins and nucleic acids. The high strength and clarity of agar gels make them suitable for varied applications; for instance, immunoelectrophoretic and isotacheophoretic study of antigenic proteins. Agarose gel electrophoresis has been particularly successful in the analysis of DNA proteins giving exceptional resolution and providing the basis for the technique known as 'DNA fingerprinting'.

3.3.3.3.4 Sephadex. The starting material, dextran, is a natural linear polysaccharide, in which the glucose residues are predominantly 1,6-linked. The individual 'dextran chains' are cross-linked with glyceryl bridges when reacted in alkaline medium with a dispersion of epichlorohydrin in an organic solvent. Though the extent of cross-linkage can be varied by controlling the relative proportions of epichlorohydrin and dextran, the position of cross-linkage is random, which results in a wide distribution of pore sizes in each gel type. Sephadex gels are commercially available with molecular weight exclusion limits of <1000 to $\sim 5 \times 10$ Da.

3.3.3.4 Thin layers. Electrophoretic studies can also be carried out on thin layers of silica, kieselguhr and alumina. As in TLC studies the studies with these materials offer the same advantages of speed and resolution when compared with paper. These materials find their greatest application in combined electrophoretic-chromatography studies in the two-dimensional separations of proteins and nucleic acid hydrolysates.

3.3.4 Techniques of electrophoresis

3.3.4.1 Low voltage. The general principle of the design of low-voltage electrophoresis (LVE) apparatus is shown in Figure 3.23. The essentials are two compartments to hold the buffer and electrodes and a suitable carrier for the support medium, such that its ends are in contact with the buffer compartments. The design of the carrier depends on the medium (strip, gel block, thin layer plate) and Figure 3.23 illustrates the use of paper strips. It will be noted that the paper does not dip into the electrode compartments, but into separate compartments connected by wicks with the anode and cathode cells. The purpose is to restrain diffusion of buffer electrolysis products along the paper, and to maintain the pH at the ends of the strip. In more recent designs a labyrinth construction replaces the wicks. The apparatus is enclosed to avoid evaporation from the paper, and provision can be made for external cooling. The strip is not supported throughout its length, but is stretched as tautly as possible across the end supports. A

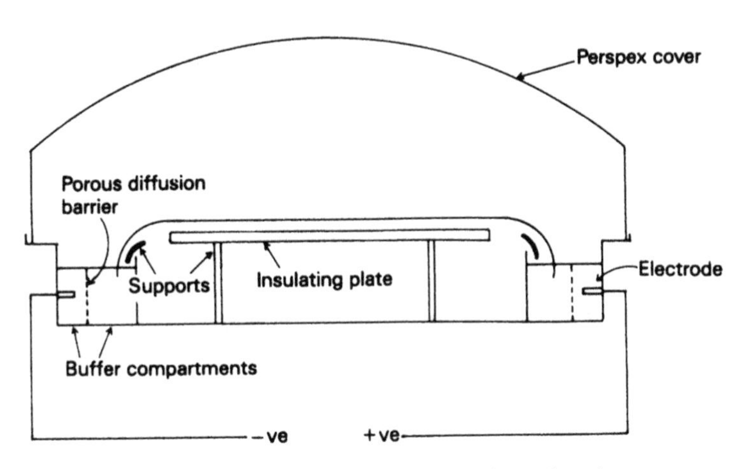

Figure 3.23 Apparatus for low voltage electrophoresis.

power pack supplying up to 500 V or even 1000 V and 0–150 mA, is needed to provide voltage gradients $\sim 5\,\mathrm{V\,cm^{-1}}$. It can be of the constant current or constant voltage type, but the former gives better results.

Horizontal electrophoresis units of similar design are available for gel slabs and thin layers of cellulose acetate. Low-voltage units for vertical electrophoresis are available for work on paper and gel slabs. Low-voltage electrophoresis can be used in principle to separate any ionic substances. In practice, its main application is in the examination of biological and clinical specimens for amino acids and proteins. The latter, until recently, were more easily separated by electrophoresis than by chromatography. There are numerous research and routine examinations of serum, plasma and other similar specimens which are done in this way. The preferred support media for these separations are cellulose acetate strips or one of the various forms of gel. The proteins are located by staining with a dye, and they can be estimated with fair accuracy with an automatic or manual scanner. Sugars can be separated in a borate buffer, in which they form complex ions. They are located with the usual chromatographic reagents.

3.3.4.2 High-voltage electrophoresis (HVE). Electrophoresis of medium to low molecular weight compounds with LVE techniques met with limited success due to the loss in resolution attributable to diffusion effects over the long development times, typically several hours. It was appreciated by many workers that improved resolution in much reduced analysis times could be achieved by using high-voltage gradients. However, though the rate of migration increases linearly with increase in voltage gradient the heat generated increases quadratically. Thus, heat dissipation for the control of evaporation was of crucial importance to the development and application

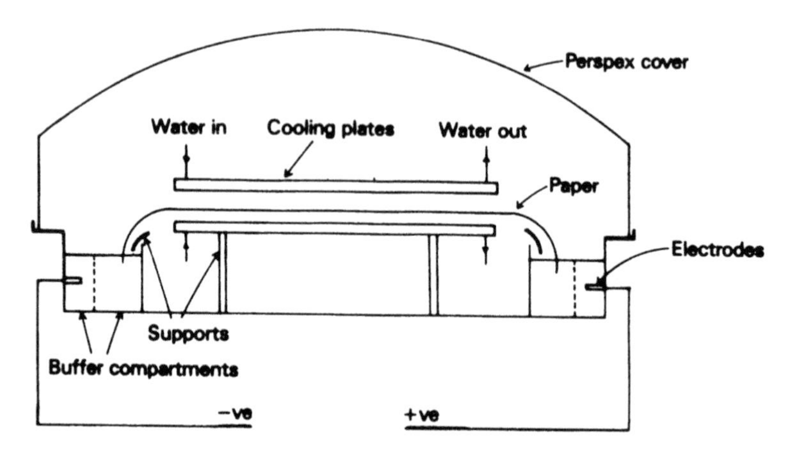

Figure 3.24 HVE apparatus.

of HVE techniques. There are three approaches. First, direct cooling systems may be incorporated into the electrophoresis unit. A typical system is shown below (Figure 3.24). The water-cooled plates are insulated from the electrophoresis medium by two polythene sheets; with this procedure it is possible to use gradients of up to $100\,V\,cm^{-1}$.

The second approach is simply to reduce the concentration of buffer solution. A reduction of one-tenth allows gradients of $30\,V\,cm^{-1}$ to be used. In the third method the sheet can be immersed in a non-conducting liquid and heat exchanger such as petroleum ether, fluorinated hydrocarbons or silicone oils. Here again, however, voltage gradients of only 30–$40\,V\,cm^{-1}$ can be used. Typical run times of less than 1 h can be achieved. This technique works best with small ions, for instance, those derived from small peptides and amino acids. In protein structure studies two-dimensional techniques have allowed fingerprinting of various protein hydrolysates. The hydrolysate is developed by high voltage electrophoresis in one direction followed by chromatography at right angles to the direction of the applied field.

3.3.4.3 Sodium dodecyl sulphate polyacrylamide gel electrophoresis (SDS–PAGE). This is a method primarily for molecular sizing by electrophoresis. SDS is an anionic detergent. The method involves treating the protein molecules in the sample with 1% (w/v) SDS and 0.1 M mercaptoethanol. The later reagent denatures the protein by destroying the disulphide bridges while the freed protein chains then the SDS binds to the protein by hydrophobic interactions. Any ionic charge on the free protein becomes swamped by the negative charges of the detergent molecules and hence all the protein–detergent 'complexes' have effectively the same size to charge ratio. The extent of migration of a complex is inversely related to the log of its molecular mass and hence by using proteins of known molecular mass and constructing a calibration graph it is possible to determine the molecular weight of a protein though not its identity nor constitution.

3.3.4.4 Isoelectric focusing. This technique, also referred to as electrofocusing, has similar underlying principles to those discussed for chromatofocusing (Chapter 4), and may be considered as a combined technique of the latter with electrophoresis. The technique is applicable to ampholytic compounds, for instance, amino acids, whose charge is determined by the matrix pH. Separation is carried out on gels on which a stable pH gradient, increasing in the direction of the cathode, has been established. As previously discussed each ampholytic compound has a pH at which it is neutral, its isoelectric point (pI). The pH gradient is achieved by impregnating the gel with polyamino-polycarboxylic acids, which are chosen such that their individual pI covers the pH range of interest. When subjected to an electric field these migrate and come to rest in order of pI, each maintains a local pH corresponding to its pI due to their strong buffering capacity.

Thus, each ampholyte migrates in the applied field until it reaches a position on the plate where the pH of the medium is equal to the component's pI. At this point the ampholyte is in its zwitterion form and is neutral and thus it loses its electrophoretic mobility and becomes focused in a narrow zone at this point. Regardless of the point of application on the plate the ampholyte always migrates to the location of its pI and then remains stationary.

As spreading of the bands is minimised due to application of the applied field and the pH gradient, high resolution can be achieved and proteins that differ by as little as 0.01 pH units can be adequately resolved. Analytical applications of IEF are concerned with the determination of pI values, assay of purity and also preparative scale procedures for the isolation of purified fractions.

3.3.4.5 Immunoelectrophoresis. The resolution of electrophoresis can further be enhanced by using the specificity of antigen–antibody reactions. Low-voltage electrophoresis is carried out in a barbitone impregnated gel for ~1–2 h. A pre-cut trough in the electrophoretogram is then filled with antisera. The antisera diffuses out of the trough laterally and on contact with the electrophoretographed antigen zones reaction occurs resulting in the precipitation of the antigen–antibody complexes, so forming curved lines of precipitation.

A number of developments to further improve the resolution and specificity of the method have been reported [78, 79], namely cross-over electrophoresis, rocket electrophoresis, two-dimensional immunoelectrophoresis, and radioimmunoassay.

3.3.4.6 Discontinuous electrophoresis. By introducing discontinuities in both buffer pH and voltage gradient in the chromatographic bed, sample ions can be concentrated as thin discs in a preliminary large pore gel prior to passage through a small pore gel where further separation is achieved in order of mass/charge ratio and also shape and size of sample components.

In discontinuous disc electrophoresis, the system used comprises a vertical cylindrical column, containing two discrete gel layers each maintained initially at a specific pH. The upper third of the column contains a wide-pore acrylamide gel in a Tris-HCl buffer of 6.7 pH units. The lower two-thirds is packed with a small-pore acrylamide gel in Tris-HCl at pH 8.9. The gels are prepared and the buffer located *in situ*.

After the sample has been applied the upper chamber of the electrophoretic apparatus is filled with Tris-glycine buffer (pH 8.3). The lower reservoir (cathode) already contains this buffer. On applying an electric field to the system the glycinate ions migrate very slowly through the spacer gel while the chloride ions are considerably more mobile. This charge separation produces a zone of lower conductivity between the leading ions (Cl^-) and the trailing ions (glycinate). This voltage gradient sweeps through the sample spacer gel and separates the components of the protein mixture into very narrow bands ($\sim$10 μm) in order of their mobilities.

As the bands move into the separating gel, the mobility of the glycinate ions which is pH dependent, increases markedly and they migrate through the protein bands to move with the chloride. The bands of protein molecules now move through the separating gel according to their electrophoretic mobility which is modified by the molecular sieving effect of the gel.

The above is a simple qualitative description. More exhaustive treatments of this and the related technique of isotacheophoresis can be found in the following references [80–82].

3.4 Capillary electrophoresis

Capillary electrophoresis is the name for a relatively new group of related techniques which have been attracting increasing interest in which separation of analyte species is achieved on the basis of differential migration in an electric field through narrow bore fused silica capillary columns (25–100 μm, i.d.). The techniques which differ significantly in operative and separation characteristics include

capillary zone electrophoresis (CZE);
capillary gel electrophoresis (CGE);
isoelectric focusing;
isotachophoresis (ITP); and
miscellar electrokinetic capillary chromatography (MECC).

The potential for capillary electrophoresis had been recognised for some time. The work of Virtanen [83] in 1977 detailed the separation Group I cations by free zone electrophoresis in small diameter tubes (200–500 μm). It was the pioneering work of Jorgenson and Lukacs [84], who performed electrophoresis studies on capillary columns (<100 μm) with an on-column fluorescence detection demonstrating efficiencies of 10^5 when using small sample volumes, which heralded the arrival of capillary electrophoresis as a high performance technique. Since the publication of that seminal work in 1981 many research groups throughout the world are studying the technique, attracted by its high separation potential and wide application range. High performance capillary electrophoresis systems have been available since the early 1990s.

3.4.1 *Overview of instrument operation*

Capillary electrophoresis differs from other electrophoretic methods in two principal ways:

- separation is achieved on very narrow capillaries (25–100 μm i.d.); and
- supporting media such as agarose or polyacrylamide gels are not used.

The basic components of a CE system are shown in Figure 3.25. The ends of a fused silica capillary column dip into two electrolyte buffer solutions containing either a platinum foil cathode or anode across which a high voltage (15–60 kV) is applied. A small volume (nl) of sample equilibrated in a suitable electrolyte is introduced at one end of the capillary and the sample components subsequently migrate through the column under the

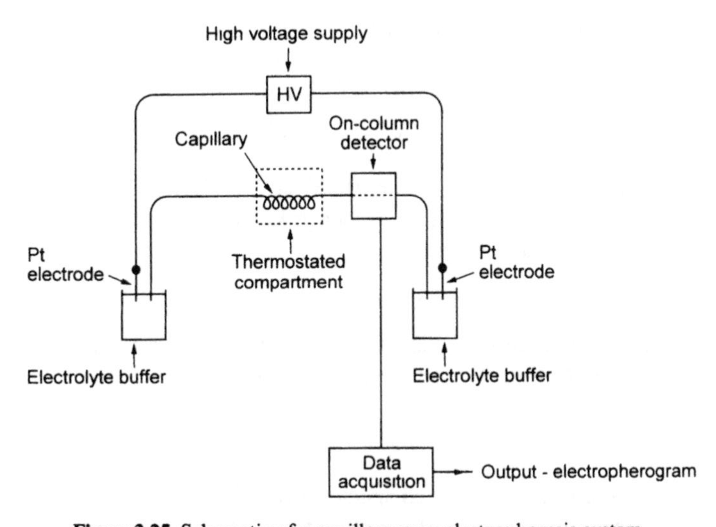

Figure 3.25 Schematic of a capillary zone electrophoresis system.

force of the applied voltage. The separated analytes are detected on column by a variety of detection devices, e.g. ultraviolet and fluorescence and the detector signal is displayed as peaks on an electropherogram.

Unlike GC or HPLC where all components migrate through the detector cell at constant flow, the signal and peak area in an electropherogram is dependent upon both analyte concentration and migration velocity—which is different and specific for each analyte.

The use of capillary tubes is an advantage in that

- the high electrical resistance minimises power dissipation and joule heating enabling the application of very high electrical fields $\sim$100–500 V cm^{-1} necessary for rapid separations;
- due to the large surface area to volume ratio of the capillary dissipation of the joule heat is very efficient thus virtually eliminating convective mixing which would result in band broadening; this feature also minimises temperature gradient across the capillary—an important consideration as electrophoretic mobility is temperature dependent. Thus capillary electrophoresis peak widths often approach the theoretical limit.

Use of large diameter columns would exacerbate these problems leading to considerable band broadening and loss of resolution. However, the use of narrow bore capillaries places stringent requirements on detectors regarding sensitivity and a compromise between performance and sensitivity is to use 25–50 μm i.d. columns with sample loadings of $\leq$10 nl. Columns employed are typically between 10 and 100 cm in length.

3.4.2 Theory and principles

The theoretical principles of capillary electrophoresis are closely related to those of electrophoresis. In electrophoresis the velocity of an ion is related to its electrophoretic mobility (Me) as follows:

$$V = \mu_e E$$

where E equals the electric field strength (V cm^{-1}). Under steady-state conditions the ion will move at constant velocity the electric force being balanced by the forces of frictional drag, i.e.

$$qE = 6\pi\eta r v$$

where q is charge, η is solution viscosity, r the hydrodynamic radius and v the velocity. The electrophoretic mobility is given by

$$\mu_e = \frac{q}{6\pi\eta r}$$

showing that small highly charged species have high mobility and vice versa. Where the extent of ionisation of the analyte is pH dependent then the actual electrophoretic mobility may be controlled or modified by varying buffer composition.

3.4.3 *Electroendosmotic flow*

Electroendosmosis or electro-osmotic flow describes the bulk flow of liquid through a fine capillary due to the effect of an applied electric field on the solution charged double layer which in this context is at the capillary wall of a fused silica column [85]. The double layer charge has its origins in

- acid–base type ionisation of the surface silanol groups (the extent of which is pH dependent); and
- adsorption/desorption of ionic species.

Counter ions, principally cations, build-up at the capillary wall to give overall charge balance forming an electrical double layer so creating a potential difference—the zeta potential. The degree of ionisation of the capillary surface and hence the magnitude of the zeta potential is pH dependent and increases with alkalinity of the buffer media (Figure 3.26).

When a voltage is applied across the capillary, cations in the diffuse area of the double layer migrate towards the cathode and because they are solvated their movement drags the bulk solution in the direction of the negative electrode. This electro-osmotic flow (EOF) and is the fundamental process which drives capillary electrophoresis. EOF is defined by

$$\nu = \frac{\epsilon \zeta E}{4\pi\eta}$$

Where ϵ is the dielectric constant, η is viscosity, ζ is the zeta potential and E the field strength. The corresponding electroendosmotic mobility of the

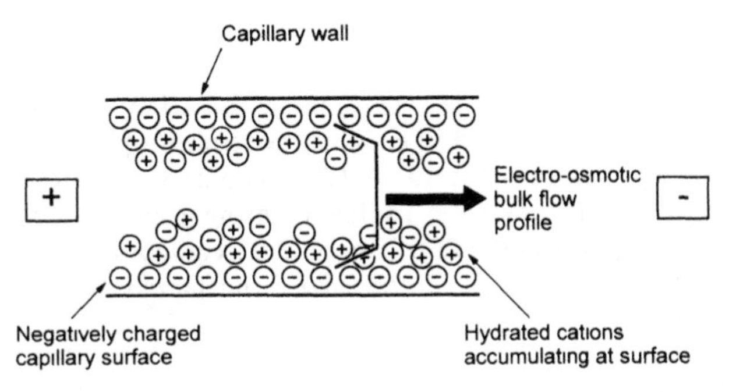

Figure 3.26 Representation of double layer at the capillary wall and subsequent electro-osmotic flow.

background electrolyte may be expressed as

$$\mu_{\text{EOF}} = \frac{\epsilon \zeta}{4 \pi \eta}$$

thus the overall mobility (μ) of a charged molecule is given by the sum of its intrinsic electrophoretic mobility and the superimposed EOF:

$$\mu = \mu_{\text{e,}} + \mu_{\text{EOF}}$$

The speed of electro-osmotic flow is typically at least an order of magnitude greater than that due to the electrophoretic mobility of the ions—and thus all species cations, neutral molecules and anions will be carried towards the cathode. The cations migrate fastest, the anions slowest with the neutral molecules being carried at the speed of the EOF. Herein lies the power and potential of CE as an analytical technique.

While EOF is beneficial and is exploited in CE it must be controlled since if it is too high the analyte species will have migrated to the cathode before separation has occurred. The EOF can be fine tuned by adjusting the following parameters.

pH	An increase causes further ionisation of the silanol groups giving a greater zeta potential and resulting in an increase in flow.
η	Decreases EOF when increased.
Ionic strength	Decreases zeta potential and EOF when increased.
Electric field	EOF increases with increase in field strength.

A unique feature of EOF is its flat flow profile in contrast to the laminar or parabolic flow which results when liquid is forced through a tube by hydrostatic pressure. As the electro-osmotic flow profile is essentially flat it does not contribute significantly to broadening of the solute zones.

3.4.4 Separation modes

Capillary electrophoresis is the generic name for a family of related techniques which have their origin in capillary zone electrophoresis (CZE) and are capillary gel electrophoresis (CGE), capillary isoelectric focusing (CIEF), miscellar electrokinetic capillary chromatography (MECC) and capillary isotachophoresis (CITP). Though the techniques differ significantly in principle of operation they can be carried out largely on the same basic instrumentation.

3.4.4.1 Capillary zone electrophoresis. The technique of CZE was first introduced by Jorgenson [86]. Operating principles have already been discussed. The capillary is filled only with buffer and separations of analytes occur on the basis of variations of mass to charge ratios. As previously detailed the EOF is usually dominant and drives all analytes—anions, cations and neutrals—in the direction of the cathode. Thus, the movement of a given species is the EOF mediated by the electrophoretic mobility. Thus, cations arrive at the cathode first in decreasing order of mobility, followed by the neutrals (no separation) and finally the anions—those of highest electrophoretic mobility arriving last.

Selectivity and separation enhancement is achieved by modification of the buffer pH and by the addition such as chiral selectors for the resolution of stereoisomers and cyclodextrins to resolve sample components by differential transient complexation.

3.4.4.2 Miscellar electrokinetic capillary chromatography. MECC first described by Terabe [87] achieves resolution of anions, cations and neutrals by the combined effects of electrophoretic migration and miscellar partitioning. Miscelles are amphiphetic aggregates of molecules known as surfactants which are long chain molecules characterised by a long hydrophobic tail and a hydrophilic head. Miscelle aggregates of surfactant molecules are formed in aqueous solution with the hydrophobic tails orientated inwards and the hydrophilic heads pointing outwards. The miscelles though negatively charged will migrate towards the cathode under the influence of EOF. During migration analyte species can interact with the miscelles through both hydrophobic and electrostatic processes. Thus, neutral molecules will partition in and out of the miscelle based on their hydrophobicity—in some ways analogous to the sorption process in reverse phase chromatography. The more hydrophobic the analyte is associated with the miscelle its migration velocity is slowed. Enhanced in separation can also be achieved by combined electrostatic hydrophilic interactions.

The most common surfactants used in MECC are anionic and cationic, for example, SDS and cetyltrimethyl ammonium bromide (CTAB). Subtle changes in the solute miscelle interactions giving enhanced resolution have been achieved by using a variety of organic modifications, e.g. chiral selectors such as cyclodextrins and crown ethers [88] and quaternary anions for ion pairing.

3.4.4.3 Capillary gel electrophoresis (CGE). The capillary can be filled with microporous gels [89] which provides molecular sizing capabilities. The underlying principles of CGE are similar to those of flat-bed (slab) or tube electrophoresis though CGE offers the advantages of on-column detection and automation, also there is less Joule heating and hence reduced zonal broadening. As with the planar electrophoretic methods, techniques such as SDS-PAGE can be employed.

A variety of polymers matrices and gel types have been used—ranging from cross-linked polymer (polyacrylamide–bisacrylamide) to linear polymers such as polyacrylamide, hydroxycelluloses, polyvinyl alcohol and dextrans. This later class can be regarded as physical gel formed by the entanglement of the polymer chains. With cross-linked polymers the polymerisation is commonly carried out in sites while for linear polymers it is possible to change the capillary with a solution of polymer.

CGE has been used extensively in molecular biology and protein chemistry and is the method of choice for DNA sequencing giving quite remarkable results—DNA fragments ranging from 100 to over 1000 base pairs have been separated [90].

3.4.4.4 Capillary isoelectric focusing (CIEF). Capillary isoelectric focusing is used to separate compounds such as amino acids, peptides and proteins, i.e. compounds which can exist as zwitterions, on the basis of their isoelectric points pI. The capillary column is filled with a solution of the carrier ampholytes on the basis of their isoelectric points pI and the sample; when a voltage is applied ampholytes which are already charged will migrate to the cathode (which is in contact with a basic solution) and those which are already charged will migrate to the anode (which is in contact with an acid solution). Migration will cease when the ampholyte attains its isoelectric point and is no longer charged. The ampholyte then maintains a local pH, corresponding to this pI, due to its strong buffering capacity. Solute, e.g. protein molecules, will also migrate either to the cathode or the anode, depending on its charge, until it encounters a pH where its net charge is zero. At this point migration halts and the solute is said to be focused.

At this point solute bands are mobilised and migrate past the detector. Mobilisation can be accomplished in either the anode or the cathode direction by either adding salt to the appropriate electrode compartment or by using pressurised flow.

EOF is eliminated in CIEF by covalent coating of the capillary wall, otherwise ampholytes and analytes would be wasted from the column before a pH gradient and analyte focusing was established. The technique has proved particularly successful for the separation of immunoglobulins and haemoglobulins.

3.4.4.5 Capillary isotachophoresis (CITP). In isotachophoresis an electric field is applied to an electrolyte system comprising a leading electrolyte and a trailing electrolyte, aliquot of sample is sandwiched between the two. The leading electrolyte is chosen such that it has greater mobility than any of the analytes and the trailing electrolyte lower mobilities. Either anion or cation mixtures can be analysed but not both on the same run. The capillary is first filled with the leading electrolyte followed by sample, the terminating electrolyte is in the other reservoir. A voltage is applied and separation of the analyte species occurs in the gap between the leading and trailing electrolytes in accordance with their relative electrophoretic mobilities.

As the electric field varies within each zone, the analyte bonds more with constant velocity through the capillary and are very sharply focused. The zones can be detected by conductivity, differential conductivity or ultraviolet detection procedures.

3.4.5 Instrumentation

3.4.5.1 Columns. Materials used for the construction of columns in capillary electrophoresis should be chemically and electrically inert, optically transparent and have good mechanical strength. Fused vitreous silica columns coated externally with a polyimide layer meet these requirements and are used almost exclusively. Column lengths range from 20 to 100 cm depending on the sample complexity. Internal diameters span the range 10–100 μm though the majority of applications are carried out on a 25–75 μm range. If the analytes are to be detected by ultraviolet–visible or fluorescence methods then a small section of the polyimide layer must be removed. Detection sensitivity can be improved by forming an extended light path which is achieved by increasing the i.d. of the section of the capillary in the detector. Before use capillaries must be thoroughly cleaned using prescribed wash procedures. Electro-osmotic flow can be eliminated if desired by coating the internal wall of the capillary. Before use the column must be thermostatically controlled to $\pm 0.1\,^{\circ}C$ to achieve consistency of sample uptake and analyte(s) migration.

Some use has been made of Teflon as it is ultraviolet transparent and also exhibits significant electro-osmotic flow. However, it is difficult to form tubing with a regular and precise internal diameter and it also suffers from adsorptive problems and poor heat transfer characteristics.

3.4.5.2 Injection. As already detailed for optimum performance narrow bore capillaries are employed, the associated column volume is of the order of 500–1000 μl and hence in order to maintain high efficiency, only minute volumes of sample (≤ 10 μl) can be loaded. Obviously neither of the traditional methods of syringe or loop injection are suitable for such small volumes.

There are primarily two methods/approaches to loading the sample into the capillary.

3.4.5.2.1 Electromigration (also known as electrophoretic) injection. The capillary inlet is immersed in the sample vial and a high voltage applied (usually three to five times lower than used for separation) for a controlled period. This causes a controlled but individually different amount of sample ions to migrate into the capillary but as shown in the equation, in direct proportion to their electrophoretic mobility. Due to this discrimination this method does not provide for optimal quantitation, though it can be used to advantage when viscous media or gels are employed and the displacement or hydrodynamic loading method is ineffective. The quantity injected, Q (g or moles) can be calculated from

$$Q = \frac{(\mu_e + \mu_{EOF})V\pi r^2 Ct}{L}$$

where μ_e is the electrophoretic mobility of the analyte, μ_{EOF} is the EOF mobility, V is the voltage, r is the capillary radius, C is the analyte concentration, t is time, and L is the capillary total length.

3.4.5.2.2 Hydrodynamic loading (HL). Samples can be successfully loaded by displacement from the sample vial by pressurising the sample vial, applying vacuum to the end of the capillary or by gravity displacement, i.e. by changing the relative heights of sample and exit vials. With pressure loading the volume of sample drawn into the capillary can be calculated using the Hagen–Poisenille equation:

$$\text{Volume} = \frac{\Delta P d^4 \pi t}{128\eta l}$$

where ΔP is the pressure difference across the capillary, d is capillary i.d., t is time of sampling, η is buffer viscosity and l is the total capillary length. The same expression can be used to determine sample volume for gravity loading by substituting the appropriate value for

$$\Delta P = \rho g \Delta h$$

where ρ is the buffer density, g is the gravitational constant and Δh is the height differential of reservoirs. This latter variant of HL is simple and extremely reproducible with the amount of sample loaded onto the capillary determined by the height difference, the period of immersion of the capillary end and by the sample viscosity. Typically, $\Delta h = 10\,\text{cm}$ and sampling time $= 10$–$30\,\text{s}$.

3.4.5.3 Detectors. A wide variety of detection methods for CE have been reported many of which are directly comparable to those employed in HPLC. Optical detectors including ultraviolet, photodiode array, ultraviolet–visible and fluorescence have proved to be the most popular and all are commercially available. With these detectors the optically opaque polyimide layer must be removed and the somewhat limited sensitivity has been significantly improved by modifying the internal geometry of the capillary to give an increased pathlength.

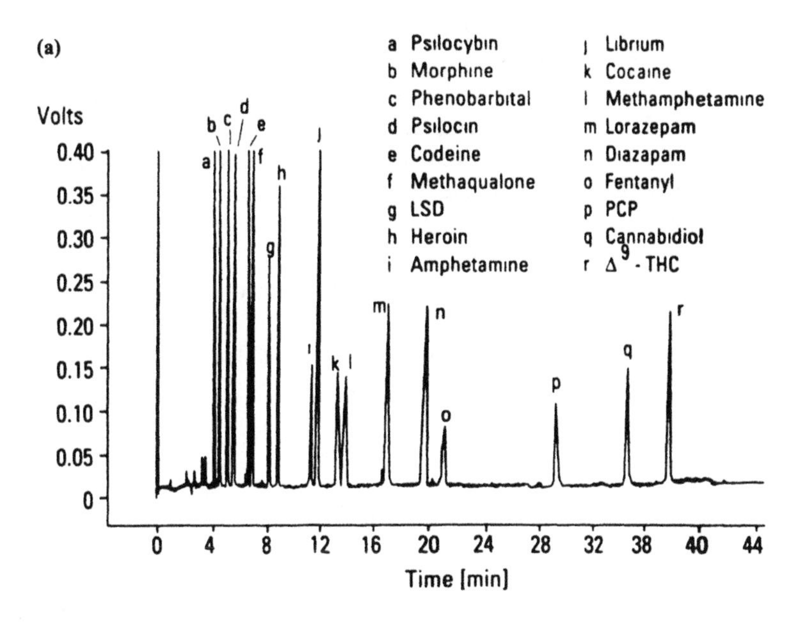

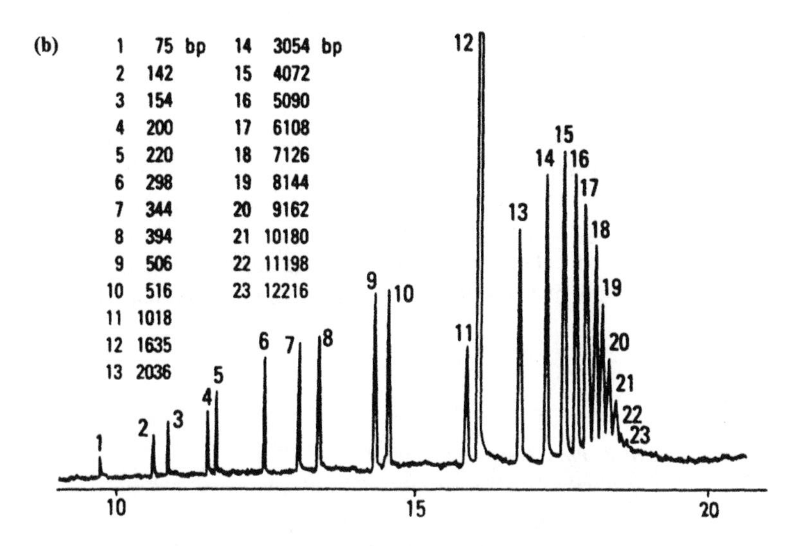

Figure 3.27 (a) MEKC forensic drug screen. Conditions: 8.5 mM borate, 8.5 mM phosphate, 85 mM SDS, 15% acetonitrile, pH 8.5, $V = 20\,kV$, $l = 25\,cm$, $L = 47\,cm$, i.d. $= 50\,\mu m$, $\lambda = 210\,nm$. CGE of 1 kbp ladder using minimally cross-linked polyacrylamide. Conditions: *Bis*-cross-linked polyacrylamide (3% T, 0.5% C), 100 mM Trisborate, pH 8.3, $E = 250$ V/cm, $i = 12.5\,\mu A$, $l = 30\,cm$, $L = 40\,cm$. (Reproduced with permission of Hewlett Packard.)

Other detection methods which have been employed with a measure of success are electrochemical (this requires the detector electrodes to be positioned very precisely on opposite sides of the capillary wall), amperometric (useful for electroactive species), conductimetric and mass spectrometry. The latter, though presenting particular difficulties in interfacing which have not been fully resolved is sensitive gives structural information and offers the potential for characterising complex biomolecules.

The mass detection limits for the above detectors range from 10^{-13} to 10^{-20} moles based on a 10 nl injection.

The principles of the above detector systems are covered in detail in other chapters of the text and an excellent review of the technique detailing the precise requirements of CE regarding interfacing to a range of detector systems is presented by Li [91].

3.4.6 Applications

The range of applications reported using CE techniques is enormous and arguably already rivals those of GC and HPLC (Figure 3.27). The efficiencies achieved are quite staggering—typically 2×10^5 theoretical plates are generated in run-times of <10 min and plate counts of 10^6 are not unusual. Separation may be achieved on the basis of differences in charge, mobility, size and hydrophobicity and compound classes where separations have been achieved include nucleotides and nucleosides, bioactive peptides, proteins, vitamins, a range of pharmaceuticals for example opiates and barbiturates and even inorganic ions.

In conclusion, capillary electrophoresis is a maturing family of techniques which has the potential to become standard in analytical laboratories. Its progress will be determined to a large extent on advances in detector design and technology and in progress in the development of facile optimisation procedures. For a fuller treatment of CE the interested reader is directed to a number of excellent texts [92, 93] which have been published recently.

References

[1] Beyerınck, M.W. *Z. Phys. Chem.*, **3** (1889) 110.
[2] Wijsman, H.P. *dE Diastase, beschouwd als mengsel van Mattese endextrinase*, Amsterdam, 1898.
[3] Izmailov, N.A. and Shraiber, M.S. *Farmatsiya*, **3** (1938) 1.
[4] Meinhard, J.E. and Hall, N.F. *Anal. Chem.*, **21** (1949) 185.
[5] Kirchner, J.G., Miller, I.M. and Keller, I.G. *Anal. Chem.*, **23** (1959) 420.
[6] Stahl, E. *Thin Layer Chromatography*. Academic Press, London, 1965.
[7] Stahl, E. *Thin Layer Chromatography* (Martini-Bettolo, G.B., ed.), p. 1. Elsevier, London, 1964.
[8] Martin, A.J.P. and Synge, R.L.M. *Biochem.*, **50** (1950) 679.
[9] Geiss, F. *Fundamentals of Thin Layer Chromatography*. Huethıg, Heidelberg, 1987.

[10] Waldi, D., Schnackerz, K. and Munter, F. *J. Chromatogr.*, **6** (1961) 61.
[11] Stahl, E. and Kaltenbach, V. *J. Chromatogr.*, **5** (1961) 351.
[12] Padley, F.H. *Thin Layer Chromatography* (Martini-Bettolo, G.B., ed.), p. 87. Elsevier, London, 1964.
[13] Bergelson, L.D., Dyatlovitskaya, E.V. and Voronkova, W.V. *J. Chromatogr.*, **15** (1964) 191.
[14] Malins, D.C. and Mangold, H.K. *J. Am. Oil Chem. Soc.*, **37** (1960) 383, 576.
[15] Stahl, E. and Trennheuser, L. *Arch. Pharm.*, **293/65** (1960) 826.
[16] Gasparic, J. and Churacek, J. *Laboratory Handbook of Paper and Thin Layer Chromatography*. J. Wiley, New York, 1979.
[17] Wortmann, B., Wortmann, W. and Touchstone, J.C. *J. Chromatogr.*, **70** (1972) 199.
[18] Neher, R. *Steroid Chromatography*. Elsevier, Amsterdam, 1964.
[19] Evans, F.I. and Kinghorn, A.D. *J. Chromatogr.*, **87** (1973) 443.
[20] Mottier, M. and Potterat, M. *Anal. Chim. Acta*, **13** (1955) 46.
[21] Bark, L.S. and Graham, R.J.T. *J. Chromatogr.*, **25** (1966) 347.
[22] Cerny, V., Joska, J. and Labler, L. *Coll. Czech. Chem. Comm.*, **26**, (1961) 1658.
[23] Blattna, J. and Davidek, J. *Experientia*, **17** (1961) 474.
[24] Davidek, J. and Blattna, J. *J. Chromatogr.*, **7** (1962) 204.
[25] Scholtz, K.H. *Dt. Apoth.-Ztg.*, **114** (1974) 589.
[26] Weill, C.E. and Hanke, P. *J. Chromatogr.*, **34** (1962) 1736.
[27] Knappe, E. and Peteri, D. *Z. Anal. Chem.*, **188** (1962) 184, 352.
[28] Kaufmann, H.P., Makus, Z. and Khoe, T.H. *Fette Siefen*, **63** (1962) 689.
[29] Kaufmann, H.P., Makus, Z. and Khoe, T.H. *Fette Siefen*, **64** (1962) 1.
[30] Honegger, C.G. *Helv. Chim. Acta*, **44** (1961) 173.
[31] Peereboom, J.W.C. and Beekes, H.W. *J. Chromatogr.*, **9** (1962) 316.
[32] Monroe, R.E. *J. Chromatogr.*, **62** (1971) 161.
[33] Vaedlke, J. and Gajewska, A. *ibid.*, **9** (1962) 345.
[34] Smith, I., Rider, L.J. and Lerner, R.P. *J. Chromatogr.*, **26** (1967) 49.
[35] Bujard, El. and Mauron, I. *J. Chromatogr.*, **21** (1965) 19.
[36] Wollenweber, P. *J. Chromatogr.*, **7** (1962) 557.
[37] Teichert, K., Mutschler, E. and Rodelmeyer, H. *Z. Anal. Chem.*, **181** (1961) 325.
[38] Randerath, K. and Struck, H. *J. Chromatogr.*, **6** (1961) 365.
[39] Tomasz, I. *J. Chromatogr.*, **70** (1973) 407.
[40] Berger, J.A., Meyniel, G. and Petit, I. *Compt. Rend.*, **225** (1962) 116.
[41] Petrovic, S.M. and Petrovic, S.E. *J. Chromatogr.*, **21** (1966) 313.
[42] Mosel, H.D. and Hermann, K. *J. Chromatogr.*, **87** (1973) 280.
[43] Davidek, J. *J. Chromatogr.*, **9** (1962) 363.
[44] Davidek, J. and Davidkova, E. *Pharmazie*, **16** (1961) 352.
[45] Hoffmann, A.F. *Biochim. Biophys. Acta*, **60** (1962) 458.
[46] Tortolani, J.G. and Colosi, M.E. *J. Chromatogr.*, **70** (1972) 182.
[47] Johansson, B.G. and Rymo, L. *Acta Chem. Scand.*, **18** (1964) 217.
[48] Morris, C.J.O.R. *J. Chromatogr.*, **16** (1964) 167.
[49] Shibukawa, M. and Ohta, N. *Chromatographia*, **13**, (1980) 531.
[50] Scott, R.P.W. *Analyst*, **103** (1978) 37.
[51] Snyder, L.R. *Chromatography*, 3rd edn (Heftmann, E., ed.). Rheinhold, New York, 1975.
[52] Wang, K.-T., Lin, Y.-T. and Wang, I.S.Y. *Adv. Chromatogr.*, **11** (1974) 13.
[53] Sander, L.C., Sturgeon, R.L. and Field, L.R. *J. Chromatogr. Sci.*, **18** (1980) 133.
[54] Rausch, R. *Recent Advances of Thin Layer Chromatography* (Dallas, F.A.A., Read, H., Ruane, R.J. and Wilson, I.D., eds), p. 151. Plenum Press, New York, 1988.
[55] Sherma, J. and Fried, B. *Handbook of Thin Layer Chromatography*. Marcel Dekker, New York, 1991.
[56] Lepri, L., Coas, V., Dessideri, P.G. and Pettini, L. *J. Planar Chromatogr.*, **6** (1993) 100.
[57] Kirchner, J.G., Miller, J.M. and Keller, G.J. *Anal. Chem.*, **23** (1951) 420.
[58] Poole, C.F. and Belay, M.T. *J. Planar Chromatogr.*, **4** (1991) 345.
[59] Poole, C.F. and Poole, S.K. *Anal. Chem.*, **66** (1994) 27A.
[60] Janchem, D.E. and Issaq, H.J. *J. Liquid Chromatogr.*, **11** (1988) 1941.
[61] Stahl, E. and Dumont, E. *J. Chromatogr. Sci.*, **7** (1969) 517.
[62] *Guide to TLC Visualisation Reagents*. J.T. Baker Chemical Co., Phillipsburg, NY, 1991.
[63] Clurczak, E.W., Murphy, W.R. and Mustillo, D.M. *Spectroscopy*, **61** (1991) 34.

[64] Goldman, J. and Goodall, R.R. *J. Chromatogr.*, **32** (1968) 24.
[65] Roberts, T.R. *Radiochromatography*. Elsevier, New York, 1978.
[66] Parrish, C.C. and Ackman, R.G. *J. Chromatogr.*, **262** (1983) 103.
[67] Sherma, J. and Fried, B. *Handbook of Thin Layer Chromatography*. Marcel Dekker, New York, 1991.
[68] Frey, R.O., Kovar, K.-A. and Hoffmann, V.J. *J. Planar Chromatogr.*, **6** (1993) 93.
[69] Consden, R., Gordon, A.H. and Martin, A.J.P. *J. Biochem.*, **38** (1944) 224.
[70] Hanes, C.S. *Canad. J. Biochem. Physiol.*, **39** (1961) 119.
[71] Tata, V.R. and Hemmings, A.W. *J. Chromatogr.*, **3** (1960) 225.
[72] Parlicek, M., Rosmus, J. and Deyl, Z. *Chromatogr. Rev.*, **6** (1964) 19.
[73] Eisenberg, F. Jr *Carbohydr. Res.*, **19** (1971) 135.
[74] Tiselius, A. *Trans. Faraday Soc.*, **33** (1937) 524.
[75] Street, H.V. and Niyogi, S.K. *Analyst*, **86** (1961) 671.
[76] Bennett, J.C. *Methods in Enzymology*, Vol. XI (Colowick, S.P. and Kaplan, N.O., eds), p. 330. Academic Press, New York, 1967.
[77] Michl, H. *Chromatography*, 2nd edn, Van Nostrand (Heftmann, E., ed.), p. 252. Reinhold, New York, 1967.
[78] Axelson, N.H., Kroll, J. and Weeks, B. *Scand. J. Immunol.*, **2**(Suppl. 1) (1973).
[79] Greenhalgh, B. *Lab. Equip. Dig.*, **Jan** (1983) 56.
[80] Davis, B.J. *Ann. NY Acad. Sci.*, **121** (1961) 404.
[81] Ornstein, L. *Ann. NY Acad. Sci.*, **121** (1964) 321.
[82] Gaal, O., Medgyesi, G.A. and Yereczkey, L. *Electrophoresis in the Separation of Biological Macromolecules*. Wiley, Chichester, 1980.
[83] Virtanen, R. *Acta Polytech. Scand.*, **123** (1974) 7.
[84] Jorgenson, J.W. and Luckacs, K.D. *J. Chromatogr.*, **218** (1981) 209.
[85] Ewing, A.G., Wallingford, P.J. and Olefrirowicz, T.M. *Anal. Chem.*, **61** (1989) 294.
[86] Lukacs, K.D. and Jorgenson, J.W. *J. High Resolut. Chromatogr.*, **8** (1985) 407.
[87] Terabe, S., Otsuka, K., Ichikawa, K., Tsuchiya, A. and Ando, T. *Anal. Chem.*, **56** (1984) 111.
[88] Kuhn, R., Stoecklin, F. and Erni, F. *Chromatographia*, **33** (1992) 32.
[89] Yin, H.F., Juergen, A. and Shomburg, G. *J. High Resolut. Chromatogr.*, **13** (1990) 624.
[90] Luckey, J.A., Drossman, H., Kostichka, A.J., Mead, D.A., D'Cunha, J., Norris, T.B. and Smith, L.M. *Nucleic Acids Res.*, **18** (1990) 4417.
[91] Li, S.F. *Capillary Electrophoresis—Principles Practice and Applications*, J. Chromatogr. Lib. Elsevier, Amsterdam, 1992.
[92] Grossman, P.D. and Colburn, J.C. *Capillary Electrophoresis—Theory and Practice*. Academic Press, San Diego, 1992.
[93] Jorgenson, J.W. *High Performance Capillary Electrophoresis*. Hewlett-Packard, 1992.

4 Liquid phase chromatography on open columns

4.1 Introduction

Open column or open tubular chromatography is the generic name for a number of different physical systems, in which a packed column is used with a liquid moving phase and in which the common feature is the practical technique used. It is distinguished from the technique of high performance liquid chromatography (HPLC) by the fact that solvent flow is gravity fed. The technique may be used with all the established sorption processes—adsorption, partition, ion exchange and gel permeation—and in addition with the techniques of affinity chromatography and chromatofocusing. The apparatus and experimental set-up shown in Figure 4.1 can be used for all the above procedures with minor modification.

Many of the qualitative uses of open column chromatography have been replaced by thin layer chromatography and HPLC, and the availability of preparative HPLC systems has further reduced the use of the technique. It does, however, find continued application for the large scale separation(s) (>10 g) of reaction mixtures encountered in synthetic organic chemistry, especially as with minor modifications to the basic apparatus, extremely inexpensive systems (*cf.* HPLC) with moderate resolution ($R_f > 0.10$) can be set up [1]. These techniques known as 'flash chromatography' and 'short path chromatography' are discussed in more detail herein. Gel and affinity chromatography are also still practised extensively in open column mode in the biosciences.

4.2 Practical aspects and considerations

4.2.1 Columns and packing procedures

Open column chromatography is commonly carried out with simple glass columns. The dimensions of the column depend on the quantity of material to be separated. The smallest columns are only a few millimetres in diameter and a few centimetres long, while the largest may be several centimetres in diameter and of correspondingly greater length. Some model experiments are described in Chapter 9, where column sizes are specified.

In order to obtain maximum efficiency the column must be evenly packed. The influence of particle size and regularity of packing on column performance has been discussed elsewhere (Chapter 2). While open tubular

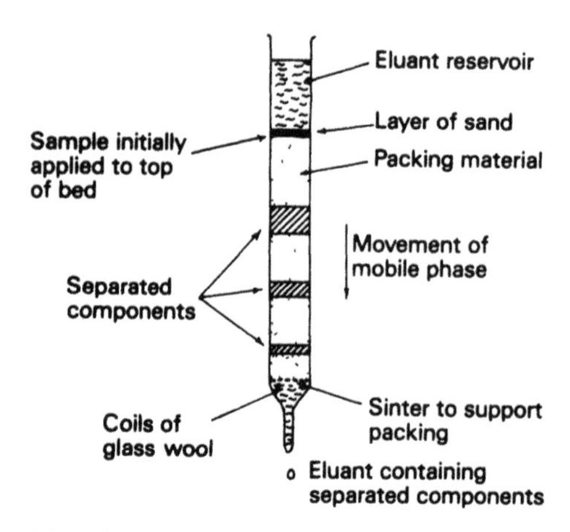

Figure 4.1 Basic experimental set-up for column chromatography.

chromatography using gravity feed of solvent is restricted to packing particles >150 μm (in order to obtain acceptable flow-rates), the column must be packed as uniformly as possible to minimise distortion of the chromatographic boundaries. Channelling is usually caused by the inclusion of air bubbles during packing. To prevent these effects, so far as possible, the packing material should be slurried with the solvent and poured as a thin stream into the tube, which should be about one-third full of solvent. If the adsorbent is allowed to settle gradually—which can be arranged by maintaining gentle agitation while there is a solvent flow through the column—reasonably homogeneous packing will result. If the particle size of the adsorbent is uniform, it is easier to get homogeneous packing. On no account should any part of the column be allowed to run dry, during packing or during a separation.

4.2.2 Sample application

The sample should be applied to the top of the column as evenly as possible, in as concentrated a solution of the eluting solvent as possible, avoiding disturbance of the column packing. The top of the column can be protected with a thin layer of sand, glass wool, filter paper or ballotini beads. When all of the sample has been adsorbed, the void can be filled with solvent and the chromatogram developed. The supply of solvent can be replenished as required.

4.2.3 Elution procedures

There are three principal elution procedures: isocratic (from the Greek isochros, meaning equal strength); stepwise (or fractional); and gradient.

4.2.3.1 Isocratic. By isocratic is meant the operation of the chromatographic column, by allowing a solvent mixture of unvarying composition to run through the column until separation is complete. The importance of mobile phase flow-rate has previously been discussed.

4.2.3.2 Fractional elution. If only one solvent is used ready elution of only some of the components of the original mixture from the column may result. To remove those which are more firmly held a stronger eluting agent will be required. Sometimes it may be necessary to use several different solvents of gradually increasing strength for the successive desorption of different components. This is known as stepwise elution. It has the advantage that sharper separations may be obtained than if only one strongly eluting solvent, capable of moving even the most firmly bound of the components of the mixture, is used—apart from the possibility of displacement development. One danger of this technique, however, is that a given compound may give rise to more than one peak by appearing in the eluates of successive steps.

4.2.3.3 Gradient elution. The technique of gradient elution analysis was first described in detail by Alm *et al.* [2]. It involves the use of a continuously changing eluting medium. The effect of this gradient is to elute successively the more strongly adsorbed substances and at the same time to reduce tailing. This means that the chromatographic bands will tend to be more concentrated and thus occupy less of the column. This desirable effect may be ascribed to the 'straightening' of the isotherms by the concentration gradient; that is, the adsorption isotherms are becoming more nearly linear as the use of a concentration gradient ensures that the tail of a particular chromatographic band is always in contact with a more concentrated solution (therefore more strongly eluting) than the front. The tail will therefore tend to move more rapidly to catch up the front.

There are available microprocessor controlled solvent delivery modules which can generate the required gradient profile, be it stepwise, linear, convex, concave or simply isocratic. Up to four solvents can be selected. A detailed discussion of the instrumentation is presented in Chapter 6. The pumping systems and accessories are equally adaptable to use with open tubular chromatography as with high pressure systems.

4.3 Modes of chromatography

The exact mode of chromatography operating in a given application is determined principally by the nature of the packing, though it must be appreciated that, while there may be one dominant mechanism, the modes are not mutually exclusive.

4.4 Adsorption chromatography

The lattice of the common porous adsorbents, e.g. alumina and silica, is terminated at the surface with polar hydroxyl groups, and it is these groups which provide the means for the surface interactions with solute molecules. (Alumina has additional structural features which can influence solute retention; these will be discussed later in the chapter.)

The eluant systems used in adsorption chromatography are based on non-polar solvents, commonly hexane, containing a small amount of a polar additive, such as 2-propanol. When the sample is applied, solute molecules with polar functionality will bond to the active sites on the packing; they will subsequently be displaced by the polar modifier molecules of the eluant as the chromatogram is developed, and will pass down the column to be re-adsorbed on fresh sites. The ease of displacement of solute molecules will depend on their relative polarities. More polar molecules will be adsorbed more strongly, and hence will elute more slowly from the column. A system as described in Figure 4.1 may be used. It was the type of apparatus first used for chromatographic separations of the kind familiar today.

After its introduction (in about 1903 by Tswett) there was little or no application of adsorption chromatography until it was again successfully employed in 1931 by Kuhn *et al.* [3] for the separation of xanthophyll pigments on columns packed with calcium carbonate. It was Tiselius and his school who made most of the major advances, such as the perfection of the elution method of separation and the devising of apparatus for the continuous analysis of the eluate by, for example, measuring changes in refractive index. Other advances, such as frontal analysis, displacement development, and gradient elution analysis, also invented by Tiselius and his co-workers, aim at the reduction or elimination of the 'tailing' always associated with adsorption chromatography, and help to achieve more efficient columns.

4.4.1 Solvents

As already indicated the solvent plays an active part in the adsorption process and competes with the sample molecules for active sites on the adsorbent. Thus, the stronger the binding of solvent molecules, the greater the amount of time the solute molecules spend in the mobile phase, and hence the faster they are eluted. Retention is therefore not so much influenced by sample solubility in the eluant as by the strength of solvent adsorption. It is advisable for a given application to choose an initial solvent of indifferent eluting power, so that stronger solvent systems can be tried subsequently. By 'strength' of the eluting agent is meant the adsorbability of the column packing.

Various workers have recorded elutropic series, e.g. Trappe [4], Snyder [5], Strain [6], Bickoff [7] and Knight and Groennings [8]. Generally, for polar adsorbents such as alumina and silica gel, the strength of adsorption increases with the polarity of the adsorbate. For carbon the order is reversed. Trappe's 'elutropic' series found that the eluting power of a series of solvents for substances adsorbed in columns such as silica gel decreased in the order:

pure water > methanol > ethanol > propanol > acetone > ethyl acetate > diethyl ether > chloroform > dichloromethane > benzene > toluene > trichloroethylene > carbon tetrachloride > cyclohexane > hexane

Practically speaking, the solvent is the controlling variable in adsorption chromatography. With the aid of the above series it is possible to select a solvent or solvent mixture with the appropriate eluting power. It should be appreciated that the eluting power of a solvent can be markedly affected by the presence of small amounts of impurity, e.g. methanol in benzene, and hence the purity of the solvents used should be as high as possible and, if necessary, further purification can be achieved by running the solvent through a column of the adsorbent to be used.

4.4.2 *Adsorbents*

Many solids have been used as adsorbents (Tswett himself tried over 100 different compounds), some of which are listed in Table 4.1, with the sorts of compounds separated with their aid. It may be noted that in the table there is little reference to inorganic separations, which are usually more conveniently carried out with the aid of ion exchange resins.

A term which is used in connection with adsorbents is 'activity'. This can relate to the specific surface of the solid; that is, the surface area measured in $m^2\,g^{-1}$ or more often to denote the strength of adsorption—the

Table 4.1 Adsorbents for chromatography

Solid	Used to separate
Alumina	Sterols, dyestuffs, vitamins, esters, alkaloids, inorganic compounds
Silica gel	Sterols, amino acids
Carbon	Peptides, carbohydrates, amino acids
Magnesia	Similar to alumina
Magnesium carbonate	Porphyrins
Magnesium silicate	Sterols, esters, glycerides, alkaloids
Calcium hydroxide	Carotenoids
Calcium carbonate	Carotenoids, xanthophylls
Calcium phosphate	Enzymes, proteins, polynucleotides
Aluminium silicate	Sterols
Starch	Enzymes
Sugar	Chlorophyll, xanthophyll

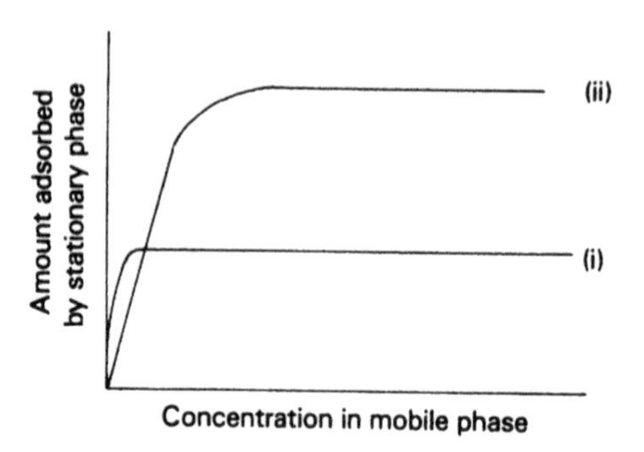

Figure 4.2 Adsorption isotherms.

chromatographic activity. This is usually the sense referred to in chromatography and is the one that will be employed here.

Often, of course, the two types of activity are found together, that is, a solid with a large surface area adsorbs tenaciously, but the situation where a substance is firmly held on the surface of a 'low area' solid is not uncommon. Thus, it might be expected that substances possessing some acidic character should be strongly adsorbed on say, lime or magnesia. As might be expected, chromatographic activity is the more specific and it is found that the strength of adsorption of polar groups on polar compounds increases in the order:

$$-CH{=}CH- < -OCH_3 < -CO_2R < {=}C{=}O < -CHO < -SH < -NH_2 < -OH < -COOH$$

This order is approximately reversed for carbon.

The shape of the isotherm associated with 'active' adsorption is illustrated in Figure 4.2(i). Since it is more strongly curved than (ii) it will give rise to more pronounced tailing, therefore the specificity mentioned above, though desirable, is often associated, in the case of the more polar compounds at least, with steep isotherms and hence pronounced tailing. Two undesirable chromatographic features may arise as consequences: first, very slow movement of the adsorbate on the column; and second, wide chromatographic bands of low concentration with a tendency to overlap. For these reasons it is often desirable to deactivate the adsorbent or to use the stepwise or gradient elution techniques mentioned above.

4.4.2.1 Preparation of Adsorbents. An indication has been given above (Table 4.1) of the wide range of materials that have been exploited as adsorbents in chromatography. Due to the breadth and detail associated with the preparation and activation of procedures, the following section will of necessity be devoted to the selective examination of the more common adsorbents.

4.4.2.1.1 Silica. Silica, silica gel and silicic acid are all terms applied to the material prepared by acidification of sodium silicate with sulphuric acid, followed by washing and drying of the gel. The final pH of the solution largely determines the specific surface area. For example, at pH 3.72 the specific surface is $830 \, m^2 \, g^{-1}$ while a pH of 5.72 gives a value of $348 \, m^2 \, g^{-1}$. Provided that the same pH is reached during preparations and the same drying procedure is followed, then the surface properties are remarkably consistent from batch to batch. The batch size has some influence, but this effect can be eliminated if an acetic acid/sodium acetate buffer is used. In this case only the final pH is important, especially if it is about 4.64, the maximum buffer capacity of the system.

The active sites on the surface consist of silanol groups which are spaced approximately 5 Å apart. The procedures used to modify the surface activity depend on the addition or removal of water. Surface adsorbed water which 'masks' the active sites is removed simply by heating; this is a reversible process and the surface can be deactivated by rehydrating the silica. Heating at higher temperatures (~400°C) leads to a permanent loss of surface activity due to the elimination of a molecule of water from two adjacent silanol groups which results in the formation of a siloxane linkage (silyl ether) which is chromatographically inactive.

The activity is tested by the relative absorbability of a number of azodyes and assigned a Brockmann number [9, 10]. The higher the grade number then the less active the surface. The surface interacts with polar solutes chiefly by means of H^- bonding and thus, for example, alkenes are more strongly retained than alkanes, and further, due to the acidic nature of the surface basic substances are held particularly strongly. The principal problem associated with silica (and adsorbents in general) is the tendency to cause peak tailing. An additional hazard is that irreversible adsorption may take place on columns and for this reason complete recovery of the adsorbates is not always achieved. Isomerisation of various compounds such as terpenes and sterols has been reported to occur on silica gel.

4.4.2.1.2 Alumina. The alumina surface is capable of exhibiting different types of solute–sorbent interaction. This may be attributed to, first, the very strong positive fields surrounding the Al^{3+}, which allow interaction with easily polarisable molecules, such that alumina would be preferred to silica for the resolution of aromatics from olefins; and second, the pressure of basic sites (probably O^{2-}) which allow interaction with proton donors.

The preparation and activation procedures for alumina are markedly different from silica in that the activity increases with the temperature of activation. Highly active alumina can be produced by heating at 400°C overnight. The surface can then be deactivated by addition of water. The surfaces can then be graded and assigned a Brockmann number as with silica.

Alumina is less widely used than silica due to its propensity to catalyse reactions with base-labile molecules and to cause rearrangements and even ring expansions in unsaturated and alicyclic compounds [11].

4.5 Partition column chromatography

In partition chromatography the solid adsorbent is replaced by a packing comprising a support material coated with a stationary phase. The stationary phase should be insoluble or at worst sparingly miscible in the mobile phase. Partition chromatography is a technique which utilises the ability of a solute to distribute itself between the two phases, to an extent determined by its partition coefficient. The basis of this method, then, is that due to differences in the partition coefficients of the various components, the mixture will be resolved.

The stationary phase is supported on a solid which is inert to the substances to be separated. The coated solid is packed into columns as in adsorption chromatography. There is, in fact, very little visible difference between the two types of column. The support material must adsorb and retain the stationary phase, and must expose as large a surface of it as possible to the flowing phase. It must be mechanically stable and easy to pack into the column when loaded with the stationary liquid, and it must not impede the solvent flow.

Needless to say, there is no support which has all these properties to the desired extent. The greatest difficulty (more or less unavoidable when a solid has to be used) is the incursion of adsorption effects. Even if the surface of the support is completely covered with liquid, adsorption effects can still make themselves felt. As complete coverage of the surface is not easy to achieve adsorption may be a major influence on the separation.

As far as liquid phase chromatography on columns is concerned it is probably true to say that the division into adsorption and partition methods is of practical, rather than theoretical, significance. The importance of adsorption varies from system to system and is mentioned briefly in connection with the different supports described below. The moving phase in partition chromatography may be a liquid or a gas, and the general principles are the same in each case.

4.5.1 Solid supports

The supports most commonly used are silica gel (sometimes referred to as silicic acid), diatomaceous earths (kieselguhr, celite, etc.) and cellulose. Other solids such as starch or glass beads have found more limited use.

4.5.1.1 Silica gel. The preparation and surface characteristics of silica gel have been discussed previously. Silica gel is almost always used with water or a buffered aqueous solution as the stationary phase. The amount of liquid held (or loading of stationary phase) is about $0.6\,cm^3\,g^{-1}$ of gel. It is fairly certain that adsorption plays a large part in all separations employing silica gel, but it is, nevertheless, extensively used. The same kinds of chemical reaction can occur in partition systems as in adsorption systems, although they may be less marked due to the diminished influence of the surface of the solid support caused by the stationary liquid. Silica gel used for chromatography is in the form of a fine white powder; a fairly narrow range of particle sizes is desirable—about the same as for adsorption column—although swelling may occur when the gel is mixed with the stationary phase.

4.5.1.2 Diatomaceous earths. These are all similar, being available commercially as kieselguhr, celite, or other proprietary products. The amount of liquid phase used with these solids is about $0.8\,cm^3\,g^{-1}$. They are usually pure enough for use, but if necessary they can be freed from iron by boiling with 3% hydrochloric acid, washed free from chloride and dried at 80°C. Kieselguhr has very little adsorptive capacity and therefore makes an ideal support for partition chromatography.

4.5.1.3 Cellulose powder. Supplied ready for use and usually requires no further treatment, not even the addition of the stationary phase, since this is acquired from the aqueous solvent. The use of cellulose in columns is an alternative to the use of cellulose in the form of thin layers coated on glass plates (Chapter 3). Cellulose columns are essential if a preparative separation is required, and they have also been found more convenient for quantitative estimations. A difference between the techniques which may affect the solvent flow, is that in a column the support and stationary phase are in contact with mobile phase before the separation starts, whereas in thin layer the mobile phase has a definite boundary which moves ahead of the solutes.

While the separation on cellulose is mainly due to partition, adsorption again plays some part and ion exchange is also possible. The extent of adsorption is uncertain, but is partly due to the polar nature of the hydroxyl groups of the cellulose molecule, and varies according to the polarity of the solutes. The ion exchange effects are also due to the hydroxyl groups and to the small number of carboxyl groups in the cellulose (Chapter 3).

4.5.2 Solvents

It is normal to choose a system so that there is a significant difference between the solvent strength parameters of the mobile and stationary phases, e.g. with water as stationary phase, pentane would be the optimum choice as eluant. However, over a long period of time the stationary phase would be stripped/washed from the column. This is referred to as solvent stripping, which led to the development of chemically modified stationary phases for HPLC (Chapter 6). This problem can be overcome by presaturating the eluant with the stationary phase before it contacts the packing. This can be achieved either by stirring the two phases together until equilibrium is achieved, or by placing a precolumn at the chromatographic column inlet. The pre-column contains a support with a high specific surface area coated with a high loading of the stationary phase to be used in the analytical column. Some typical applications and separations achieved with partition columns are shown in Table 4.2.

Table 4.2 Some typical separations on partition columns

Separation	Support	Stationary phase	Mobile phase
C_1–C_4 alcohols	Celite	Water	$CHCl_3$ or CCl_4
C_2–C_3 fatty acids	Silica gel	Water (buffered)	$CHCl_3$/BuOH
C_1–C_2 fatty acids	Silica gel	Water	Skellysolve/Bu_2O
Acetylated amino acids	Silica gel	Water	$CHCl_3$/BuOH
Acetylated amino acids	Kieselguhr	Water	$CHCl_3$/BuOH
Amino acids	Starch	Water	PrOH or BuOH/HCl
Proteins (ribonuclease)	Kieselguhr	Water	$(NH_4)_2SO_4$/H_2O/cellosolve
Purines	Starch	Water	PrOH/HCl
17-oxo-Steroid glucuronides	Silica gel	Aqueous sodium acetate	$CHCl_3$/EtOH/AcOH
Corticosteroids	Celite	Water	EtOH/CH_2Cl_2[a]/40–60 petrol/ CH_2Cl_2[a]
Methoxy aromatic acids	Silica gel	0.25 mol dm^{-3} H_2SO_4	BuOH/$CHCl_3$
Phenols	Cellulose	Water	MeOH/BuOH/$CHCl_3$
DNP amino acids	Chlorinated rubber	Butanol	Aqueous buffers
17-Oxo-steroids	Silica gel	Water	CH_2Cl_2/petrol[a]
Inorganic	Cellulose	Water	Acetone/HCl
Dibasic acids	Silica gel	Water	BuOH/$CHCl_3$ (stepwise in three mixtures)
Alkanes and cycloalkanes	Silica gel	Aniline	*iso*-PrOH/benzene
Organic acids	Silica gel	Aqueous sulphuric acid	$CHCl_3$/BuOH[a]
Lanthanides	Kieselguhr	Tributyl phosphate	(i) 15.8 mol dm^{-3} HNO_3[b] (ii) 15.1 mol dm^{-3} HNO_3 (iii) 11.5 mol dm^{-3} HNO (iv) Methylene bis-(di-*n*-hexyl)-phosphine oxide
Lipids	Silica gel	Water	Various

[a] Gradient elution.
[b] Stepwise elution.

4.6 Ion exchange chromatography

Ion exchange is a process wherein a solution of an electrolyte is brought into contact with an ion exchange resin and active ions on the resin are replaced by ions (ionic species) of similar charge from the analyte solution.

A cation exchanger is one in which the active ions on the ion exchange material are cations and the exchange process involves cations. The polar groups in cation exchangers are acidic, commonly SO_3H, CO_2H, OH or PO_4H_3. They are attached to the polymer molecule in a regular way and are accessible to the solution containing the ions to be removed or separated. The polar groups in anion exchangers are tertiary or quaternary ammonium groups ($-CH_2-NR_2$ or $-CH_2-NR_3^+$) and they function in an analogous manner. Anion exchangers are usually supplied in the chloride form rather than the hydroxide, because of the former's greater stability.

Ion exchange resin materials are based on styrene–divinylbenzene polymers and polyacrylates (Table 4.3). The synthesis of these materials is easily controlled and gives materials of the necessary chemical and physical stability, in terms of uniformity of particle size and shape, porosity and chemical composition. These resins find wide application in demineralisation, water treatment and ion recovery from wastes.

The insoluble polymeric resin is synthesised by the suspension-radical copolymerisation of styrene with divinylbenzene (DVB) (see Figure 4.3). The polymerisation results in a three-dimensional column framework which is porous in character and can thus allow diffusion of ions to take place through it. By conducting the polymerisation in an aqueous medium beads of definite size can be produced. The extent of cross-linkage and hence pore size can be varied by altering the proportion of DVB.

Sulphonic acid (SO_3H) exchange units are introduced in a controlled fashion by reacting the solvent swollen resin with chlorosulphonic acid, which results in mainly *para*-substitution of the benzene rings. The sulphonic acid resin is described as a strong cation exchanger, is fully dissociated over a

Table 4.3 The general characteristics of various ion exchange resins

	Type	Exchanging group[a]	Effective in the pH range	Exchange capacity[b]
Cation exchange	Strong acid	$-SO_3H$	1–14	$4\,mmol\ H^+\ g^{-1}$
	Weak acid	$-CO_2H$	5–14	$9–10\,mmol\ H^+ g^{-1}$
Anion exchange	Strong base	$-CH_2\text{-}\overset{+}{N}R_3$	1–15	$4\,mmol\ OH^- g^{-1}$
	Weak base	$-CH_2\text{-}NR_2$	1–9	$4\,mmol\ OH^- g^{-1}$

[a] $R = -CH_3$ (usually).
[b] The capacity quoted is for the dry resin.

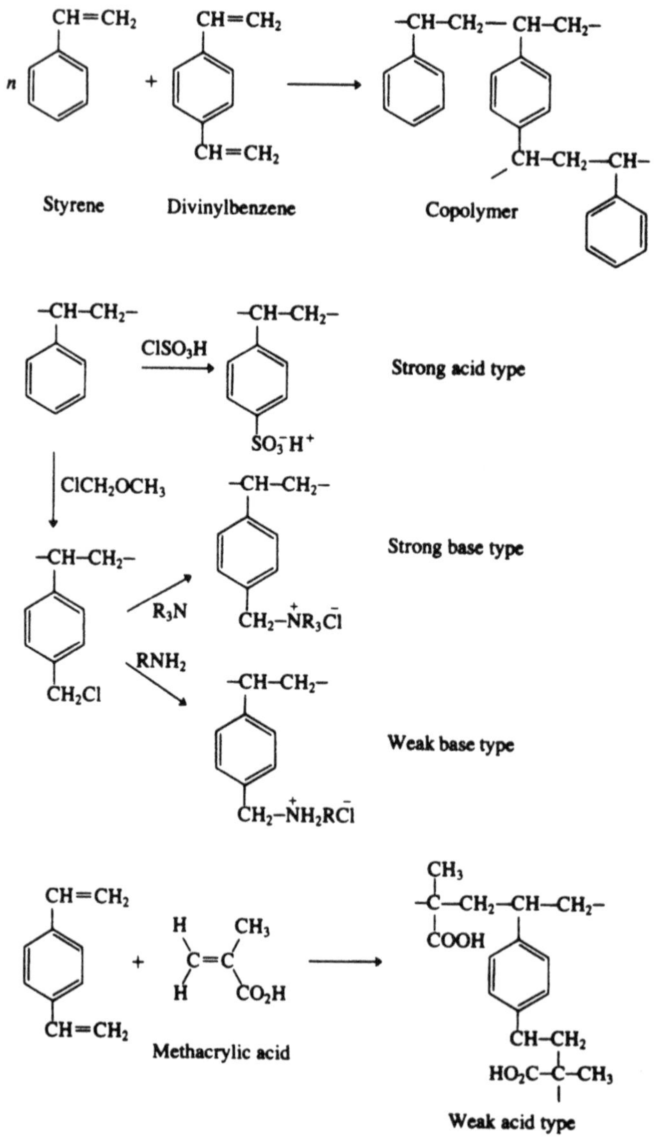

Styrene Divinylbenzene Copolymer

$ClSO_3H$ Strong acid type

$ClCH_2OCH_3$ Strong base type

R_3N RNH_2 Weak base type

Methacrylic acid Weak acid type

Figure 4.3 Synthetic routes to ion exchange resins.

wide range of pH, and will exchange its protons for other cations under a wide range of pH.

Weak acid exchangers (COOH) are restricted in use to $5 < pH < 14$. The exchange groups are introduced directly by the polymerisation of DVB and methacrylic acid. Phosphate groups commonly bonded to cellulose form ion exchangers of intermediate strength.

Common exchange units on anion exchange resins are

$-CH_2-\overset{+}{N}-(CH_3)_3X^- \; X^-=Cl, OH, NO_3$ etc.	quaternary ammonium resin
$-CH_2-N-(CH_3)_2$	tertiary amino resin
$-CH_2-NH-CH_3$	secondary amino resin
$-CH_2-NH_2$	primary amino resin

The resins are prepared by chloromethylating the polymer matrix, followed by treatment with the appropriate amine. The quaternary resin in both hydroxyl and salt forms is ionic, hence it may be described as a strongly basic resin which may be regarded as an insoluble polymeric cation associated with an equivalent number of active exchangeable hydroxide or halide ions. Anion exchangers are normally supplied in the chloride form rather than the hydroxide, due to the former's greater stability.

The amino resins are weakly basic and are generally too weak to attract a proton from water to form a stable cation. They are therefore not commonly found in the hydroxide form. Salt forms are obtained by interaction of the base with an acid and these are ionic containing the stable ammonium type ion which is associated with an equivalent number of anions.

The wet form of the resin is denser than water and so the resin beads may be packed into a column and conveniently eluted with water or aqueous solutions. Column operation is the usual procedure, though batch operation is possible where the resin and aqueous solution are stirred together to effect exchange.

4.6.1 Properties desirable in resins

For most applications using ion exchange resins an important factor is the accessibility of the exchange sites to the ions in solution. The exchange takes place partly in the thin film of solvent adsorbed on the surface of the beads, and partly within the resin matrix; it is normally assumed that for small ions (including all metallic ions and simple inorganic anions, and many small organic ions), all the sites are equally accessible to the displacing ions in solution. This accessibility within the lattice depends partly on the degree of cross-linking of the polymer chains, and if that varies in different parts of the resin, the exchange properties will be variable also, and that is not desirable for efficient chromatography.

With small amounts of cross-linkage, exchange equilibria are established more quickly, due to the extra swelling of the resin, so that the diffusion of ions becomes more rapid. The degree of cross-linking is controlled in the manufacturing process; a larger range of cross-linking can be achieved in the polystyrene–DVB acidic resins than in basic or polymethacrylate resins. By altering the degree of cross-linkage in the polymerisation the resin pore size may be varied and the resin made selective for ions of a given size. If the pore size is small—a high degree of cross-linkage—then the resin becomes selective towards smaller ions, the larger species being excluded from the resin. For acceptable chromatography a resin must also possess the following properties.

1. It must possess mono-functional exchange groups. There is no difficulty about this with modern resins, but earlier products made from phenol were polyfunctional (OH and COOH groups) and their exchange properties depended on the pH of the solution in which they were immersed. They were not, on this account, suitable for chromatography.
2. It must have a controlled degree of cross-linking; 4–8% is best for chromatography.
3. The range of particle sizes must be as small as possible.
4. The particle size must be as small as practicable.

4.6.2 *Ion exchange capacity*

The ion exchange capacity of a resin is a quantitative measure of its ability to take up exchangeable ions. The total capacity is defined as the amount of charged and potentially charged groups per gram of dry resin. The available capacity is the actual capacity obtained under specified experimental conditions and will be influenced by accessibility to functional groups, eluant concentration, ionic strength and pH, the nature of the counter ions, and the strength of the ion exchanger and its degree of cross-linkage.

For a cation exchanger the available capacity is determined by converting the resin to the hydrogen form, then using a neutral solution of a sodium salt to displace H^+ ions which are then titrated as free acid by a standard solution of sodium hydroxide. Then if: wt of dry resin = W g; NaOH titre = T ml; molarity of NaOH = A; dry weight capacity = $T.A/W$ m eq gm^{-1}.

The available capacity of an anion exchanger is determined by converting it to the chloride form and using a neutral nitrate or sulphate solution to displace the chloride, which are then estimated using a standard silver nitrate solution. As previously mentioned resins absorb water, causing the beads to swell and so ion exchange capacities are determined on the basis of dry weight.

4.6.3 *Selectivity of resins*

The affinity between a resin and an exchangeable ion is a function both of the resin and the ion. Ion exchange is an equilibrium process which for a cationic process can be represented by the equation

$$n(R^-H^+) + M^{n+} \rightleftharpoons (R^-)_n M^{n+} + nH^+$$

where R represents the resin matrix. The equilibrium distribution coefficient (K_d), also known as the selectivity coefficient, is given by

$$K_d = [M^{n+}]_R[H^+]^n/[M^{n+}][H^+]^n_R$$

where $[M^{n+}]$ and $[H^+]_R$ are the concentrations of the exchanging ion and hydrogen ion within the structure. The greater the affinity for a particular ion, relative to hydrogen, the greater the value of K_d. For solutions which depart from ideality, then strictly concentration should be replaced by activity. However, there is no entirely satisfactory way of measuring the activities of ions on a resin.

To illustrate how different cations may be separated let us consider a column which is packed with a cation exchanger in the hydrogen form. As a solution containing M^+ ions flows through the column the M^+ ions will replace H^+ ions on the resin according to the value of K_d. If only a small amount of the solution is used and it is washed down the column with pure water, all the M^+ ions will eventually replace hydrogen ions, and will form a stationary adsorbed band. The distribution of the ions within this band will depend on the value of K_d, if large, the band will be narrow and concentrated, if small, wide and diffuse. Thus, if a few ml of 0.1 mol dm^{-3} sodium chloride are placed on a column which is then washed with distilled water, the sodium ions will remain in a more or less narrow band near the top of the column and an equivalent amount of hydrochloric acid will be liberated to be eluted from the column.

In order to make the adsorbed band of ions (M^+) move down the column, water is ineffective and it is obviously necessary to elute with a solution of an acid (or a solution containing another cation), so that exchange can take place. The M^+ ions will then be washed out of the column, leaving it in its original form. The rate at which the band of ions moves will depend on the pH of the eluting acid and the value of K_d. Thus, two ions having different affinities for the resin will move at different rates down the column and a separation will be achieved.

4.6.4 Nature of the resin

The selectivity coefficient (K_d) is influenced by a number of factors operating in the resin; most obvious is the acid or base strength, or in other words, the polarity of the exchange groups. Since in cation exchangers the exchange groups are either SO_3H or CO_2H, there is little scope for differences within each class (strong or weak). The selectivity can also be affected by the degree of cross-linking. Small amounts reduce the selectivity while large amounts may totally or partly exclude larger ions—particularly organic ones. Selectivity is also slightly influenced by temperature.

The affinity of cations for the resin in dilute aqueous solution varies mainly as their charge and increases with it; for cations of the same charge the affinity is inversely proportional to the radius of the hydrated ion. This may be explained in terms of the polarising power of the cation, i.e. the greater the polarising power of the cation, the greater its affinity for the resin and the more strongly will it be retained. (The polarising power of a cation is proportional to Z^2/r where Z is the charge on the ion and r is the radius of the hydrated ion.)

Some affinity sequences are given below:

$Na^+(aq.) < Ca^{2+}(aq.) < Al^{3+}(aq.) < Th^{4+}(aq.)$
$Li^+(aq.) < Na^+(aq.) < NH_4^+(aq.) < K^+(aq.) < Ti^+(aq.) < Ag^+(aq.)$
$Mg^{2+}(aq.) < Ca^{2+}(aq.) < Sr^{2+}(aq.) < Ba^{2+}(aq.)$
$Al^{3+}(aq.) < Sc^{3+}(aq.) < Lu^{3+}(aq.) < \ldots$ lanthanides... $< La^{3+}(aq.)$

The orders above, however, are only true for solutions whose strength is about 0.1 mol dm^{-3} in the exchanging ion. At higher solution strengths the tendency is for the most concentrated ion to be most firmly bound, the others remaining in the same relative order. It is difficult to achieve separations of mixtures of multivalent cations solely on differences in affinities. Mixtures of this type are usually resolved by taking advantage of the ions' ability to form complexes.

Anion exchange resins, however, allow for more variation because the exchange groups are either $-NR_3^+$ or $-NR_2$ and the nature of the R group can be changed. The ionic form of a resin will also have some influence; the nitrate form of an anion resin will behave slightly differently (though not markedly so) from the chloride form; a series of K_d values is valid for only one specific ionic form of the resin.

In dilute solution the affinity of anions for an anion exchange resin depends on the degree of polarisation of the anion by the exchanger cation; i.e. the greater the degree of polarisation the more firmly shall the anion be held; and the greater the affinity of the anion for the resin. Polarisation of the anion increases with increasing negative charge and increasing size of the anion. In general, polyvalent anions have greater affinity than monovalent anions and for anions of the same charge the larger have the greater affinity.

$$F^- < HCO_3^- < Cl^- < HSO_3^- < CN^- < Br^- < NO_3^- < I^- \ll SO_4^{2-}$$

In terms of elution of anions from the resin the anions of least affinity will be eluted first.

4.6.5 *Separation methods*

The strong acid and strong base resins find much wider application than the weak varieties, because of the wide pH range over which they retain their exchange properties. The weak acid and weak base types are highly ionised only when in a salt form, which is why their operation is restricted to the pH ranges indicated in Table 4.4. They may be preferred for certain separations, however; for instance, polymethacrylic acid has more than twice the exchange capacity of the sulphonic acids at pH values greater than 7, and both weak acid and weak base resins may show a greater selectivity in certain

Table 4.4 Some commercially available resins suitable for chromatography

Name	Type	Functional group	Bead sizes (mesh)	% cross-linking	Exchange capacity g^{-1} (dry resin)	Form supplied	Working pH range	
Zeo-Karb 225	Strong acid	$-SO_3H$	14–52, 52–100 100–200, >200[a]	4, 8, 12 20	4.5–5.0 mmol H^+	Na^+	1–14	
Amberlite CG 120	Strong acid	$-SO_3H$	100–200, 200–400 400–600[b]	8	5.0 mmol H^+	Na^+	1–14	
Zeo-Karb 226	Weak acid	$-CO_2H$	14–52, 52–100[a]	2.5, 4.5	9–10 mmol H^+	H^+	6–9	
Amberlite CG 50	Weak acid	$-CO_2H$	100–200, 200–400 400–600[b]		10.0 mmol H^+	H^+	5–14	
Deacidite FFIP	Strong base	$-CH_2\overset{+}{N}R_3$	14–52, 52–100[a]	2–3, 3–5, 7–9	4.0 mmol OH^-	Cl^-	1–14	R = alkyl
Amberlite CG 400	Strong base	$-CH_2\overset{+}{N}R_3$	100–200[b] 200–400	8	3.8 mmol OH^-	Cl^-	0–12	R = alkyl
Amberlite CG 45	Weak base	$-CH_2\overset{+}{N}R_2$	100–200[b] 200–400		5.0 mmol OH^-	OH^-	0–9	R = alkyl

[a] British Standard Screens.
[b] US Standard Screens.

circumstances. A weak base resin will adsorb strong acids but not weak ones and, conversely, weak acid resins will adsorb strong bases but not weak ones. Again, weak acid resins show greater selectivity for certain divalent ions.

Exchange isotherms measured in dilute solutions are approximately linear; that is, the amount of a particular ion held by a resin is directly proportional to the concentration of the ions in the solution in contact with it. At higher concentrations the uptake by the resin begins to fall off, with the result that the isotherm becomes concave to the concentration axis in a manner similar to that depicted in Figure 4.2. Tailing can therefore be expected when the elution technique is used to separate concentrated solutions, otherwise symmetrical peaks, familiar in partition chromatography, will be obtained (Chapter 9).

In chromatography the solutions used are mostly dilute, hence the elution technique is much employed and it frequently gives highly satisfactory separations. In addition, all the other methods described in Chapter 1 have been used, although frontal analysis is not often encountered.

A modification of the elution technique, already referred to in section 4.6.3, is also used to good advantage. It depends on the alteration of the activities of the ions being separated by means of an eluting agent which will complex with them. A simple example is provided in experiment 4 in Chapter 9, where copper(II) and iron(III) ions are separated by elution with phosphoric acid. If this separation is tried with hydrochloric acid as the eluant, the copper(II) ions tend to move more rapidly down the column than the iron(III), because they carry a smaller charge and are therefore not so firmly held by the resin. A good separation will not be obtained. If, however, phosphoric acid is used, the iron(III) ions are rapidly removed from the column, and a sharp separation results. Subsequently, the copper(II) ions can be removed with hydrochloric acid. Clearly the phosphate ions form a much more stable complex with iron(III) ions, which are rendered colourless, than with copper(II). Complex formation is undoubtedly an important factor in other types of chromatography, particularly in inorganic separations on paper, but in no other technique has it been exploited to quite the same extent as in ion-exchange chromatography.

One of the earliest and most spectacular successes of ion exchange chromatography was the separation of the lanthanides on a strong acid resin with a buffered citrate solution for elution. Straightforward elution with hydrochloric acid brings about little separation, but the citrate ions complex with the M^{n+} ions, thus reducing their activity, which will now depend largely on the stability of the various citrate complexes. The equilibrium constants for the formation of the complexes (stability constant K_s) vary more than the affinities (K_d) of the free ions for the resin. Separation is therefore largely due to differences in K_s rather than in K_d. Similar operations have been carried out on the actinides with Dowex 50 in the ammonium form and elution with ammonium lactate or ammonium-hydroxy-isobutyrate. In

Table 4.5 Ion exchange resins: inorganic separations

Separation	Ion exchanger	Method
Lanthanides	Sulphonated polystyrene	Elution with citrate buffers
Lanthanides	Dowex 1-X10 in nitrate form—325 mesh	Stepwise and gradient elution with $LiNO_3$ solution
Actinides	Dowex 50 in ammonium form	Elution with ammonium lactate or ammonium α-hydroxy-isobutyrate
Lanthanides from actinides	Dowex 1-X8	Elution using 10 mol dm^{-3} LiCl in 0.1 mol dm^{-3} HCl
Ca, Al, Fe	Dowex 2 in citrate form	Stepwise elution with (a) water, (b) conc. HCl, (c) dil. HCl. (Also involves backwashing with conc. HCl.)
Ca, Sr, Ba, Ce	Dowex 50-X8	Stepwise elution with ammonium α-hydroxy-isobutyrate
Zr, Hf	Dowex 1-X8	Elution with 3.5% H_2SO_4
Zr, Ti, Nb, Ta, W, Mo	Dowex 1-X8 200–400 mesh	Stepwise elution with mixtures involving HCl, oxalic acid, H_2O_2, citric acid and ammonium citrate
Many simple mixtures of metal ions	Dowex 50W-X8 100–200 mesh (in polythene columns)	Stepwise elution with dil. HF, followed by a mineral acid such as HCl or HNO_3

Table 4.6 Ion exchange resins: organic separations

Separation	Ion exchanger	Method
Amino-acids	Dowex 50-X4	Stepwise and gradient elution with citrate and citrate/acetate buffers
Phosphate esters	Dowex 1-X2 200–400 mesh in chloride or formate form	pH gradient elution (HCl)
Folic acid analogues	Diethylaminoethyl cellulose 100–250 mesh	Gradient elution 0.1–0.4 mol dm^{-3} phosphate
Chlorophenols	Dowex 2-X8 200–400 mesh in acetate form	Gradient elution: acetic acid in ethanol
Aminobenzoic acid, aminophenols, and relates substances	Dowex 1-X10 200–400 mesh in chloride form	Stepwise elution: water; 1 mol dm^{-3}, 5 mol dm^{-3}, and 8 mol dm^{-3} HCl
Proteins	Diethylaminoethyl cellulose	Elution
Separation of acids from bases, both of which are freely water soluble	Sulphonated polystyrene in H form	Stepwise elution: water to remove acids: dil. HCl for bases

Tables 4.5 and 4.6, some typical chromatographic separations performed on ion exchange resins are listed.

4.7 Inorganic ion exchangers

Considerable work has been carried out over the years to develop inorganic ion exchange materials. Apart from clays and zeolites, which were in fact the first ion exchange materials to be investigated, substances studied include heteropolyacid salts, zirconium salts, tin(IV) phosphate, tungsten hexacyanoferrate, and others. A general account of inorganic ion exchangers was published by Amphlett [12] and chromatographic aspects were reviewed by Marshall and Nickless [13].

One of the reasons for the interest in inorganic materials is that resinous ion exchangers are susceptible to radiation damage and are therefore not really suitable for use with highly active solutions, although they have been successfully used for the separation of, for example, mixtures of actinides. Inorganic materials possess other advantages, including a much greater selectivity for certain ions such as rubidium and caesium, and the ability to withstand solutions at high temperatures. In addition, inorganic ion exchangers do not swell appreciably when placed in water and there is no change in volume when the ionic strength of the solution in contact with them is changed. On the other hand, certain of the inorganic materials possess disadvantages such as solubility or peptisation at certain pH values at which resins are normally stable, or they may be soluble in solutions in which resins are insoluble. Zirconium molybdate, for example, dissolves in EDTA, oxalate and citrate solutions. Again, they may exist in microcrystalline forms which are not very convenient for packing columns, since they tend to impede the flow of mobile phase, although there are ways of overcoming this problem.

Inorganic ion exchangers fall roughly into two groups: crystalline, such as the ammonium salts of the 12-heteropolyacids, and amorphous, such as zirconium phosphate. One example from each group will be described.

Of the crystalline variety, perhaps the most thoroughly investigated compound is ammonium molybdophosphate, $(NH_4)_3PO_4.12MoO_3$(AMP). AMP can only be used in neutral or acid solutions because of its solubility in dilute alkalis. In acid solution (pH less than 2) only metal ions which form heteropolyacid salts insoluble in water exchange significantly with the ammonium ion, namely K^+, Rb^+, Cs^+, Ag^+ and Tl^+ [14]. Those ions are therefore selectively adsorbed from solutions containing other ions and very sharp separations are possible. In addition, very good separations of, for example, sodium from potassium, potassium from rubidium, and rubidium from caesium have been carried out, and more complicated separations are undoubtedly possible. Under acid conditions polyvalent ions do not exchange readily, but at pH 2–5, they are fairly strongly adsorbed, especially if a suitable buffer solution is used [15]. Simple group separations are possible, for example Sr^{2+} from Y^{3+}, a quite strong acid solution being necessary to elute the Y^{3+}.

Zirconium phosphate, which is easily prepared by the addition of a solution of zirconyl nitrate or chloride to a phosphate solution is extremely insoluble and will not dissolve in solutions of high or low pH. It is an inorganic polymer of composition $Zr(H_2PO_4)_2.H_2O$ and a relative

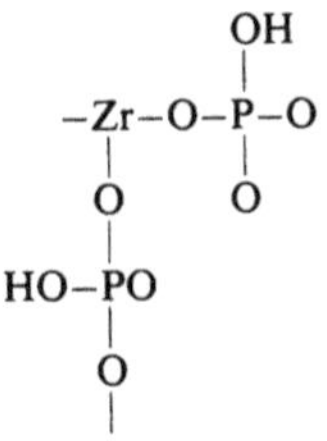

```
          OH
          |
  -Zr-O-P-O
   |      |
   O      O
   |
HO-PO
   |
   O
   |
```

Figure 4.4 Partial structure of zirconium oxide.

molecular mass of about 880 [16] with the repeating unit shown in Figure 4.4.

It has a high exchange capacity—about 4 mmol $OH^{-} g^{-1}$ in alkaline solution—and behaves as a weak acid cation exchanger. It can be prepared in bead form but does not show the very high selectivity of AMP. For example, the separation factor for caesium and rubidium on zirconium phosphate (α_{Rb}^{Cs}) is 1.3–1.5; in the case of AMP the corresponding figure is 26. Nevertheless, zirconium phosphate shows adsorptive properties for a large number of cations and has been recommended for separations involving radioactive solutions.

4.8 Gel ion exchangers

The separation of biological molecules from complex mixtures requires the use of high resolution techniques. It is, however, not normally practical to separate such molecules on conventional ion exchange materials because the molecules cannot penetrate the resin matrix, even if cross-linking is as little as 1%. Most of the exchange sites are therefore inaccessible and, since the molecules do not use the full exchange capacity of the material, the separation is very inefficient.

Ion exchange, using gels modified with ionogenic groups [17], has overcome this problem and made possible the ion exchange separation of large polar and charged biological molecules. In many cases it is unclear whether an anion or cation exchanger is required for a separation, but some important macromolecules of biological origin are amphoteric electrolytes (such as proteins). At a pH above the isoelectric point they are in the anionic form, and are separated on an anion exchanger; below the isoelectric point they are cationic, and a cation exchanger is used. The pH of the buffer is chosen according to the nature of the substances to be separated, and this then dictates the gel type to be employed. If, at the pH to be selected, the ampholyte has a high affinity for the exchange group of the resin, the appropriate weak ion exchanger should be used.

Ion exchange gels are used and handled in much the same way as the unmodified gels, combined with some of the techniques used for conventional

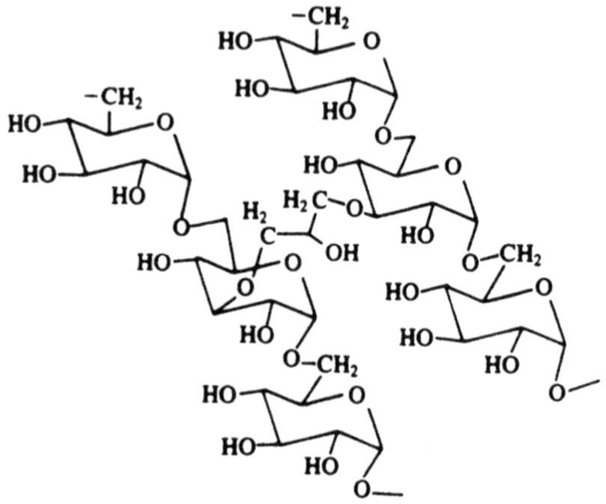

Figure 4.5 Partial structure of a cross-linked dextran (Sephadex).

ion exchange materials. Allowance must be made for changes in the gel condition when the ionic strength of the eluant is changed. Ion exchanger gels are supplied in the sodium form (cation exchangers) or the chloride form (anion exchangers) and they are usually converted, before a separation is started, into the H_3O^+ or the OH^- form, respectively.

4.8.1 Dextrans

One of the first commercially available gel ion exchangers was produced by introducing ionogenic units onto a cross-linked dextran (Figure 4.5). In the dextran ethers the –OH groups are converted into ethers where the alkyl group contains either an acidic or basic ionogenic group. Some of the properties of ion exchange dextrans are listed in Table 4.7.

Gel ion exchangers based on dextran have proved particularly useful since, due to their hydrophilic character and low adsorptivity, they have little

Table 4.7 Commercially available dextran ion exchangers

Name	Type	Functional group	Exchange capacity mmol $H^+(OH^-)g^{-1}$ (approx.)
SP–Sephadex	Strong acid	Sulphoxyl	2.3
CM–Sephadex	Weak acid	Carboxymethyl	4.5
QAE–Sephadex	Strong base	Diethyl-(2-hydroxypropyl)aminoethyl	3.0
DEAE–Sephadex	Weak base	Diethylaminoethyl	3.5
DEAE–Agarose	Weak base	Diethylaminoethyl	1.5
CM–Agarose	Weak acid	Carboxymethyl	2.5
DEAE–Sephacel	Weak base	Diethylaminoethyl on cellulose	1.4

tendency to denature or adsorb biological molecules. Furthermore, due to the high degree of substitution of the dextran matrix with ion exchange groups, high sample capacities are obtained, and finally, the material has good packing characteristics as the gel is bead-formed.

4.8.1.1 Physical properties. Ion exchange dextrans are physically very similar to ordinary dextrans, and have similar chemical stability, however the extent of swelling in aqueous buffers is very much more dependent on the ionic strength of the solute. When the ionic strength of the solution is lowered, the gels swell considerably; they shrink again when the ionic strength is raised. This is of practical importance in the operation of a column since if a considerable shrinkage occurs when a buffer is changed, cavities and irregularities may appear in the bed. Again, excessive swelling may clog the column. Swelling is most likely to happen during regeneration, and in consequence it is often better to regenerate the resin in a Buchner funnel, and then to repack the column.

Ion exchange dextrans are supplied in two degrees of porosity, the less porous being recommended for separations where M_r is less than 3×10^4 or greater than 2×10^5 (in the latter case the molecules will be totally excluded from the gel and little ion exchange will take place). The more porous gel is recommended for separations between the two M_r limits stated. The bead size in all cases is 40–120 μm.

4.8.1.2 Applications. The number of reported uses of ion exchange dextrans is increasing rapidly. Examples are the use of DEAE–Sephadex for the separation of mucopolysaccharides by stepwise elution with NaCl [18] in dilute HCl, soluble RNA [19], and the separation of myoglobin and haemoglobin [20]. Cation exchange dextrans have been used for the separation of α- and β-globulin [21], alkaloids [21], and for the purification of enzymes [22].

4.8.2 Agarose gels

Agarose is a mixture of charged and neutral polysaccharides the agarose repeating subunit is shown below (Figure 4.6). A gel is formed from the neutral fraction of agar and the agarose units cross-linked with 2,3-dibromopropanol. This process gives a bead-form gel with a very rigid structure arising from the alignment of the polysaccharide chains, which are stabilised by hydrogen bonding, while further rigidity is conferred on the chain by intrachain cross-linking.

Ion exchange capability is conferred by the introduction of carboxymethyl and diethylaminoethyl functionality, these groups being bonded to the monosaccharide subunit via an ether linkage. The above preparation process gives a highly rigid gel material with a similar porosity, and molecular weight

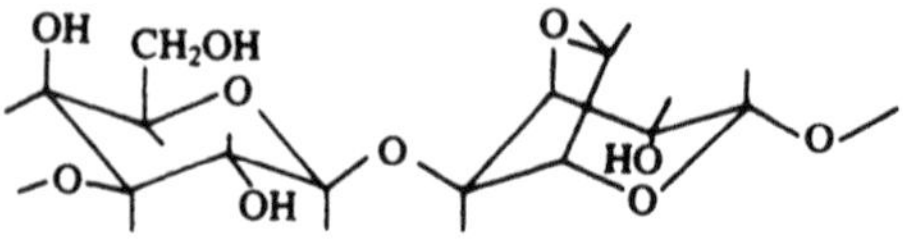

Figure 4.6 (a) Partial structure of agarose; (b) partial structure of Sepharose, a cross-linked agarose.

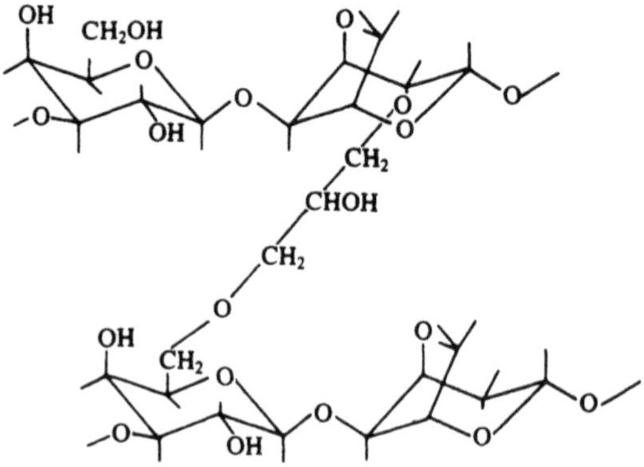

exclusion limit as dextran gels, but with improved mechanical strength and capacity. Further benefits deriving from the rigid structure are consistency of bed volume, which is relatively unaffected by changes in the ionic strength and pH of the eluant; and the improved flow properties and high flow-rates which can be achieved. Table 4.7 lists the properties of some agarose ion exchange gels.

4.8.3 Cellulose

The first ion exchangers designed [23] for the separation of biological molecules utilised a cellulose matrix. Though cellulose, due to its hydrophilic properties, had little tendency to denature proteins, these packings suffered from low sample capacities and poor flow characteristics, both defects stemming from the irregular shape of the particles. It was not until the mid-1960s that cellulose gels were produced in the optimal bead-form. In the production of commercial gels the polysaccharide gel is broken down, and during the regeneration bead process it is cross-linked for added strength with epichlorohydrin. The resulting macroporous bead (30–120 μm diameter) has good hydrolytic stability with an exclusion limit of 1×10^6 for proteins.

The only commercially available material, DEAE–Sephacel can be used in the pH range 2–12, however, hydrolysis can occur in strongly acidic solutions while strongly alkaline mediums can cause breakdown of the macromolecular structure. These packings have excellent flow characteristics and increased physical strength and stability arising from the cross-linked bead structure. Re-equilibration is also facilitated as the bed volume is stable over a wide range of ionic strength and pH. The above material is used in the ion exchange separation of proteins, nucleic acids, hormones and other biopolymers.

4.9 Gel chromatography

Adsorption studies on silica gel and active carbon had shown molecular sieve effects with materials of high relative molecular mass, and in 1954 Mould and Synge [24] showed that separations based on molecular sieving could be performed on uncharged substances during electro-osmotic migration through gels. This formed a basis for separations based on the relative sizes of molecules, and the systematic use of the principle was introduced in 1959 by Porath and Flodin [25], who used the term 'gel filtration' to describe their method of separating large molecules of biological origin in aqueous systems by means of polysaccharide gels. In a pioneering paper on non-biochemical uses, Moor [26] used the term 'gel permeation chromatography' (GPC). Both of these terms are still used in their respective fields, and others have been proposed, but in 1964 Determann [27] suggested that 'gel chromatography' was the most general name for the technique, and this is the one that will be used here.

The stationary phase is a porous polymer matrix whose pores are completely filled with the solvent to be used as the mobile phase. The pore size is highly critical, since the basis of the separation is that molecules above a certain size are totally excluded from the pores, and the interior of the pores is accessible, partly or wholly, to smaller molecules.

The flow of mobile phase will cause larger molecules to pass through the column unhindered, without penetrating the gel matrix, whereas smaller molecules will be retarded according to their penetration of the gel. The principle is illustrated in Figure 4.7.

The components of the mixture thus emerge from the column in order of relative molecular mass, the larger first. Any compounds which are completely excluded from the gel will not be separated from each other, and similarly, small molecules which completely penetrate the gel will not be separated from each other. Molecules of intermediate size will be retarded to a degree dependent on their penetration of the matrix. If the substances are of a similar chemical type, they are eluted in order of relative molecular mass.

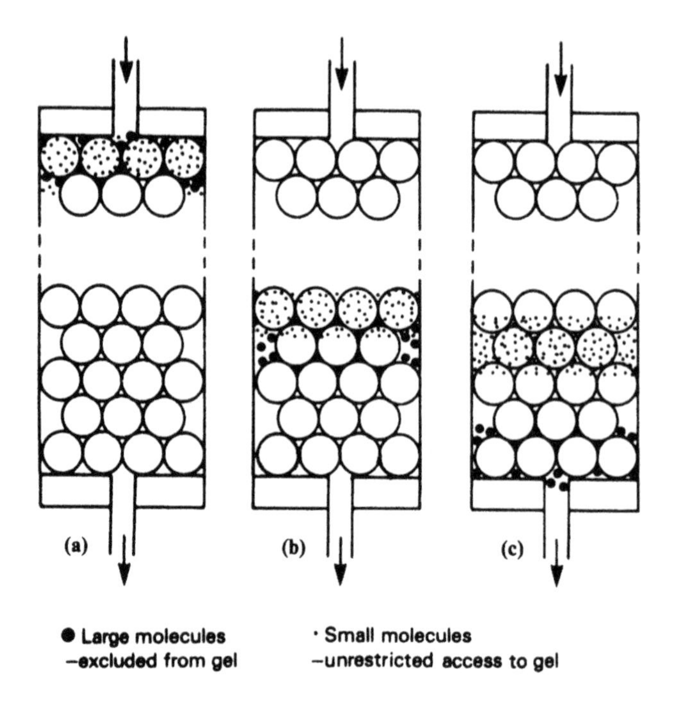

Figure 4.7 Principle of gel chromatography: (1) mixture applied to the top of the column; (2) partial separation; (3) complete separation; excluded substance emerges from the column.

Adsorption effects of the surface of the gel particles can usually be ignored, and thus gel chromatography can be looked upon as a kind of partition chromatography. The liquid stationary phase is the liquid within the gel matrix, and the mobile phase is the flowing eluant which fills the rest of the column. We have, in other words, a partition column where the two liquid phases, mobile and stationary, are of the same composition.

The model proposed above suggests that the differential exclusion of the solute molecules was achieved on the basis of their hydrodynamic volumes, that is, their size and shape. This model has been extended by Porath [28], who considered the pores in dextran gels to be conical in nature. A further approach [29], which considers the gel to be composed of randomly arranged rigid rods, shows good correlation between the molecular radius and retention volume of the solute.

Gel chromatography was originally used for separation of biological materials, because the earliest gel media, cross-linked dextrans, were suitable for use only with an aqueous system. In 1964 [26], cross-linked polystyrene gels suitable for use with organic solvents were first produced, and this made possible the extension of gel chromatography to the separation and characterisation of synthetic polymers. Application of the method, not only to a large variety of separations but also the determination of relative molecular mass, proceeded rapidly.

Early accounts of the use of gel chromatography were given by Porath [30], Tiselius *et al.* [31], Gelotte [32] and Granath [33]. In addition, the manufacturers and suppliers of many commercially available gel media provide an extensive information service.

4.9.1 Column parameters and separations

A number of theoretical treatments of gel chromatography have been published, including applications of the theoretical plate concept similar to those outlined in Chapter 2. It is appropriate here to mention only a few simple parameters of most value in practical work.

A column is made up by pouring a slurry of swollen gel particles in the solvent used to swell the gel into a suitable tubular container. The total volume of the column, V_t (which can be measured), is the sum of the volume of liquid outside the gel matrix, V_o, the volume of liquid inside the matrix, V_i, and the volume of the gel matrix, V_m (Figure 4.8); that is

$$V_t = V_o + V_i + V_m$$

V_0 is also known as the void or dead volume; it is the volume of mobile phase which will elute a totally excluded molecule (A). The volume required to elute a particular molecule is the elution volume (or retention volume), V_e. The use of these volumes is not very satisfactory in describing the behaviour of particular solutes, because although they are characteristic of the gel and the solutes they also depend on the volume and packing density of the column. Separation parameters for gel columns are shown in Figure 4.8.

As in other forms of partition chromatography the elution volume can be related to the column dead volume (V_e) and the volume of stationary phase

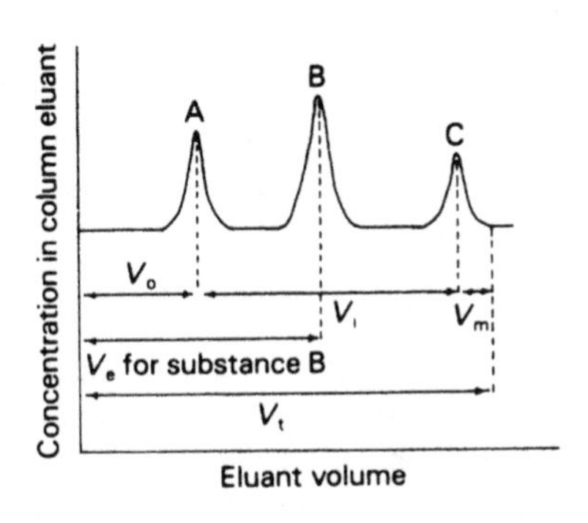

Figure 4.8 Separation parameters for gel columns.

(V_s) through the following equation:

$$V_e = V_o + K_d \times V_i$$

where K_d is the volumetric distribution coefficient. Thus

$$K_d = (V_e - V_o)/V_i$$

K_d is a constant for a given solute and gel under constant operating condition of eluant composition and temperature, hence the determination of K_d characterises the retention behaviour of the solute independently of the dimensions of the chromatographic bed.

K_d represents the fraction of the gel volume that is accessible to the molecule concerned, and is thus zero for a totally excluded molecule ($V_e = V_o$), and unity for a molecule which has access to the gel equal to that of the solvent ($V_e = V_o + V_i$). The relationship between these various parameters is shown in Figure 4.8.

In practice, not all these volumes are readily determined. V_o can be measured as the volume required to elute a completely excluded solute, and V_e for a particular solute is equally readily measured. On the other hand, the methods which can be used for determining V_i (requiring, for example, observation of the elution of tritiated water) give only approximate results, since exchange reactions on the Sephadex gel cause the ^3H peak to emerge after the bed volume [34]. K_d is often, therefore, replaced by an alternative distribution coefficient, K_{av}, which is defined [35] as

$$K_{av} = (V_e - V_o)/(V_i + V_m)$$

that is

$$K_{av} = (V_e - V_o)/(V_t - V_o)$$

where ($V_i + V_m$) approximates to the stationary phase volume, V_g. Although all the volumes in this expression can be determined without difficulty, K_{av} has the disadvantage (because of the inclusion of a term for the gel volume, V_m) of not approaching unity for small molecules. Fractionation ranges are shown in Figure 4.9 and choice of gels are shown in Figure 4.10.

If K_d or K_{av} is plotted against the logarithm of the molecular weight for a series of solutes similar in molecular shape and density, an S-shaped curve is obtained. Over a small range of K_{av}, the fractionation range, the curve approximates to a straight line

$$K_{av} = A + B \log M$$

The former plot shows the fractionation range of the gel, which is defined as the approximate range of M_r, within which a separation can be expected, provided that the molecules concerned are in different parts of the range. The upper end of the range ($K_d \rightarrow 0$) is the exclusion limit, the M_r of the smallest molecule which cannot penetrate the pores of the matrix. The plot

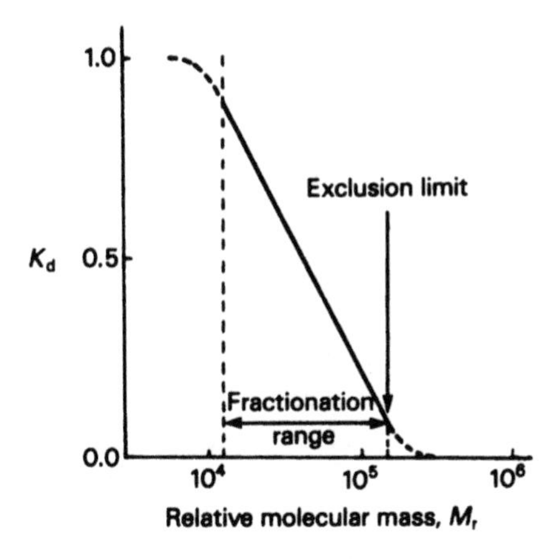

Figure 4.9 Fractionation ranges.

of K_d against $\log M_r$ diverges from linearity as K_d approaches 0 or 1, and the fractionation range is the range of M_r over the linear part of the curve. For most gels the lower limit of the range ($K_d \to 1$) is about one-tenth of the exclusion limit.

Since M_r is not directly related to the size or shape of the molecule, it is necessary to state, in publishing fractionation ranges for gels, what type of molecule has been used in the determinations. Most commercial gels are made in a number of grades of different fractionation range. A gel is selected such that the M_r values of the substances to be separated lie on the straight part of the curve. This is illustrated in Figure 4.10, where A ($M_r \sim 5 \times 10^4$) and B ($M_r \sim 6 \times 10^3$) are the substances to be separated. It can be seen that

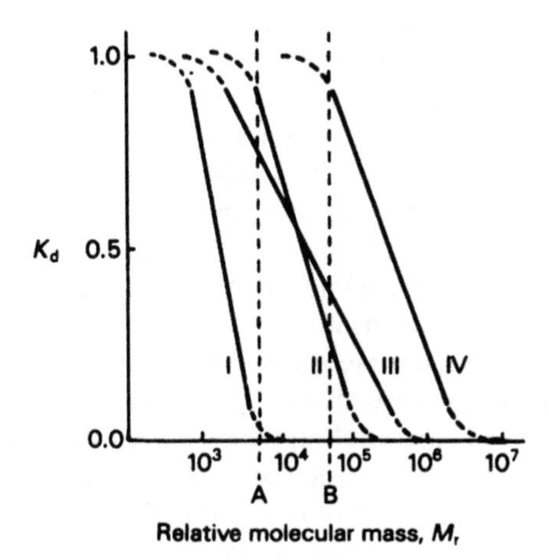

Figure 4.10 Choice of gel.

either of the gels II or III could be used. It would be better to use gel II because the K_d values are in that case further apart, and the fractions of the eluate containing A and B would be further apart also. More complex mixtures would require the use of gels of wider fractionation range, but in general it is best to use a gel of as narrow a range as possible.

Sorption effects by the gel matrix may alter the values of K_{av} and K_d for certain solutes, and the use of K_d against $\log M_r$ plots to select a gel, takes into account neither those effects nor zone spreading in the column, and variations in flow-rate (which could arise, for example, from differences in viscosity of sample and eluant). Taken together these may have an important effect on the separation, and hence on the choice of the best gel for a particular purpose. The behaviour of small molecules on tightly cross-linked gels is profoundly influenced by adsorption, ion exclusion, ion exchange and various other processes which dominate the size differences between solutes.

4.9.2 Nature of the gel

Separation by gel chromatography is influenced by the properties of the pores in the three-dimensional network, and the nature of the pore forming material itself. The gel must be chemically inert, mechanically stable and have a carefully formed and reproducible porous structure, with a fairly uniform particle size. The fractionation range is governed by the range of pore sizes in a given gel. A wide range of pore sizes gives rather poor resolution, though a narrow range, while giving improved resolution, has a more limited molecular weight range of application. Normally, a series of sorbents is used which separates solutes over a characteristic molecular weight range, commonly one-order of magnitude.

The chemical stability of commercially available gels is good, though they are not resistant to strongly acidic or alkaline conditions. For high flow-rates the sorbents must have good physical stability. However, the large pore structure contains only moderate amounts of the solid matrix; the gels are thus mechanically weak and are distorted even when moderate pressure is applied, thus the attainable flow-rates are limited. As in other modes of chromatography the optimum particle shape is spherical. The bead size should be as small as possible while still affording the desired flow-rate, the limit of which is effectively determined by the rigidity of the structure—compressible gels would give impracticably low flow-rates. Some examples of commercially available gels are given in Table 4.8.

Two main types of gel material may be distinguished: xerogels and aerogels. Xerogels are gels in the classical sense; they consist of cross-linked polymers which swell in contact with the solvent to form a relatively soft porous medium, in which the pores are the spaces between the polymer chains in the matrix. If the liquid is removed the gel structure collapses, although it

Table 4.8 Some commercially available media for gel chromatography

Name	Type	Chemical nature	Eluant (mobile phase)	Maximum exclusion limit M_r	Calibration	Number of fractionation ranges	Bead size (μm)	Notes
Sephadex G	Xerogel	Dextran	Aqueous	6×10^5	Peptides/proteins	12	10–40	Superfine grade for very high resolution and thin layers
				2×10^5	Dextrans	8	40–120	
							50–150	
							100–200	
Sephadex LH	Xerogel	Dextran hydroxypropyl ether	Polar organic	5×10^3		1	25–100[a]	Exclusion limit varies according to solvent
Bio-Gel P	Xerogel	Polyamide	Aqueous	4×10^3	Peptides/proteins	10	50–100[a]	
							100–200	
							200–400	
Enzacryl Gel K	Xerogel	Polyacryloyl-morpholine	Aqueous	1×10^5	Polyethylene glycols	2		Little swelling in lower alcohols
			Polar organic					
Bio-Beads SX	Xerogel	Polystyrene	Organic	14×10^3	Polystyrenes	7		
Styragel	Hybrid	Macroporous polystyrene	Non-polar organic	4×10^8	Polystyrenes	12		
Bio-Gel A	Hybrid	Agarose	Aqueous	15×10^7	Dextrans	6	50–100[a]	1, 2, 4, 6, 8, 10% agarose
							100–200	
							200–400	
Sepharose	Hybrid	Agarose	Aqueous	4×10^7	Proteins	3	60–250	2, 4, 6% agarose
				2×10^7	Polysaccharides		40–190	
							40–210	
Merck-O-Gel	Hybrid	Polyvinyl acetate	Polar organic	1×10^6	Polystyrenes	6		
Porasil	Aerogel	Silica	Aqueous	2×10^4	Polystyrenes	6		Calibration in organic solvents
			Organic					
Bio-Glas	Aerogel	Glass	Aqueous	9×10^4	Polystyrenes	5		Calibration in toluene
			Organic					

[a] US Standard Screens.

can sometimes be restored by replacing the liquid. Aerogels, on the other hand, are rigid materials which are not really gels at all; they are porous solids which are penetrated by the solvent, and they do not collapse when the solvent is removed; porous glass and porous silica are examples. Some gel materials, such as polystyrene, are xerogel–aerogel hybrids. These have a fairly rigid structure, but swell to some extent on contact with solvents, in the same way that ion exchange resins do. An exception is agarose (see below), which behaves in an unusual way.

Much greater care is needed in handling and using the relatively soft xerogels and agarose than in handling the much more rigid polystyrene hybrids and aerogels, particularly in controlling column conditions to prevent particle breakdown and coagulation of the particles, both of which will retard the flow of mobile phase.

4.9.2.1 Dextran gels. The original gel chromatography medium, which is still widely used, was Sephadex G, a cross-linked dextran. The starting material, dextran, is a natural linear polysaccharide, that is, a polyglucose, in which the glucose residues are predominantly α-1,6 linked; it is insoluble in aqueous media.

The individual polysaccharide chains of dextran can be cross-linked, with glyceryl bridges between the hydroxyl groups, on reaction under alkaline conditions with a dispersion of epichlorohydrin in an organic medium (Figure 4.5, p. 138). The resulting product is a water insoluble solid in bead-form, which is hydrophilic due to the many residual hydroxyl groups, a feature which causes the gel to swell in water/aqueous solutions; the water regain is 1–20 ml g^{-1} of dry resin. The extent of cross-linkage can be varied by controlling the relative proportions of epichlorohydrin and dextran. Thus, a range of Sephadex gels of varying porosity have been obtained, which are useful over different molecular weight ranges. The random distribution of cross-linkages also produces a wide distribution of pore sizes in each type of gel, thus providing a wide linear working range. The Sephadex gels are insoluble and stable in water, salt and alkaline solutions and organic media.

Little swelling occurs in non-polar solvents, and thus dextran gels are only useful in aqueous media. Traces of carboxyl (10–20 mmol $H^+ g^{-1}$ of dry gel) remain, which may affect the separation of some polar species; this property is exploited in aromatic adsorption studies. High chemical stability is conferred on the gels due to the covalent cross-linking of the individual polysaccharide chains; they can be safely used between pH 2 and 12, though at very low pH values hydrolysis of the glycosidic linkages may occur.

4.9.2.2 Dextran hydroxypropyl ether. Sephadex LH-20 is the hydroxypropyl ether of Spehadex G-25 (the number indicates the porosity of the gel). It will form a gel in both aqueous and organic solvents, and is particularly effective in solvents such as dimethyl sulphoxide, pyridine and dimethyl formamide.

4.9.2.3 Polyacrylamide gels. Bio-Gel P is a wholly synthetic polyacrylamide gel made by suspension co-polymerisation of acrylamide and the cross-linking agent *N,N'*-methylenebisacrylamide [36, 37]. The pore size is regulated by variation of the proportions of the monomers. Polyacrylamide gels behave in a similar way to dextran gels, with a water regain of 1.5–18 ml g^{-1} of dry resin. These gels are less resistant to alkali than Sephadex and hydrolysis of the amide residues to carboxylic acid groups occurs on exposure to solvents of extreme pH.

4.9.2.4 Polyacryloylmorpholine. Suspension co-polymerisation of acryloylmorpholine and *N,N'*-methylenebisacrylamide gives the medium known as Enzacryl Gel. This material swells to a gel in water, and also in pyridine and chloroform, although not in lower alcohols.

4.9.2.5 Polystyrene. Styrene–divinylbenzene polymers, as used in ordinary ion exchange resins, but without ionising groups, will form gels in less polar organic solvents. One form of this polymer is Bio-Beads S. Unfortunately, this type of polymer has a very small pore size, which rather restricts the range of uses.

If styrene and divinylbenzene are co-polymerised in a solvent mixture in which the polymer is sparingly soluble, a macroporous form of the polymer bead is obtained. One such material is Styragel, which can be used very effectively in the gel chromatography of organic polymers in organic solvents. Introduction of a few $-SO_3H$ groups into a macroporous resin (Aquapak) makes the gel hydrophilic, and thus usable in aqueous solvents, although the presence of the strongly acidic group may be a disadvantage.

4.9.2.6 Polyvinyl acetate. Co-polymerisation of vinyl acetate and 1,4-divinyloxybutane gives a material (Merck-O-Gel OR) which forms a gel in polar organic solvents, including alcohols.

4.9.2.7 Agarose gels. All the xerogels described are characterised by extreme softness of the swollen gel when the pore size is large. Larger pore sizes with mechanical stability are obtainable in the aerogels glass and silica, but with a penalty of incursion of adsorption effects. To overcome this adsorption difficulty, and to obtain larger pore sizes, agarose gels were developed.

Agarose is a polysaccharide (alternating 1,3-linked β-D-galactose and 1,4-linked 3,6-anhydro-α-L-galactose units) obtained from seaweed (Figure 4.6). Above 50°C it dissolves in water, and if the solution is cooled below 30°C a gel is formed, which is insoluble below 40°C. Above that temperature the gel 'melts' or collapses; freezing also causes irreversible changes in the structure of the gel. The chemical stability of agarose gels is similar to that of dextran gels, and is derived from the helical formation between individual polymer molecules. Residual sulphate groups may be eliminated by alkaline hydrolysis of the cross-linked agar and agarose under reducing conditions [38]. The charged hydroxyl groups can be reduced to the corresponding alcohol on reaction with litium aluminium hydride [39].

Several forms of agarose are obtainable commercially (Sepharose, Bio-Gel A, Gelarose, Sagavac), which indicates the usefulness of this gel. It can be made with a very large pore size, and it is mechanically much more stable than a dextran gel of similar pore size.

4.9.2.8 Porous silica. Silica is commercially available in bead form (Porasil; Merck-O-Gel Si) with a range of porosities. This is an aerogel, with a very rigid structure. It can be used in some organic solvents, but it is best used in water. It is rather highly polar, and can tend to retard polar molecules by adsorption.

4.9.2.9 Porous glass. Various grades of granular porous glass (Corning; Bio-Glas) are available. Glass can be used for gel chromatography in both aqueous and organic media, but there may be undesirable adsorption effects, as with silica, due to the presence of negatively charged sites. These groups can be temporarily 'capped' with polyethylene oxide [40] or polyethylene glycol [41]. The gels are derived from borosilicate by heat treatment, followed by acid leaching; this process gives a gel of accurately controlled pore size and very narrow distribution. Porous glasses have numerous advantages accruing from their rigid structure. They have good mechanical strength, are heat stable and can be used in aqueous and non-aqueous solvent systems.

4.9.3 Methodology

4.9.3.1 Column preparation. Columns for gel permeation studies are generally packed with a suspension of the gel material in the appropriate solvent. The gel suspension is adjusted to the consistency of a thick slurry, though sufficiently mobile to allow trapped air to bubble freely through it. The 'fines' can be decanted at this stage and then the well mixed slurry poured into the column. The gel-bed is then equilibrated by passing through two to three volumes of the eluant, at a flow rate slightly higher than that to be used in the experiment. The column is then tested by running through a totally excluded coloured solute, commonly Blue Dextran 2000, in a sample volume <10% of the column volume. If the column is properly packed and stabilised the coloured substance should be eluted as a compact horizontal band. This test also gives a measure of the void volume, V_o, that is, the volume required to elute the unretained coloured substance.

During packing of the column and development of the chromatogram the top of the column must not be allowed to run dry and must not be disturbed. Generally, the most crucial factor in operation of the column is control of the flow rate. Operating the column at pressure greater than specified (in the manufacturers' literature) can lead to diminution of the flow rate due to compression of the chromatographic bed which arises from fracture and deformation of the gel particles.

4.9.3.2 Sample preparation. Solid particles and any substances which may be strongly adsorbed on the gel should be removed from the sample. The mass and concentration of the solutes are only important in so far as they affect the viscosity. If the sample viscosity is more than about twice that of the eluant, then this leads to poor resolution. The sample capacity, in terms of volume, is also important.

The sample volume is also important. The largest volume which can be handled for complete separation in one run through a column is $V_{e1} - V_{e2}$, where these are the elution volumes of the two most closely related components (from the chromatographic point of view). For group separations where $V_{e1} - V_{e2}$ is large, the sample volume can be 10–30% of the bed volume, V_t. For separation of closely related components ($V_{e1} - V_{e2}$ small) the sample volume should only be 1–3% of V_t. No useful gain in resolution is obtained by further reduction of the sample volume. The smaller the sample volume, the greater will be the reduction of the component concentration in the eluate; this dilution effect may have to be taken into account in deciding upon column and sample sizes. For very small samples, in the cubic millimetre region, gel chromatography may well be more effective on gel thin layer plates than on columns (Chapter 3). The thin layer technique can often be used in a 'pilot' experiment to determine the best conditions for a larger scale separation on a column.

With simple laboratory columns the sample is applied to the top of the gel bed by any of the usual methods. Application of sample to the drained bed, though requiring little specialist equipment, is difficult to achieve without disturbing the surface of the packing. A technique suitable for samples which are denser than the eluant, such as blood serum, involves sample application through a fine capillary to the top of the column under a layer of the eluant. If required, the sample density can be increased by the addition of glucose, buffer salts or other suitable inert material.

The sample can best be applied to the column using three-way valves in conjunction with a loop or syringe. In the former, the loop can be isolated from the eluant flow and filled with sample. The loop contents are then

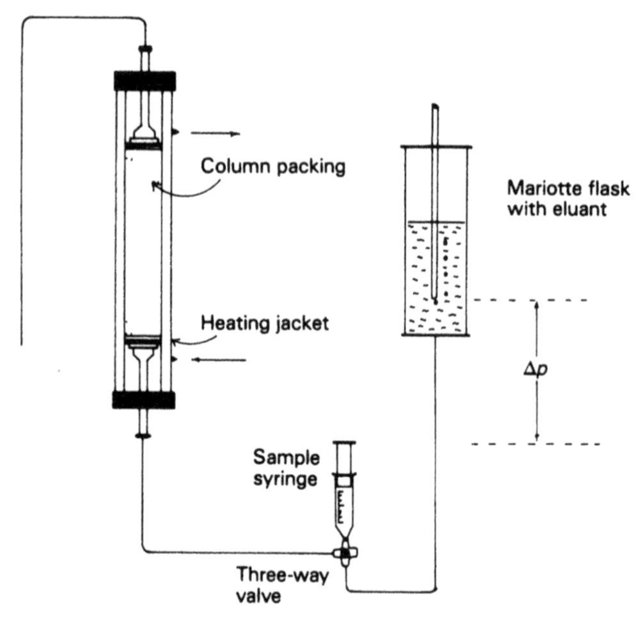

Figure 4.11 Experimental set-up for gel chromatography. The operating pressure (Δp) is the height between the air inlet and the eluant outlet. The three-way valve connects either the sample or the eluant to the column.

flushed onto the column by switching the valve position, so bringing the loop in series with the solvent delivery from the reservoir. Loop-valve systems afford very reproducible sample volume application. Syringe-valve systems, as illustrated in Figure 4.11, necessitate interruption of the solvent delivery during sample application. This latter technique, however, allows for variation in the sample loading. The eluant flow can be generated either by gravity feed or by a pump. In the system illustrated (Figure 4.11) the sample is applied from a syringe via a three-way valve to the bottom of the column. This system provides a constant hydrostatic pressure, which corresponds to the vertical distance between the lower end of the flask air-inlet tube and the free end of the column outlet tubing.

Development of the chromatogram is strikingly simple. Only isocratic techniques are used and the solutes are eluted in order of decreasing molecular weight and/or size. The number of solutes which can be resolved is normally <10 as the maximum elution volume in the absence of adsorption forces can be no greater than the total volume of mobile phase ($V_o + V_i$) present in the column. Resolution in gel chromatography is subject to the same theoretical considerations as other forms of liquid chromatography. The important variable for control of resolution are column length (L), particle size (dp) and homogeneity of packing. The influence of flow-rate on column performance has previously been discussed. While the column performance increases as the $\sqrt{L}$, the column length which can be used in practice is limited due to compressibility of gel. Alternatively, the resolution can be improved by using a smaller particle size, though again this is limited by the back pressure and the restriction on flow.

Homogeneity of column packing is essential for good column performance, as irregularities in packing will cause additional peak broadening. Peak resolution is generally little affected by solvent composition, however, where the solutes are electrolytes then retention behaviour can be modified by varying the ionic strength of the eluant. The effect is especially marked where the solute configuration is influenced by electrolyte solutions. Furthermore, partition can be controlled by adding small amounts of materials such as dextran, polyvinylalcohol and polyethyleneglycol to the eluant. Small amounts of these additives have a significant effect on the K_{av} values. Where adsorptive interactions occur between solute molecules and the gel packing then, of course, eluant strength has a pronounced effect on resolution.

4.9.4 Applications

Since 1959, literally thousands of papers have been published on the use of gel chromatography and related applications. This section will therefore be restricted to a broad but necessarily selective examination of the areas of application, and only some instructive examples will be described.

4.9.4.1 Group separations and desalting. The simplest application of GPC is the separation of two groups of solutes of widely differing sizes. By choosing a suitable gel, high-molecular-weight material may be totally excluded ($K_{av} \rightarrow 0$) and will wash through in the void volume, while low-molecular-weight materials ($K_{av} \rightarrow 1$) will be strongly retained. For example, low-molecular-weight materials, such as urea, sugars and peptides, liberated during chemical modification of proteins can be separated from the high macromolecular molecular weight products. The particular application where large molecules of biological origin are separated from inorganic or ionisable species is known as desalting, a typical example is the separation of haemoglobin and sodium chloride on a Sephadex gel (Figure 4.12) [42].

An interesting application of GPC involved the isolation of a mixture of porphyrins prepared from the acid catalysed acetylation of haematoporphyrin. Reversed phase HPLC showed the reaction mixture to comprise more than 30 components comprising several monomeric porphyrin derivatives though the active component comprised highly aggregated or higher molecular weight porphyrins; this active component was termed HPD and it has been used with some success in the photodynamic therapy of solid tumours.

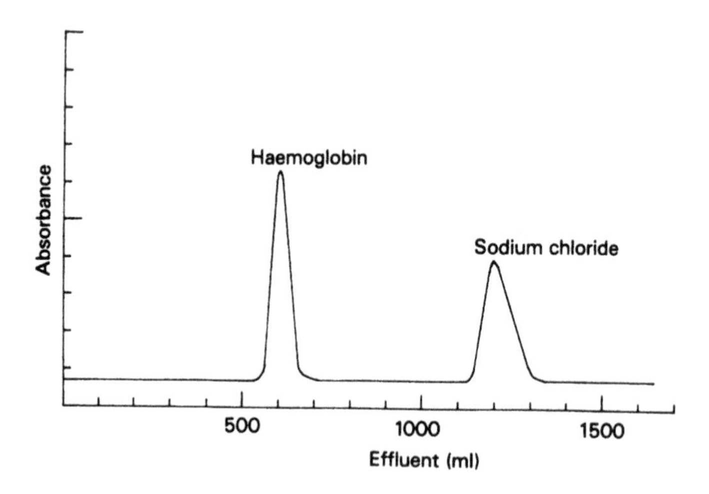

Figure 4.12 Group separation of haemoglobin and sodium chloride [42].

Reverse phase HPLC was successful in separating the bulk of the monomeric material from DHE with the exception of protoporphyrin. This component was present in the broad elution profile of the active fraction and could not be separated from HPD by reverse phase HPLC (Figure 4.13). However, the active component could be isolated quantitatively by gel filtration using aqueous elution on a Sephadex (G25) or a polyacrylamide column (nominal exclusion limit 20 000 Da). Non-aqueous fractionation using a tetrahydrofuran/methanol/aqueous buffer eluant on Sephadex-LH20 has also allowed the isolation of the active component from the above mixture free of any monomeric species [43]. This column material, -LH20, has also been used to successfully remove excess cuprous acetate present in the preparations of copper amine derivatives of protoporphyrin [44].

4.9.4.2 Fractionation of mixtures. Fractionation of mixtures of biopolymers is the most widespread application of gel chromatography and can be carried out on the analytical and preparative scale. When the molecular weight and sizes of the components of the sample do not differ much then a gel of the appropriate fractionation range should be chosen. Choice of suitable gels allowed various peptides [45], nucleic acid [46], enzyme [47] and polysaccharide [48] mixtures to be separated and purified.

4.9.4.3 Molecular weight determinations. Gel chromatography was first used to separate large molecules of biological origin such as proteins, polysaccharides, nucleic acids and enzymes. The initial observation that substances were eluted in order of decreasing molecular weight initiated

Hydroxyvinyldeuteroporphyrin (HVD) | Protoporphyrin (PP) | Haematoporphyrin (HP)

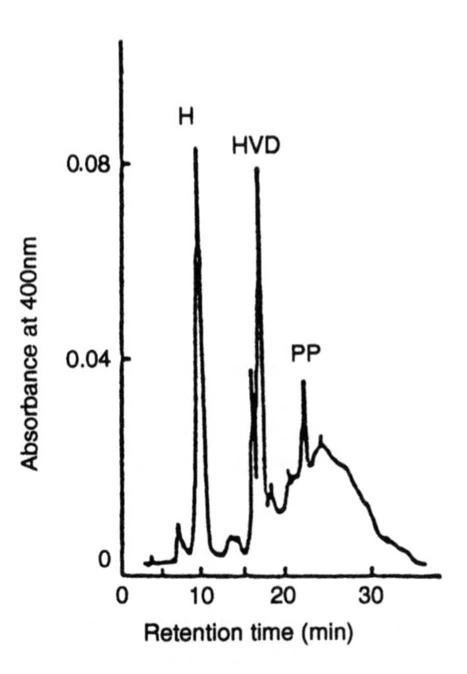

Figure 4.13 HPLC: reverse phase chromatogram of HPD on a C_{18} column eluted with methanol and phosphate buffer.

studies exploring the relationship between the chromatographic behaviour of the molecules and their molecular weight (M) [25].

On Sephadex the relationship:

$$\log M = A - B(V_e/V_o)$$

has been reported, where A and B are constants. Before applying such a relationship to the estimation of molecular weights, the column must be calibrated using several materials of similar type and of known molecular weight. Where the molecules concerned are chemically related the fractionation curve can be used as the basis for determining the M_r and M_r ranges. Though the studies were initially of naturally occurring macromolecules [49, 50], the separation and examination of synthetic polymers [51] has been added to the field of application, and the method has become an important part of polymer technology.

4.9.4.4 Protein–ligand binding. A recent area of study of increasing interest/importance is the examination of the equilibrium binding of small ligands to proteins (R) [52, 53]. A sample of protein equilibrated with the ligand is applied to a gel column containing free ligand. The normal processes of molecular sieving disturb the ligand–protein equilibrium concentrations of the applied sample and elution results in the appearance of a protein front followed by a mixture of free and bound material. The binding ratio can be estimated from peak area measurements obtained from the chromatogram.

4.10 Affinity chromatography

Ordinary chromatographic techniques depend for their effectiveness on differences (often small) in adsorption, partition, ionic charge, or size, between solute molecules of similar chemical character. A modified gel technique in which use is made of the high specificity of biochemical reactions is affinity chromatography [54, 55].

4.10.1 Principles

A ligand, which exhibits a specific binding affinity for a particular compound, is covalently bonded to the gel matrix and the material is packed into the column in the normal manner. The sample is applied to the top of the column in a suitable buffered eluant. On development, components with a specific affinity for the ligand become bound (though not irreversibly) and are retarded. The unbound material washes straight through the column (see Figure 4.14).

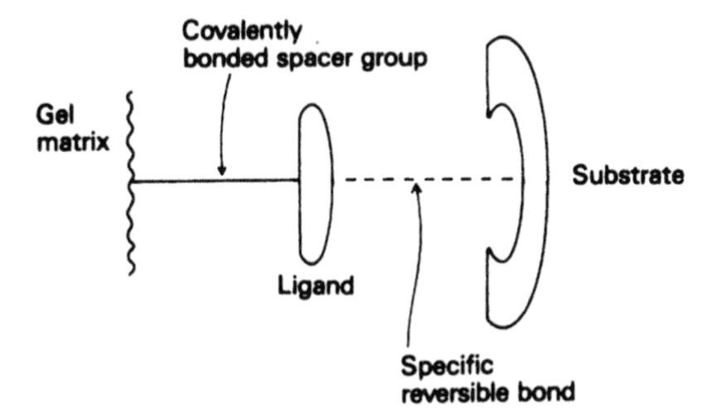

Figure 4.14 Principle of affinity chromatography.

The composition or pH of the eluant is then altered to weaken the component–ligand binding, thus promoting dissociation and facilitating elution of the retained compounds. The conditions which must be fulfilled are, therefore, the gel–ligand bond must be stable under the conditions of the experiment, the ligand–substrate bond must be specific and reversible under the conditions of the experiment and the gel–ligand bonding must be such that the specific adsorption properties of the ligand are not interfered with by the gel matrix.

4.10.2 Column materials

Sepharose, a bead-formed gel derived from agarose, possesses many of the properties necessary for a matrix for affinity chromatography. The residual hydroxyl groups on the gel can be readily derivatised by reaction with cyanogen bromide (CNBr) [56]. The CNBr activated matrix can be coupled with a range of ligands giving packings with various specificities. The open pore structure makes the interior of the matrix available for ligand attachment and ensures good binding capacities, in addition Sepharose exhibits extremely low non-specific adsorption and thus separation is achieved principally by the specific interactions of affinity chromatography.

Several intermediates are available commercially; for instance, CNBr–Sepharose, Sepharose and Bio-Gel P (polyamide) intermediates with reactive spacer groups attached, and Sepharose already coupled to a ligand specific for polysaccharides and glycoproteins. An outline of the reactions developed to couple ligands to Sepharose is illustrated in Figure 4.15; dextran and cellulose matrices can be derivatised in a similar manner.

To accommodate steric interferences, associated with the binding between the ligand and substrate, it has been found necessary for retention of ligand activity to insert short alkyl chains between the ligand and gel matrix, thus facilitating interaction of the ligand with the active site of the substrate.

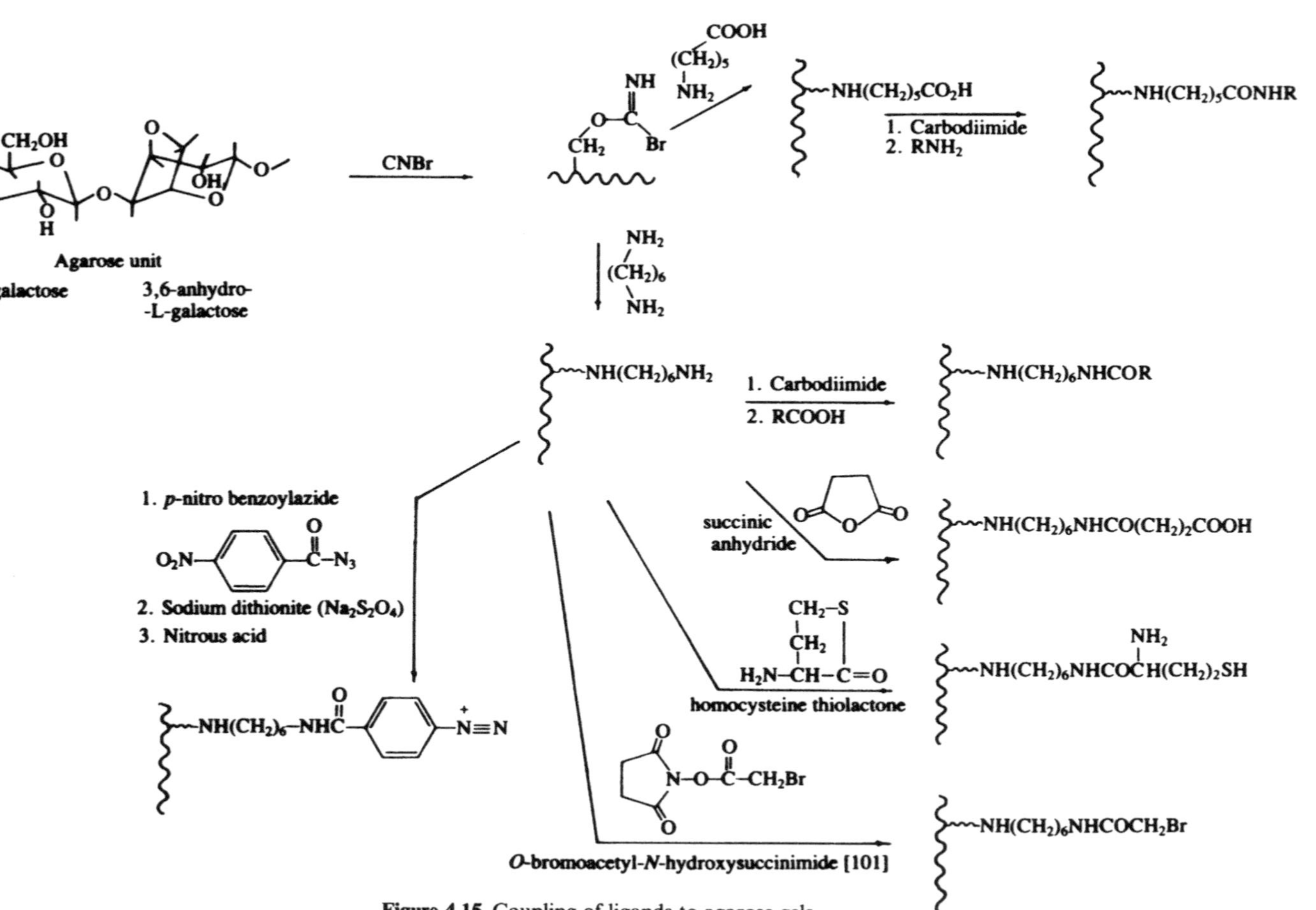

Figure 4.15 Coupling of ligands to agarose gels.

Bifunctional amines, for instance hexamethylenediamine, coupled to CNBr activated sepharose, are useful reagents in this respect and also furnish a free amino group, often suitable for bonding the affinity ligand. These 'spacers' can be lengthened by attaching succinic acid via its anhydride. The resulting free carboxyl group can then be bonded to affinity compounds containing an amino group via the intermediacy of *N,N′*-disubstituted carbodimides. The latter have proved an important synthetic intermediate in promoting condensation between free amino and carboxylic acid groups.

4.10.3 Applications

The complexity and chemical similarity of many compounds of biological interest, such as nucleic acids and proteins, makes affinity chromatography, because of its high specificity and its concentrating effects, an indispensable separation technique in biochemical research.

Affinity chromatography has considerable potential and has already found extensive application in biochemical and chemical studies requiring the purification of antigens, enzymes, proteins, viruses and hormones. It provides an elegant method for the high yield purification in a single step under mild conditions of pH and ionic strength. The selectivity of the technique derives both from the ligand interactions and the use of selective elution conditions. An example of the use of affinity chromatography is the isolation of whole genes. A further example uses CNBr activated Sepharose to immobilise DNA, which can then be employed to bind specific enzymes and proteins. A detailed discussion of the applications is outwith the scope of the text and the interested reader is directed to the papers cited and Ref. [57].

4.11 Covalent chromatography

Covalent chromatography is a technique used for the specific isolation and purification of thiol containing proteins and peptides. Sepharose is

$$\bigcirc\!-O-CH_2-\underset{\displaystyle OH}{\underset{|}{C}H}-CH_2-SH \quad + \quad \text{(2-pyridyl)}-S-S-\text{(2-pyridyl)}$$

$$\longrightarrow \bigcirc\!-O-CH_2-\underset{\displaystyle OH}{\underset{|}{C}H}-CH_2-S-S-\text{(ring)}$$

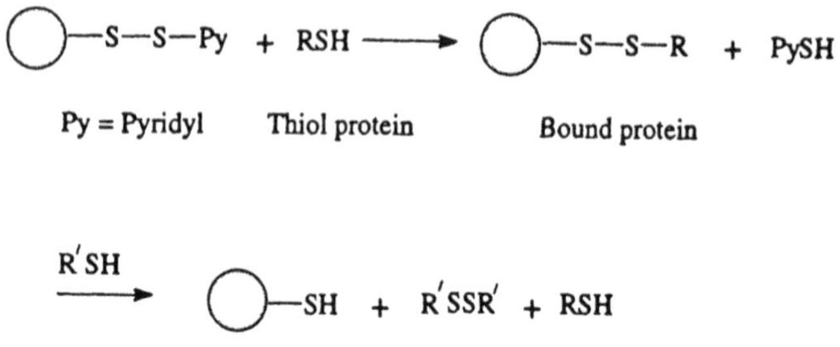

$$\bigcirc\!\!-S-S-Py + RSH \longrightarrow \bigcirc\!\!-S-S-R + PySH$$

Py = Pyridyl   Thiol protein   Bound protein

$$\xrightarrow{R'SH} \bigcirc\!\!-SH + R'SSR' + RSH$$

Figure 4.16 Reaction scheme showing binding and liberation of thiol protein.

commonly used as the gel matrix and the packing is prepared in a two-step synthesis and involves, for instance, the coupling of a thiopropylsepharose derivative with 2,2′-dipyridyl disulphide to give pyridyldisulphide; the hydroxylpropyl residue is effective as a spacer arm.

On application of the sample to the column, thiol containing components become bound to the column. The disulphide linkage can then be reduced in a controlled fashion, normally with an excess of the activated thiol, thus freeing the thiol compounds which can be collected in the eluant. The technique thus allows the separation of thiol-containing proteins from non-thiol and disulphide functionalities (Figure 4.16).

4.12 Chromatofocusing

Chromatofocusing is a unique column chromatographic method for the separation of proteins on ion exchange resins according to their isoelectric points. The technique, first described by Sluyterman and Elgersma [58], proposed that a pH gradient could be produced internally in an ion exchange column by taking advantage of the buffering action of the ion exchanger and running a buffer initially adjusted to one pH, through a column initially adjusted to another pH. If such a pH gradient is used to elute proteins bound to an ion exchanger, the proteins are eluted in order of their isoelectric points. This technique has a number of advantages, for instance, a protein is not subjected to a pH more than its isoelectric pH value, in addition focusing effects result in band sharpening, sample concentration and very high resolution. The pH range is chosen such that the isoelectronic points of the proteins of interest fall in the region of the middle of the pH gradient.

For a descending pH gradient, that is, where the column is at a higher pH than the eluant, a basic buffering group is required on the ion exchanger and thus an anion exchanger is used. As the eluant, which contains a mixture of differently charges species, migrates down the column, the most acidic components bind to the anion exchanger. The pH of the eluant leaving the column is gradually lowered to that of the eluting buffer solution.

4.12.1 Ionic properties of amino acids and proteins

Amino acids have two distinct pK_a values, the actual ionic form of any amino acid depends on the pH of the solution. The ionic forms of glycine which predominate at various pH values are illustrated in Figure 3.18. Because of the presence of the charged groups, an amino acid in solution will have a positive charge at low pH. As the pH is raised, a point will be reached where the amino acid exists exclusively in the dipolar form, known as the zwitterion. This pH is known as the isoelectric point (pI).

In a system using a descending pH gradient, the pH of the initial eluant being less than the isoelectric point ensures that the amino acid carries a positive charge and thus will migrate down the column of anion exchanger in the eluant. However, as it migrates down the column, the pH of the solvent surrounding the protein/amino acid increases. Eventually the pH is greater than the protein's pI, the equilibrium readjusts, the protein reverses its charge and binds to the ion exchanger. The protein remains bound until the passing eluant pH is less than the compound's pI. The process is repeated continuously until the amino acid is washed from the column at its pI.

The theoretical considerations indicating the possibility of producing focusing effects, as in electrophoretic methods, in the ion exchange chromatography of proteins in a pH gradient were soon confirmed experimentally. In chromatofocusing protein molecules at the rear of the zone (i.e. acidic pH) are repelled from the ion exchanger and migrate more rapidly than proteins at the front. If a second sample of the same protein is applied to the column after development of the first has commenced, it will catch up and co-elute with the first. If a second sample of greater pI is applied, it will overtake the first zone and elute before it.

The techniques and practice relating to column packing, sample preparation and sample application are similar to that described for affinity chromatography, as indeed is the apparatus employed. Chromatofocusing is a generally applicable technique for separation problems in biochemistry, for instance, the genetic variants of haemoglobin (responsible for several blood disorders, such as sickle cell anaemia) can be readily resolved from carboxyhaemoglobin by chromatofocusing over a narrow pH interval (pH 8–7).

A more detailed treatment of the principles and practice of chromatofocusing is outwith the scope of this text and the interested reader is directed to the handbook by Pharmacia [59].

4.13 Flash chromatography

Flash chromatography (FC) is a simple, low cost and rapid column chromatographic technique used principally for the separation and purification of mixtures of organic compounds (Chapter 9, Experiment 15). The technique, first described by Still [1] uses a smaller and narrower range particle size of packing than conventional preparative column chromatography and as such has to be carried out with a slightly increased over-pressure in order to achieve adequate flow-rates.

The technique is extremely efficient, gives high recoveries and the separation of mixtures containing components with retardation factors differing by less than 0.1 can be effected in less than 15 min. It is suitable for sample loads ranging from less than 10 mg to 10s of grams depending on the differences in retardation factors.

FC uses glass columns and glass/Teflon fittings with slight modification so that positive pressure, usually nitrogen or air, can be applied at the top

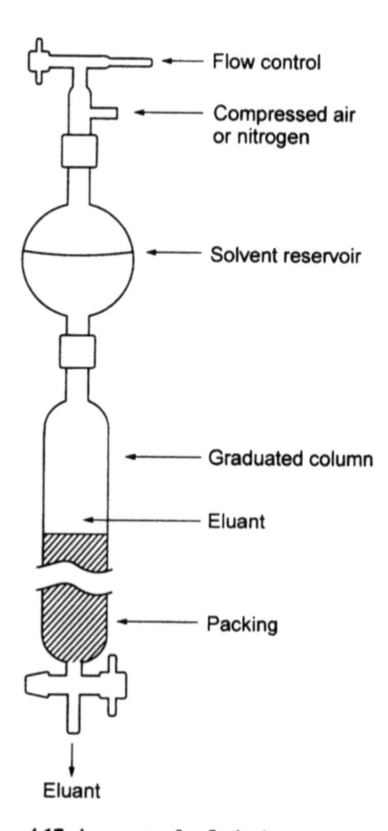

Figure 4.17 Apparatus for flash chromatography.

of the column (Figure 4.17). Column diameters range from 10 to 100 mm in diameter though they are no more than 40 cm in length and are fitted with modified stop-cocks and valves to minimise dead volumes. The columns may be either dry or wet packed the latter giving better results with larger columns. With the packing techniques described the best results are obtained with 40–63 μm particle size; there is no advantage in using a smaller dp. Prior to running a flash column, thin layer chromatography is used to identify the eluant system to be used—this should give an $R_f \geq 0.3$ for the principal analyte. Optimum column performance is achieved using a linear flow-rate of $5\,cm\,min^{-1}$, and thus all analytes of interest should be eluted within two bed volumes.

A wide variety of supports—silica, reverse phase silica, cellulose, alumina and amino bonded and chiral stationary phases—of the required particle size are now available.

References

[1] Still, W.C., Kahn, M. and Mitra, A. *J. Org. Chem.*, **43** (1978) 2923.
[2] Alm, R.S., Williams, R.J. and Tiselius, A. *Acta Chem. Scand.*, **6** (1952) 826.
[3] Kuhn, E., Winterstein, A. and Lederer, E.Z. *Physiol. Chem.*, **197** (1931) 141.
[4] Trappe, W. *Biochem. Z.*, **305** (1940) 150.
[5] Snyder, L.R. *Principles of Adsorption Chromatography*. Dekker, New York, 1968.
[6] Strain, H.H. *Chromatographic Adsorption Analysis*. Interscience, New York, 1942.
[7] Bickoff, E.M. *Anal. Chem.*, **21** (1942) 20.
[8] Knight, H.S. and Groennings, S. *Anal. Chem.*, **26** (1954) 1549.
[9] Hernandez, R., Hernandez Jr, R. and Axelrod, L.R. *Anal. Chem.*, **33** (1961) 370.
[10] Brockmann, H. and Schodder, H. *Chem. Ber.*, **74** (1941) 73.
[11] Frew, A.J., Proctor, G.R. and Silverton, U.V. *J. Chem. Soc.*, (1980) 1251.
[12] Amphlett, C.B. *Inorganic Ion-Exchange Materials*. Elsevier, Amsterdam, 1968.
[13] Marshall, G.R. and Nickless, G. *Chromatogr. Rev.*, **6** (1964) 18.
[14] Smit, J. van R., Robb, W. and Jacobs, J.J. *J. Inorg. Nucl. Chem.*, **12** (1959) 10.
[15] Smit, J. van J. and Robb, W. *J. Inorg. Nucl. Chem.*, **26** (1964) 509.
[16] Bactslé, L. and Paelsmaker, J. *J. Inorg. Nucl. Chem.*, **21** (1961) 124.
[17] Porath, J. and Linden, E.B. *Nature*, **191** (1961) 69.
[18] Schmidt, M. *Biochem. Biophys. Acta*, **63** (1962) 346.
[19] Kawade, Y., Okamoto, T. and Yamamoto, Y. *Biochem. Biophys. Res. Commun.*, **10** (1963) 200.
[20] Gondko, R., Schmidt, M. and Leyko, W. *Biochim. Biophys. Acta*, **86** (1964) 190.
[21] Bjorling, C.O. and William-Johnson, B. *Acta Chem. Scand.*, **317** (1963) 2638.
[22] Mathews, C.K., Brown, F. and Cohen, S.S. *J. Biol. Chem.*, **9** (1964) 2957.
[23] Peterson, E.A. and Sober, H.A. *J. Am. Chem. Soc.*, **78** (1956) 71.
[24] Mould, D.L. and Synge, R.L.M. *Biochem. J.*, **58** (1954) 571.
[25] Porath, J. and Flodin, P. *Nature*, **183** (1959) 1657.
[26] Moore, J.C. *J. Polym. Sci.*, **A2** (1964) 835.
[27] Determann, H. *Angew. Chem.* (Int. Edn), **3** (1964) 608.
[28] Porath, J. *J. Pure Appl. Chem.*, **6** (1963) 233.
[29] Laurent, T.C. and Killander, J. *J. Chromatogr.*, **14** (1964) 317.
[30] Porath, J. *Pure Appl. Chem.*, **6** (1963) 233.
[31] Tiselius, A., Porath, J. and Albertsson, P.A. *Science*, **141** (1963) 3.
[32] Gelotte, H. *New Biochemical Separations* (eds James, A.T. and Morris, L.J., eds), p. 93. Van Nostrand, New York, 1964.

[33] Granath, K. *New Biochemical Separations* (James, A.T. and Morris, L.J., eds), p. 110. Van Nostrand, New York, 1964.
[34] Marsden, N.V.B. *J. Chromatogr.*, **58** (1971) 304.
[35] Laurent, T.C. and Killander, J. *J. Chromatogr.*, **5** (1961) 103.
[36] Lea, D.J. and Schon, E.H. *Can. J. Chem.*, **40** (1962) 159.
[37] Hjerten, S. *Archiv. Hiochem. Biophys.*, (Suppl. 1) (1962) 147.
[38] Porath, J., Janson, J.-C. and Lååls, T. *J. Chromatogr.*, **60** (1971) 167.
[39] Lååls, T. *J. Chromatogr.*, **66** (1972) 347.
[40] Hiatt, C.W., Shelokov, A., Rosenthal, E.J. and Galimore, J.M. *J. Chromatogr.*, **56** (1971) 362.
[41] Hawk, G.L., Cameron, J.A. and Dufault, L.B. *Prep. Biochem.*, **2** (1972) 193.
[42] Flodin, P. *J. Chromatogr.*, **5** (1961) 103.
[43] Dougherty, T.J., Boyle, D.G., Weishaupt, K.R., Henderson, B.W., Potter, W.R., Beellnier, D. and Wityk, K.E. *Porphyrin Photosensitisation* (Kessel, D. and Dougherty, T.J., eds), Chapter 1, pp. 3–15. Plenum Press, New York, 1983.
[44] Forbes, E., Keir, W.F., MacLennan, A.H., Moore, J.V. and Truscott, T.G. *Cancer Lett.*, **67** (1992) 175.
[45] Kusnir, J. and Meloun, B. *Coll. Czech. Chem. Commun.*, **38** (1973) 143.
[46] Petrovic, S.L., Petrovic, I.S. and Markovic, R.A. *et al.*, *Prep. Biochem.*, (1974) 509.
[47] Gelotte, B. *Acta Chem. Scand.*, **18** (1964) 1283.
[48] Barth, H.G. and Smith, D.H. *J. Chromatogr.*, **206** (1981) 410.
[49] Zeichner, M. and Stern, R. *Biochemistry*, **16** (1977) 1378.
[50] Ansati, A.A. and Mage, R.G. *J. Chromatogr.*, **140** (1977) 98.
[51] Heitz, W. and Ullner, H. *Macromol. Chem.*, **120** (1968) 58.
[52] Colman, R. *Anal. Biochem.*, **46** (1972) 358.
[53] Wood, G.C. and Cooper, P.F. *Chromatogr. Rev.*, **12** (1970) 88.
[54] Friedberg, F. *Chromatogr. Rev.*, **14** (1971) 121.
[55] Turkova, J. *J. Chromatogr.*, **91** (1974) 267.
[56] Porath, J., Axen, R. and Ernback, S. *Nature*, **215** (1967) 1491.
[57] Turkova, J. *Chromatographic and Allied Methods* (Mikes, O., ed.), John Wiley, New York, 1979.
[58] Sluyterman, L.A. and Elgersma, O. *J. Chromatogr.*, **150** (1978) 17.
[59] *Affinity Chromatography—Principles and Methods*. Pharmacia Fii Chemicals, Uppsala, 1984.

5 Gas chromatography

5.1 Introduction

It was nearly 50 years from Mikhail Tswett's description of chromatographic separations (1906) to the development of gas–liquid chromatography by Martin and James (1952). Since that time gas chromatography (GC) has developed rapidly, particularly during the 1960s, producing sweeping changes in analytical chemistry and in many areas of research and development. Subsequently, capillary or open tubular columns were produced which increased considerably the separating capabilities of GC so that complex samples such as flavours and perfumes and environmental residues could be successfully resolved into in some cases over a hundred components. Hybrid techniques, particularly using a mass spectrometer as a detector, GC–MS, have added a further dimension to GC analyses enabling separated compounds to be readily identified. It is now possible to separate, quantify, and subsequently characterise, components in a mixture, from permanent gases, and hydrogen isotopes to essential oils in perfumes, fatty acids in vegetable oils and waxes and pollutants in water and soils. Although direct analysis is limited to compounds with a molecular weight up to 400–500, derivatives of non-volatile materials and metal complexes can readily be made which have some measure of volatility without thermal degradation. Preparative-scale gas chromatography offers a method of obtaining very pure compounds in sufficient quantity for further analytical and preparative work. It is also possible to obtain physicochemical measurements using GC, with regard to surface properties, kinetics and thermodynamics of separation and adsorption processes which have application in, for example, the development of catalysts.

5.2 Principles of gas chromatography

The general principles of chromatography have been discussed in Chapter 2. However, a more detailed account in the context of GC is presented herein; the reader is also referred to the Glossary for additional detail and explanations of the terminology used. The treatment of theoretical aspects is at a level which should enable the analyst to obtain maximum efficiency and performance from the system and application.

There are broadly three modes in which GC is carried out:

1. gas–liquid chromatography (GLC) using a packed column with the liquid stationary phase coated onto inert support particles;

2. capillary column GC where open tubular columns are used with the liquid or solid stationary phase coated onto the inner walls of the column tubing referred to as wall coated, porous layer and surface coated open tubular columns (WCOT, PLOT and SCOT); and
3. gas–solid chromatography (GSC) using a packed column with the solid surface of the particles forming the stationary phase, for example, alumina or a cross-linked polymer.

In each case we have a gaseous mobile phase transporting components as a vapour over a stationary phase, separation being effected by interaction of the individual components with the stationary phase resulting in retardation according to their distribution ratios (K):

$$K = \frac{C_{SP}}{C_{MP}}$$

where C_{SP} is the concentration of the component in a unit volume of stationary phase and C_{MP} is the concentration of the component in a unit volume of mobile phase.

In GC the distribution ratio is dependent on the component vapour pressure, the thermodynamic properties of the bulk component band and affinity for the stationary phase. The equilibrium is temperature dependent and, therefore, the stationary phase column must be precisely and accurately maintained.

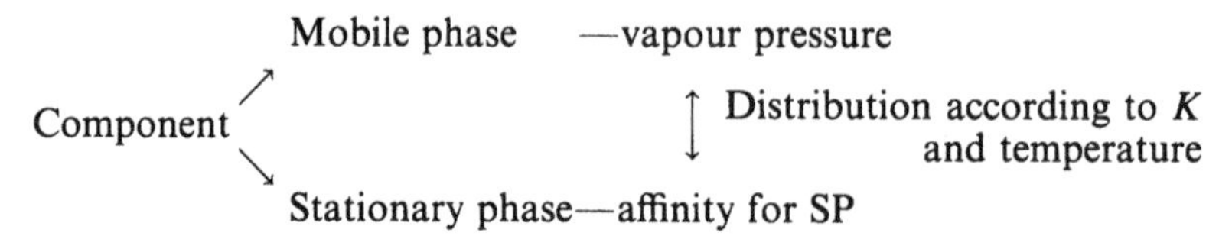

The relationship between the vapour pressure, p^0, and temperature is described by the Clausius Clapeyron equation:

$$\ln(p^0) = -\frac{\Delta H_V}{RT} + C$$

where ΔH_v is the molar heat of vaporisation of the component, R the gas constant, T the column temperature and C a constant. Retardation of a component and therefore its specific retention volume, V_g, will vary inversely with temperature. Also, when the activity coefficient of the component is unity $\Delta H_V + \Delta H_S = 0$, where ΔH_S is the molar heat of solution, the above equation can be modified by substituting for $\ln(p^0)$ and ΔH_v;

$$V_g \propto \frac{1}{p^0} \quad \text{and} \quad \ln(V_g) \propto -\ln(p^0)$$

$$\text{also} \quad \Delta H_S = -\Delta H_V$$

$$\text{therefore} \quad \ln(V_g) = -\frac{\Delta H_S}{RT} + C'$$

Specific retention volume is the net retention volume (V_N) of component per unit weight of stationary phase, W_S:

$$V_g = \frac{V_N}{W_S}$$

It follows that the net retention volume and therefore the adjusted retention volume (V'_R) and retention volume (V_R) vary as the logarithm of $1/T$. Similarly, adjusted retention time (t'_R) and retention time (t_R) will decrease logarithmically as temperature increases.

$$\text{therefore,} \quad \ln(V_R) \propto -\frac{\Delta H_S}{RT} \quad \text{and} \quad \ln(t_R) \propto -\frac{\Delta H_S}{RT}$$

The relative retention factor, α, for an adjacent component pair A and B will also depend on the column temperature.

$$\ln(\alpha) = -\frac{(\Delta H_{SB} - \Delta H_{SA})}{RT} + C$$

The component with the longer retention time, B, will have the higher heat of solution in the stationary phase. Relative retention will therefore decrease as the stationary phase column temperature increases and the components will elute closer together. Hence the importance of selecting the best column temperature and the value of temperature programming in optimising a separation. Figure 5.1 shows the isothermal separation of a homologous series of alcohols at 100°C and the improvements in the separation obtained by temperature programming in reducing the retention time and band broadening of the later eluting components. Temperature programming is used routinely for the analysis of complex mixtures particularly when using capillary columns.

5.3 Resolution

An ideal chromatogram is obtained when all the analytes in a sample mixture are separated with base line resolution in the minimum analysis time. In practice the chromatographic parameters are selected to obtain the best practical separation. The most appropriate stationary phase and column system are chosen after considering the polar characteristics of the analytes, their volatility range and selecting the optimum mobile phase velocity and column temperature programme. Resolution is a function of retention characteristics of the components, column efficiency (band broadening) and the selectivity or separating capabilities of a column. These are reflected by the terms retention factor (capacity factor), k, column efficiency, N or N_{eff},

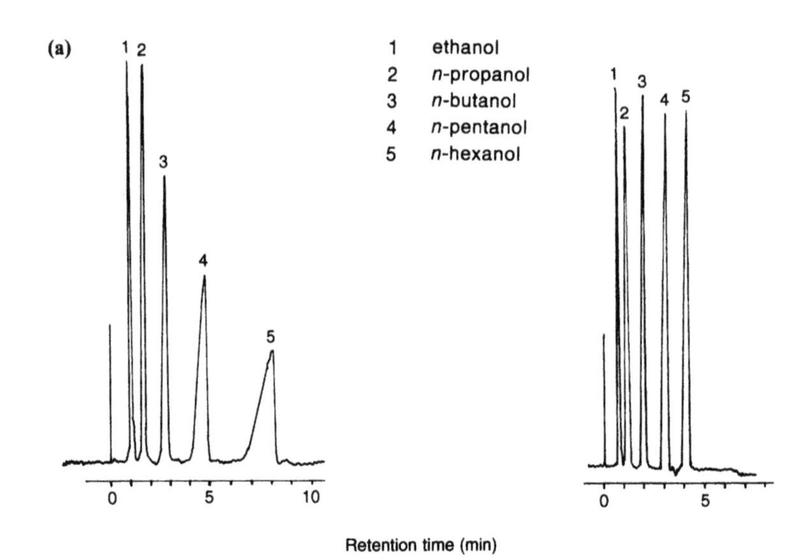

Figure 5.1 (a) Separation of alcohols. Isothermal and temperature programmed separation at 100°C and 100–150°C at 15°C min^{-1} respectively. Column: 2m × $\frac{1}{8}$in. Carbowax 20; carrier gas: nitrogen, 30 ml min^{-1}; injection: 1 μl, neat 1:1:1:1:1 mixture; detector: FID.

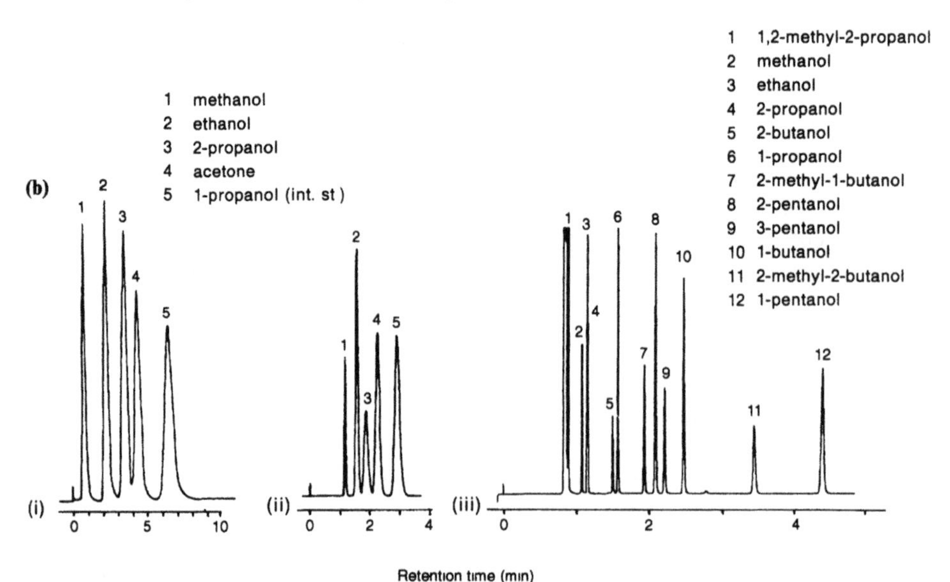

Figure 5.1 (b) Separations on packed gas solid column, wide bore and narrow bore capillary WCOT columns. (i) Column: 2m × $\frac{1}{4}$in. Porapak Q; Carrier gas: He 30 ml min^{-1}; injection: 1 μl, 200mg% (aq.); detector: FID; column temperature: 150°C. (ii) Column: 10 m × 0.53 mm 2 μm film HP M, 50% phenylmethylsiloxane; carrier gas: He 3.5 ml min^{-1}; injection: 1 μl, 200 mg% (aq.) 20:1 split; detector: FID; column temperature: 50°C. (iii) Column: 25 m × 0.25 mm, 0.25 μm film Carbowax-20 M; carrier gas He 1 ml min^{-1}; injection, 1 μl, split ratio 200:1; detector: FID; column temperature: 60°C.

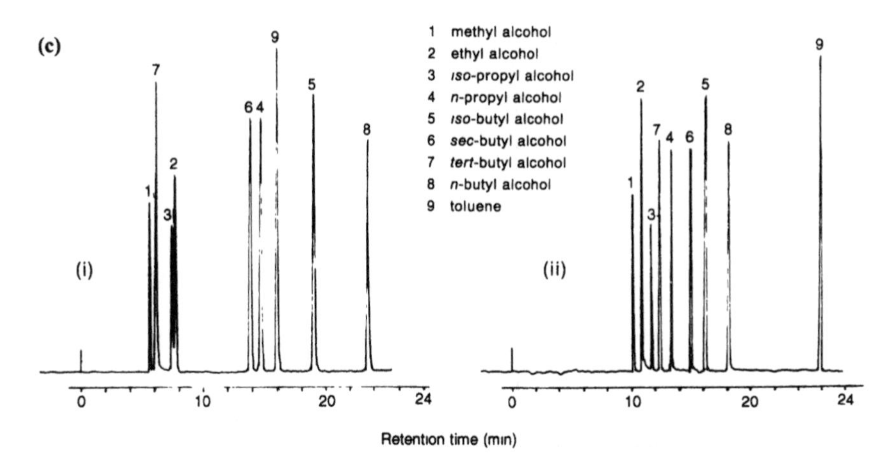

Figure 5.1 (c) On (i) polar and (ii) non-polar capillary WCOT columns illustrating 'like attracts like' polar/non-polar retention properties and two column analysis to confirm components present. Toluene is used as the reference marker. 1, Methyl alcohol; 2, ethyl alcohol; 3, *iso*-propyl alcohol; 4, *n*-propyl alcohol; 5, *iso*-butyl alcohol; 6, *sec*-butyl alcohol; 7, *tert*-butyl alcohol; 8, *n*-butyl alcohol; 9, toluene (reference marker). Polar column: Carbowax 20M (25 m × 0.32, 0.25 μm film). Non-polar column: HP-1, dimethylsiloxane (25 m × 0.32, 0.25 μm film). Both columns: carrier gas He 1 ml min^{-1}; injection: 0.5 μl split 20:1; detector: FID; column temperature: programmed 50–200°C at 5°C min^{-1}.

and separation factor, α (Figure 5.2):

$$\text{where} \quad k = \frac{t'_R}{t_M} = \frac{t_R - t_M}{t_M}$$

$$\alpha = \frac{k_B}{k_A}$$

$$N = \frac{L}{H} = 16\left(\frac{t_R}{w_b}\right)^2 = 5.54\left(\frac{t_R}{w_h}\right)^2$$

$$\text{and when } t_M \text{ is significant} \quad N_{eff} = 16\left(\frac{t'_R}{w_b}\right)^2 = 5.54\left(\frac{t'_R}{w_h}\right)^2$$

where t_M is dead time; t_R and t'_R are retention time and corrected retention time, respectively; w_b and w_h are base width of a peak and width at half height; N and N_{eff} are column efficiency and effective column efficiency; k_A and k_B are retention factors for components A and B; H is HETP, height equivalent to a theoretical plate or equilibrium step height; and L is column length.

For a given column length, optimum column efficiency is obtained when the equilibrium step or plate height is at a minimum, that is, the column band broadening processes described by the van Deemter equation (see section 2.5.1) are minimised by selecting the optimum velocity, μ_{opt}, of the mobile phase. Figure 5.3 shows the van Deemter plot of H against μ for

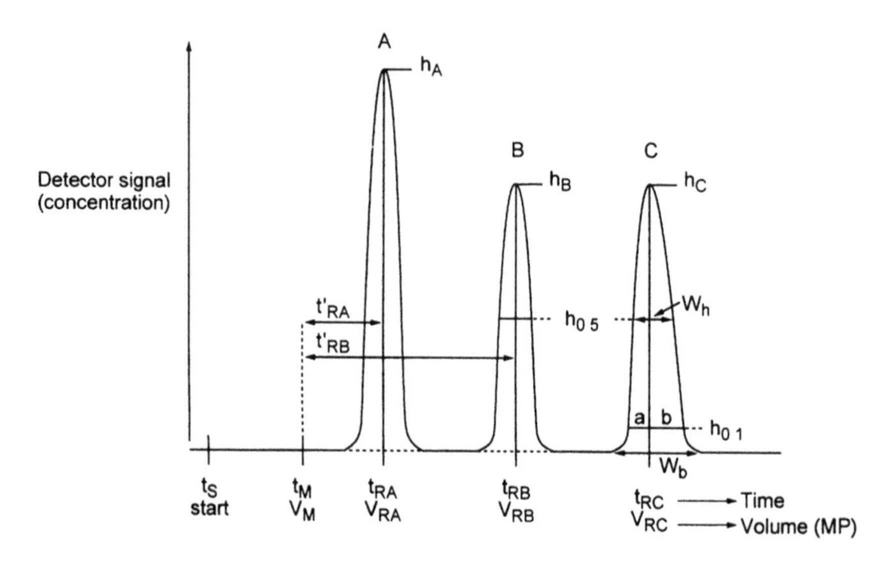

Figure 5.2 Data and symbols on chromatograms of A, B and C: t_M, dead time, time for the solvent or mobile phase to pass through the system; V_M, dead volume, volume of mobile phase in the system; t_{RA}, retention time for A, the time for A to pass through the system; t'_{RA}, corrected retention time of A; V_A, retention volume of A; V'_A, corrected retention volume of A; t_{RB}, retention time of B; t'_{RB}, corrected retention time of B; V_B, retention volume of B; t_{RC}, retention time of C; t'_{RC}, retention time of C; V_C, retention volume of C; A, peak area, A_A for component A, A_B for B, A_C for C; h, peak height, h_A for component A, h_B for B, h_C for C; $h_{0\,5}$, peak half height; $h_{0\,1}$, 10% peak height; w_b, width at the base of a peak; w_h, width of a peak at half height; F_C, flow-rate of mobile phase; a, b, the forward part and rear part of a peak at 10% h.

packed and WCOT columns and when using nitrogen, helium and hydrogen carrier gases with WCOT columns [1]. Nitrogen is frequently used in packed columns where the flow-rate of carrier gas is high (30–50 ml min^{-1}) and small changes in velocity do not significantly affect H and column efficiency. However, in WCOT capillary columns the optimum efficiency for nitrogen is at a low gas velocity and efficiency rapidly decreases as the velocity increases. If hydrogen or helium are used the plot is flatter so that higher gas velocities than the theoretical minimum may be used to optimise a separation and improve analysis time without significant loss of efficiency.

Adjacent peaks are considered resolved if there is baseline separation, that is, if the sum of their half widths is less than the separation between the peak maxima. The ratio of peak separation to half widths is a measure of the resolution (Figure 5.2):

$$R_S = \frac{(t_{RB} - t_{RA})}{\frac{1}{2}(w_{bB} + w_{bA})} = \frac{2\Delta t}{(w_{bB} + w_{bA})} \approx \frac{\Delta t}{w_{bB}}$$

$$\text{also} \quad R_S = \frac{(t_{RB} - t_{RA})}{(w_{hB} + w_{hA})} \approx \frac{\Delta t}{2w_{hB}}$$

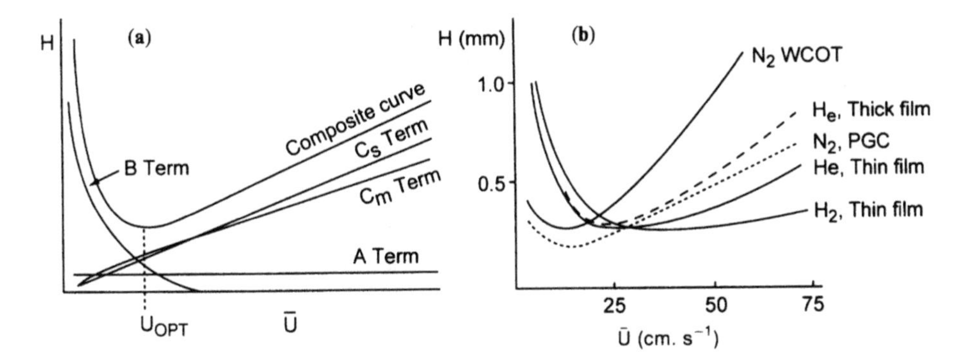

Figure 5.3 Van Deemter plots of plate height, H, against average linear velocity, $\bar{U}$, of the mobile phase. (a) The contribution of each term to the composite curve. (b) Plots for WCOT GC column using N_2, He, H_2 carrier gas and N_2 for packed column GC.

Thus, for satisfactory resolution the peak maxima separation, Δt, should be at least the width of the second peak. If $R_S = 1.0$ (peak width 4σ) there is a 2.3% overlap and for $R_S = 1.5$ a 0.2% overlap. For an acceptable separation without too long an analysis time R_S should be between 1.2 (corresponding to a peak overlap of 1%) and 1.8 (see section 2.6). Note that baseline resolution is achieved at $R_S \approx 1.5$. The resolution can be directly obtained from a chromatogram. A further factor which affects resolution is when a sample is analysed and one component of an adjacent pair is found to be in much greater concentration than expected. In this case greater resolution may be required for adequate separation for peak area measurements and quantitation.

Resolution can be expressed in terms of column efficiency and step height, retention characteristics and separation properties and is useful for column selection and for determining the optimum carrier gas velocity.

$$R_S = \frac{\sqrt{N}}{4}\left(\frac{k}{k+1}\right)\left(\frac{\alpha-1}{\alpha}\right)$$

and $$R_S = \frac{1}{4}\sqrt{\frac{L}{H}}\left(\frac{k}{k+1}\right)\left(\frac{\alpha-1}{\alpha}\right)$$

Resolution is a function of the square root of N, large changes are therefore necessary for a marked effect, although in WCOT columns where lengths of 10–100 m are used column length is an important factor to consider. Efficiency is more easily improved by

- selecting a critical pair of component peaks from the chromatogram;
- selecting the stationary phase and film thickness to give satisfactory retention characteristics, k, and separation factors, α; changes in k are only significant for early eluting peaks with $k < 5$ (see section 5.6.3 and Figure 5.7 below);

- determining the optimum temperature programme and optimising α; and
- setting the carrier gas velocity for the best time effective separation, optimising H.

5.3.1 10% valley resolution

An alternative rapid method of assessing resolution, the 10% valley method, uses the valley between adjacent peaks, h_V. For a satisfactory separation with <2.3% peak overlap, the ratio of h_V to mean peak height $\bar{h}$ of the two peaks should be $\leq$10% (Figure 5.2):

$$h_V \leq 0.1\left(\frac{h_A + h_B}{2}\right) \leq 0.1\bar{h}$$

$$R_S \approx 0.1\left(\frac{\bar{h}}{h_V}\right)$$

5.3.2 Separation number

Separation number (SN) is a practical, alternative means of describing the separating capabilities of a capillary column. SN shows the ability of a given column and stationary phase to separate a mixture into its components with a resolution of 1.0 between the peaks. SN is calculated from retention times and peak widths of two consecutive homologous n-alkanes having Z and $Z + 1$ carbon atoms and indicates the number of component peaks that can be resolved ($R_S \approx 1.0$) between the two alkanes in a given part of the chromatogram.

$$SN = \left(\frac{t_{R(Z+1)} - t_{RZ}}{w_{h(Z+1)} + w_{hZ}}\right)$$

5.3.3 Asymmetry

Resolution is compromised if the normally bell-shaped Gaussian peaks are asymmetric, that is the peak is fronting or tailing. Asymmetry (A_s) is the result of a number of factors which affect normal partition of the analytes between the vapour phase and stationary phase, for example, fronting caused by poor injection technique or overloading the stationary phase and tailing from dead volume effects or retention via adsorption or hydrogen bonding. Asymmetry A_s is measured at 10% peak height as the ratio of the tailing and leading parts of the peak (Figure 5.2):

$$A_s = \frac{b}{a} \quad \text{at } 10\%h$$

A good, uniformly packed and WCOT column should give an asymmetry ratio of between 0.9 and 1.2.

5.4 Columns and stationary phase

The key to good GC separations is to use the most appropriate stationary phase and column at the optimum mobile phase velocity and column temperature. The main factors to consider when specifying the system for analysis of a given sample mixture are

- to note the boiling point range and vapour phase characteristics of the components and number of components;
- to identify the polar and non-polar characteristics of the components in the mixture and their functional groups;
- selection of a stationary phase/column system and stationary phase film thickness to give the required selectivity and separation factors;
- optimising the carrier gas velocity for the fastest analysis time and minimum dead time set at the elution temperature of the critical pair of peaks, helium should be used as a carrier gas if at all possible;
- optimising the temperature programme or selecting the best temperature for isothermal analysis; and
- testing the system using a standard mixture either representative of the sample(s) to be analysed or a mixture which includes compounds of differing polar character and boiling points.

5.4.1 Gas chromatography and column oven

The column is perhaps the most important feature of a GC system. It contains the stationary phase and thus effects the separation of components in a mixture. The column may be made of glass or metal and typically 2–6 mm i.d. and 1–3 m in length when packed with stationary phase material or 0.2–0.7 mm i.d. and 10–100 m long if in the form of a capillary column. Columns are formed into a coil of between 4 and 8 in (10–20 cm) in diameter and specially designed end fittings are used to connect the columns to the injector and detector with minimum dead volumes. A packed column contains solid particles of uniform size coated with stationary phase in GLC or uses uncoated particles as the stationary phase in GSC (Figure 5.4).

Aluminium or copper tubing can be used for the column, however, an active oxide film may form on the inner surface, which can initiate catalytic reactions in some sensitive components and are therefore generally avoided. Stainless-steel or glass tubing is therefore used for most columns. Capillary

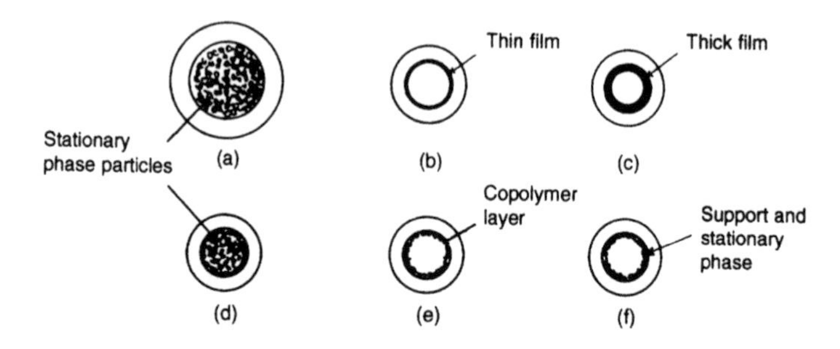

Figure 5.4 Cross-section of GC columns. (a) $\frac{1}{8}$in. packed column; (b) thin film WCOT column; (c) thick film WCOT column; (d) $\frac{1}{16}$in. micropacked column; (e) PLOT column; (f) SCOT column.

or open tubular columns consist of long narrow tubing coated on the inner surface with a 0.2–5.0 μm film of stationary phase and are known as WCOT columns. Columns are usually stainless-steel or glass (Pyrex) for packed columns and silica for capillary columns.

5.4.2 *Column oven and column temperature*

The separation process occurring in the column involves an equilibrium established by the component between the stationary and the mobile phases. The distribution coefficient and hence retardation is dependant on vapour pressure and retention properties of the bulk solute and is a function of temperature as discussed earlier.

Packed columns with relatively large amounts of stationary phase on the support material require oven temperatures 20–50°C above their boiling point to obtain satisfactory peak profiles and elution times commensurate with the low stationary phase loadings. However, decreasing the amount of stationary phase or reducing column temperature results in the peaks first eluted having poorer resolution. A balance of stationary phase loading and temperature is therefore required. Loading of 7–15% (w/w) is used for small molecules up to C_7–C_8 and 3–10% (w/w) for C_9–C_{22}. WCOT capillary columns have a stationary phase film thickness of 0.2–5 μm, satisfactory separations being achieved with shorter analysis times than for packed columns at temperatures often 20–50°C lower.

Once a column has been chosen, the two variables which can be modified to optimise the separation of components are temperature and carrier gas flow-rate. The latter is often preset although the newer microcomputer controlled instruments allow the flow rate and temperature to be changed between runs for automatic optimisation. Accurate control of column temperature is important in order to obtain reproducible chromatograms, retention times and peak area to height ratios. Oven temperatures can be set to operate from about 10°C above ambient, up to 450°C with reproducibility of better than 0.1°C. Lower temperatures are possible if cryogenic cooling is employed using, for example, a liquid nitrogen cooled coil in the oven. The temperature of the heating element is controlled by an electronic proportional controller and the oven carefully designed so that the temperature throughout is uniform. Accurate control is important in isothermal analyses where retention times are being measured, and also in automatic and capillary GC systems for the separation of closely related components within a narrow boiling range.

Column temperature can be progressively increased while the flow rate is kept constant to avoid problems of the least volatile components in a mixture taking too long to elute and hence forming broad tailing peaks. The rate of temperature increase is pre-set and typically involves an initial-hold stage when the starting temperature is held, perhaps long enough for the solvent to elute, a temperature ramp stage where the temperature is increased at a selected rate, usually between 1 and 40°C min^{-1}, and a final stage where the upper temperature is held for a pre-selected time. At the end of the cycle the oven is cooled before a new analysis is carried out. Again, microcomputer controlled instrument functions enable temperature increase profiles other than a linear increase to be programmed and also many more stages can be included in the overall cycle. An example of the effect of temperature programming is shown in Figure 5.1.

5.4.3 Packed columns

The packed column consists of 1/4 or 1/8 in (6.35 or 3.175 mm) o.d., 2–4 m long glass or metal tubing packed with either adsorbent particles or stationary phase-coated particles, typically 60–80 (0.25–0.2 mm) or 80–100 (0.2–0.15 mm) mesh size range (Figure 5.4). Both empty metal and glass columns can be obtained already formed to the configuration required to fit various instruments, and packed as described below. Straight metal tubing can be used and carefully coiled to shape before packing. Alternatively, pre-packed columns can be purchased. A good packed column has 1000–2000 steps or plates m^{-1}, in contrast to capillary columns, where 2000–5000 plates are obtainable. Therefore, a typical packed column will contain 2000–10 000 plates in which to achieve the separation, while a capillary column may have over 100 000 plates. The purpose of a support material is to provide a uniform, inert support for the stationary phase with sufficient mechanical strength to avoid crushing of the particles. Although many materials have been studied, only diatomaceous materials are commonly used. These can be divided into two groups.

1. Pink diatomaceous materials derived from crushed firebrick, for example, Chromosorb P. The material is used for high performance stationary phases due to its high surface area ($4.0\,m^2\,g^{-1}$) and support for high stationary phase loadings, up to 35% (w/w). It is particularly suitable for alkanes but must be deactivated by silanisation for polar compounds.
2. White diatomaceous materials prepared from calcined diatomite, for example, Chromosorb W, a fragile packing of lower surface area ($1.0\,m^2\,g^{-1}$) than Chromosorb P, suitable for polar compounds. Chromosorb G is the hardest Chromosorb and is twice as dense as Chromosorb W. Maximum stationary phase loading is approximately 5% (w/w), equivalent to 12% loading on Chromosorb W because of its higher density.

Chromosorb T is a polytetrafluoroethylene (PTFE) support material. It is inert and hydrophobic and therefore suitable for analysis of small polar molecules such as water, organic acids, phenols, amines and acidic gases (HF, HCl, SO_2, NO_x). Chromosorb T can be coated with stationary phases such as polyethylene glycol, Apiezon or a fluorocarbon oil (Kel-F, Fluoropak-80) but can be difficult to pack.

The surface of the diatomaceous support materials can contain mineral impurities which may promote thermal decomposition of components as they move through the column. Surface silanol (Si–OH) groups form polar sites on the surface leading to tailing of polar compounds and formation of hydrogen bonds with suitable components interfering with the equilibrium processes and results in tailing. The materials are therefore thoroughly acid washed (AW) by treating with hydrochloric acid to remove the minerals and silanised using dimethyldichlorosilane (DMDCS)/methanol or hexamethyldisilylazane (HMDS) to block the Si–OH groups with methylated siloxane bonds, Si–O–Si–Me.

$$\text{Si-OH} + \text{Cl}_2\text{SiMe}_2 \rightarrow \text{–Si-O-Si(Cl)Me}_2 + \text{MeOH} \rightarrow \text{–Si-O-SiMe}_3 + \text{HCl}$$

$$\begin{array}{c}\text{Si-OH}\\ | \\ \text{O} \\ | \\ \text{Si-OH}\end{array} + \text{Cl}_2\text{SiMe}_2 \rightarrow \begin{array}{c}\text{–Si-O}\\ | \\ \text{O} \\ | \\ \text{–Si-O}\end{array}\!\!\!\!\!>\text{SiMe}_2$$

$$\text{Si-OH} + \text{Me}_3\text{Si-N=N-SiMe}_3 \rightarrow \text{–Si-O-SiMe}_3 + \text{N}_2$$

HMDS is useful for on-column silanisation since the by-product of the reaction is nitrogen, which will not harm the metallic surfaces of the column, end fittings or detectors. Several useful reviews of solid supports and their properties have been published [2, 3]. In some applications it is necessary to deactivate the column before use by repeated injection of a substance with similar properties to the sample. For example, analysis of barbiturates on an Apiezon column requires priming with barbitone (see experiment 18, Chapter 9). Further examples are priming by phosphoric acid for organic acids and phenols and potassium hydroxide for amines. This process is sometimes referred to as 'tail reducing'.

5.4.4 Stationary phases for packed columns

Separation of components occurs by partition between the mobile phase and a suitable stationary phase. The stationary phase therefore needs to be thermally stable, unreactive and have negligible volatility over the operating temperature range to give a reasonable lifetime. The maximum column temperature at which each stationary phase may be used is therefore specified to minimise stationary phase bleed, that is slow loss of stationary phase, which contributes to detector noise and baseline drift. The maximum operating temperature is often restricted to about 20°C less than the upper temperature limit at which degradation or slow vaporisation become significant. Although a large number of materials are available, few are necessary to achieve the separation of a wide range of gases and volatile chemical compounds. Many analytical laboratories achieve 90% of their analyses on as few as four or five different stationary phases, e.g. Apiezon; OV101/SE30; OV3/Dexil 300; OV17; Carbowax 20M (Table 5.1).

Table 5.2 is a list of common stationary phases for packed columns, squalane an alkane being the reference non-polar material. Polar character (PI) increases down the table. OV101 and similar dimethylsilicone oils are commonly used non-polar stationary phases and Carbowax 20M, polyethylene glycol, the most used polar stationary phase. Silicone oils are particularly useful due to their thermal stability and high maximum temperature whilst higher column temperatures, up to 500°C can be used with carborane methyl silicones, the Dexsil range of stationary phases. McReynold's numbers are a method of describing retention characteristics of a stationary phase for a series of compounds of varying polar character. A detailed discussion is included in section 5.5.

Columns can be purchased already filled with packing material coated with stationary phase at a suitable loading of between 3 and 15% (w/w) or coated support material can be purchased for packing columns. Alternatively coated packings may be prepared by weighing out sufficient support material to fill the column and the appropriate weight of stationary phase to give the desired weight/weight ratio. The stationary phase is completely dissolved in a suitable volume of a dry solvent such as acetone, ethyl acetate, methanol or toluene and added to the support material. The solvent is removed on a

Table 5.1 Basic stationary phases

Stationary phase	Material	Structure
Apiezon L	Branched-chain alkane grease, mp 43°C	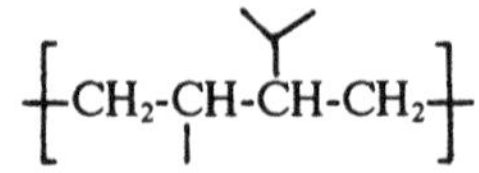 $-[CH_2-CH(CH_3)-C(CH_3)_2-CH_2]-$
OV101	Dimethyl silicone	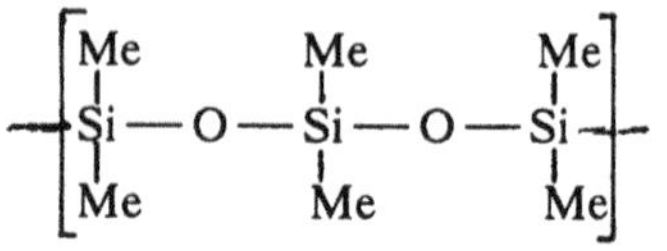 $-[Si(Me)_2-O-Si(Me)_2-O-Si(Me)_2]-$
OV3	5% Phenyldimethylsilicone	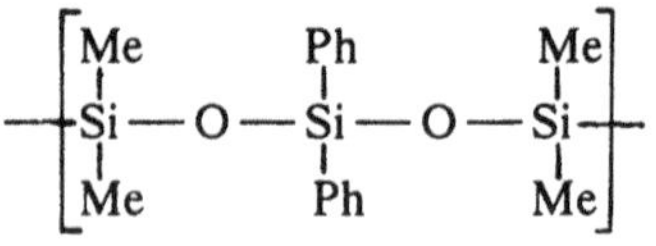 $-[Si(Me)_2-O-Si(Ph)_2-O-Si(Me)_2]-$
OV17	50% Phenyldimethylsilicone	
Dexil 300	Carborane dimethyl silicone	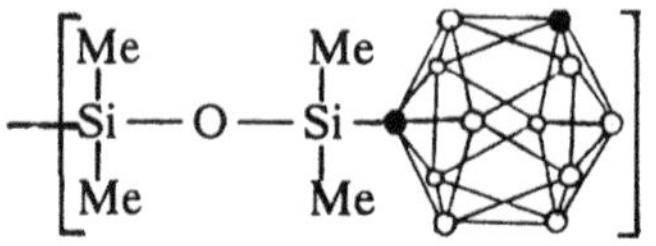 $-[Si(Me)_2-O-Si(Me)_2-(\text{carborane})]-$ • = C, carbon, ○ = borane, B–H
Carbowax 20M	Polyethylene glycol	$HO-[-CH_2-CH_2-O-]_n-H$

rotary evaporator to ensure an even distribution of stationary phase on the surface of the support material. Columns are packed by first plugging one end with silanised glass wool and then introducing the packing into the other end via a funnel whilst connecting a vacuum pump to the plugged end. The column should be coiled into its final configuration before packing and the coils arranged horizontally during the packing. The column is gently vibrated to ensure uniform packing without voids.

5.4.5 *Specialised stationary phases*

A number of stationary phases/column systems have been developed or optimised for specific applications, particularly for analysis of a complex mixture such as samples from environmental monitoring of atmospheric or water pollution, for example:

- 1% SP1000 (Supelco Inc.), a polyethylene glycol substituted terphthalic acid, on 100–120# inert diatomite (AW, DCDMS) support for US EPA method 612 for chlorinated hydrocarbons;

Table 5.2 GC stationary phases, retention indices and McReynold's numbers[a]

Stationary phase	Minimum/maximum temperature (°C)	X	Y	Z	U	S	P_{SUM}	PI (P/5)
Squalane hexamethyltetracosane	20/120	0	0	0	0	0	0	0
Apiezon L alkane grease, mp 43°C	50/280	32	22	15	32	42	135	27
OV101, SE30 100% methyl silicone	50/300	15	53	44	64	41	215	43
SE54 1% vinyl 5% phenyl methyl silicone	50/300	33	72	66	99	67	335	67
OV3 10% phenyl methyl silicone	0/350	44	86	81	124	88	425	85
Dexsil 300 carborane methyl silicone	50/550	47	80	103	148	96	475	95
Dexsil 400 carborane phenyl methyl silicone	40/400	72	108	118	166	123	590	118
OV7 20% phenyl methyl silicone	0/350	69	113	111	171	128	590	118
OV1701 14% cyanopropyl methyl silicone	0/250	67	170	153	228	171	795	159
Dinonylphthalate, DNP	0/150	83	183	147	231	159	805	161
OV17 50% phenyl methyl silicone	0/380	119	158	162	243	202	885	177
Dexsil 410 carborane cyanoethyl methyl silicone	50/400	72	286	174	249	171	950	190
OS124 polyphenylether	0/210	176	227	224	306	283	1215	243
Tricresyl phosphate	20/130	176	321	250	374	299	1420	284
OV210 50% trifluoropropyl methyl silicone	0/280	146	238	358	468	310	1520	304
OV225 cyanopropylmethyl phenylmethyl silicone	0/280	228	369	338	492	386	1815	363
Carbowax 20M polyethylene glycol	40/220	322	536	368	572	510	2310	462
FFAP (free fatty acid phase) terephthalic acid modified Carbowax	50/250	340	580	397	602	627	2545	509
Carbowax 1000 polyethylene glycol	30/150	347	607	418	626	589	2585	517
Diethylene glycol succinate, DEGS	0/200	496	746	590	837	835	3505	701

[a] PI = polarity index, calculated from $(X + Y + Z + U + S)/5$, $P_{SUM} = (X + Y + Z + U + S)$, X = retention index for benzene, Y = butanol, Z = pentanone, U = nitropropane, S = pyridine. Values measured at 120°C and a 20% (w/w) loading.

- 10% Carbowax 20M/2% KOH on 80–100# Chromosorb W (AW) for US EPA method 607 for nitrosamines.

Examples of specific stationary phases include silver nitrate dissolved in polyethylene glycol which forms loose adducts with alkenes and is highly specific for such compounds [4]. *N*-Dodecylsalicylaldimines of nickel, palladium and platinum are able to selectively retain those molecules which can act as ligands to transition metals such as amines, ketones, alcohols, and C=C and C≡C containing compounds [5]. Tri-*o*-thymotide dissolved in tritolyl phosphate will selectively retain straight chain organic compounds relative to those with branched chains and dicarboxylrhodium(II) trifluoroacetyl(+) camphorate has been used in squalane for purity determinations and separation of olefins [6]. Dimeric rhodium benzoate in squalane interacts reversibly with lone pair electrons, for example, in ethers, ketones and esters, leading to increased attraction and selectivity. Other materials investigated include vanadium(II) and manganese chlorides and chiral stationary phases for separating enantiomers.

5.4.6 Micropacked columns

Chromatographers have always sought to improve on the analytical capabilities of a GC system particularly by improving the performance of the column. In principle the efficiency of a packed column can be improved by increasing the number of separation steps (theoretical plates) in the column either by using a longer column or by reducing the particle size of the packing material (Figure 5.4). Micropacked columns have typical efficiencies between conventional packed columns and wall coated open tubular capillary columns. Micropacked columns have an internal diameter of less than 1 mm and a particle size, d_P, to column internal diameter, d_C, ratio of less than 0.3, typically between 0.1 and 0.3.

$$\frac{d_P}{d_C} \leq 0.3$$

Suitable packing materials have a particle size between 0.2 and 0.08 mm (80 and 200#). Scott first proposed high efficiency packed columns in 1958, but it was Carter and Virus in 1963 who described the preparation and applications of micropacked columns, 2 m long with internal diameters of 0.5 mm, packed with 0.04 mm glass beads coated with 10% squalane [7]. Column efficiencies of over 60 000 steps have been obtained using column lengths up to 15 m and d_P/d_C ratios down to 0.03. High carrier gas pressures are required causing sample introduction problems and a need for special instrumentation. However, micropacked columns can be easily prepared using 1/16 in (1.59 mm) o.d. stainless-steel or 1/8 in (3.18 mm) o.d. PTFE tubing of suitable internal diameter and fitted into packed column GC instruments using the appropriate end fittings. Glass and stainless-steel columns, 2 m in length, with an internal diameter of 0.75 mm, are available from Supelco Inc. and may be fitted in place of conventional 1/4 in (6.35 mm) o.d. glass or 1/16 in (1.59 mm) o.d. stainless-steel columns. These columns are available pre-packed with a carbon molecular sieve for the analysis of permanent gases and C_1–C_4 hydrocarbons. Micropacked columns can be prepared with conventional coated support materials of 100–120 or 120–150#, using the packing procedure described in section 5.4.4 with the additional use of an ultrasonic bath to assist in packing the particles uniformly.

The advantages of micropacked columns have been overtaken by the development of thick film wide bore WCOT capillary columns with internal diameters of 0.5–0.7 mm and up to 5.0 μm stationary phase films (section 5.5).

5.5 Choice of stationary phase

The suitability of a stationary phase for a particular application depends on the selectivity and the degree to which polar compounds are retarded relative to what their retardation would be on a completely non-polar stationary phase. Since retention time is a function of temperature, flow-rate, stationary phase type and loading or film thickness it cannot be used to relate the retention characteristics of one column to another. Various retention index methods have been described such as evaluating the partition and separation properties of solute–stationary phase systems. Kovats (1958) devised a system of indexing chromatographic retention properties of a stationary phase with respect to the retention characteristics of *n*-alkanes, alkanes being used as reference materials since they are non-polar, chemically inert and soluble in most common stationary phases [8–10]. The retention index (RI) for the *n*-alkanes is defined as

$$\mathrm{RI} = 100Z$$

where Z is the number of carbon atoms present in the alkane, for example,

$$\mathrm{RI}_{\mathrm{hexane}} = 600, \qquad \mathrm{RI}_{\mathrm{octane}} = 800$$

Retention time varies logarithmically with vapour pressure. A graph of log of retention time against carbon number, Z, for a homologous series will be linear since there is an almost uniform increase in boiling point between members of a given series. A graph of log of corrected retention time against RI for several alkanes with retention times similar to the compound is constructed. Log t'_R of the compound is then noted on the graph and the RI determined (Figure 5.5). For example, propanol may have an RI of 650 on a Carbowax column, but only 500 on a OV101 column. This implies that the Carbowax is the more polar column by 150 RI units.

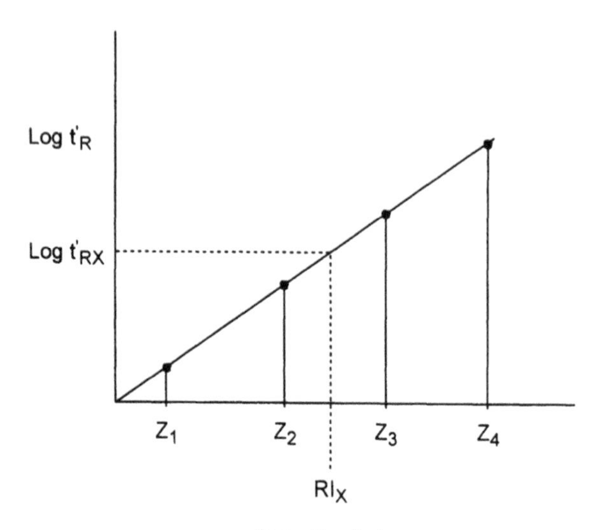

Figure 5.5 Determination of retention index for X, $Z_1 \ldots Z_4$ are RI values of alkanes.

Alternatively, the RI for an individual compound, RI_C, may be calculated using the following equation:

$$RI_C = 100Z + 100\left(\frac{\log t'_{RC} - \log t'_{RZ}}{\log t'_{R(Z+1)} - \log t'_{RZ}}\right)$$

where t'_{RC}, t'_{RZ} and $t'_{R(Z+1)}$ are the corrected retention times for the compound and the alkanes eluting before and after the compound, respectively.

Rohrschneider and McReynold extended the RI system to predict a PI for various stationary phases measured at a column temperature of 120°C with a 20% (w/w) loading, to minimise retention contributions from the diatomite support [11, 12]. A set of five reference compounds were selected to reflect a range of polar characteristics and functional groups: benzene, X; butanol, Y; 2-pentanone, Z; nitromethane, U; and pyridine, S. Squalane, 2, 6, 10, 15, 19, 23-hexamethyltetracosane ($C_{30}H_{62}$), is used as the reference stationary phase as it is a readily available completely non-polar, non-volatile liquid, bp = 176 at 0.05 mm. The values of X, Y, Z, U and S represent the relative affinities of the reference compounds for the stationary phase, calculated as the differences, ΔRI, between the RI of the reference on a chosen stationary phase compared to the RI on squalane. The polarity index, PI, is the mean of the RI values (Table 5.2).

$$PI = P/5 = (X + Y + Z + U + S)/5 \quad \text{or} \quad P_{SUM} = (X + Y + Z + U + S)$$

Although PI gives an indication of the polar character of a stationary phase it is necessary to consider the individual values for each reference compound to obtain a true picture of the selectivity of the stationary phase. Note that dinonylphthalate, PI = 161 and OV17, PI = 177 have a similar polarity index but benzene (RI = 83 and 119) has a greater retention on OV17 whereas butanol (183 and 158) is more retarded by DNP. McReynold's constants are quoted by suppliers and in reference texts to indicate column polarity and selectivity for classes of compounds, for example, analysis of drugs, stimulants, antihistamines, plasticisers, solvents [13, 14]. The RI and PI values in Table 5.2 were determined on a packed column, however, values for capillary columns are relatively similar although a different range of stationary phases are used. A more detailed discussion is included in section 5.6.

A method to select the appropriate stationary phase for analysis of a sample mixture is to consider the polar characteristics of the analytes and select a stationary phase of similar polarity. An analyte with similar polar character to the stationary phase will be well retained, the principle of *like attracts like* applies, and useful retention is then likely to occur leading to adequate selectivity and separation of the analytes primarily on the basis of volatility. Conversely, if the solute is immiscible with the stationary phase then little or no retention difference will be obtained. Further useful indication of retention characteristics may be obtained by analysing a sample on a non-polar and polar column, for example, a dual column GC fitted with Apiezon/OV101 and Carbowax 20M columns temperature programmed 50–220°C at 10°C min^{-1} with a final hold of 10 min. The chromatogram will indicate the polarity of stationary phase required for the components and the analysis can be repeated with columns of differing polarity, e.g. OV17, OV1701. Tables of RIs for various classes of compounds have been published, mainly for squalane, the reference non-polar stationary phase and Apiezon L, with RI values for Carbowax 20M as a reference polar stationary phase [10].

5.6 Capillary column gas chromatography

The theory and application of capillary columns was first expounded by Golay [15]. The theory predicts a very high efficiency for a column which is simply a long length of tubing, 10–100 m long, 0.1–0.7 mm internal diameter, whose inner wall is coated with a thin layer of stationary phase. No column packing material is present (Figure 5.4). The early developments have been reviewed [16–20]. Capillary columns have been constructed of copper, stainless-steel and glass but developments in vitreous silica materials which produce flexible inert columns are the preferred types today. Metal columns are generally avoided since catalysed reactions with the analytes may occur and produce modifications to the chromatography although silica lined metal capillary columns have recently become available.

Capillary columns, and in particular WCOT columns have very high efficiencies compared to packed columns, mainly because of their length and the unrestricted path for the mobile phase. Stationary phase loading is lower, a 0.2–5.0 μm film compared to a 3–30% (w/w) loading in packed columns, giving a more favourable phase ratio for rapid separation. Columns are now available where the stationary phase is chemically bonded to the column wall and is laid down as films of varying thickness. Columns with high thermal stability are produced enabling separations to be carried out over a wide temperature range. Rapidly eluting peaks have a low retention ratio, k, requiring high column efficiencies for good resolution, conversely well retained high boiling point compounds have higher k values, $k > 7$, therefore require high temperatures to achieve elution in a reasonable analysis time. High column efficiencies, a wide temperature programming range and good selectivity enable complex mixtures to be readily analysed. A comparison of the features of WCOT and packed columns are summarised below:

- the path of the carrier gas is unrestricted in open tubular columns, permeability is therefore very high so that long column lengths can be used and high column efficiencies and separating power obtained;
- capillary columns have high efficiencies, containing up to 500 000 theoretical plates, packed columns have up to 10 000 plates/columns;
- samples introduced onto capillary columns are in the nanogram (10^{-9} g) range due to the much reduced retention volume, μg samples are used on packed columns;
- column temperatures are generally 20–50°C lower than required for packed columns because of the more favourable phase ratio in capillary columns;
- there is greater flexibility of carrier gas flow-rates when helium or hydrogen are used due to the flatter curves obtained for the van Deemter plots of H (HETP) against linear velocity of carrier gas (μ);
- retention times are less than for packed columns due to high column efficiencies and minimum band dispersion; the linear velocity of the carrier gas can be significantly higher than the optimum value when helium is used; analysis times are therefore shorter by up to a factor of 10;
- development of pure vitreous silica materials enables chemically inert columns to be produced with thermally stable bonded stationary phases, complex mixtures with a wide boiling point range can be separated that are impractical on packed columns;
- generally only four or five types of stationary phase are required to enable most sample mixtures to be analysed, especially when temperature programming is used. For example:

analytes with non-polar groups	OV101, Apiezon
analytes with medium polar groups	OV17, OV1701
polar analytes	Carbowax 20M

- the limit of detection is approximately the same as for packed columns even though much less sample is introduced onto the column, less band broadening occurs leading to narrow sharp peaks; and
- capillary column chromatography can be used for fingerprint/identification purposes, especially if a two-column system is used.

WCOT capillary columns are the most commonly used GC columns, their high separating efficiencies enabling routine samples and complex mixtures to be analysed more rapidly with less band broadening and asymmetry and at lower temperatures than packed columns. The chromatograms in Figure 5.1 illustrate these attributes. Although the WCOT columns are longer than packed columns absence of a column packing material enables more separation steps to be accommodated in longer columns without excessive band broadening. The two main observations obtained from comparison of the chromatograms are as follows.

1. An analysis carried out on a packed column could be achieved in about one-tenth the time using a capillary column.
2. A more complex sample containing many more (at least five times as many) components can be separated on a capillary column than in an equivalent packed column analysis time; alternatively longer analysis times can be used to obtain the maximum separating capability for complex samples.

5.6.1 *Band broadening and analysis time*

The high separating capabilities of WCOT columns is attributable to much reduced band broadening and rapid elution of components, reflected by reduced peak widths, w_b, and retention times. Since resolution, R_S, is a function of peak separation, Δt_R and peak width, sharp narrow peaks allow resolution to be maintained even though more peaks are separated in a given analysis time:

$$R_S = \frac{\Delta t_R}{w_{bB}} = \frac{\Delta t_R}{2w_{hB}}$$

The stationary phase is supported on the wall of the column, the carrier gas therefore has an uninterrupted flow through the column. The absence of column packing means that the unequal pathways term, A term, in the van Deemter equation (see section 2.5.1) is zero and band broadening is due to the effects of longitudinal diffusion, B term, and mass transfer, C term. The reduced equation is referred to as the Golay equation [15].

van Deemter equation $$H = A + \frac{B}{\mu} + C_M\mu + C_S\mu$$

Golay equation for WCOT columns $H = \frac{B}{\mu} + C_M\mu + C_S\mu$

and column efficiency, $N = \frac{L}{H}$

where H is the equilibrium step height or HETP and μ is the average linear velocity of the carrier gas.

Note that the C term varies directly with carrier gas velocity and the B term varies inversely. Dispersion due to the mass transfer term, C, is a function of the component/stationary phase interaction and is the critical dispersion term. There are two components, mass transfer in the mobile phase, C_M, and in the stationary phase, C_S. Figure 5.3 shows the van Deemter/Golay plots for packed and WCOT columns and the effect of film thickness.

Dispersion due to the B term decreases as the carrier gas velocity increases and is effectively time dependent. Thus, for minimum dispersion attributable to B, analysis times should be optimised to achieve separation of the critical components and elution of the last component in minimum practicable time. The retention factor, k_Z, for the last component to elute will determine the analysis time, that is, the retention time for this component, t_{RZ}.

$$k = \frac{t_R - t_M}{t_M}$$

$$\text{rearranging,} \quad t_R = t_M(k+1) \quad \text{but} \quad t_M = \frac{L}{\mu}$$

$$\text{therefore} \quad t_R = \frac{L}{\mu}(k+1) \quad \text{or} \quad t_{RZ} = \frac{L}{\mu}(k_Z+1)$$

Analysis time is therefore directly proportional to the retention factor and column length and inversely proportional to the carrier gas velocity.

5.6.2 Optimum practical gas velocity

Optimum practical gas velocity (OPGV) is the average carrier gas velocity that gives the best separation for a particular sample in the shortest practical analysis time whilst maintaining satisfactory resolution between peaks. It is important therefore to determine the optimum practical gas velocity, μ_{PRACT}, for the analysis rather than the optimum value, OGV, at μ_{OPT}. Figure 5.6 is the van Deemter/Golay plot for a typical 25 m, 0.25 mm internal diameter BP-1 WCOT column. Note the increase in H (about 30%) if a value for μ_{PRACT} is chosen which is twice μ_{OPT}; there would also be a corresponding decrease in column efficiency (N). In practice μ_{PRACT} is between 1.5 and $2.0 \times \mu_{OPT}$; therefore, to determine the OPGV for a specific analysis a value of $2 \times \mu_{OPT}$ is used initially and then gradually reduced until the most time efficient separation is obtained.

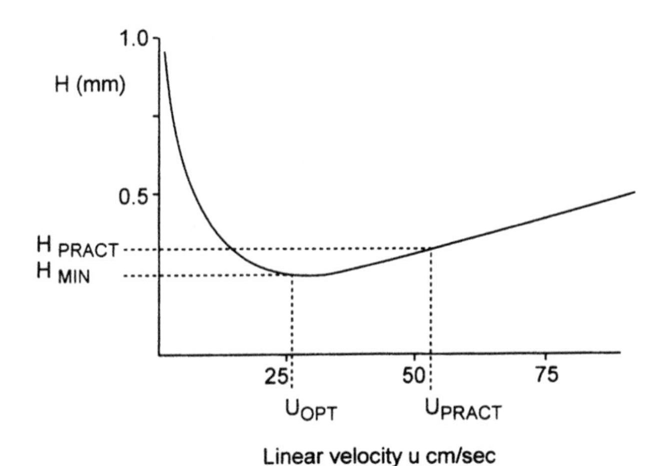

Figure 5.6 Optimum practical gas velocity (OPGV) approximate values for a 25 m × 0.25 mm dimethylsiloxane WCOT column using He carrier gas.

5.6.3 *Column efficiency*

Maximum column efficiency occurs at the OGV (Figure 5.6). The value of H_{MIN} can be calculated using the equation proposed by Golay from the development of the theory for open tubular columns [15]:

$$H_{MIN} = r\sqrt{\frac{(11k^2 + 6k + 1)}{3(1 + k)^2}}$$

where $r = d_C/2$, the internal radius of the column and $H \propto r$. Table 5.3 shows the typical values of H_{MIN} (mm) and N (steps/plates m^{-1}) calculated using Golay's equation and retention factors of $k = 3$ and $k = 5$ for a 25 m, 0.25 i.d. column. Approximate carrier gas flow rates are also listed for $\mu_{OPT} \approx 35\,cm\,s^{-1}$ of helium as a guide to setting up a column.

μ_{OPT} is calculated from $\mu_{OPT} = L/t_M$, and t_M, the system dead time is readily determined by analysing an unretained component such as methane

Table 5.3 H_{MIN} and column efficiency for WCOT columns

Internal diameter, d_C (mm)	H_{MIN} (mm)		Column efficiency ($N\,m^{-1}$)		He flow-rate ($ml\,min^{-1}$) at μ_{OPT}
	$k = 3$	$k = 5$	$k = 3$	$k = 5$	
0.18	0.141	0.152	7092	6578	0.35
0.25	0.196	0.210	5102	4762	0.75
0.32	0.251	0.269	3984	3717	1.4
0.53	0.416	0.446	2404	2242	2.6
0.75	0.588	0.631	1701	1585	5.2

using mains gas or propane from a camping gas cylinder. μ_{OPT} for hydrogen is approximately twice the value for helium. Carrier gas flow-rate is calculated from

$$F_C = \frac{\pi r^2 L}{t_M} \text{ ml min}^{-1}$$

where r is the internal radius of the column (cm), and L is the column length (cm).

The separating capability and efficiency of a column is a function of the distribution ratio, K, and the retention ratio, k, of each component, each increasing as the retention time increases. Column efficiency, N, therefore varies with the retention characteristics of each component.

$$N = 16\left(\frac{t_R}{w_b}\right)^2 \quad \text{and} \quad t_R = t_M(k+1)$$

$$\text{therefore} \quad N = 16\left(\frac{t_M(k+1)}{w_b}\right)^2 = 16t_M^2\left(\frac{k+1}{w_b}\right)^2$$

Figure 5.7 shows the relationship between column efficiency, N, and k. $N \propto (k+1)^2$ therefore decreases rapidly up to $k \approx 3$. Above $k \approx 5$, column

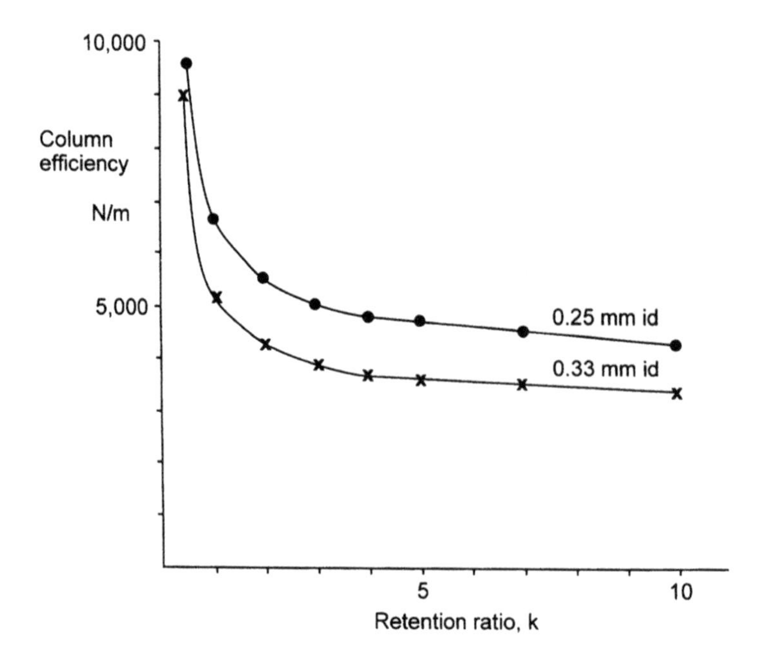

Figure 5.7 Relationship between column efficiency and retention ratio for a 25 m thin film dimethylsiloxane WCOT column.

efficiency remains almost the same. The data plotted was calculated using the Golay equation and $N = L/H$ for a 25 m, 0.25 i.d. column.

5.6.4 *Coating efficiency*

Coating efficiency (CE, %) is a convenient term used to indicate the actual separating efficiency of a column compared to the theoretical efficiency, it is not a measure of how efficiently the column wall is coated with stationary phase.

$$CE = \frac{H_{MIN}}{H_{PRACT}} \times 100$$

H_{MIN} can be calculated using the Golay equation, and H_{PRACT} from column efficiency, $N_{eff} = L/H$ (Figure 5.6). CE reflects the additional factors, such as injector characteristics, dead volumes and uniformity of the stationary phase that contribute to the resolution of the peaks and which result in a decrease in column efficiency. Typical CE values range from 60 to 80%.

5.6.5 *Wall coated open tubular columns*

A successful analysis is obtained when the components in a sample are adequately separated. This is achieved by optimising the retention characteristics that is, K and k, by selecting the most appropriate variables:

- select the most suitable stationary phase by noting the molecular structure of the components and hence their polar character, remember *like attracts like*;
- optimise the temperature programme cycle;
- determine the OPGV for optimum analysis time; and
- consider alternatives:
  —column lengths,
  —columns with different internal diameters, or
  —stationary phase film thickness.

When WCOT columns were first introduced there were basically two methods of coating the inner wall with stationary phase. A solution of stationary phase in a volatile solvent was passed through the column until an equilibrium film had been deposited on the wall. Alternatively, a plug of neat stationary phase was forced through the column. Eventually these techniques were refined so that uniform films were obtained. SCOT columns with higher sample capacity were also developed by depositing a thin layer of diatomite onto the inner wall, giving a larger surface area, which was coated with stationary phase. This was achieved in a one step process using a stationary phase coating solution containing a suspension of the support material [21]. SCOT columns have been largely replaced by thick film

Table 5.4 Analysis of glasses used for capillary columns showing the metal content (%w/w)

Column material	SiO_2	Al_2O_3	Na_2O	CaO	MgO	B_2O_3	BaO
Soda glass	68	3	15	6	4	2	2
Borosilicate glass	81	2	4	<0.1	<0.1	13	<0.1
Synthetic fused silica	100	<0.02	<0.02	<0.02	<0.02	<0.02	<0.02

WCOT columns. The stationary phase coating was susceptible to column bleed and damage by solvents and high boiling contaminants. PLOT columns with alumina, molecular sieve or copolymer surface coatings have been developed particularly for the separation of C_1–C_6 alkanes and alkene and atmospheric gases [22–24].

The adsorption of even trace amounts of sample on the walls of a capillary column can cause the loss of a significant proportion of any polar compounds. Stainless-steel columns are therefore unsatisfactory for many applications and although glass columns have less active surfaces, problems still occur due to the metal content, particularly of the strong Lewis acids, boron, calcium and magnesium. Pure vitreous silica columns overcome many of these problems [25]. They are inert, have a high tensile strength and flexibility due to the physical properties and molecular structure of silica and the low metals content (Table 5.4). Inert fused silica columns also have higher temperature stability of the stationary phase, lower bleed levels and longer lifetimes. They are also suitable for the preparation of chemically bonded stationary phases, thus providing a range of tough, inert and versatile columns. There are two methods for the manufacture of silica columns. The silica may be prepared from natural quartz crystals by melting at 1900°C under vacuum. The metal content is typically 10–100 ppm. A second process uses purified silicon tetrachloride vapour which is introduced into a flame where it is hydrolysed by the water formed in the flame from combustion of methane.

$$SiCl_4 + 2H_2O \rightarrow SiO_2 + 4HCl$$

The silica is collected on a substrate as fused silica, is extremely pure (<1 ppm of impurities) and has about 0.1% hydroxyl groups. These silanol groups are valuable sites for the preparation of bonded stationary phases and may also be reduced by heating, when water is eliminated and siloxane bridges formed. The final structure is a three-dimensional slightly distorted Si–O tetrahedral lattice giving great strength to capillary columns. Many silica columns are further strengthened by coating the external wall with a polymer such as a polyimide which is thermally stable up to 350°C.

Current column technology utilises the surface properties of pure silica tubing to immobilise the stationary phase producing extremely stable inert columns which are not degraded by solvents and have minimum column bleed up to their maximum operating temperatures. Use is made of the surface silanol groups to bond silicone polymers to the tube wall after first cleaning and etching the surface. Thick film stationary phases are obtained by cross-linking the stationary phase polymer via reactive functional groups such as vinyl moieties or substituents present in the polymer. Cross-linking can be initiated using ultraviolet photolysis through the silica tubing (pure silica is transparent to ultraviolet radiation) or via free radicals to give a three-dimensional film. Finally, the column is given added strength by an outer coating of polyimide high temperature polymer.

5.6.6 *Stationary phases for wall coated open tubular columns*

Many stationary phases are simply substituted methyl silicone polymers the substituent groups giving varying degrees of polar character. They have excellent thermal stability up to 300–350°C. Two values are usually given for upper temperature limits. The first is the isothermal limit at which the column can be continuously used, the second is the upper limit which can be used for a few minutes in a temperature programming cycle. Exceeding these limits will accelerate thermal degradation and polymer pyrolysis and therefore loss of the stationary phase. Strong inorganic acids and bases and oxygen also accelerate the degradation process. This is apparent on a chromatogram as tailing of peaks, loss of separating efficiency and column bleed. Column bleed is the normal background signal caused by very slow loss of the stationary phase resulting in a small detector signal. Normally the signal will start to increase just before the isothermal temperature limit, rise at the limit and stabilise by the upper limit (Figure 5.8). The

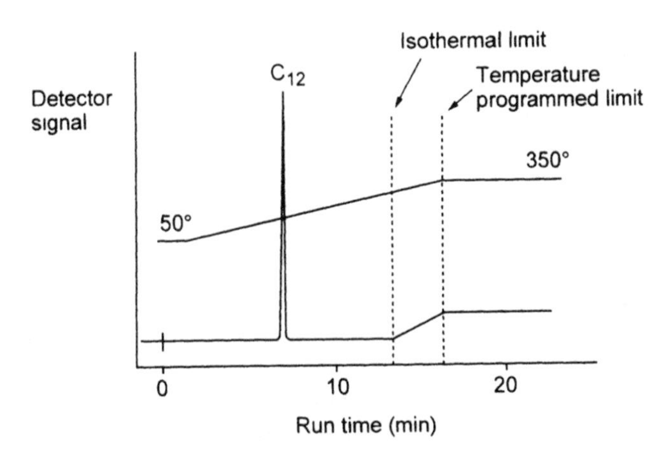

Figure 5.8 Column bleed, 25 m × 0.25 mm WCOT column dimethylsiloxane stationary phase, temperature programmed 50–350°C at 20°C min^{-1}.

Table 5.5 Stationary phases for WCOT columns

Stationary phase	Packed column equivalent	Structure R-groups	Polarity	Applications
X-1	OV101, SE30	100% methyl	Non-polar	Solvents, petroleum products, VOCs, environmental samples drugs, amines
X-5	SE54	5% phenyl 95% methyl	Non-polar	Aromatics, PAHs, perfumes, flavours, environmental samples, drugs
X-1701 X-10	OV1701	14% cyanopropyl 86% methyl	Medium polar	Pesticides, alcohols, phenols, esters, ketones
X-17 X-50	OV17	50% phenyl 50% methyl	Medium polar	Drugs, esters, ketones, plasticisers, organochlorine samples
X-200 X-210	OV210	50% trifluoropropyl 50% methyl	Polar	Selective for loan pair electrons, steroids, esters, ketones, drugs, alcohols, Freons
X-WAX	Carbowax 20M	polyethylene glycol	Highly polar	Alcohols, methyl esters of fatty acids, solvents, fatty acids, amines
Prefix X BP® Rtx® SPB® DB® HP® CP-SIL®	Supplier SGE Restek Supelco J & W Hewlett Packard Chrompak	Silicone structure R | Wall–Si–O– | R		Notes: VOCs = volatile organic compounds PAHs = poly-aromatic hydrocarbons

condition of the stationary phase and column bleed is checked by introducing a high boiling alkane, such as *n*-decane ($C_{10}H_{22}$), *n*-dodecane ($C_{12}H_{26}$) or *n*-decanol ($C_{10}H_{21}OH$), noting the retention time and peak shape and the baseline profile in a temperature programmed run.

Table 5.5 lists five commonly used stationary phases their properties and applications. Since WCOT columns have high column efficiencies over 90% of analyses can be carried out with these stationary phases using columns of varying length, internal diameter and film thickness. The equivalent liquid stationary phases are included. Reference to Table 5.2 will give their polarity index and McReynold's retention indices.

Figure 5.9 shows a number of chromatograms obtained using columns with a range of internal diameters, stationary phases, and film thickness. Note the carrier gas velocity, column temperature, sample size and detector used, also changes in elution order, retention times and analysis time reflecting the column parameters.

Columns of 0.53 mm i.d. are often referred to as widebore or megabore WCOT columns. They can be used in capillary column mode with carrier gas flow-rates of 1–3 ml min^{-1} and a film thickness of up to 5 μm. Such columns have a large sample capacity accepting up to 2 μg per component without overloading the stationary phase, permitting direct splitless injections of up to 2 μl using a microlitre syringe. Alternatively, widebore columns can also be used in packed column mode with a carrier gas flow-rate of 10–15 ml min^{-1}. Resolution is usually better, peak tailing of polar components less pronounced and analysis times are shorter than packed columns. However, concentrated solutions may require dilution since the sample capacity is less than packed columns and a special connector and injector liner may be required to accommodate direct injections.

Columns are produced by various manufacturers, their proprietary prefixes being included in Table 5.5. A range of over 100 different columns are available from numerous manufacturers and suppliers, some columns having been developed for specific applications, for example,

- 50 m, 0.25 mm i.d. column with a 5 μm film of poly(dimethylsiloxane) for petroleum products analysed to ASTM test method D3710;
- 30 m, 0.25 mm i.d. column with 0.25 μm film of poly(diphenyl 5% dimethyl 95% siloxane) for US EPA methods for water pollutants, that is, the following methods:
  - —610 and 8100 for polynuclear aromatic hydrocarbons;
  - —612, 8120, chlorinated hydrocarbons;
  - —680, 8800, pesticides and polychlorinated biphenyls;
  - —1653, acyl derivatives of phenols.

Chiral WCOT columns suitable for the enantiomeric separation of small molecules such as esters, ketones, alkanes and alcohols have recently been introduced. One example of a chiral column has a stationary phase film consisting of a non-bonded mid-polar, 35% phenyl 65% methyl siloxane modified by embedding in the film permethylated α- or β-cyclodextrin. Elution characteristics are modified by the cyclodextrin content; DEX 110 contains 10% cyclodextrin and DEX 120, 20% [127, 128].

5.6.7 *Phase ratio, film thickness and column internal diameter*

Phase ratio, β, is defined as the volume occupied by the mobile phase (carrier gas) relative to the volume occupied by the stationary phase per unit length. In WCOT columns β is the ratio of the internal radius, r (mm), to the

Figure 5.9 Chromatograms illustrating GC analyses using a range of column types and detectors. Note the polarity of the stationary phase, analysis times of the components and sequence in which they are eluted. Polarity of stationary phases are given in Table 5.5. (Chromatograms reproduced by permission of Restek Corporation.)

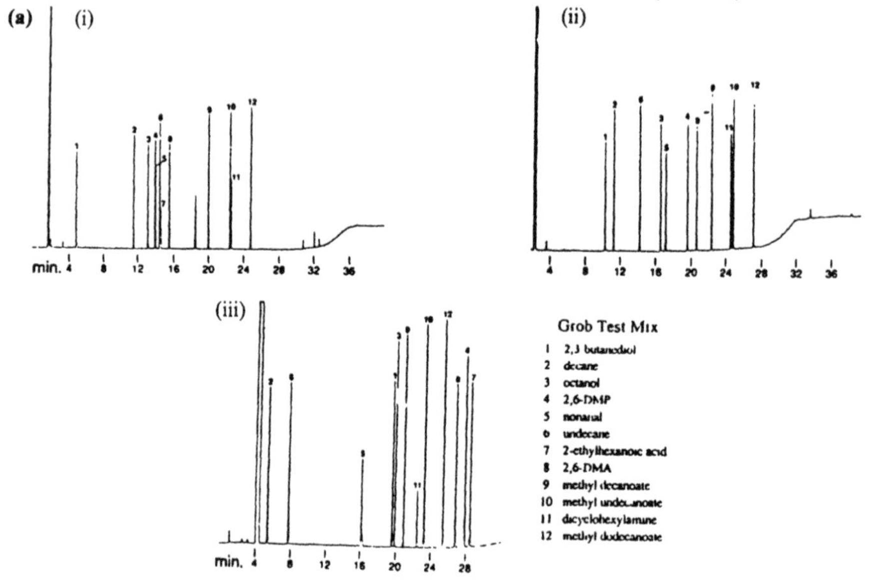

Figure 5.9 (a) Analysis of a grob test mix on (i) non-polar Rtx®-1, (ii) polar Rtx®-1701 and polar Stabilwax® columns. All 30m, 0.32mm i.d., 1.0μm, 40°–210°C, 6°Cmin⁻¹, to 320°C, 15°Cmin⁻¹, He.

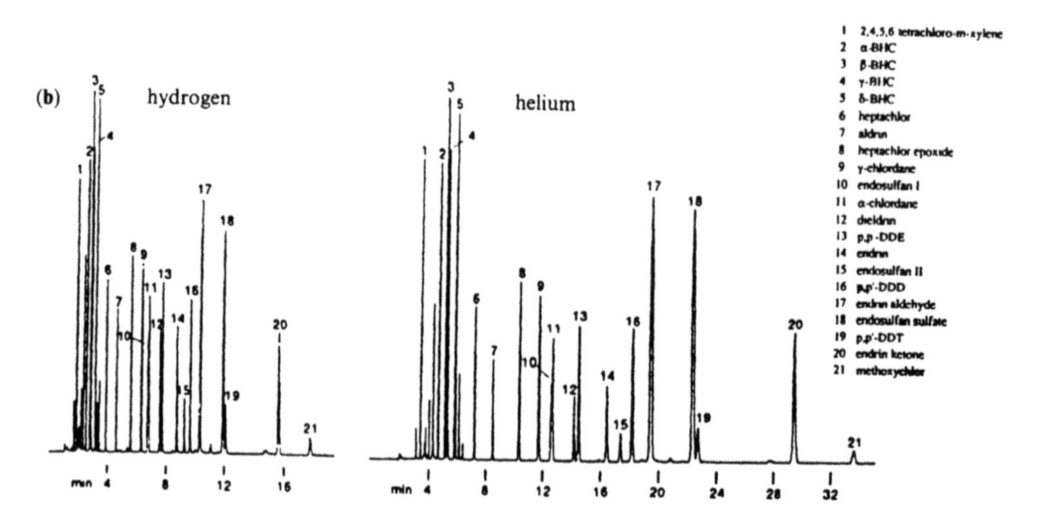

Figure 5.9 (b) EPA Method 608 pesticides using an ECD, illustrating the different carrier gas properties of hydrogen and helium. Rtx-5, 30m, 0.25mm i.d., 0.25μm, 210°C, methoxychlor: H_2 17 min, He 34min.

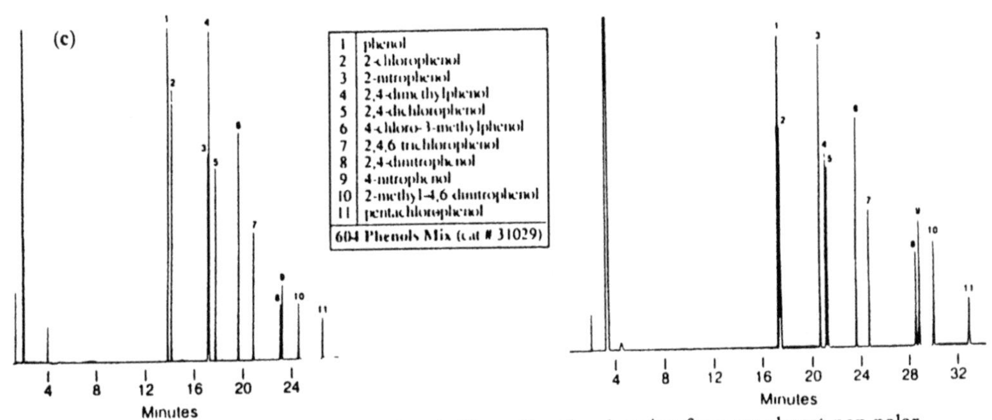

Figure 5.9 (c) EPA Method 604 phenols illustrating the changing from an almost non-polar Rtx®-5 column to a mid-polar Rtx®-50. All 30m, 0.53mm i.d., 1.5μm, 40°C, 6°Cmin^{-1}, to 250°C, 10°Cmin^{-1}, He.

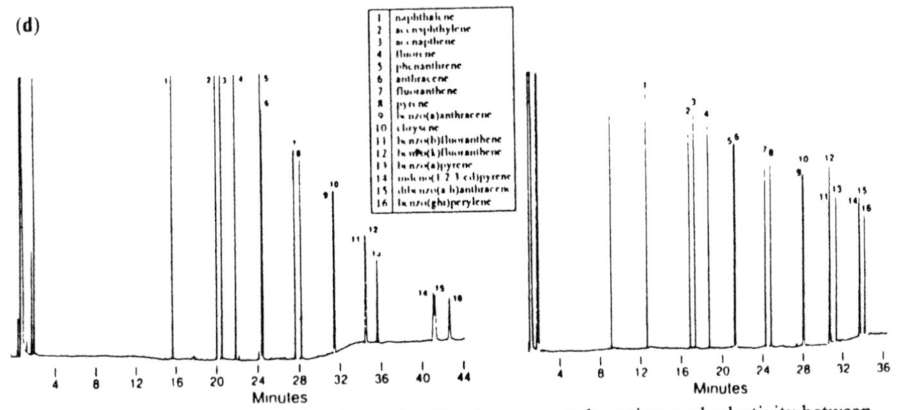

Figure 5.9 (d) EPA Method 610, illustrating the change in analysis time and selectivity between thick film wide bore and thin film narrow bore columns. Rtx-5, 30m, 0.53mm i.d., 1.5μm, 40°C, 6°Cmin^{-1}, to 300°C, 10°Cmin^{-1}, H_2.

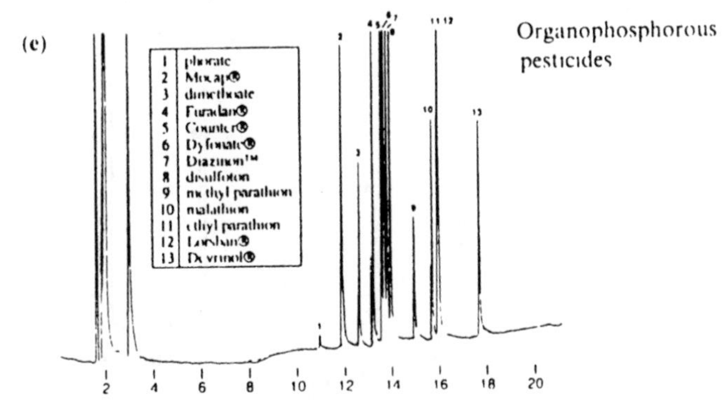

Figure 5.9 (e) Organophosphorous pesticides using a nitrogen/phosphorous detector and split-less injection onto a wide bore column. Rtx-5, 30m, 0.53mm i.d., 0.5μm, 100°C, 2°Cmin^{-1}, to 280°C, 10°Cmin^{-1}, He.

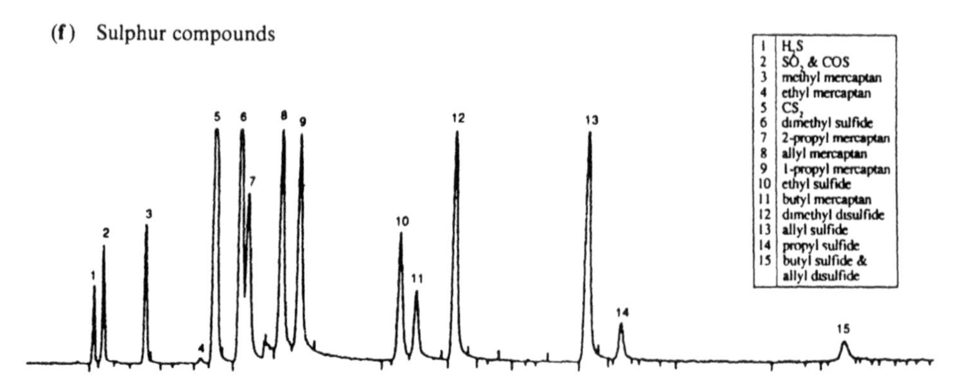

Figure 5.9 (f) Sulphur compounds using a flame photometric detector and direct injection onto a wide bore thick film column. Rtx-1, 60m, 0.53mm i.d., 7.0μm, 50°-200°C, 15°Cmin-1, He.

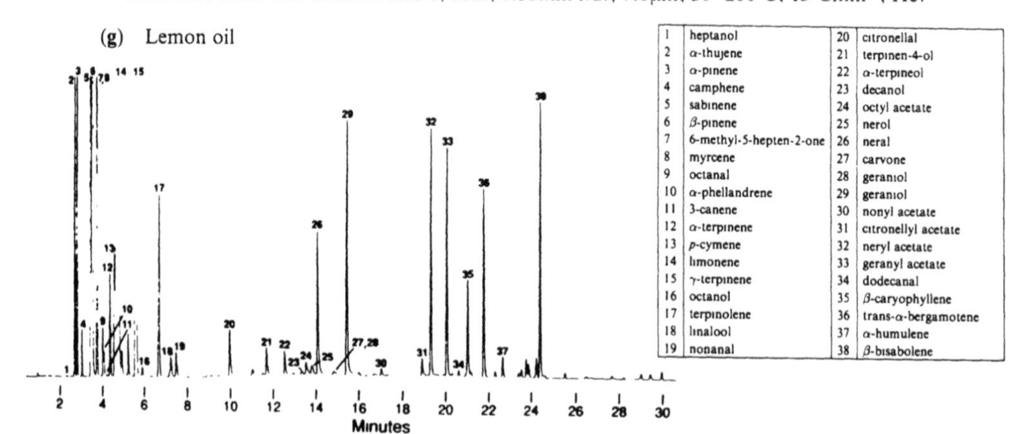

Figure 5.9 (g) Analysis of lemon oil using hydrogen as carrier gas and a thin film almost non-polar column. Rtx-5, 30m, 0.32mm i.d., 75°C, 8min, to 250°C, 4°Cmin-1, H_2.

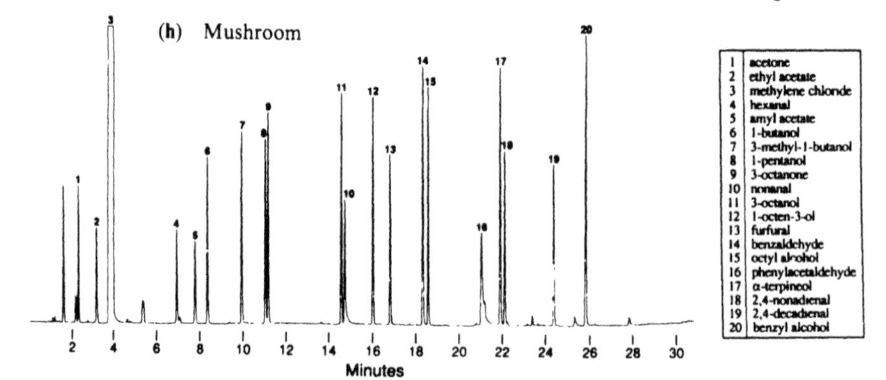

Figure 5.9 (h) A mushroom aroma using hydrogen as carrier gas and a polar column with split injection. Stabilwax, 30m, 0.32mm i.d., 1.0μm, 40°C-200°C at 6°Cmin-1, H_2.

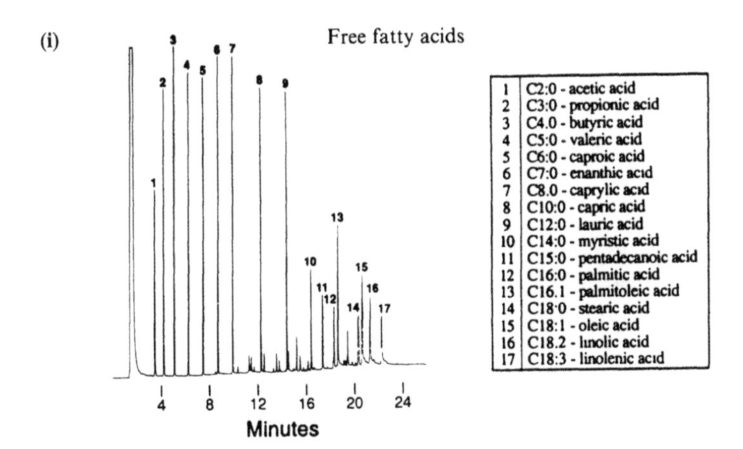

Figure 5.9 (i) Free fatty acids analysed on a wide bore very polar thin film column using helium carrier gas and direct injection. Stabilwax-DA, 30m, 0.53mm i.d., 0.25μm, 100°C, 2min, to 250°C, 8°Cmin^{-1}, He.

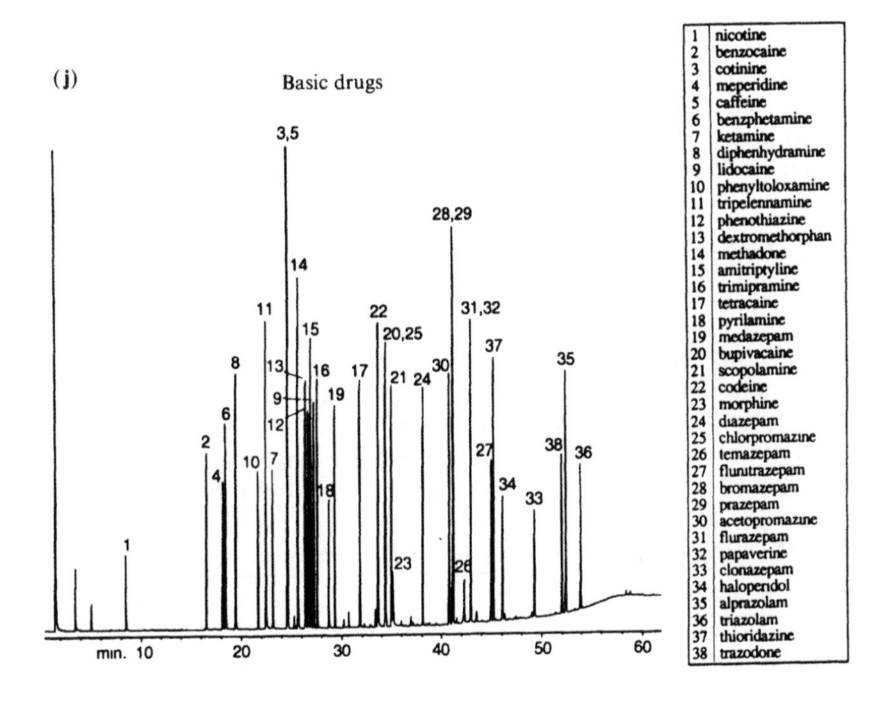

Figure 5.9 (j) Basic drugs analysed on a narrow bore thin film column using helium and polar stationary phase. Rtx-200, 30m, 0.25mm i.d., 0.25μm, 100-325°C, 4°Cmin^{-1}, He.

stationary phase film thickness d_f (mm):

$$\beta = \frac{r}{2d_f}$$

$$k = \frac{K}{\beta} \quad \text{(see section 2.3.2)}$$

$$\text{therefore} \quad k = K\left(\frac{2d_f}{r}\right)$$

Since the distribution ratio, K, remains constant for a given stationary phase, temperature and carrier gas velocity the retention ratio, k, will vary directly with film thickness and inversely with the internal diameter ($d_C = 2r$). Retention times can therefore be manipulated by selecting alternative i.d. columns or a different film thickness, therefore retention time (i) increases as column i.d. decreases; and (ii) increases as film thickness increases.

One disadvantage of moving to smaller column i.d. and thinner stationary phase films is that the amount of sample that can be introduced onto a column also decreases. Exceeding the sample capacity results in asymmetric peaks (fronting) due to overloading of the first few plates or equilibrium steps in the column. Sample capacity increases as the phase ratio decreases and is higher for a component that has a high affinity for the stationary phase—the maxim *like attracts like* again! Sample capacity ranges from approximately 80–100 ng on 0.18 and 0.25 mm, 0.25 μm film columns to 200 ng on a 0.32 mm, 0.5 μm column and 2000 ng on a 0.32 mm i.d., 5.0 μm film column. Table 5.6 shows typical phase ratios and sample capacities for various columns and stationary phase films.

5.6.8 *Performance of wall coated open tubular columns*

The fundamental criteria for assessing a column's capabilities relate to its separating capabilities in terms of the retention characteristics, column efficiency and the stationary phase at OPGV, and temperature cycle. Retention characteristics of a component are indicated by its retention ratio, k ($k = t'_R/t_M$). Dead time, t_M, is an indication of the OPGV and is readily determined by introducing a sample of methane onto the column. Column efficiency, N or N_{eff}, can be calculated using data from the chromatogram

Table 5.6 Phase ratios for various stationary phase film thickness and column i.d.

SP film thickness (μm)	0.25	0.5	1.0	1.5	3.0	5.0
Phase ratio						
0.18 mm i.d.	180	90	45	30	15	9
0.25 mm i.d.	250	125	63	42	21	13
0.32 mm i.d.	320	160	80	53	27	16
0.53 mm i.d.	530	265	128	88	43	27

Table 5.7 Standards included in the Grob mixture

Standard	bp (°C)
2,3-Butanediol	88
Decane	174
Octan-1-ol	195
2,6-Dimethylphenol	212
Nonanal	192
Undecane	193
Dodecane	215
2-Ethylhexanoic acid	228
2,6-Dimethylaniline	214
Methyldecanoate	224
Methylundecanoate	241
Dicyclohexylamine	256
Methyl dodecanoate	262

and CE determined as indicated in section 5.6.4. The chromatography is evaluated using a test mixture which includes a marker to use as a quick check for retention characteristics and peak asymmetry and then for further calculations of N_{eff} and CE (Figure 5.1(c)). The mixture should also reflect the components being routinely analysed. Alternatively a standard test mixture such as that proposed by Grob can be used [27]. The mixture contains alkane markers, polar alcohols, mid-polar esters, relatively high boiling components to check resolution and acids and amines to detect any tailing due to deterioration of the column. Standards included in the Grob mixture are shown in Table 5.7.

Figure 5.9 includes chromatograms obtained using the Grob mixture on different stationary phases. Note the elution sequence, although GC is often referred to as a boiling point separation elution of the components is not in boiling point order. The chromatograms illustrate the relative retention characteristics, that is, the principle of *like attracts like*. A column which has deteriorated may be improved (temporarily) by deactivation of the polar sites by silylation, injecting a solution of a chloromethylsilane or HMDS. An alternative procedure is film deactivation using a polar stationary phase. This method uses a short packed pre-column of 5% Carbowax 20M on an inert support which is placed into the heated injection port of the gas chromatograph and connected directly to the column. The precolumn is maintained at about 260°C and the column at 250°C for 5–10 h [28].

5.6.9 Porous layer open tubular columns

Porous layer open tubular (PLOT) columns have been developed primarily for the analysis of low molecular compounds, atmospheric gases and C_1–C_6 hydrocarbons which have very low retention ratios typically <1.0 and down to <0.01 on WCOT columns. Retention ratio can be increased by increasing the distribution coefficient K by replacing the liquid stationary phase film with a solid phase adsorbent to produce a more active retention mechanism. PLOT columns consist of silica capillary columns with a thin coating of solid phase adsorbent particles on the inner wall (Figure 5.4). PLOT columns were developed in the early 1960s but de Nijs in 1981 first successfully used alumina PLOT columns [29–32]. Alumina has an active surface due to the Al–O–Al matrix, however, the surface can be modified by alkali salts to give selective retention properties for molecules that have π electrons. An extensive study has shown that sodium and potassium salts are the best modifiers [33, 34]. Modified alumina PLOT columns are now available, for example, as 25 m, 0.3 mm i.d. and 50 m, 0.53 mm i.d. Al_2O_3/KCl and Al_2O_3/Na_2SO_4, for the analysis of C_1–C_6 hydrocarbons. Figure 5.9 includes a chromatogram of such an analysis using a Micro GC-TCD (thermal conductivity detector) system.

Molecular sieve 5 Å packed columns have been used for the analysis of atmospheric gases for many years. However, the high capacity of a zeolite adsorbent layer has added an extra dimension to gas analysis. A 25 m, 0.53 i.d., 50 µm layer PLOT column can successfully separate helium, argon, neon, oxygen, nitrogen and methane at 30°C using a TCD.

Carbon layer open tubular columns provide a mixed adsorption/partition retention mechanism since the active layer is a graphitised carbon material coated with a liquid phase. The columns can be used from subambient temperatures to 300°C and have a flat Golay/van Deemter curve so that high carrier gas velocities can be used to optimise analysis times with little loss in efficiency. A 25 m, 0.53 mm i.d., 25 µm film column programmed from 30 to 110°C is capable of separating hydrogen, oxygen, nitrogen, carbon monoxide, methane, carbon dioxide and C_2 hydrocarbons using a TCD [35].

Recent developments in coating technology have enabled cross-linked styrene divinylbenzene porous copolymers to be successfully immobilised on the silica wall of a capillary column. These PLOT columns have similar retention properties to the packed column equivalents but with higher resolution. PoraPLOT columns with Porapak Q, S or U stationary phases are general purpose columns suitable for the analysis of C_1–C_4 hydrocarbons, chlorofluorocarbons and atmospheric pollutants [35, 36].

5.7 Gas–solid chromatography

GSC is a complementary technique to GLC. Separation is based on the adsorption properties of a solid surface which gives a more active stationary phase and higher retention characteristics than liquid film stationary phases. The distribution ratio, K, and the retention ratio, k, are therefore higher than for GLC columns so that gases and highly volatile compounds which cannot be easily separated on liquid phases can be readily analysed, for example, methane is not retained on packed GLC and WCOT columns and is used to determine t_M, but a molecular sieve or carbosieve column is capable of separating atmospheric gases and C_1–C_6 hydrocarbons. The instrumentation and techniques required for GSC are the same as GLC, with the additional requirement for a gas sample valve or gas syringe for introducing samples on to the column. Active solids discussed herein include alumina, silica gel, molecular sieves, carbon and porous polymer materials.

Separation of components in GSC is determined by their relative affinity for the adsorbent as described by the distribution ratio, K:

$$K = \frac{C_S}{C_M} = \frac{\text{amount adsorbed on the solid surface}}{\text{partial pressure in the gas phase}}$$

Isotherms are used to describe the adsorption process and also represent the distribution ratio, K and are obtained by plotting the amount of component adsorbed against concentration in the gas phase (Figure 5.10). Curve I shows a typical Langmuir adsorption isotherm which is approximately linear at low concentrations of adsorbate, but at high concentrations is non-linear giving rise to peaks with sharp fronts and diffuse tailing. Curve II is a typical GLC partition isotherm where the tendency at higher concentrations is to peak fronting. A linear isotherm, curve III, represents the distribution ratio for an ideal distribution of the component between the two phases giving symmetrical peaks. The success of GSC depends on the low concentrations of components so that the relative amounts in the gas phase and adsorbed

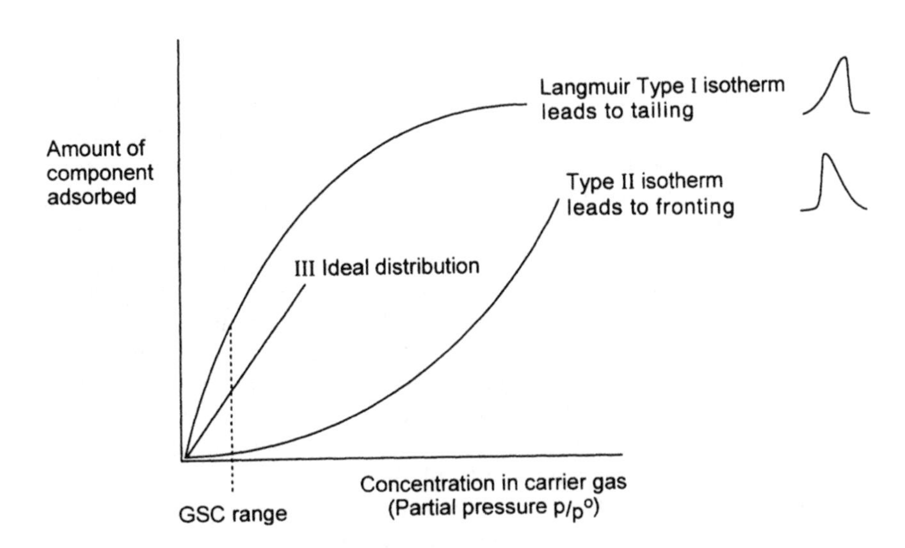

Figure 5.10 Isotherms and GSC.

on the stationary phase are such that the chromatographic equilibrium is within the linear part of the isotherm and thus gives symmetrical sharp peaks.

Adsorption results from two properties of the adsorbent, polarity and microporosity. Polarity describes the normal van der Waal's type interactions due to permanent electric effects and may be supplemented by hydrogen bonding. Microporosity describes the presence within the adsorbent macro structure of pores with diameters in the range 50–300 nm (5–30 Å). Many active solids are also efficient catalysts and may result in the irreversible adsorption or complete or partial conversion of an analyte giving rise to artefacts which will show up in the chromatogram. There are three main requirements for successful GSC.

1. Separation must be carried out at temperatures which are high relative to the boiling point of the analytes, GSC is therefore appropriate for separation of gases and highly volatile substances.
2. Adsorbents must have suitable surface characteristics for reversible interaction with the analyte, for example, alumina modified with inorganic salts, silica modified by hydrothermal treatment and silanisation, zeolites, graphitised materials and porous polymers.
3. The surface should be pre-loaded with a liquid phase to modify the non-uniform surface, for example, squalane on carbon black, alternatively a carrier gas may be modified to interact with the surface, for example, using water vapour with alumina or silica columns.

5.7.1 *Alumina adsorbents*

Alumina has found wide application in GSC. It is a highly polar material with a typical surface area of 250 $m^2 g^{-1}$. It interacts strongly with polar molecules such as water, and also has some catalytic activity, for example, it may convert acetone to diacetone alcohol or dehydration may occur. Much of the early work on alumina was carried out by Scott who subsequently concentrated his research on surface modified materials [37, 38]. He measured polarity in terms of the retention of ethylene relative to non-polar ethane and propane. Activation by heating at temperatures up to 500°C increased the polarity by loss of water and adsorption of water reduced polarity until a minimum was reached when the amount required for a monolayer had been adsorbed. Further water continued to reduce the activity (by reducing the surface area) but increased the polarity [39–41]. Scott later extended his work on alumina to substances modified with sodium hydroxide and obtained results similar to those obtained for the water-modified alumina without the need to pre-saturate the carrier gas. Retention of benzene relative to heptane increased markedly on alumina modified with sodium salts in the order $OH^- >$ $Cl^- > Br^- > I^-$. A recent comparative study of the modifier effects alkali metal salts showed that 2% (w/w) of NaCl on alumina gave the best separation of C_2–C_5 alkenes and alkanes [34, 40]. Other interesting modifiers reported are CuCl, $AgNO_3$ and CdI. Alumina and magnesium silicates modified with various salts, for example, alumina coated with 10% Na_2SO_4, Na_2MoO_4, and $Al_2(SO_4)_3$, were used for the separation of *cis–trans* isomers, chlorobenzene and dichlorobenzenes [39–41]. Wall coated solid phase capillary columns and PLOT columns have been discussed in section 5.6.9.

5.7.2 *Silica gel*

Porous silica beads have been used in GSC for separating mixtures of gases and volatile organic hydrocarbons (VOCs). Retention characteristics arise from adsorption by the micropores, 50–500 nm in diameter, and polar properties of the surface silanol groups. Retention of VOCs decreases as the pore size decreases and surface area increases. Hydrothermal treatment with steam at 850°C for up to 24 h is used to increase pore size to over 1000 nm and chemical modification by silanisation of the silanol groups reduces polar character. Esterification of the silanol groups with methanol and ethanol has also been used.

Porasil is a porous silica bead material available as 80–100 or 100–120# particles with a pore size of 150 nm and surface area of 185 $m^2\,g^{-1}$ (type *B*) and 300 nm pores, 100 $m^2\,g^{-1}$ (type *C*). Porasil can also be used as an inert support for GLC with stationary phase loadings of 1–4% (w/w) of, for example, OV101, SE30, OV17, DNP, DEGS. Durapak is a range of chemically modified Porasil with bonded stationary phases. The uniform beads, bonded liquid film and zero column bleed provide a retention mechanism that leads to less band broadening than with diatomite supports. Sharp symmetrical peaks are obtained and separation efficiencies are higher than conventional columns, N_{eff} up to 3000 plates m^{-1}. Bonded phases available include polar Carbowax 400 and non-polar *n*-octane for the separation of C_1–C_6 esters, alcohols and hydrocarbons.

5.7.3 *Molecular sieves*

Conventional molecular sieves consist of an aluminium silicate skeleton with pores of regular size and shape formed by the stacking of polyhedra made up from $-SiO_4$ and $-AlO_4$ tetrahedra. Adsorption properties arise from the large internal surfaces of the crystal lattice and the effective pore diameters. Molecular sieves are excellent drying agents for the carrier gas having a high affinity for water, which is adsorbed in preference to practically any other molecule. Pelletised material of 60–80 or 80–100# is used in GSC for the separation of permanent gases, although carbon dioxide like water is irreversibly adsorbed at room temperature. Hydrogen, nitrogen and oxygen are easily separated on 5 Å molecular sieve at 100°C and argon may be separated from oxygen at −78°C on a 1 m column or at 30°C with a 5 m column. Separations are carried out at relatively high temperatures compared with the boiling points of the gases, and hence the isotherms are effectively linear. Other separations depend on the complete or nearly complete exclusion of some molecules but not others; for example, ethane (0.4 nm diameter) is adsorbed by 4 Å sieve but butane (0.5 nm) is not. Type 5 Å sieve will take up butane and unbranched alkanes but not branched chain compounds or cyclic compounds such as benzene (0.63 nm diameter). However, benzene is adsorbed on 10X and 13X in preference to unbranched alkanes because of π electron interaction. Analyses are therefore possible in which a small molecule such as methane may be extracted from a mixture in a small pre-column of molecular sieve prior to analysis.

5.7.4 *Carbon molecular sieves and graphitised carbon*

Carbon molecular sieves have a carbon skeletal structure formed by the pyrolysis of proprietary polymeric materials. Further processing provides a pelletised form with a surface area of 1000–1500 $m^2 g^{-1}$. Carbon molecular sieves such as Carboxen and Carbosieve (Supelco Inc.) are designed for the separation of C_1–C_3 hydrocarbons and permanent gases. Higher molecular weight material is frequently adsorbed irreversibly.

Active carbons are difficult solids to prepare in a reproducible form. The surface of normal active carbon black is generally slightly polar due to the presence of chemisorbed oxygen, sulphur or nitrogen. The slightly polar microporous nature of the carbon surface results in strong adsorption properties. Graphitisation reduces both polarity and microporosity to give an adsorbent with a highly homogeneous surface. Graphitised carbon blacks are non-porous, non-polar, inert materials suitable for analysis of atmospheric and permanent gases. Carbopack adsorbents (Supelco Inc.) have a surface area of 10 or 100 $m^2 g^{-1}$ (type *B* or *C*) and are available as 60–80$^{\#}$ packings that can be used up to 500°C.

5.7.5 *Microporous polymers*

Co-polymerisation of divinylbenzene (DVB) with styrene or other monomers yields a porous polymer. Retention is by surface adsorption accompanied by some partition due to interaction with the polymer itself behaving as a 'solid stationary phase'. As with most adsorbents for GSC, porous polymers can only be used in the analysis of substances of relatively low molecular weight and boiling points <250°C. Separations are carried out at temperatures 50–100°C higher than with conventional packed columns. Higher molecular weight materials may be irreversibly retained or move slowly along the column appearing as 'column bleed' on the chromatogram.

Microporous polymers are used as adsorbents in sample collection tubes for monitoring gases and vapours in ambient and workplace air. The tubes are subsequently thermally desorbed into a GC for analysis of the retained substances (see section 5.9.8).

Basic microporous polymers are produced by the copolymerisation of DVB with other monomers such as styrene, acrylonitrile and ethylene glycol dimethacrylate producing copolymers with a wide range of properties and applications [42]. The DVB content controls pore size through cross-linking to produce surface areas of 50–500 $m^2 g^{-1}$ and pore diameters of 10–300 nm. The polymers produced are non-polar and hydrophobic, little adsorption of polar compounds occurs, therefore water, amines, glycols and other hydrogen bonding compounds are rapidly eluted with symmetrical peaks. Maximum temperatures are generally 250°C and no column bleed occurs thus giving good baseline stability. All the polymers are susceptible to oxidation above 200°C and require conditioning at 200–250°C for 4–6 h under nitrogen or helium before packing into a column or adsorption tube.

Porapak (Waters Associates) is a generic title for a commercial range of microporous co-polymers; a summary of the monomers used in their preparation, the resulting surface areas and typical applications are given in Table 5.8. Retention indices are included as a reference for benzene (bp = 80°C) and *n*-butanol (bp = 118°C) representing the retention characteristics of aromatic, non-polar alkane chains and highly polar groups [42]. Table 5.9 is a list of retention times relative to *n*-hexane for a wide range of compounds on various Porapak copolymers.

Chromosorb Century Series (Johns Manville Corp.) is a range of microporous polymers formed from the co-polymerisation of DVB and styrene, additional vinyl monomers are included to produce a range of co-polymers with varying surface areas and polar characteristics suitable for analysis of a wide range of samples. Table 5.10 summarises the composition of the co-polymer, properties and applications.

Tenax TA is a microporous polymer based on 2,6-diphenyl-*p*-phenylene oxide and has largely replaced Tenax GC for adsorption tubes and GC columns. It has excellent thermal stability up to 350°C and a surface area of 35–40 $m^2 g^{-1}$ and pore size of approximately 200 nm [43, 44]. Tenax TA has lower retentive properties than Porapak Q or Chromasorb 101, 102 and 106 with a low affinity for water and so is suitable for packing adsorbent tubes including those used in purge and trap applications and for analysis of solvent mixtures, alcohols, esters, ethanolamines, etc. Tenax GR is a graphitised form of Tenax containing up to 30% graphitised carbon incorporated in the co-polymer during the polymerisation process. GR has less affinity for water than TA. It also has higher breakthrough volumes and produces highly symmetrical GC peaks and high column efficiencies making it more suitable than TA for adsorption tubes. A rather unusual

Table 5.8 Properties and applications of Porapak microporous polymers

Porapak	Polymer	Surface area ($m^2 g^{-1}$)	Polarity	RI benzene	RI 1-butanol	Applications
P	DVB styrene	150–200	Non-polar	765	560	General purpose, alcohols, glycols, ketones, esters
P–S	Surface silanised form of P	150–200	Non-polar	NA	NA	As P, less tailing for alcohols, glycols
Q	DVB ethylvinylbenzene polymer	550–600	Non-polar	630	638	General purpose, for water, aqueous samples, C_2–C_6, hydrocarbons, esters, ketones, NO_x
Q–S	Surface silanised form of Q	550–600	Non-polar	6625	525	As Q, less tailing for water and carboxylic acids
R	DVB/4-vinyl-2-pyrollidone	400–500	Mid-polar	645	545	Ethers, ketones, esters, aldehydes, HCl Cl_2 from water
S	DVB/4-vinyl-pyridine	300–450	Mid-polar	645	550	Normal and branched chain alcohols
N	DVB/ethylene glycol dimethacrylate	250–350	Polar	735	605	High water retention C_2, C_3 hydrocarbons and CO_2, NH_3, water
T	Ethylene glycol dimethacrylate polymer	250–350	Highly polar	750	675	High water retention aldehydes, HCHO in water, carboxylic acids and esters

Table 5.9 Retention times relative to *n*-hexane using Porapak copolymers (Reproduced by permission of Phase Separations Ltd)

Sample	P–S	P	Q	Q–S	R	S	N	T
Water	0.408	0.467	0.056	0.082	0.131	0.109	0.135	0.188
Methanol	0.475	0.542	0.127	0.134	0.180	0.168	0.193	0.244
Formaldehyde	0.475	0.517	0.134	0.127	0.190	0.172	0.195	0.172
Acetaldehyde	0.475	0.542	0.169	0.170	0.190	0.187	0.222	0.259
Ethanol	0.592	0.666	0.218	0.230	0.307	0.291	0.367	0.462
Formic acid	0.717	0.717	0.225	0.189	0.368	0.386	0.819	0.187
Acetonitrile	0.792	0.934	0.287	0.286	0.358	0.348	0.497	0.670
Propylene oxide	0.666	0.784	0.314	0.327	0.336	0.329	0.406	0.444
Propionaldehyde	0.750	0.808	0.338	0.343	0.376	0.383	0.476	0.543
Acetone	0.758	0.850	0.343	0.349	0.390	0.391	0.544	0.666
Isopropanol			0.351					
Methylene chloride	0.960	0.950	0.373	0.403	0.407	0.438	0.510	0.545
Acrylonitrile	0.892	1.00	0.388	0.404	0.474	0.475	0.660	0.853
Acetic acid	0.926	1.03	0.419	0.379	1.31	1.91	1.34	1.90
Methyl acetate	0.800	0.883	0.419	0.434	0.445	0.438	0.598	0.735
Propanol	0.883	0.984	0.479	0.478	0.660	0.641	0.862	1.06
Pentane	0.666	0.684	0.501	0.536	0.481	0.469	0.490	0.467
Isobutyraldehyde	0.934	1.05	0.598	0.623	0.670	0.676	0.905	1.04
Butyraldehyde	1.10	1.22	0.711	0.710	0.776	0.802	1.09	1.28
2-Butanone	1.18	1.26	0.734	0.730	0.820	0.846	1.20	1.41
Chloroform	1.28	1.35	0.753	0.718	0.854	0.791	0.966	1.24
Ethyl acetate	1.13	1.22	0.812	0.852	0.864	0.862	1.20	1.44
Isobutanol	1.26	1.36	0.902	0.900	1.24	1.21	1.76	2.07
Propionic acid	1.53	1.67	0.909	0.843				4.27
Hexane	1.00 (0.944 min)	1.00 (0.945 min)	1.00 (4.93 min)	1.00 (5.50 min)	1.00 (4.67 min)	1.00 (4.97 min)	1.00 (6.94 min)	1.00 (4.84 min)
Butanol	1.48	1.58	1.07	1.07	1.47	1.46	2.08	2.50
Benzene	1.69	1.86	1.16	1.16	1.24	1.25	1.42	1.67
Carbon tetrachloride	1.53	1.53	1.16	1.14	1.07	1.07	1.17	1.34
Isopropyl acetate	1.47	1.46	1.33	1.43	3.48	1.44	2.04	2.38
Propyl acetate	1.85	1.83	1.72	1.83	4.20	1.83	2.64	3.19
Isopentanol	2.29	2.27	2.10	2.09	2.91	2.85	4.25	5.10
Heptane	1.64	1.58	2.28	2.28	2.18	2.05	2.18	2.20
Pentanol	2.61	2.63	2.46	2.46	3.38	3.35	4.93	5.86
Toluene	2.92	3.18	2.71	2.69	2.43	2.85	3.24	3.65

There are eight types of Porapak, basically porous polymer beads, modified to give varying retention characteristics.
1 m × 2.3 mm i.d. column.
Temperature: 175°C.
Flow-rate: 25 ml min^{-1}.
Detector: FID (except TC for water).

application of Tenax-GC was in the Viking Mars Lander mission in the 1970s for the analysis of Martian Soil to detect the presence of organic compounds and thus provide evidence for the existence of life forms [43, 44]. The column consisted of 60–80 mesh Tenax-GC coated with poly(methylphenyl ether) and was selected for effective separation of water, carbon dioxide and elution of compounds over a wide boiling point range at ng levels.

Table 5.10 Properties and applications of Chromasorb Century Series of copolymers

Chromasorb	Polymer	Surface area ($m^2 g^{-1}$)	Polarity	RI benzene	RI 1-butanol	Applications
101	DVB/styrene	300–350	Non-polar	745	565	General purpose, alcohols, glycols, ketones, esters, fatty acids and esters
102	DVB/styrene	300–350	Slightly polar	650	525	Gases, VOCs, C_2–C_6 hydrocarbons, solvents, alcohols
103	Cross-linked polystyrene	300–350	Non-polar	720	575	Amines, basic compounds, esters, ketones, not glycols
104	DVB/acetonitrile	100–200	Highly polar	684	735	Nitrogen oxides, sulphur gases, CO_2, vinyl chloride
105	Polyaromatic		Mid-polar	635	545	Aromatic, therefore has different selectivity to 101, 102, C_1–C_{10} hydrocarbons, VOCs solvents, aldehydes
106	Cross-linked polystyrene	700–800	Non-polar	605	505	Selective for non-polar compounds, C_3–C_8 hydrocarbons, solvents and VOCs
107	Cross-linked acrylic ester	400–500	Polar	660	620	General mid-polar, solvents, esters, aldehydes, ketones
108	Cross-linked acrylic ester	100–200	Polar	710	645	Selective for polar compounds, HCHO aldehydes, carboxylic acids and esters, glycols
T	Polymerised tetrafluoro-ethylene (Teflon-6)	200–300	Non-polar	NA	NA	Water, amines aldehydes, acid gases, SO_2, HF, HCl, chlorosilanes, carboxylic acids

5.8 Gas chromatography instrumentation

The instrumentation for GC incorporates the features common to all forms of chromatography, namely a mobile phase, sample introduction system, a stationary phase and a detection system. GC has a gaseous mobile phase, the carrier gas, a column containing the stationary phase, with sample introduction by syringe injection. The detectors generate a minute signal current requiring an amplifier for output to produce the chromatogram. Column temperature, instrument variables and signal processing are controlled by a microprocessor. A GC system therefore consists of six sections:

1. carrier gas supply and controls;
2. sample introduction/injector system;
3. chromatographic column and oven;
4. detector;
5. amplifier and signal processing and control electronics; and
6. integrator and chromatogram printout.

The role of the column and requirements of a column oven have been discussed in section 5.4.2. A dual column gas chromatograph is shown in Figure 5.11.

5.8.1 *Carrier gas supply and control*

The carrier gas acts as the mobile phase and transports the sample component through the column to the detector, retardation occurring due to interaction with the stationary phase. The individual partition and adsorption properties of the components determine the rate at which they move through the system.

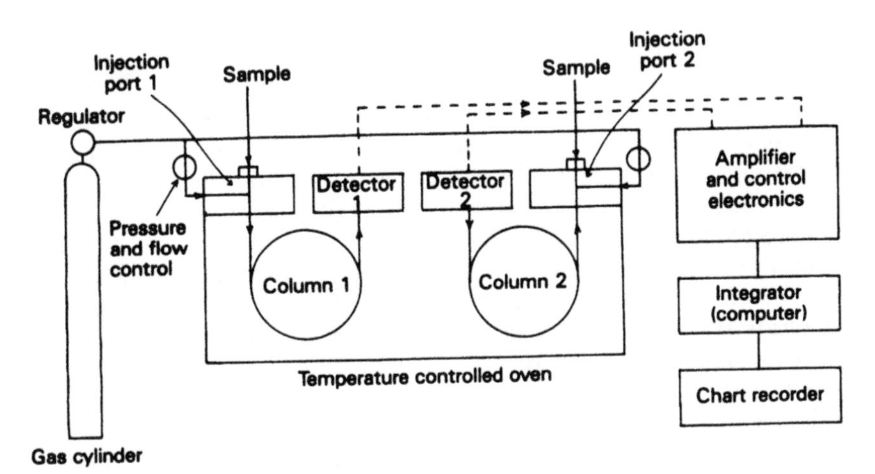

Figure 5.11 Schematic diagram of a dual column GC.

Table 5.11 Viscosity and thermal conductivities of GC gases at 100°C

Gas	Molecular weight	Viscosity, $\eta \times 10^{-6}$ (μp)	Thermal conductivity (cal s^{-1} cm^{-2} °C^{-1} cm^{-1}) $\times 10^{-6}$
CO_2	44	189	42
Ar	40	269	44
O_2	32	256	66
N_2	28	219	64
He	4	228	369
H	2	108	459
Methane	16	—	86
Ethanol	46	—	35

Selection of the best carrier gas is important, because it affects both the column separation processes and detector performance. The carrier gas has to be inert to the column materials and sample components. Gases with the smallest diffusion coefficients, hydrogen and helium, will give better separation efficiencies than higher molecular weight gases such as nitrogen, carbon dioxide and argon. The ratio of viscosity to diffusion coefficient should be a minimum for rapid analysis and therefore hydrogen and helium are the preferred carrier gases (Table 5.11).

Impurities in the carrier gas such as air, water vapour and trace gaseous hydrocarbons can cause sample reactions, column deterioration and affect detector performance. For example, water is often retained for longer than the sample components and therefore is eluted as a rather broad flat peak, affecting the baseline zero. In practice, a compromise is employed and N_2 is used with packed columns and helium or hydrogen with WCOT capillary columns. Ancillary gases are required for the detectors; air and hydrogen for flame ionisation detector (FID) and nitrogen–phosphorous detector (NPD), argon or helium for electron capture detectors (ECD) and hydrogen or helium with katherometer detectors. The gases are supplied from high pressure gas cylinders, being stored therein at pressures up to 3000 psi (20.7 MPa). Carrier gas purity should be better than 99.99% and 99.999% (referred to as five 9s or 99999 grade) is often used. The gas is further purified by passing through a series of traps to remove contaminants, molecular sieve 5 Å for water vapour, charcoal for hydrocarbons and silica gel to produce dry oxygen-free mobile phase. In many laboratories the use and storage of hydrogen cylinders is restricted due to safety regulations. An alternative supply of hydrogen can be obtained from the electrolysis of water using commercially available hydrogen generators. These can supply up to six FIDs at 0–60 psi (0–414 kPa) pressure and 0–250 ml min^{-1}. Nitrogen and air/oxygen generators are also available. Small variations in the carrier gas flow-rate will affect column performance and retention times. Therefore, to achieve reproducible separations, it is necessary to maintain a constant flow-rate. Pressure regulators and flow-rate controllers are built into the carrier gas

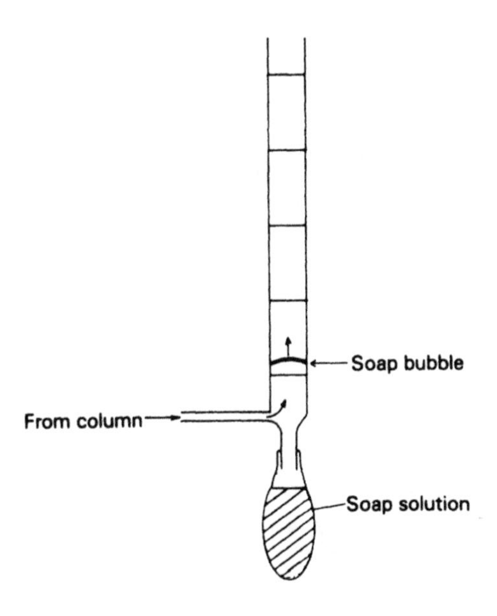

Figure 5.12 Soap bubble flow meter.

lines at the cylinder and in the instrument to obtain a pulse-free supply at pre-set pressures and flow-rates. Flow control is important to counteract the changes in flow-rate occurring during temperature programming since at constant pressure the flow-rate will change, as the viscosity of most gases increases with temperature. Flow controllers maintain the flow-rate accurately over all the full operating temperature range of 0–450°C. Controllers may be in the form of restrictors, consisting of fine bore capillary tubing in the gas supply line, needle valves or automatic regulation under the control of the instrument microprocessor and on-board computer which sets the method parameters. Column flow-rates are generally set using an electronic meter with direct readout or using a soap bubble flow meter and a stop watch (Figure 5.12). The flow meter is connected to the column outlet by the side arm and soap bubbles are introduced into the gas stream by squeezing the bulb at the bottom. The bubble moves up the calibrated tube and the time taken for a bubble to travel between the calibration marks (e.g. 10 ml) is measured. The flow per minute can thus be calculated.

5.8.2 Column switching

Column switching enables analyses to be carried out which are either impossible to achieve on a single column system or do not yield the required information using conventional methods. Switching of the carrier gas is achieved using multiport gas switching valves. These are used to switch the

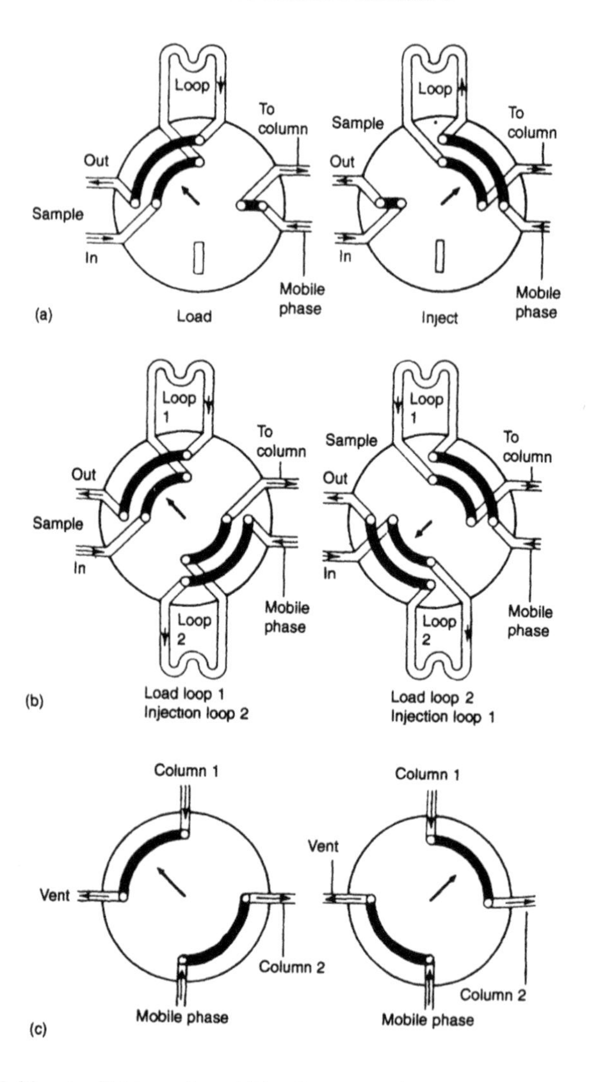

Figure 5.13 Multiport switching valves: (a) loading sample loop and injecting sample onto the column; (b) two loop sample injection system; (c) four port valve for column or solvent switching and heart cutting.

carrier gas through sample loops as described later, and also for venting, back-flushing, peak or heart cutting dual column operations and to achieve column switching (Figure 5.13).

Frequently the sample to be analysed contains major components, such as solvents which can lead to column overloading affecting the separation process, masking of minor peaks and overloading the detector. Venting or fore-flushing permits the rapidly eluting solvent to be removed from the system before the analytical column by using a pre-column in series with the main column. The major components are vented after the pre-column and when completely eluted the valve is turned to direct the components on to the main column. Similarly, when the main components have been eluted from the main column any slow moving and high boiling components which would either prolong the analysis time or might accumulate on the column, can be flushed backwards off the column. Column switching is also used for peak or heart cutting. Separation of components is mainly influenced by polarity of the stationary phase and column temperature and these are characteristic of various classes of compounds. It is therefore useful to use one column for the main separation of a mixture and to separate further poorly resolved peaks by directing the column eluant containing this fraction on to a second column of different polarity to effect a better separation. Heart cutting refers to the transfer of a group of components which may be poorly separated on the first column onto a second column where a good separation is obtained. Column switching is used to direct the components eluting from the first column before and after the components of interest to an alternative detector or to waste. Column switching is particularly useful in capillary GC, since the small column capacities can easily be overloaded with a solvent band.

5.8.3 Sample inlet systems

The introduction of a sample into a GC is the first stage in the chromatographic process and its efficiency is reflected in the overall efficiency of the separation procedure and the accuracy and precision of the qualitative and quantitative results. The sample may be introduced as a gas using a gas sampling valve or as a liquid introduced into the chromatograph via an injection port. Liquid sample injectors contain a self-sealing septum to retain the high pressure carrier gas. The resealing capability of the septum depends on temperature, flexibility of the silicone rubber, sharpness of the syringe needle and design of the injector. The septum holder usually incorporates a needle guide to reduce the mechanical damage (Figure 5.14). High injection block temperatures can gradually lead to loss of flexibility of the silicone septum due to degradation, particularly at temperatures over 300°C. This is accompanied by septum bleed of low molecular weight and depolymerised material which results in ghost peaks and baseline drift. Another phenomena with similar results is the memory effect, a consequence of the ability of silicon rubber to absorb compounds which are later desorbed. These problems can be reduced by the use of PTFE-coated septa. Typically, septa are formed from 0.09 to 0.125 in thick (2.3–3.2 mm) silicone rubber with a 0.01 in (0.26 mm) PTFE film on the inner surface. Septum purge, that is, flushing the surface of the septum with 2–3 ml min^{-1} of carrier gas, is used to reduce artefacts due to septum effects.

The sample should be introduced on to the beginning of the column with the minimum of dispersion in the mobile phase, so that it enters the column

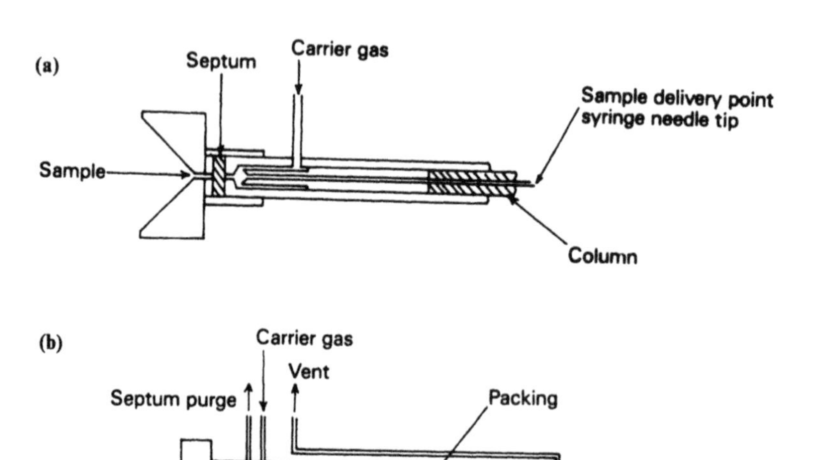

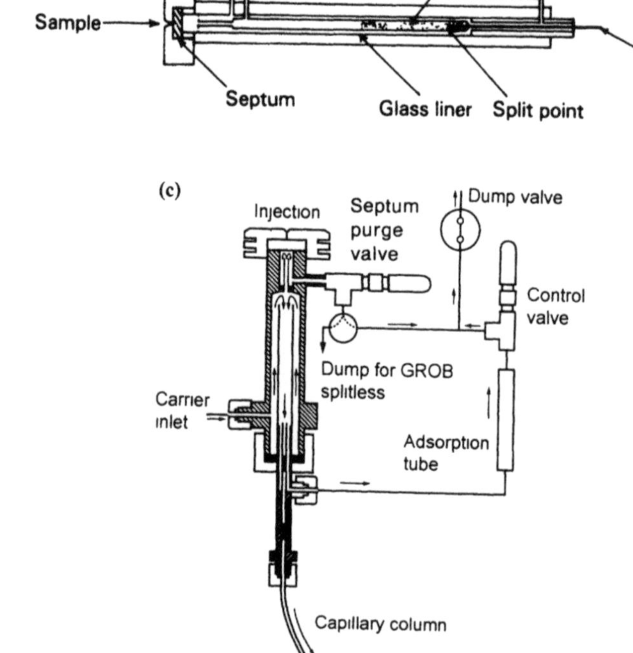

Figure 5.14 GC injectors for packed and capillary columns. (a) On-column injector; (b) split/splitless injector; (c) multipurpose split/splitless injector for WCOT columns. (Reproduced by permission of SGE Ltd.)

as a narrow band containing all the sample components. Some features of inlet systems are

- rapid clean switching or injection of the sample into the mobile phase with no tailing or dispersion of sample;

- correct inlet temperatures high enough to vaporise instantaneously all components in a sample without decomposition and condensation;
- minimum dead volumes to avoid diffusion of the sample in the mobile phase;
- design of the overall inlet system for good precision (better than ±1%);
- no contamination of samples or catalytic effects;
- no loss of retention of sample in the inlet system; and
- no septum bleed or leak.

5.8.4 Sample preparation

Gaseous samples can either be pumped into a gas sampling valve or introduced from a pressurised container. Liquid and solid samples can be introduced directly but are often contained in solution, which may be made up directly or obtained from an extraction process. The solvent needs to be carefully chosen to avoid problems in the chromatography. Some of the solvent requirements are

- no reaction with the sample or stationary phase;
- must completely dissolve the sample and must be fully miscible with it;
- there should be no co-elution of solvent and sample peaks, that is, should not have similar retention times;
- no non-volatile material should remain in the column; and
- overloading the column with large solvent volumes should be avoided.

5.8.5 Gaseous samples

Gaseous samples are generally introduced on to the column using a multiport switching valve (Figure 5.13). The sample is introduced into the sample loop while the valve is in the LOAD position. It is important that excess volume of sample is used to ensure that the sample loop is completely flushed through. The sample may be drawn through the sample loop, injected into it using a gas-tight syringe or introduced under pressure. Typical sample loop volumes are 0.1–10.0 ml and the loops are readily interchangeable. The overall inlet system is carefully designed to minimise carrier gas pressure fluctuations as the valve is turned to the INJECT position when the carrier gas is directed on to the column via the sample loop. An alternative method of gas sampling is to use a gas-tight syringe available for injection volumes ranging from 1 to 50 ml. The sample is injected into the carrier gas via the rubber septum of the liquid injection port. Great care is required to avoid mishaps due to back pressure of carrier gas in the syringe.

5.8.6 Liquid sample inlet systems

It is important to use the correct injector and injection technique for a particular application to obtain maximum performance from a GC system. Most present-day instruments incorporate interchangeable injectors so that a conventional glass-lined injection port can be replaced with an on-column capillary column injector or autosampler injection port. Flash vaporisation injectors are mounted in a heated block which is maintained at a temperature 20–50°C higher than the column itself. Most are of straight-through design which incorporates a replaceable glass liner. The liner can be used to collect non-volatile residues and also provides an inert non-catalytic vaporising surface for metal sensitive compounds. The glass liner can be replaced by an extension to the end of a glass column, which can also contain column packing material thereby minimising sample dispersion problems. Capillary columns require much smaller on column samples than packed columns, 10–1000 times less, due to the small column loading capacity. It is technically difficult to introduce such small samples on to the end of the column and therefore a modified interchangeable injector is used, which allows split or splitless techniques to be employed (Figure 5.14). The capillary column projects into the glass liner providing an all-glass system with minimum dead volume. The reduced sample size is achieved by splitting the injected sample allowing only a small fraction on to the column, typical split ratios of 1:100 to 1:1000 are used.

Samples are introduced into the injection port using a syringe with a capacity between 1 and 10 μl. A sample of 0.1–5 μl is normally used for packed columns with the components of interest typically 1% of the injected sample in solution. A 'needle barrel' syringe is used where the stainless-steel microbore needle forms the sample reservoir. A sample is pulled into the needle and expelled by a stainless-steel plunger wire which just protrudes from the end of the needle in the empty position. Both needle and barrel are accurately constructed and since almost all the sample is expelled from the syringe there is a minimum dead volume and sample retention, thus avoiding as far as possible cross-contamination of samples. The syringes are fully loaded, carefully checked to make sure that no air bubbles are present, and adjusted to the required volume. The volume used should be between 20 and 70% of the syringe's full capacity for best reproducibility ($<\pm 2\%$) of the volume injected. The needle is introduced into the injection port through a self-sealing elastomeric septum, the sample rapidly expelled and the syringe quickly withdrawn to avoid vaporisation of additional sample which may be retained in the needle. The syringe is flushed out with solvent between samples. A check on the purity of all solvents used in GC is essential and, as in any analysis, a blank should be analysed.

5.8.7 Capillary column gas chromatography injectors

Ideally a sample should be introduced onto the first few plates of the column. However, capillary columns have a very low gas volume, in the order of 10 μl per plate, which means that equally small liquid samples are needed otherwise the column would be saturated and column efficiency decreased. Specially designed glass-lined split/splitless injectors are used where the proportion of sample introduced on to the column can be varied from splitless mode where all the sample goes onto the column to split mode and where a predetermined ratio, e.g. 1:200 is set so that 1/200 of the gaseous volume enters the column, the rest being vented through the gas control valves (Figure 5.14) [45, 46]. The split ratio required will depend on the WCOT column used and the analyte concentrations in the sample.

Splitless injection mode is used when all the sample needs to go onto the column such as in trace analysis where the analytes may be in ppm ($\mu g\,ml^{-1}$) concentrations. The vent valve is closed so that all the rapidly vaporised sample is carried on to the column. The solvent is selected so that it is eluted before the analytes and so does not interfere with the separation process. An alternative approach which minimises band broadening effects due to the injection process is 'cold trapping'. The sample is vaporised as normally but the column is at such a temperature (150°C below the analyte boiling points) that the analytes are trapped on the first meter or so of the column and the solvent is not retained. The column is then heated by a temperature programmed cycle to effect separation of the analytes. The Grob splitless method uses a solvent such as octane, bp 126°C, which elutes after the analytes [46]. A column starting temperature is selected such that on injection of a sample the analytes separate as usual but the solvent is retained on the first part of the column until near the end of the temperature programme.

In split mode the proportion of the sample that goes onto the column is predetermined to achieve the best column efficiency and to avoid overloading the capacity of the stationary phase. The split ratio is controlled automatically by the GC microprocessor or manually by adjusting infinitely variable microneedle valves so that the major proportion of vaporised sample is vented or dumped. Capillary injectors also have a vented stream of carrier gas that purges the septum area to prevent accumulation of traces of solvent or sample and cross contamination. Silanised glass beads or glass wool may be placed in the liner to reduce dead volume and band broadening. This also produces a homogeneous sample at the splitting point just above the end of the column and serves as a clean-up trap for dirty samples. Split ratios of 1:100 to 1:1000 are typically used.

5.9 Sampling techniques

5.9.1 Derivatisation of samples

There are many compounds which cannot be readily analysed by GC, either because they are not sufficiently volatile or because they tail badly and are too strongly attracted to the stationary phases. Derivatisation before analysis to form volatile products enables many more compound classes to be analysed [47, 48]. The main reasons for derivatisation are

- to increase the volatility of the sample;
- to reduce thermal degradation of the sample by increasing thermal stability;
- to increase detector response by incorporating into the derivative functional groups which produce a higher detector signal such as CF_3 groups for electron capture detectors; and
- to improve separation and reduce tailing.

Derivatisation methods may be classified into three groups according to the reagents used and the reaction achieved, namely silylation, acylation, and esterification or alkylation. In many cases the derivatives are formed as soon as the sample and reagent are mixed in a solvent. Alkylation usually occurs quickly, but many reactions require heating for the reaction to occur. Thus, derivatisation time will vary and in order to find if the reaction has gone to completion a sample is analysed at selected time intervals until no further increase in the derivative peak(s) is detected. Most derivatives are thermally stable, although trimethylsilyl derivatives may be decomposed on the stainless steel of an injector port at >210°C. The hydrolytic stability varies considerably with acid derivatives being the most stable and those of amines the least stable. Solvents with active hydrogen atoms such as water; alcohols and enol-forming ketones, cannot be used and non-polar solvents such as hexane tend to produce slow reactions. Pyridine is a suitable solvent, it also acts as an acid scavenger and basic catalyst if required. Dimethylformamide (DMF), toluene and methanol are also used. All solvents are specially purified and stored under nitrogen.

5.9.2 *Silylation*

Silylation is the most widely used derivatisation technique. It involves the replacement of an acidic hydrogen on the analyte molecule with an alkylsilyl group, for example, $-SiMe_3$. The derivatives are generally less polar, more volatile and more thermally stable. Two examples are shown below:

1. using trimethylchlorosilane, TMS

$$ROH + Cl{-}SiMe_3 \rightarrow R{-}O{-}SiMe_3 + HCl$$

2. using hexamethyldisilylazane (HMDS)

$$2ROH + Me_3Si{-}N{=}N{-}SiMe_3 \rightarrow 2(R{-}O{-}SiMe_3) + N_2$$

Silylation reactions generally proceed very rapidly (within 5 min) with pyridine being the most frequently used solvent. GC columns used for analysis of silyl derivatives are conditioned by HMDS before use to block any acidic sites and avoid possible reactions with silyl derivatives. Many varied and improved silylation reagents have been developed [47, 48]. Examples are the substituted acetamides:

BSTFA (*N*,*O*-bis(trimethylsilyl)trifluoroacetamide)

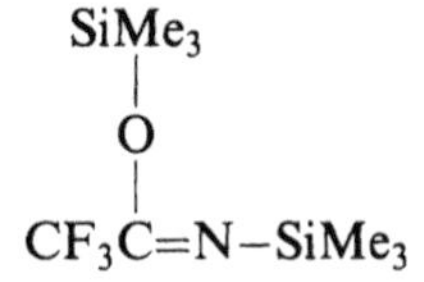

$$\begin{array}{c} SiMe_3 \\ | \\ O \\ | \\ CF_3C{=}N{-}SiMe_3 \end{array}$$

BSA (*N*,*O*-bis(trimethylsilyl)acetamide), the non-fluorinated analogue of BSTFA

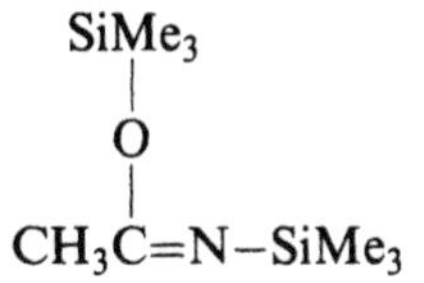

$$\begin{array}{c} SiMe_3 \\ | \\ O \\ | \\ CH_3C{=}N{-}SiMe_3 \end{array}$$

Both react rapidly and quantitatively under mild conditions using pyridine or DMF as solvent, forming esters, ethers, or *N*-TMS derivatives. The main advantage of BSTFA over BSA is that the by-products are more volatile and often elute with the solvent front.

$$R{-}OH + BSTFA \rightarrow R{-}O{-}SiMe_3 + CF_3CO{-}NH{-}SiMe_3 \text{ or } CF_3{-}CONH_2$$

5.9.3 *Acylation*

Acylation is used to form perfluoroacyl (trifluoroacetyl from trifluoroacetic anhydride) derivatives of alcohols, phenols or amines, for enhanced detector performance using an electron capture detector as well as increased volatility.

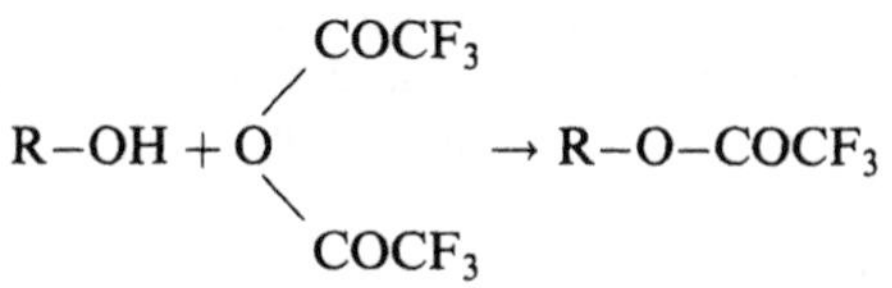

$$R{-}OH + O\begin{array}{l} \diagup COCF_3 \\ \diagdown COCF_3 \end{array} \rightarrow R{-}O{-}COCF_3$$

N-Fluoroacyl-imidazoles react readily with hydroxyl groups and secondary or tertiary amines to form acyl derivatives. No acids are produced which could hydrolyse the products. The imidazole produced as a by-product is relatively inert. A reaction using *N*-trifluoroacetylimidazole is shown below:

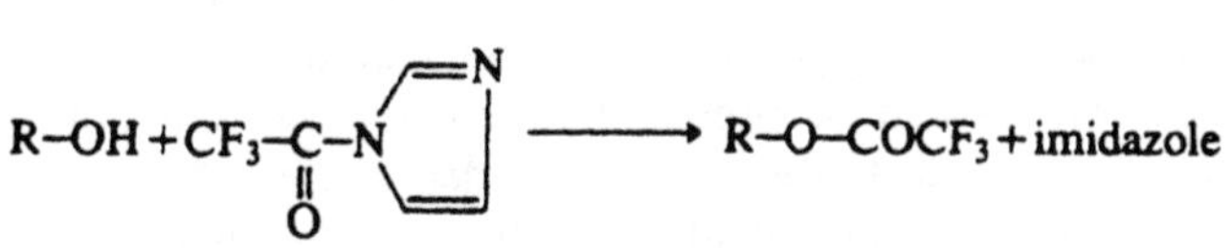

$$R{-}OH + CF_3{-}\underset{\underset{O}{\|}}{C}{-}N\langle\text{imidazole ring}\rangle \longrightarrow R{-}O{-}COCF_3 + \text{imidazole}$$

5.9.4 *Alkylation (esterification)*

Alkylation is the addition of the alkyl group to an active functional group, formation of methyl esters being the most useful derivatisation reaction. A number of reagents are available but boron trifluoride in methanol is most commonly used.

$$\mathrm{RCOOH + BF_3/MeOH \rightarrow RCOOMe_3}$$

Methylation can also be achieved using $MeOH/H_2SO_4$ or MeOH/NaOH, for example, formation of fatty acid methyl esters (FAMEs) [49]. Flash alkylation has been developed, in which the high temperature of the injection port is used to form the derivative on injection of the sample together with an appropriate reagent. Two major classes of alkylation reagents have been used, the quaternary alkylammonium hydroxides such as tetramethyl and tetrabutylammonium hydroxide as a 0.2 M solution in methanol which is used mainly for low-molecular-weight acids and the general purpose reagent trimethylanilinium hydroxide (TMAH). TMAH is also used where normal methylation might form derivatives similar to naturally occurring methyl derivatives in biological systems.

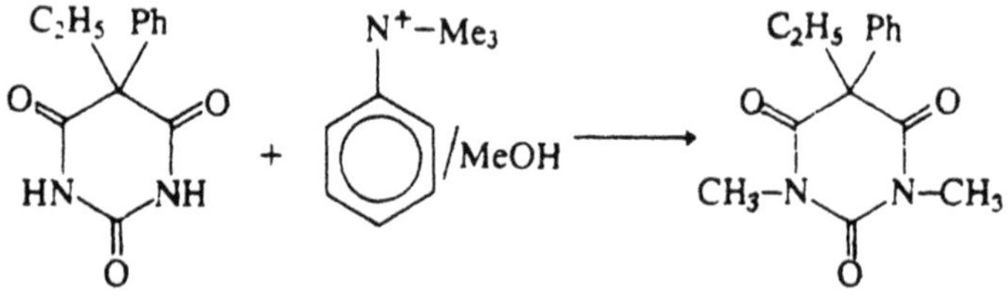

Phenobarbitone

Other reagents commonly used include pentafluorobenzyl bromide developed for the analysis of acids, amides and phenols using an electron capture detector for enhanced sensitivity. Organics in surface waters have been successfully analysed by this procedure [47, 48].

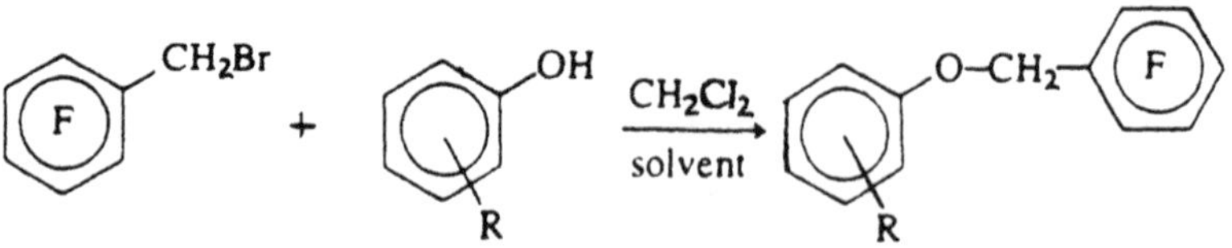

Dialkylacetals of DMF react instantaneously and quantitatively with acids, amines, amides, barbiturates either in solution or directly by on-column derivatisation using a mixture of the analyte and reagent. Methyl and butyl derivatives can be prepared. There is a considerable range of derivatisation reagents available and the reader is referred to the literature and the reference texts quoted [47–49]. In addition, many suppliers of chromatography materials include details of reagents and methods of preparation in their catalogues.

5.9.5 *Headspace sampling*

Static headspace sampling is the analysis of the vapours above a sample contained in a sealed vial [50–52]. Volatile compounds in almost any matrix can be analysed without the need for extractions, dissolving samples or dilutions. Since the headspace sampling relies totally on volatisation to extract the analytes from the sample, extraction, clean-up and preconcentration are not necessary. Samples can be sealed immediately they are taken in sample vials suitable for analysis in the automated instruments available. The sample vial is maintained at a constant temperature for sufficient time to allow the vapour and liquid or solid sample phase to equilibrate. Sampling is achieved by transferring a predetermined volume of the vapour phase on to the column. The technique avoids problems due to non-volatile materials being carried into the injector or on to the column (Figure 5.15). The sampling process involves three stages:

1. sample vial is introduced into a thermostatted oven where the vapour-sample equilibrium is established;
2. after a pre-programmed time the vial is pressurised to column head pressure by the sampling needle; and
3. the valves are switched so that the carrier gas flow to the column is interrupted and sample vapours from the vial pass directly onto the column, sample volume is determined by the time the valves are in the sampling position.

Automated dedicated headspace GC instruments such as the Perkin Elmer HS 40 can analyse up to 40 samples unattended and run continuously for 24 h. Temperature and pressure conditions are precisely controlled for each sample so that overall precision is better than $\pm 1\%$. Figure 5.15 includes examples of headspace analysis. A more complete discussion of applications of headspace analysis is given later in the chapter.

5.9.6 *Sample introduction by pyrolysis*

Many solid materials have been qualitatively analysed by fingerprinting the products from controlled pyrolysis of a sample of polymeric materials, plastics, rubbers and paint flakes [53, 54]. The sample is placed in a glass or platinum tube and inserted into a small heating element; alternatively a solution is used and the solvent evaporated leaving a film of sample on the element. The sample probe is inserted into a GC injection port and the temperature of the sample raised rapidly to a predetermined level, usually in the range 450–800°C, in about 1–5 s. The temperature is accurately

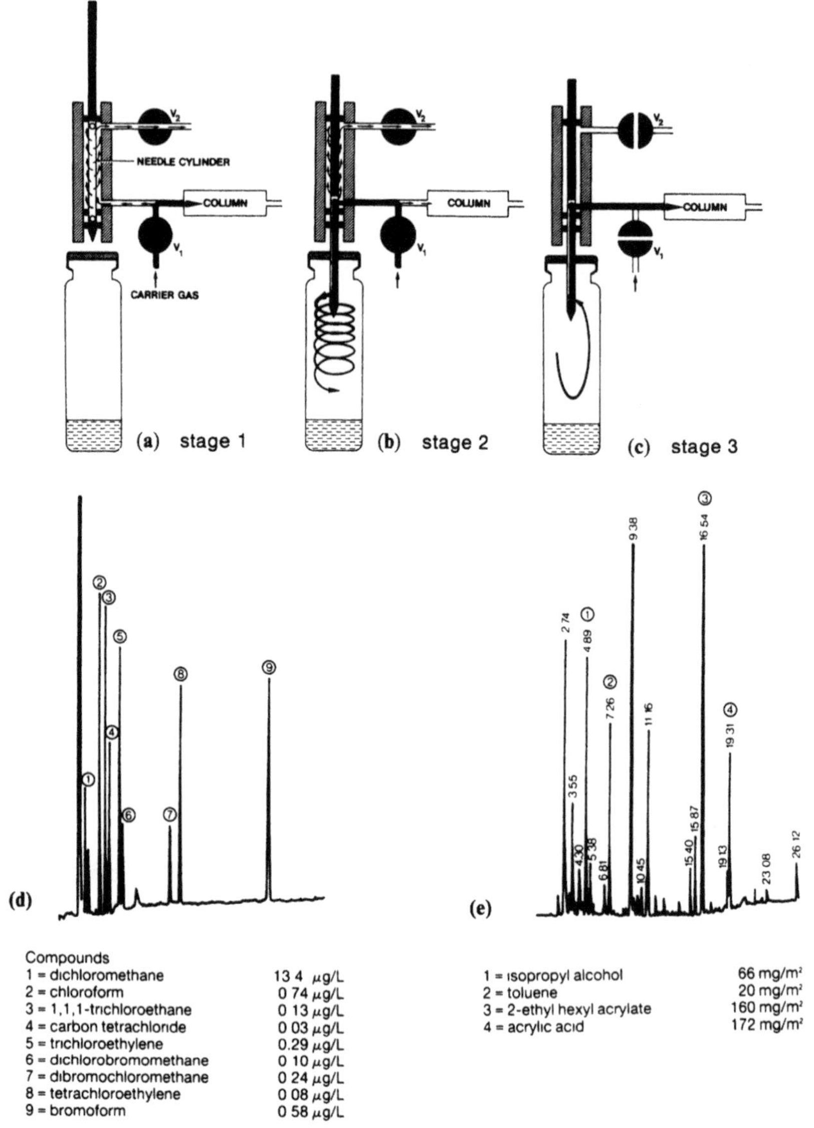

Figure 5.15 Headspace analysis showing the sampling sequence for the Perkin Elmer HS40 and sample chromatograms. (Reproduced by permission of Perkin Elmer Ltd.). The HS40 employs a unique sampling technique – a pneumatic pressure balanced system. The headspace sample is introduced onto the column without resorting to a gas syringe, thus avoiding fractionation due to pressure changes in the syringe. Since the needle is sealed, there is no loss of headspace gas during transfer. (a) In this figure the HS40 sampling needle is shown in the upper position as it is during sample thermostatting. Carrier gas passes solenoid valve V1 to the column and at the same time the needle cylinder is flushed. (b) The sampling needle has moved downward to a lower position. This occurs at the end of the sample thermostatting time. Carrier gas passes through solenoid valve V1 and branches to the column and the vial. The vial is now pressurised to the column head pressure. (c) For sampling, solenoid valve V1 interrupts the carrier gas supply

controlled so that reasonable reproducibility is possible. Thorough method development is required to determine optimum pyrolysis conditions so that the characteristic volatile molecular fragments, monomers and entrained solvents are produced and transferred on to the column. An alternative heating method uses curie point tubes or needles which attain the curie point temperature of the metal in 20–30 ms by induction heating effected by a high frequency electric field. The ferromagnetic material heats up to the curie point for that particular alloy. At this point the energy absorption decreases rapidly as the ferromagnetism disappears, thus limiting the temperature. Different temperatures are obtained by choosing suitable alloys such as iron–nickel (40:60 alloy), curie point 590°C, pure iron 770°C, pure nickel 358°C. The advantages of the curie point pyrolyser are two-fold: firstly, heating is extremely rapid, and secondly very small samples can be handled.

5.9.7 *Automated sample introduction*

Automatic sample introduction is regularly used in routine analyses. Automation of sample handling and sampling have enabled maximum use to be made of the data handling and control capabilities of microcomputer controlled instrumentation. The entire process of sample transfer, sample injection, collection of chromatographic data and calculation of results is handled by microcomputer controlled instruments. All the sample introduction methods discussed above (with the possible exception of pyrolysis) have been automated. Automatic gas sampling valves for on-line monitoring of gases and vapour use pressurised or pumped samples introduced via a pneumatically operated or solenoid multiport valves, the sample interval being predetermined and automatically controlled. Headspace analysis has been successfully automated and blood alcohol levels are measured using such a system. The samples are contained in a thermostatically controlled oven and automatically move to the injection needle position where they remain while the injection cycle shown in Figure 5.15 is carried out. Problems due to blood residues remaining in the injection port and on the column are thus avoided. Automatic liquid samplers also use sealed sample vials which

for the preset injection period. The vial is at the column head pressure and acts as a reservoir of carrier gas and delivers the sample to the column. At the end of the sampling period valve V1 is reopened and the normal carrier gas supply returns. The cycle is completed by the sampling needle returning to the position shown in (a). (d) Halocarbons in water by capillary/ECD-headspace analysis. 50 m × 0.32 mm fused silica capillary, permaphase PVMS/54, 2 μm film; 140 kPa helium carrier gas, ECD: ×1; splitless sampling; sample: 5 ml water, 60 min at 60°C. (e) Residual monomers and solvents in an adhesive tape. 50 m × 0.32 mm fused silica capillary, SP-1000 liquid phase, 1 μm film; 190/120 kPa high pressure sampling, helium carrier gas, splitless sampling; sample: 15 cm^2, 60 min at 140°C.

are transported to the sampling position. The sample may be taken up into a single, automatically activated syringe (the syringe is pumped several times to eliminate sample cross-contamination and air bubbles) and then transferred into the injection port. An alternative method is to flush the sample through a sample loop attached to a motorised multiport valve which functions in an analogous manner to a gas sampling valve. Liquid and volatile solid samples can also be automatically injected by measuring, weighing and loading the samples into small aluminium capsules which are placed into a magazine. The capsules are pushed by a piston through a gas interlock into the special heated injection port, where they are pierced by a needle and the sample flushed out by carrier gas.

5.9.8 Sampling by adsorption tubes

Monitoring of the atmosphere and workplace for trace pollutants (substances) and compounds of environmental interest is now a common routine requirement to meet environmental legislation such as the Control of Substances Hazardous to Health (COSHH) regulations in the UK. Substances to be analysed are present at concentration levels down ppbv, $\mu g\,m^{-3}$ so that pre-concentration techniques are necessary. The most effective method is to collect the substances using an adsorbent tube, although trapping in an appropriate solvent may also be used.

VOCs are collected on an adsorbent by drawing a known volume of air through a tube packed with an adsorbent (Figure 5.16). Small personal sample pumps are available which have an adjustable flow-rate of 10–200 ml min^{-1} and can run for a typical working day of 8 h. A standard volume of 1–10 litres is sampled and the tube capped for return to the laboratory. The stainless-steel or glass tubes are approximately 10 cm in length, 6.3 mm o.d. and 2–4 mm i.d. packed with a co-polymer, charcoal or carbon molecular sieve.

Desorption is by two procedures, thermal desorption being the preferred and more universal method.

1. *Solvent desorption.* Samples are collected on glass tubes containing 100–200 mg of 20–40$^{\#}$ granular charcoal adsorbents made from coconut shells. Solvents such as carbon disulphide, dichloromethane or hexane (2–5 ml) are used to desorb the analytes followed by capillary column analysis. Many applications using charcoal adsorbent tubes are reported in the literature [55, 56].
2. *Thermal desorption.* Substances collected on to the adsorbent are analysed by thermal desorption at a temperature up to 300°C using dry helium carrier gas to transport the substances into a small volume (approximately 500 μl) 'cold trap'. The cold trap is then flash heated, the desorbed sample being transferred to a GC for analysis using a WCOT column.

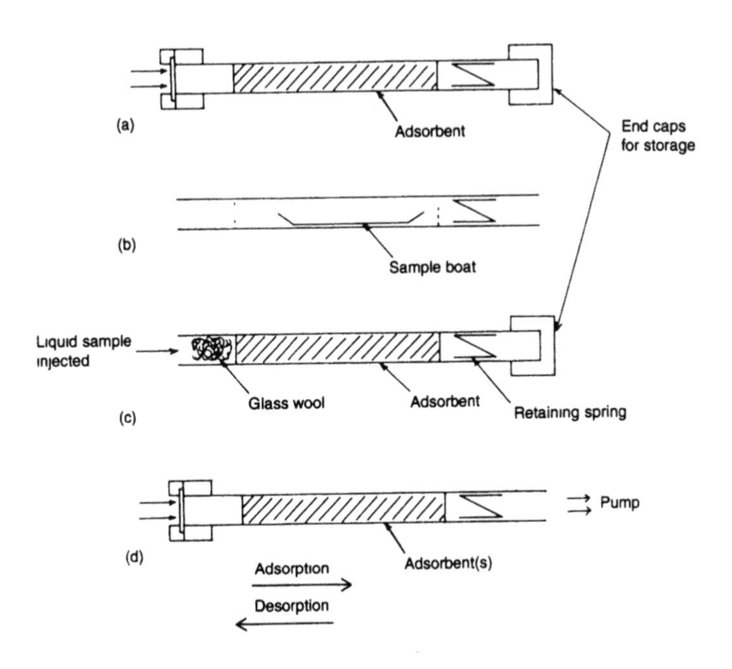

Figure 5.16 Thermal desorption. (a)–(d) Automatic thermal desorption sample tubes: (a) diffusive sampling, (b) solid sampling, (c) liquid sampling, (d) pumped active sampling.

Microporous co-polymers such as Tenax TA and GC (polymer of 2,6-diphenyl-*p*-phenylene oxide), Chromosorb 102 and 106 (polymer of divinylbenzene and styrene) and carbon molecular sieves are used as adsorbents for sampling VOC, more details can be found in sections 5.7.4 and 5.7.5 and Tables 5.8 and 5.10. The co-polymers have differing adsorption properties and breakthrough volumes and therefore are used to sample different molecular weight ranges of VOC. Tenax is a weaker adsorbent than Chromosorb 102 and 106 and the molecular sieves are the highest adsorbent properties. The co-polymers are hydrophobic and so are suitable for air sampling VOC without retention of water vapour. As a general guide Tenax is most suitable for substances with a boiling point range of 100–250°C, Chromosorb 106 for 30–150°C and carbon molecular sieves for −30 to 80°C.

Tenax TA and Tenax GC are a recently introduced porous polymeric material originally developed as a GCS column packing. It is hydrophobic and is excellent for adsorbing volatiles from the atmosphere at room

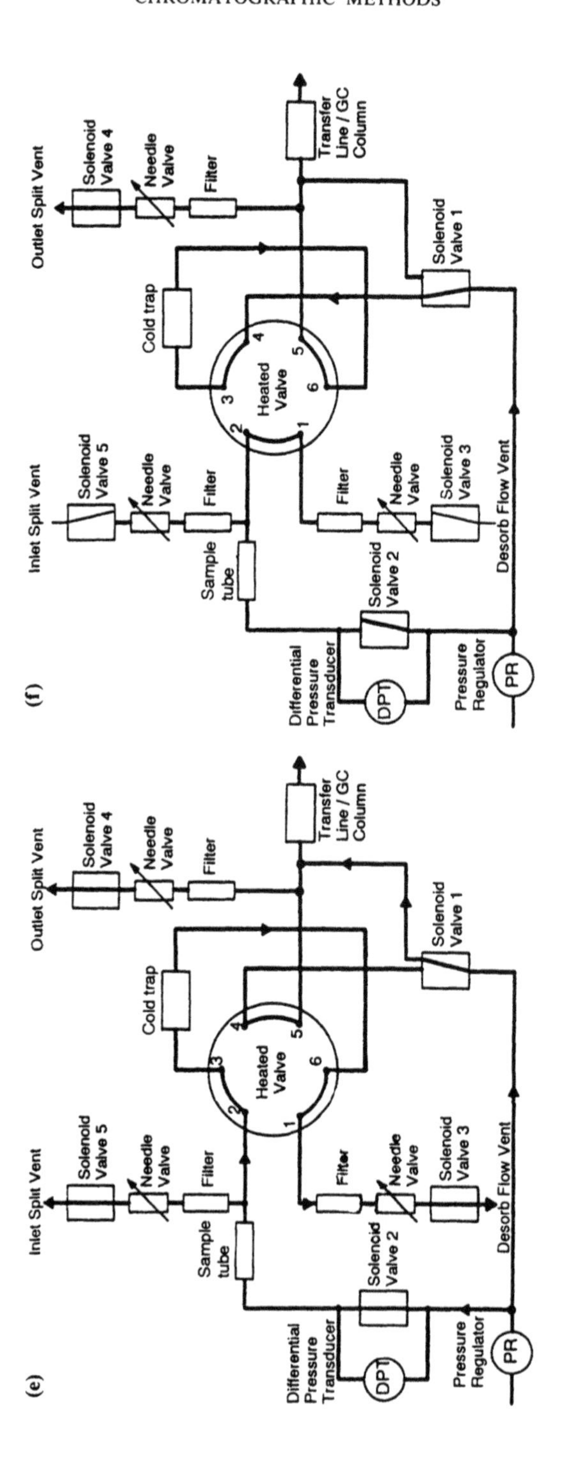

Figure 5.16 *cont.* (e), (f) Perkin Elmer ATD 400. Operation for a two stage thermal desorption of samples collected on an absorbent tube. (e) First stage desorption of the sample tube onto the cold trap; (f) second stage desorption of the cold trap onto the GC column.

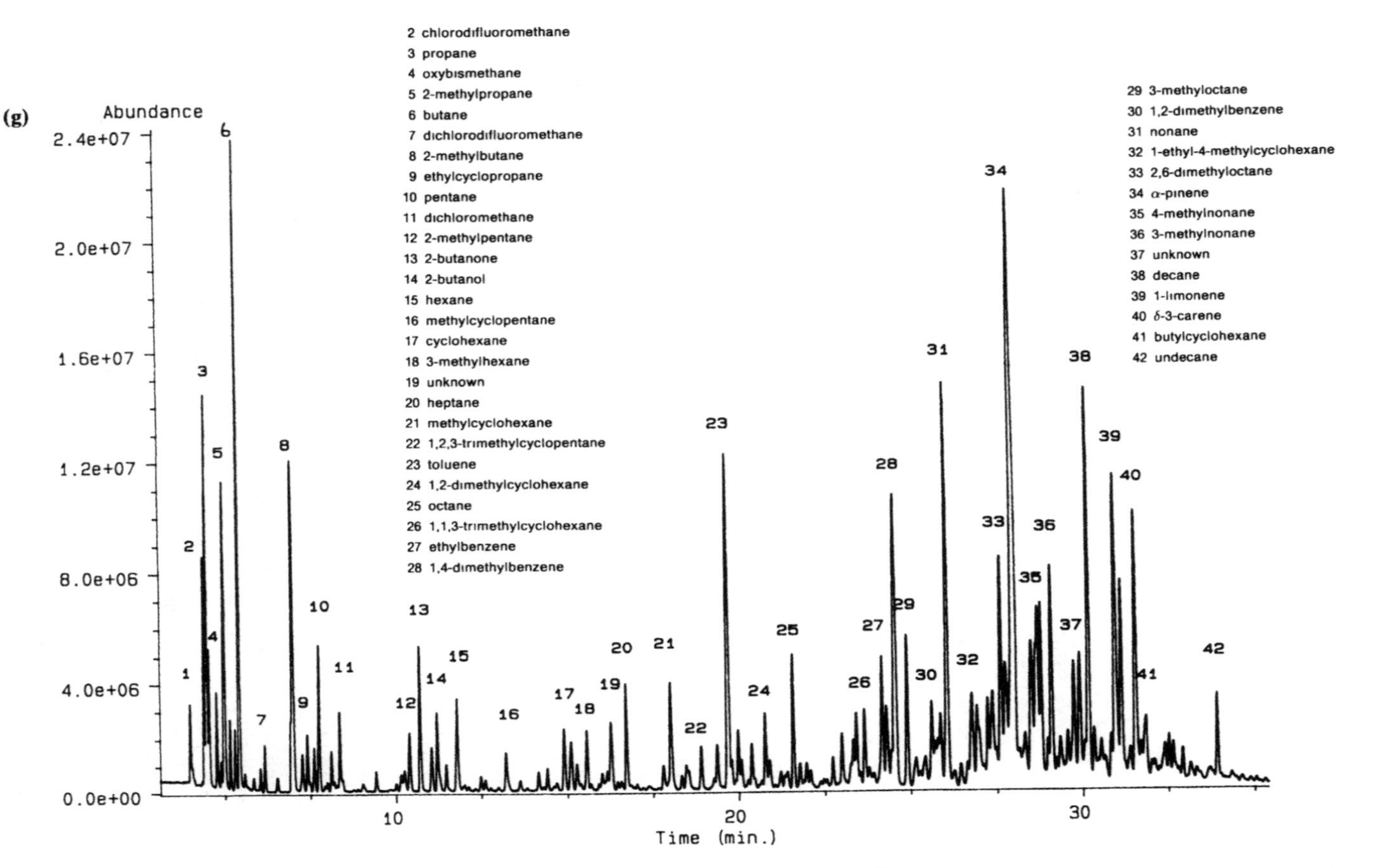

Figure 5.16 *cont.* (g) Chromatogram of a landfill gas sample collected on an ATD tube at 50 ml min^{-1} for 20 min, analysis on a 60 m × 0.32 mm 1.5 μm film Rtx column at 35–300°C programme at 5°C min^{-1}, He carrier gas 1 ml min^{-1}, HP-MSD detector. (Analysis by M. Allen, A. Braithwaite, C. Hills.)

temperature yet is stable enough at high temperatures, up to 300°C and more, for efficient thermal desorption. Many studies and applications are described in the literature, including studies of the storage and stability of samples on Tenax, analysis of solvent vapours and pollutants in the atmosphere [57–59]. The importance of environmental and personal monitoring is demonstrated by the specialised equipment developed and marketed for this purpose. An example is the Perkin Elmer automated thermal desorption instrument the ATD400 which is capable of analysing 50 sample tubes unattended using pre-programmed desorption methods. Sampling may be carried out by pumped or passive sampling of vapours or volatiles from gaseous, liquid or solid samples. Direct desorption of liquid samples injected on to the adsorbent and direct analysis of volatiles from solid samples placed in a sample tube can also be carried out. The ATD400 automatically moves each tube in turn into the desorption oven where thermal desorption takes place for the time and temperature specified (100–280°C for 5–20 min) in either a single or two stage process. In the former method the desorbed vapours are swept directly on to the GC WCOT column via a heated transfer line whereas in the two-stage process the desorbed material is collected in a small trap packed with Tenax TA and cooled to −30°C by a Peltier device. The analytes are therefore further concentrated. Rapid ohmic heating of the trap to 280°C results in a narrow band of vaporised material being introduced on to the column. Figure 5.16 includes an example of ATD-GC-MS analysis. Chromatograms, particularly of environmental samples, can contain over a hundred peaks, however, GC detectors do not identify the components so a mass spectrometer is used to aid identification of components. GC–MS is discussed in Chapter 8.

5.10 Detectors for gas chromatography

The purpose of a detector is to monitor the carrier gas as it emerges from the column and generate a signal in response to variations in its composition due to eluted components. Chromatography is a separation technique, detectors such as the flame ionisation, electron capture and thermal conductivity detectors do not identify the components. Additional techniques such as mass spectrometry are used to assist in identification of the eluted components. The chromatogram is a plot of detector signal against time to form a concentration profile of the molecules present in the carrier gas as it passes into the detector. Most are proportional or differential detectors, that is, a zero base line signal is produced by the carrier gas alone. When a component is eluted the signal produced is proportional to the concentration or mass of that component present (Figure 5.17). Commercial instruments use mass flow proportional detectors which respond to the

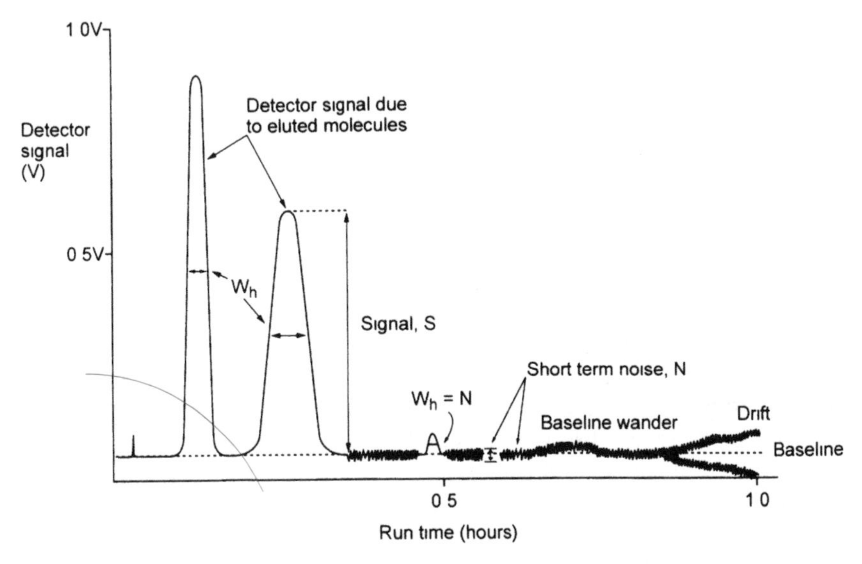

Figure 5.17 Signal to noise ratio and determinable signal levels.

concentration of the components eluted. The following detectors will be discussed:

- flame ionisation detector, FID;
- electron capture detector, ECD;
- thermal conductivity detector, TCD (katherometer);
- nitrogen–phosphorous detector, NPD;
- flame photometric detector, FPD;
- photoionisation detector, PID; and
- miscellaneous detectors.

5.10.1 Detector performance

The requirements of a detector for GC are exacting, and include adequate sensitivity to monitor the eluted sample components present in very low concentrations in the major eluting component, the carrier gas, and a rapid response to the changing concentration of the minor components. There are four criteria to consider:

1. sensitivity, defined as the smallest concentration of analyte in the detector that produces a signal of at least twice the noise level of the background signal;
2. response, the magnitude of the signal produced for a unit concentration of analyte in the carrier gas; response can also be defined in terms of the amount of an analyte measured as peak area;

3. response time, that defines the detector's ability to produce a signal which accurately follows the change in analyte concentration, that is, how rapidly the detector can respond to a change in analyte concentration; and
4. linear range, that defines the range of concentration of analyte over which the detector responds with the same sensitivity.

5.10.2 Signal noise and detection limits

Noise present in the detector signal may have two components, long-term noise and short-term noise. The former causes a slow baseline wander measured over a 1 h period and may be attributed to fluctuations in temperature, column stationary phase bleed, flow rate variation, or pneumatic leaks. Short-term noise is observed as small, sharp spikes of shorter duration than component peaks and usually arises in the detector. Most integrators smooth the signal so that noise is not apparent unless a direct plot mode is selected. It is important to establish the mean noise level, the baseline, in order to determine the limit of detection. The time period of a peak is most conveniently described by the peak width at half height and the noise, N, is measured as the variation between maxima and minima of the noise peaks over the time period. The contribution of noise to the total component signal should be less than 1% (Figure 5.17).

The minimum detectable amount (MDA), also referred to as the limit of detection, is the analyte concentration level or amount of analyte that produces a signal level, S, that is at least twice that of the noise range, N, that is, the signal to noise ratio, S/N, is at least 2. In practice this corresponds to a minimum peak height, h_{MIN}, being at least twice the noise signal.

$$S/N \geq 2 \quad \text{or} \quad S = \geq 2N \quad \text{and} \quad h_{MIN} \geq 2N$$

MDA is usually expressed in $\mu g\,ml^{-1}$ or $g\,ml^{-1}$.

Linearity or linear dynamic range (LDR) is the sample concentration range over which the detector response is linear, that is from the MDA to the upper concentration level which produces a deviation from linearity of about 5%. MDA and LDR for a specific analyte may be determined by plotting peak area against concentration in the carrier gas or in the sample analysed (Figure 5.18). LDR corresponds to the concentration where the deviation from linearity is greater than 5% and MDA is the concentration which corresponds to $2 \times N$ where N is the intercept which corresponds to the noise signal, obtained by extrapolating the line. The method detection limit (MDL), also referred to as the limit of quantitation, is generally defined as 10 times the baseline noise, i.e. $10N$. The MDA and LDR of the common detectors is given in Table 5.12.

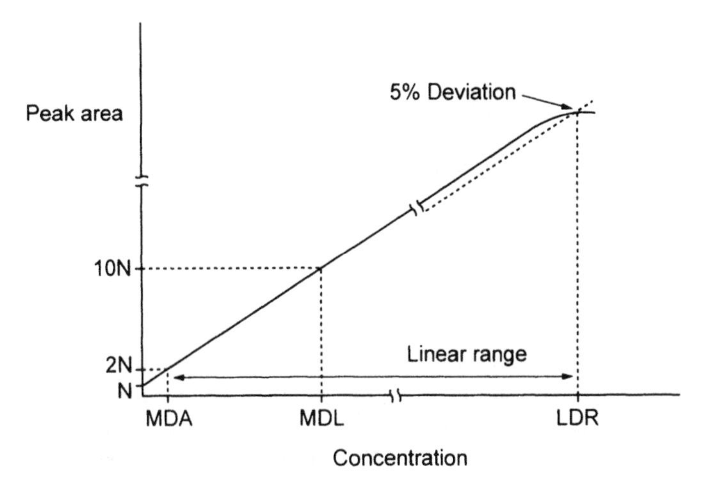

Figure 5.18 Method detection limit (MDL), minimum detectable amount (MDA) and linear dynamic range (LDR).

5.10.3 Detector response factors

Detectors produce an output signal in response to the concentration of analytes passing through, however, different analytes can produce varying signals for the same concentration level. Thus the signal observed will need correcting to determine accurately the amounts of analytes in a mixture. The total amount of a component in a sample is given by the total or integrated signal, measured as peak area, produced as the band of analyte passes through the detector. In order to obtain the correct ratios the different response signals for a given detector to each of the compounds being analysed have to be obtained for each GC instrument, detector and detector settings. Typical FID detector response factors (D_{RF}) for the *n*-alcohols with respect to ethanol are given in Table 5.13.

The corrected peak area is given by the ratio of area from the chromatogram adjusted for detector response (see experiment 16, Chapter 9).

$$A_{CORR} = \frac{A_{CHROM}}{D_{RF}} \quad \text{therefore, for butanol,} \quad A_{CORR} = \frac{A_{CHROM}}{1.63}$$

Table 5.12 MDA and LDR of the common detectors

Detector	MDA ($g\,ml^{-1}$)	LDR
FID	10^{-12}	10^{7}
ECD	10^{-14}	10^{4}
TCD	10^{-7}	10^{4}
NPD	10^{-14}	10^{5}
FPD	10^{-11}	10^{4}
PID	10^{-12}	10^{5}

Table 5.13 Detector response factors (D_{RF}) for *n*-alcohols

n-Alcohol	D_{RF}
Ethanol	1.00
Propanol	1.41
Butanol	1.63
Pentanol	1.79
Hexanol	1.95

Detector response factors need to be accurately determined for good reproducible quantitative analyses [60]. They may be conveniently obtained by the constant volume method, that is, approximately 10 repeatable volumes of a sample containing an equal amount of the analytes are injected and the mean determined. D_{RF} values are calculated by normalising each peak area to that of the peak to be used as the reference.

$$D_{RF_x} = \frac{A_x}{A_{REF}}$$

The main disadvantage is the difficulty of achieving good reproducibility of sample injection volumes.

5.10.3.1 Internal standardisation. A second method of standardisation uses an internal standard as a marker. A known amount of the standard is added to a standard mixture of all the components to be analysed and the D_{RF} values relative to the internal standard calculated. The accurate amount of internal standard is then added to each sample and the sample analysed. The amount of each component is then calculated using the equations below (a more detailed account is given in section 2.7):

$$D_{RF_x} = \frac{A_x}{A_{IS}} \times \frac{C_{IS}}{C_x} \quad \text{and} \quad C_{UNKNOWN} = \frac{A_x}{A_{IS}} \times \frac{C_{IS}}{D_{RF}}$$

where A is the peak area, C is the concentration, subscript x refers to a component, subscript IS refers to the internal standard and $C_{UNKNOWN}$ refers to the amount of the analyte in the sample.

5.10.4 Flame ionisation detector

The FID is regarded as the universal GC detector and hence is used for routine and general purpose analyses. The outstanding features are

- high sensitivity to virtually all organic compounds;
- little or no response to water, carbon dioxide and carrier gas impurities, and hence gives a zero signal when no analyte is present;

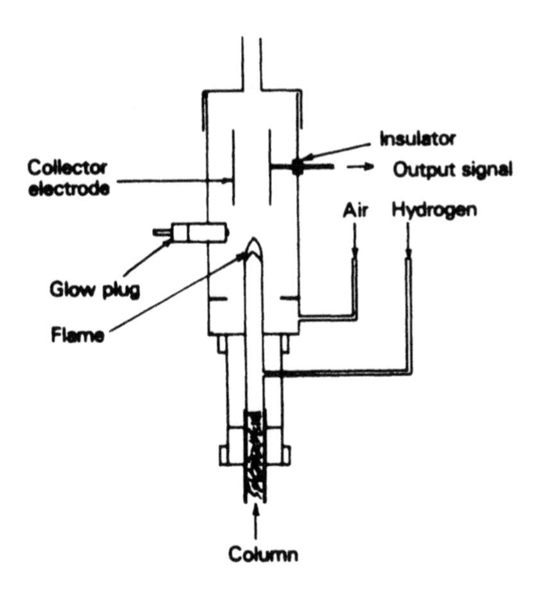

Figure 5.19 Flame ionisation detector.

- a stable baseline; it is not significantly affected by fluctuations in temperature or carrier gas flow-rate and pressure; and
- good linearity, LDR, over a wide sample concentration range.

The detector consists of a minute hydrogen-air flame burning at a small metal jet, with an electrode located above the flame to collect the ions formed from the analyte molecules (Figure 5.19). The hydrogen is introduced into the column eluant and thoroughly mixes with the carrier gas before emerging at the jet into the air stream to form the flame. The flame processes are complex and direct ionisation only forms a small contribution to the overall ionisation process [61, 62]. The organic molecules undergo a series of reactions including thermal fragmentation, chemi-ionisation, ion molecule and free radical reactions to produce charged-species. As organic compounds enter the flame the thermal energy available causes cracking and stripping of protons and terminal groups. A pure hydrogen–air flame contains $H^{\cdot}$, $O^{\cdot}$, $OH^{\cdot}$ radicals and excited species, but no ions. However, when organic molecules are present in the flame, ionisation occurs, the amount being proportional to the number of carbon atoms present and hence the number of molecules. The main flame processes initially involve formation of $CH^{\cdot}$ from the organic molecules which immediately reacts with oxygen radicals:

$$-CH^{\cdot} + O^{\cdot} \rightarrow -CHO^{+} + e^{-}$$

The chemical nature of the organic molecule influences the flame response

Table 5.14 ECNC of organic molecules

Atom	Type of molecule	ECNC
C	Aliphatic	1.0
C	Aromatic	1.0
C	Alkene	0.95
C	Carbonyl	0.0
O	Ether	−1.0
O	Primary alcohol	−0.6
O	Secondary alcohol	−0.75
N	Primary amines	−0.6
N	Secondary amines	−0.75
Cl	Aliphatic	−0.12

and is corrected using the effective carbon number contribution (ECNC) (Table 5.14).

The ions travel to the collector electrode which is maintained at a negative potential (about −150 V) with respect to the flame jet. Thus, the electrical current observed (about 10^{-14} A) is due to the concentration of the charged species present in the flame and the chemical structure of the molecules. The sensitivity is generally in the region of 0.015 coulombs g^{-1} (carbon) with a linear dynamic range of 10^7, and overall response varies slightly for a given type of compound and carbon number. The signal is amplified and conditioned by an electrometer amplifier with a high input impedance to produce an output signal typically over a 0–10 mV or 0–1 V range, enabling a chart recorder, integrator or computer interface to be easily used to produce the chromatogram and data. Materials not detected by the FID include H_2, O_2, N_2, $SiCl_4$, SiF_4, H_2S, SO_2, COS, CS_2, NH_3, NO, NO_2, N_2O, CO, CO_2, H_2O, Ar, Kr, Ne, Xe; HCHO and HCOOH have a very small response.

5.10.5 Electron capture detector

The ECD is most frequently used for analysis of trace environmental pollutants such as chlorinated pesticides and herbicides due to its high sensitivity for compounds containing electronegative elements. The ECD responds to changes in electrical conductivity of gases in an ionisation chamber due to the presence of electron acceptor molecules [61–66]. Construction of the detector is usually one of two forms, the plane parallel or concentric cylindrical cell design (Figure 5.20). The cylindrical design is preferred because of the ease of construction, greater sensitivity and smaller dead volume, typically 0.2–0.5 μl.

The cell consists of two electrodes, an outer source electrode and a central collector electrode. The cylindrical source electrode consists of a β-radiation emitter. The high energy electrons produced by the radioactive decay processes interact with the carrier gas (nitrogen or argon) to produce large quantities of thermal electrons which are collected by the positively polarised

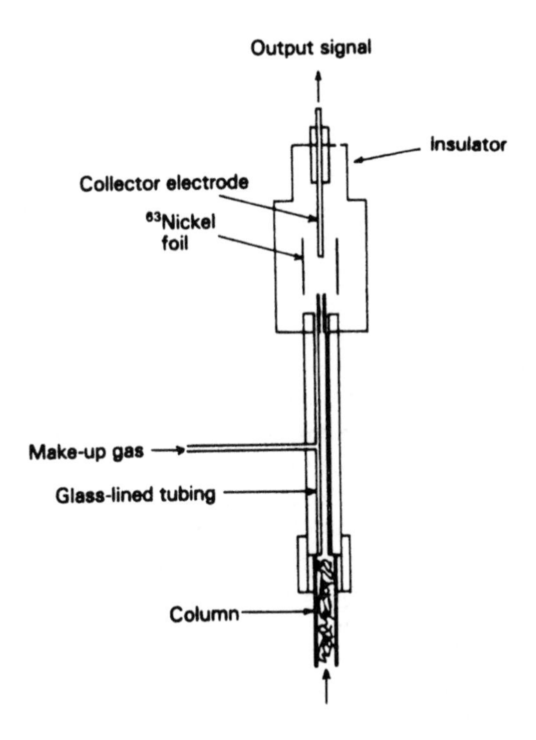

Figure 5.20 Electron capture detector.

central electrode, producing the standing current or baseline signal.

$$N_2 + \beta^- \rightleftharpoons N_2^+ + e^-$$

$$Ar + \beta^- \rightleftharpoons Ar^* + Ar^+ + e^-$$

$$AB + e^- \rightarrow AB^+ \quad \text{and} \quad A^\bullet + B^-$$

Molecules, AB, with high electron affinities capture the highly mobile thermal electrons as they pass through the detector to form slow secondary ions and radicals thus reducing the detector current and producing a decreased detector output signal. The signal is subsequently electronically processed to form the chromatogram. Any contaminants which can quench the electron flux such as oxygen (air), water vapour and stationary phase bleed absorb electrons producing a background signal and decreased sensitivity. A further problem is a build up of electrical charge in the cell (space charge) and on the electrodes when a continuous potential is applied to the collector. This can be avoided by operating the detector in pulse mode. A short duration positive potential is applied to the collector electrode (+50 V for 1.0 μs at 100 μs intervals) to collect the mobile electrons but not the slower negative ions. This allows maximum time for capture of the randomly moving thermal electrons by the sample molecules between the pulses, increasing sensitivity and enabling reproducible responses to be obtained. The detector response follows a Beer's Law type linear relationship:

$$I = I_0 \, e^{(-ack)}$$

where I is the current when electron capturing material is present; I_0 is the current when no electron capturing material is present; a is the electron capture cross-section of material; c is the concentration; k is the proportionality constant related to the geometry of the cell and the operating conditions.

Although a number of radiation sources may be used, only tritium (^3H) and nickel (^{63}Ni) are commonly used. Tritium adsorbed on a titanium foil is a weak β-emitter (18 keV) and has a high flux of radiation, but short effective range of approximately 2.0 mm for efficient ionisation of the carrier gas. ^{63}Ni on the other hand is a more durable source, the β-radiation (67 keV) can penetrate at least 5.0 mm but the flux is lower. A ^{63}Ni source is therefore less efficient but this is offset by the longer active lifetime and higher maximum operating temperature up to 450°C as opposed to that of ^3H, 250°C. For safety reasons instruments have sealed ECDs which can only be serviced by the manufacturers. Nitrogen or argon is frequently used as the carrier gas, however, metastable ions and cross-section effects can produce anomalous results. The addition of 5–10% (v/v) of an alkane, usually methane, overcomes these problems. Methane molecules undergo inelastic collisions with the slow β-electrons and metastable ions, which lose energy until thermal equilibrium is established. Thus, the thermal electrons rather than the higher energy β-electrons are captured by the electrophilic molecules. ECDs have high sensitivity to electrophilic molecules, in the order of $10^{-12}\,g\,s^{-1}$, but have a rather small linear range of 10^4. Thus, careful preparation of samples is required to ensure that the sample component concentrations fall within the operating range. At higher concentrations the detector may operate in an ionisation mode. Relative response factors have been determined for a range of compound types [67]. These show enhanced responses, relative to n-butyl chloride, of 10^2–10^4 for polyhalogenated, polynuclear aromatic and nitro compounds, anhydrides, conjugated carbonyls, and sulphur compounds as shown in Table 5.15.

An approximate order of response for the halogens is $F < Cl < Br < I$. Applications include trace analysis of environmental samples for chlorinated solvents, plasticisers, chlorofluorocarbon gases, pesticides and herbicides (DDT, γBHC/lindane, aldrin), SF_6 tracer gas in flue gases and mine atmospheres, organometallics (lead tetra-alkyls), polynuclear aromatic carcinogens, NO_x and SO_2 in chimney-stack gases.

5.10.6 *Thermal conductivity detector*

The TCD or katherometer is a bulk property detector, that is, it responds to some overall physical property, the thermal conductivity, of the carrier gas.

Table 5.15 ECD responses

Type of compound	ECD response relative to *n*-butyl chloride
Chloroalkane	1
Dichloroalkane	10^2
Bromoalkane	10^3
Dibromoalkane	10^5
Chloroform	10^5
Carbon tetrachloride	10^6
Benzene	10^{-1}
Bromobenzene	10^3
Polynuclear aromatics	$10–10^3$
Aliphatic alcohols, esters, ethers	1
Butan-2,3-dione	10^5

The response to the carrier gas forms the baseline signal and any change in composition of the eluant will produce a change in the overall physical property being monitored by the detector, and hence a change in the detector signal. The detector is also sensitive to variations in parameters that affect the specific property being measured, for example, temperature, flow-rate, pressure and carrier gas purity.

Devices to measure thermal conductivity have been demonstrated since the 1920s, but it was in the 1950s that TCDs were developed as universal detectors for GC. A study of the properties of carrier gases established that hydrogen and helium give the highest detector signal due to their high thermal conductivity (459 and 369 $(\text{Cal s}^{-1}\,\text{cm}^{-2}\,{}^\circ\text{C}^{-1}\,\text{cm}^{-1}) \times 10^6$, respectively, Table 5.11, p. 210), and therefore maximum signal differences when organic compounds (thermal conductivity 30–90) and inorganic gases are eluted. An excellent account of the theory and practice of TCDs is given by Littlewood [68].

Thermal conductivity is the flow of heat from a body at a higher temperature to an accepting material. The flow of heat is dependent on the cross-sectional area, temperature, temperature coefficient and geometry of the donor body, and the nature and cross-sectional area of the accepting material. The TCD signal is therefore highly dependent on cell design and is also directly proportional to the concentration, C, and velocity, V_{MOL}, of the analyte molecules in the detector, the temperature difference, ΔT, between the hot filament and the cell, and is inversely proportional to the specific heat, C_{V}, of each analyte:

$$S_{\text{TCD}} \propto \frac{C V_{\text{MOL}} \Delta T}{C_{\text{V}}}$$

A cylindrical cell geometry is used with a platinum, tungsten or tungsten–rhenium wire filament, 0.02 mm in diameter coiled along the axis of the cell (Figure 5.21). The filaments have a high temperature coefficient of resistance and are heated by a low voltage, constant mA current, 100 mA

(a)

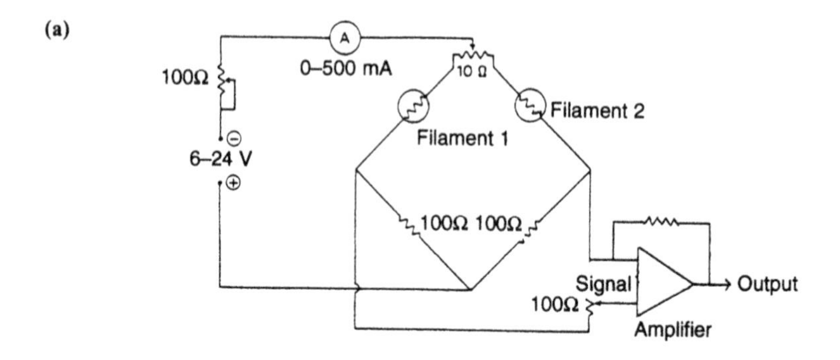

(b)

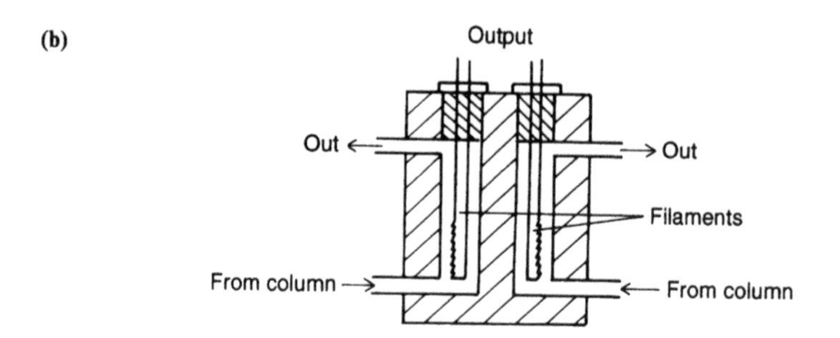

(c)

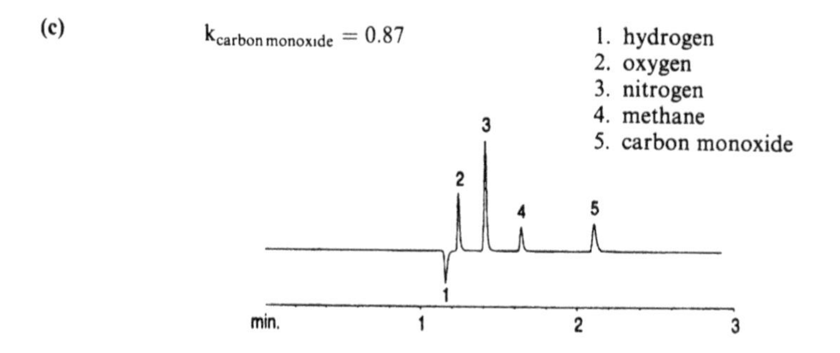

30m. 0.32mm ID Rt-Msieve™ 13X PLOT column
15μl split injection of permanent gases (hydrogen spiked)

Oven temp.: 40°C isothermal
Inj./det. temp.: 200°C/200°C **Detector:** microcell TCD
Carrier gas: helium
Linear velocity: 44cm/sec. set @ 40°C (2cc/min.)
Detector sensitivity: 50mV full scale **Split ratio:** 15 : 1

Figure 5.21 Thermal conductivity detector: (a) wheatstone bridge network; (b) twin filament cell; (c) analysis of permanent bases and hydrogen using a PLOT column and helium carrier gas. Note the negative peak for hydrogen. (Reproduced by permission of The RESTEK Corp.)

at 6–8 V. The temperature of the filament and hence its resistance is determined by the supply current and thermal conductivity of the ambient gas. Changes in composition of the carrier gas as a component elutes from the column causes a change in thermal conductivity and therefore a change in temperature of the filament. This results in a change in resistance and since the filament forms one arm of a Wheatstone bridge network an out-of-balance current is produced which is amplified to form the recorder signal (Figure 5.21). The filament is housed in a detector block which acts as a thermal sink to minimise temperature fluctuations. In order to improve the stability and reduce baseline drift due to flow and pressure fluctuations an identical filament is also housed in the detector block with pure carrier gas only passing through. This also forms part of the Wheatstone bridge. Four cell detector blocks are also used to enhance the temperature stability of the detector each filament forming an arm of the Wheatstone bridge network. Often one of the reference cells is used as an additional detector to monitor a second analytical column in a two column GC, both columns are fitted with injector blocks and either column may be used for analysis. In practice the filaments are balanced using potentiometers or automatic electronic circuitry.

The TCD responds to all types of organic and inorganic compounds including those not detected by the FID. Figure 5.21 includes an example of GC–TCD analysis of permanent gases using a molecular sieve PLOT capillary column and helium carrier gas. Hydrogen has a higher thermal conductivity (see Table 5.11) than helium, hence the negative peak. It does not destroy the eluted components and therefore is suitable for use with fraction collectors for trapping of the separated components for preparative work. Cell dead volume varies from about 1.0 ml for packed columns to 20 µl for capillary columns. The TCD is less sensitive than the FID by a factor of 10^5 (see section 5.10.3). Relative response factors are similar for a homologous series with most compounds having a response similar to ethanol (±20%).

Halocarbons have lower response factors, and low molecular weight compounds such as methane have higher response factors (Table 5.11). Thermistor metal oxide beads which have a negative temperature coefficient of resistance are used as an alternative to filaments. They are efficient over only a small temperature range (−20 to +50°C), but are robust and therefore are used in portable air pollution monitoring systems.

5.10.7 Nitrogen–phosphorus detectors

Since its introduction in 1964 the thermionic alkali bead detector has been successfully used for the specific detection of phosphorus compounds, particularly pesticides, and has been developed for the analysis of nitrogen-containing compounds such as drugs [62, 69–72]. The NPD was originally

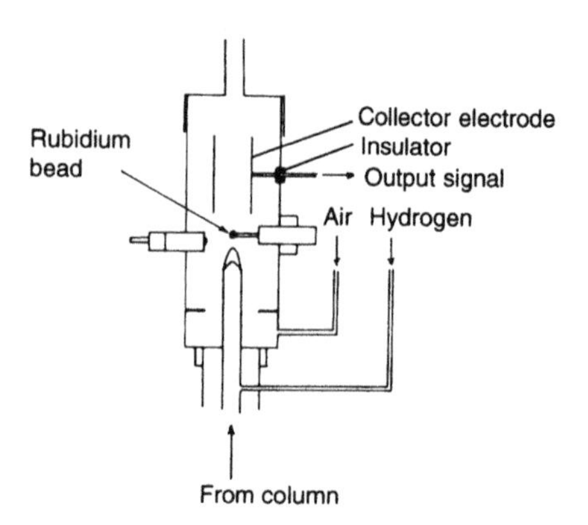

Figure 5.22 Nitrogen–phosphorus detector.

a modified FID with a bead of an alkali metal salt mounted between the flame tip and the collector electrode with a flame temperature sufficient to vaporise the alkali metal salt and generate a stable population of alkali metal ions necessary for the thermionic process. The combustion products of nitrogen and phosphorus compounds interact with the alkali metal ions by a complex series of reactions, which produce thermionic electrons. These are collected and give rise to the increase in current. The sensitivity of the detector is dependent on the alkali ion concentration, and therefore on the flame temperature. For good reproducible results, careful control of the flame, carrier gas and oven temperature are required.

NPDs now use an electrically heated temperature controlled glass bead which contains the alkali as a rubidium silicate which is thermally stable and results in a long working life time (Figure 5.22). By changing the electrical polarity of the jet and the carrier gas flow rate the detector can be made specific either to phosphorus containing compounds only, or to those containing both nitrogen and phosphorus. The bead is polarised at about −150 V with respect to the collector while the jet polarity is switched for the phosphorus only or nitrogen and phosphorus modes of operation. In the P mode a hot flame with a high hydrogen flow-rate of 40–50 ml min^{-1} is used. Electrons from combustion of non-P-containing compounds are grounded via the jet which has a positive potential relative to the bead. Combustion products from P-containing compounds react on the surface of the bead to produce thermionic electrons which are captured by the collector electrode producing the P specific signal. In the N–P mode the hydrogen flow rate is reduced to 1–5 ml min^{-1} and the jet polarised to the same potential as the bead. No flame is formed at the jet, the hydrogen 'burns' on the surface of the bead to produce a plasma. N-containing compounds produce CN radicals probably as a result of pyrolysis, which react with the alkali atoms in the plasma to form an ion pair. The cyanide ions migrate to the collector electrode to form the detector signal. Alternatively, the cyanide ion may be 'burnt' in the oxidising region of the flame producing electrons which travel to the collector. A similar process occurs for the formation of PO and PO_2 intermediate radicals.

When compared with a standard flame ionisation detector the NP detector is approximately 50 times more sensitive to nitrogen compounds and 500 times more sensitive to phosphorus compounds, a chromatogram of pesticides analysed using a NPD detector is included in Figure 5.9. Normal FID type response is suppressed so compounds that do not contain nitrogen and phosphorus do not produce a measurable signal. The NPD detector has a linear range of 10^5 and an MDA of $10^{-14}\,\mathrm{g\,ml^{-1}}$.

5.10.8 *Flame photometric detector*

The FPD is another type of flame detector. It is compatible with the FID and offers specificity for phosphorus and sulphur containing compounds with similar levels of sensitivity. Although first developed in 1966, the FPD has been progressively developed to take advantage of technological developments [62, 73–76]. The FPD measures the radiation emitted by excited species in the flame, the process observed in the simple flame tests familiar to all undergraduates. In a hydrogen flame, sulphur and phosphorus form excited species which emit radiation at 394 and 526 nm, respectively, and it has been reported that some nitrogen-containing compounds form emissions at 690 nm [77]. A narrow band-pass filter selects the wavelength to be observed and a photomultiplier detects and amplifies the emissions to form the detector signal (Figure 5.23). Fibre-optics are used to transmit the radiation to a photomultiplier located away from the heated detector block, giving improved stability.

In contrast to the oxygen-rich flame of FIDs, the FPD uses a hydrogen-rich flame which is cooler. This enhances production of the two reactive species of interest, HPO^* and S_2^*, which give off the characteristic emissions at 526 and 394 nm, respectively. The mechanisms for formation of HPO^* are not fully understood, however, the detector response for phosphorus containing compounds is linear. In contrast, response to sulphur containing compounds varies as the square (approximately) of the concentration.

$$\mathrm{R{-}P} + x\mathrm{O_2} \rightarrow \mathrm{HPO^*}$$

$$2\mathrm{R{-}S} + (x+2)\mathrm{O_2} \rightarrow x\mathrm{CO_2} + 2\mathrm{SO_2} \rightarrow 4\mathrm{H_2O} + \mathrm{S_2} \rightarrow \mathrm{S_2^*}$$

$$\text{detector signal} \propto [\mathrm{HPO}] \quad \text{or} \quad \propto [\mathrm{S}]^x \quad \text{where } x = 1.84\text{–}2.0$$

Careful calibration for sulphur is required using a range of standards to establish the response curve. Microcomputer controlled instruments can provide on board routines to produce a linear response based on a factor

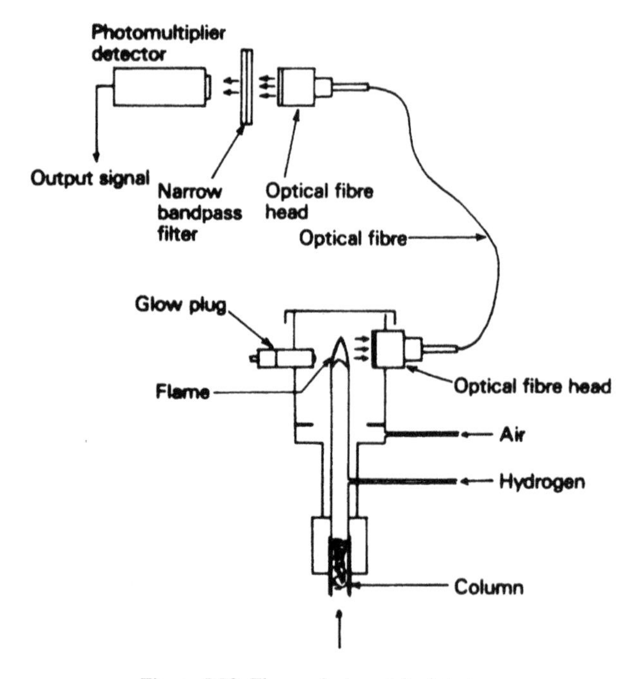

Figure 5.23 Flame photometric detector.

of $x \approx 1.84$. The overall response is mass/flow rate sensitive. Selectivity of both modes over hydrocarbons is $10^4:1$, minimum detectable amount is of the order of $10^{-12}\,\mathrm{g\,ml^{-1}}$ with a linear range of 10^4 for phosphorus and 10^3 for sulphur. Discrimination between S- and P-containing compounds is only about 4:1 due to the spectral interference of HPO^* and S_2^* emissions. It is therefore necessary to separate all the S-containing compounds from the P compounds. The FPD has been used in a number of applications, particularly in environmental and food additive analyses. S- and P-containing pesticides and their residues are of particular concern and the subnanogram sensitivity of the FPD over the FID and electron capture detector have enabled trace analyses to be readily carried out. Similar benefits have been utilised in the analyses of dyes and flavour additives in foods, soft drinks and organosulphur compounds in beer. Gaseous sulphur compounds such as thiophenes, mercaptans, disulphides, H_2S, SO_2 and the sulphurous content of fuel oils, petroleum and coal products have been analysed.

5.10.9 Photoionisation detector

PIDs were originally developed in the 1960s but it was 1976 before the first commercial detectors were available. Considerable developments have

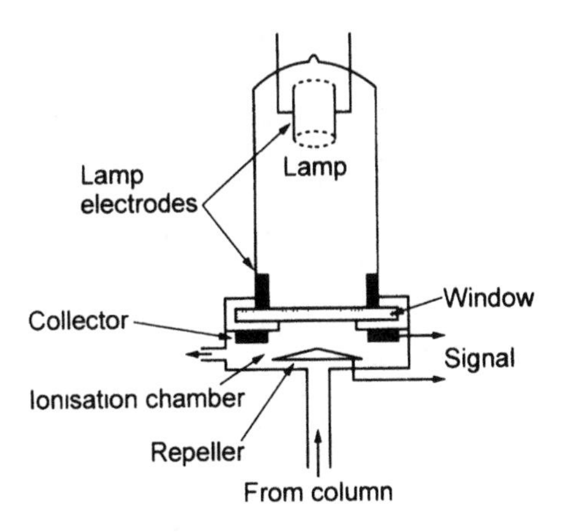

Figure 5.24 Photoionisation detector.

taken place over the last 10 years particularly in lamp design and micro ionisation chambers. The PID is now recognised as a sensitive detector with a wide range of applications particularly in environmental analyses [62, 78–80].

The photoionisation detector uses the energy of a vacuum ultraviolet lamp to ionise the analytes as they elute from the column. Excitation of the analyte molecules takes place in a small ionisation chamber attached to the window of a gas discharge lamp, where the molecules are ionised by the energy emitted from the lamp. The lamp contains a rare gas such as krypton or argon excited by a direct current at a potential of about 1000 V or a radio frequency of 100 kHz. Radiation emitted by the krypton lamp has 10.2 eV energy corresponding to a wavelength of 121 nm, argon emission is at 11.7 eV (104 nm). The window is made of magnesium fluoride (cut-off at 112 nm) or lithium fluoride (cut-off 105 nm) transparent to the emissions. The ionisation chamber has a small dead volume of 50–100 μl and contains a repeller electrode at a potential of 100–200 V, which directs the ions to a shielded collector electrode. The ion current formed between the electrodes produces the detector signal (Figure 5.24).

Ionisation is initiated by absorption of a photon, $h\nu$, by the analyte molecule, A. Direct ionisation results in a one step formation of ions and indirect ionisation involves initial excitation of the carrier gas, G, which then transfers energy to the analyte molecules forming ions:

Direct ionisation $A + h\nu \rightarrow A^{+} + e^{-}$

Indirect ionisation $G + h\nu \rightarrow G^{*} + A \rightarrow A^{+} + e^{-}$

$A + h\nu \rightarrow A^{*} \rightarrow A_{1}^{+} + e^{-}$

Successful ionisation of analytes depends on their ionisation potential and the energy of the lamp. The most useful source is the krypton lamp of

10.2 eV which has sufficient energy to ionise alkenes and aromatics and some alkanes (>C_6), alcohols, esters and aldehydes having low ionisation potentials. Ionisation potentials of most classes of organic compounds and inorganic gases have been published [81–83]. The argon 11.7 eV lamp has sufficient energy to ionise most organic compounds. The 10.2 eV lamp can therefore be used to selectively detect aromatic compounds in the presence of alkanes and other organic materials.

5.10.10 Miscellaneous detectors

The use of spectroscopic techniques as detectors for GC is discussed in Chapter 7. The detectors already mentioned are those commonly used in present day instruments, however, there are a large number that have been developed over the years either for specialised applications or in the quest for the ideal universal detector. Some of these are outlined below. The reader is referred to specialised texts for further details of these and other detectors [61, 62, 68].

5.10.10.1 Gas density balance. The gas density balance, designed by Martin and James, is a universal detector which appears to be ideal for GC [84–86]. However, problems of sensitivity and stability compared to ionisation detectors have prevented its more widespread use. Interest has been maintained since direct determination of molecular weight is achieved. The operation depends on detecting minute gas flows in what amounts to a supersensitive anemometer/TCD (Figure 5.25).

The reference gas (pure carrier gas) enters the cell at A, divides and passes over both detector filaments. Column effluent enters at B with a flow rate

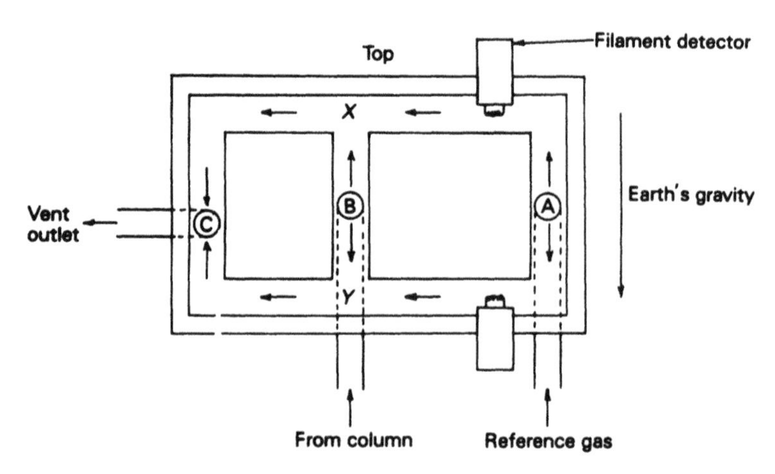

Figure 5.25 Gas density balance detector, mounted as shown.

approximately 20 ml min^{-1} less than that of the reference gas, thus preventing back diffusion to the detectors. If the overall density of the gas mixture at B is higher than the reference gas at A, and hence at X and Y, then the flow of gas from A is less on the lower arm since more of B moves downwards than proceeds to the upper arm. The resulting differing flow-rate over the detector elements causes imbalance in a Wheatstone bridge network, thus producing the detector signal and chromatogram (Figure 5.21). Negative responses are produced if less dense components are eluted. The response is a linear function of concentration and relative molar mass of the component.

$$A = \frac{Wk(M - m)}{M}$$

where A is the peak area; W is the amount (weight) of component; k is a constant; M is the molecular weight of component; m is the molecular weight of carrier gas.

Hydrogen and helium are not usually used as carrier gases because of their high diffusion rates. In order to determine molecular weights a standard of known molecular weight and the unknown component(s) are analysed using two different carrier gases. The equation below is solved for the molecular weight of the unknown.

$$\frac{A_U^1(M_S - m_1)}{A_S^1(M_U - m_1)} = \frac{A_U^2(M_S - m_2)}{A_S^2(M_U - m_2)}$$

terms are as in the previous equation with u being an unknown, s a standard, 1 a carrier gas of lower molecular weight, and 2 a carrier gas of higher molecular weight.

5.10.10.2 Conductivity detectors. Electrolytic (Hall) conductivity detectors convert the eluted components into ionic species in solution, which are then monitored in a d.c. conductivity cell. The chemical reactions used in the conversion process determine the specificity of the detector. Trace amounts of compounds containing nitrogen, chlorine and sulphur can be readily determined by firstly converting these to NH_3, Cl^-, S^{2-}, respectively [62, 87, 88]. Typically, the effluent is mixed with hydrogen and hydrogenated over a nickel catalyst at 800°C, forming NH_3, HCl, H_2S, H_2O and saturated low-molecular-weight hydrocarbons. Unwanted gases are removed in scrubbers such as $Ca(OH)_2$ or $Sr(OH)_2$ for acidic gases, and the remaining gases pass into a microelectrolytic cell for measurement. An alternative arrangement is to pass an oxygen-enriched effluent over a platinum catalyst where C, H and O containing compounds react. A constant amount of oxygen is continuously added to the effluent and after catalytic combustion any unreacted oxygen is determined in an oxygen-sensitive electrochemical cell. The resulting signal is therefore proportional to the amount of oxygen consumed and hence to the amount of eluted components. A similar method using hydrogenation and a hydrogen-sensitive cell has also been reported. Although capable of detecting nanogram samples and having a 10^4 linear range, the above detector systems suffer from a time lag in their operation, which can cause problems unless suitable separation times can be achieved.

5.10.10.3 Helium and argon ionisation detectors. Helium and argon ionisation detectors operate by formation of excited carrier gas atoms (He or Ar) which collide with eluted component molecules in a small cylindrical cell of similar construction to that of the ECD [89–91]. The ionisation energy of excited helium atoms (19.8 eV) is sufficient to ionise most other molecules in ion-molecule collision processes. The ions formed are collected on a central electrode under a polarised electrical field of up to 1 kV. Helium presents a number of problems, particularly the purity requirement of the carrier gas and small linear range. Argon is less sensitive to sample saturation and due to the lower ionisation energy of the excited atoms (11.7 eV) is less sensitive to contamination of the carrier gas by the permanent gases. Excitation of helium or argon is achieved using a high energy β-radiation source such as tritium on titanium foil. Both detectors are sensitive to contamination due to stationary phase bleed and therefore are frequently used with gas solid chromatography. Analysis of permanent gases, H_2, O_2, N_2, CO, CO_2, CH_4, NO, N_2O, H_2SO_2, and low-molecular-weight hydrocarbons have been achieved. Sensitivity is $10^{-13}\,g\,ml^{-1}$ over a 10^4 linear range.

5.10.10.4 Solid state detectors. Detectors for portable environmental monitoring equipment frequently use amorphous or thin film metal oxide detector elements, for example, based on doped zinc or tin oxides. The elements operate at 200–350°C and involve a chemisorption/chemioxidation process [92]. The adsorption of the active compounds results in a change in the distribution of electrons and charge carriers on the solid surface which in turn produces a change in electrical conductivity. The oxides used are semiconductors because of the ability of the metal ion to exist in different oxidation states. All the oxides are non-stoichiometric and either contain an excess (*n*-type) or deficiency (*p*-type) of metal ions. It is these oxidation states and the electrical neutrality of the oxides which are ultimately modified by the complex chemisorption processes. Recent developments using mixed metal oxides have produced detectors with rapid response times and improved sensitivities ($10^{-8}\,g\,ml^{-1}$) [48]. Metal oxide detectors can be fabricated to respond to a wide range of inorganic and low molecular weight organic compounds and therefore show promise as GC detectors for pollution monitoring.

5.10.10.5 Organic semiconductor detectors. Adsorbed gases have a marked effect on the electrical conductivity of organic semiconductors such as metal complexes of phthalocyanine, merrocyanine and porphyrins [94, 95]. The changes in electrical conductivity are due to the chemisorption of the gas onto the surface of a thin film (200 nm) of the organic semiconductor and subsequent electron transfer, the direction of transfer depending on the electronegativity of the gas and work function of the semiconductor. The change in conductivity can be several orders of magnitude with sensitivity of the order of 10^{-9} g ml^{-1} and a linear range of 10^5. These detectors are inexpensive and easy to make and show potential in environmental pollution monitoring systems.

5.10.10.6 The silicon chip GC. Fabrication of a thermal conductivity detector on a silicon chip has received continuing attention since the first reports of a complete capillary GC system with TCD detector on a slice of silicon, 5 cm^2 [96]. The capillary column is a spiral 1.5 m long and 200 μm by 40 μm cross-section etched into the surface of the silicon. The seating for the inlet valve and mounting for the TCD detector, a thin film metal resistor, fabricated on a separate chip, are also etched on the chip. The silicon slice is bonded to a Pyrex glass wafer to form the final instrument. The small size has led to proposals for development of five parallel capillary columns lined with different materials and five detectors all on one chip. The different retention times for the different columns enable an instrument with multiple detector outputs and microcomputer decoding of the data to identify readily over one hundred polluted gases. Recently, a miniature GC system for gas analysis has become commercially available [97]. The injector and TCD type detector are made by micromachining, an etching technique based on photolithography, a process used to manufacture semiconductors. The injector is a pneumatically operated microvalve which moves only 30 μm when opening and closing. The detector is a four 'filament' two channel TCD with a 200 nanolitre dead volume with a claimed sensitivity approaching that of an FID. Both injector and detector are compatible with micro WCOT and PLOT columns. Figure 5.21 includes a TCD chromatogram of permanent gases obtained using a molecular sieve PLOT column and a microcell TCD.

5.11 Supercritical gas chromatography

The greater selectivity of adsorption chromatography for compounds of different chemical types may be extended using supercritical fluids (SFs) as the mobile phase [96–99]. In normal GC the carrier gas does not interact significantly with the stationary phase, but when it is above it's critical temperature and pressure strong interactions occur. This makes the analytes being separated appear more volatile due to enhanced molecular interaction with both the mobile phase and stationary phase which has surface

Table 5.16 Critical values for various gases

Gas	Critical temperature (°C)	Critical pressure (bar)
Carbon dioxide	31.1	72.7
Ammonia	132.5	112.5
Pentane	196.6	33.3
Diethylether	192.6	35.6
i-Propanol	235.3	47.0
n-Hexane	234.2	29.9
Water	374.2	218.3

properties modified by the mobile phase. Supercritical fluids have liquid and gaseous states that are indistinguishable and are formed when both the critical temperature, T_C, and critical pressure, P_C, are exceeded. The critical values for various gases are shown in Table 5.16.

The critical values are high, requiring specialised equipment to withstand the temperatures and pressure involved. Supercritical water behaves as a covalent material dissolving organic solvents, but not inorganic salts. It is no longer hydrogen bonded and is therefore miscible with many permanent gases, e.g. N_2, O_2 (air), a property with commercial potential for extracting organic compounds from a difficult matrix. Supercritical hydrocarbons have equally interesting properties and potential, but also present toxic and fire hazards. The properties of supercritical carbon dioxide have been extensively studied since it is used commercially for extracting low-molecular-weight organic material from water and for decaffeinating coffee. It is inexpensive, not too hazardous and has therefore been developed for supercritical chromatography. It is a mobile non-polar solvent with properties similar to hexane and can be used from almost ambient temperatures (>31.1°C). Increasing the temperature and decreasing the pressure decreases the density of CO_2 and the chromatograms produced change from LC-like to GC-like. Ammonia has been used as the mobile phase for the analysis of C_1–C_4 hydrocarbons in synthetic fuel fractions.

One of the major problems with packed columns is the pressure drop across the column, CO_2 is the only practical SF for these systems. In contrast, a wider range of SFs can be used with open tubular capillary columns, e.g. ammonia and hydrocarbons. SF chromatography requires equipment capable of generating mobile phase pressures of up to 10 000 psi (69 MPa), together with suitable detectors. Ultraviolet detectors are used with carbon dioxide since it is transparent in this region of the spectrum.

5.12 Applications of gas chromatography

The range and variety of applications to which G has been applied demonstrate the versatility and scope of the technique. The ever-changing complexity of present-day products (e.g. drugs, foodstuffs, consumer products, environmental monitoring of air, water and legislation), the increased sensitivity of the technique and data collection and processing have all contributed to the steadily increasing use of GC. The aim of this section is to present an overview of the range of applications and where appropriate include references for further reading. Reviews frequently appear in the chromatography and analytical chemistry journals and the Royal Society of Chemistry (Cambridge UK) now publish *Chromatography Abstracts*, with excellent monthly abstracts of the literature. Chromatograms illustrating various applications are included in Figure 5.9.

5.12.1 Headspace gas chromatography analysis

Headspace GC (HSGC) analysis employs a specialised sampling and sample introduction technique, making use of the equilibrium established between the volatile components of a liquid or solid phase and the gaseous/vapour phase in a sealed sample container [50–52]. Aliquots of the gaseous phase are sampled for analysis. In many cases the detection limits for a particular component in a mixture are improved and the 'clean' nature of the sample makes HSGC ideal for automatic repetitive quantitative and qualitative analyses, at the same time prolonging the effective lifetime of a column. An example of a quantitative application is the forensic analysis of blood and urine alcohol (ethanol) levels, particularly in connection with driving offences. The technique eliminates the need for a precolumn or regular re-packing of the first section of the column, rendered ineffective by sample residues, since only volatile compounds are introduced on to the column. The blood sample is homogenised and an aliquot taken and diluted with an aqueous solution of an internal standard (*n*-propanol or isopropanol). Aqueous solutions of ethanol (80 mg per 100 ml and 200 mg per 100 ml) are used as standards to calibrate the internal standard. The sealed vials containing the diluted samples are loaded into an autosampler and are allowed to equilibrate at 40°C for 10 min. The vapour samples are then taken as described in the section on injection techniques and analysed on a polar column such as PEG 20M or Porapak Q GSC column, see Exercise No. 17, Chapter 9 for further details on the chromatography and calculations. In addition to ethanol, methanol, acetone, acetaldehyde and other volatiles which may be present in the blood can also be observed. HSGC of volatile constituents in a sample can be extended to the detection of volatile solvents and materials such as white spirits (turpentine substitute), paraffin, petrol, diesel fuel in case samples and glue solvents in the blood, to assist with the investigations in arson, explosions, robberies and accident cases. Quality and production control of beer constituents and other brewery products can be effectively carried out by HSGC to determine the ethanol content and the ratios and amounts of the fusil oils, higher alcohols (*n*-propanol, iso-propanol, *n*-butanol, 2-methyl and 3-methylbutan-1-ol), esters (ethyl acetate), acetaldehyde, etc., using a Carbowax column [54]. Aromatic flavours and trace volatiles in beer (26 have been observed), foods and soft drinks are also readily analysed. Plastics packaging and consumer products have been analysed for residual monomer and other process residues such as carbon dioxide, water, solvents, acetates, aliphatic alcohols, printing inks and solvents [55]. These compounds present problems of film stability, porosity and contamination of foods. Analysis of volatile free fatty acids produced by bacteria, particularly anaerobic bacteria, enables a fingerprint of the particular microorganism(s) to be obtained, which assists in the identification of the bacteria [56].

5.12.2 Food analysis

Analysis of foods is concerned with the assay of lipids, proteins, carbohydrates, preservatives, flavours, colorants and texture modifiers, and also vitamins, steroids, drugs and pesticide residues and trace elements. Most of the components are non-volatile and although HPLC is now used routinely for much food analysis, GC is still frequently used [57]. For example, derivatisation of lipids and fatty acid to their methyl esters (FAMEs), of proteins by acid hydrolysis followed by esterification (*N*-propyl esters) and of carbohydrates by silylation to produce volatile samples suitable for GC analysis. Thus, GC quality control analysis of food products can confirm the presence and quantities of the analytes to ensure the product complies with company and legislative requirements. For example, fruits, fruit derived foodstuffs, vegetables and soft drinks, tea and coffee, are analysed for their polybasic and hydroxy acid content (citric, maleic acids) as TMS derivatives using an OV17 or equivalent WCOT column; flavanols and caffeine are analysed on an OV101 or equivalent column whilst the carbonyl-containing volatiles, e.g. vanillin, are analysed using a more polar Carbowax 20 M column. Dairy products are analysed for volatile components (aldehydes, ketones) to determine age and rancidity, fatty acids (as methyl ester derivatives) and milk sugars (as TMS derivatives). The use of GC in the analysis of butter, cheese and yoghurt is not only for the butter fat content, but also for added colour and flavourings which are monitored to determine the quality and quantity of permitted additives and brand or type of product [57]. The composition of volatile components in a wide range of natural foods, fruits, vegetables, flavourings and beverages (tea, coffee, cocoa), which give them the characteristic flavours and odours has received considerable interest, particularly for the preparation of artificial flavourings and perfumes. The volatile nature of the flavours and essential oils are particularly suitable for WCOT column analysis since most of the above contain a complex mixture of a large number of these compounds, e.g. rose perfumes (more than 55), essential oils such as juniper (more than 100), peaches and figs (more than 40). A chromatogram of mushroom aroma is included in Figure 5.9.

5.12.3 Drugs

Analysis of drugs by GC is now complemented by HPLC. However, there are still numerous GC applications involving both quantitative and qualitative identification of the active components and possible contaminants, adulterants or characteristic features which may indicate the source of the particular sample [111–113, 127, 128]. Forensic analysis frequently uses GC to characterise drugs of abuse, in some cases the characteristic chromatographic fingerprint gives an indication of the source of manufacture of the sample or worldwide source of a vegetable material such as cannabis [114]. The drugs may be contained in commercial preparations, illicit drug samples or samples of blood, urine or stomach contents and therefore appropriate sample preparation and extraction are required. Although some classes of drugs can be analysed directly after extraction or dissolution in a solvent, many require derivatisation, usually silylation, acetylation or methylation. Analytical procedures, chromatographic methods and retention data are published for over 600 drugs, poisons and metabolites [111–113, 127, 128]. These data are extremely useful for forensic work and in hospital pathology laboratories to assist in identifying drugs.

5.12.4 Pyrolysis gas chromatography

Pyrolysis GC (PGC) is used principally for the identification of non volatile materials, such as plastics and natural and synthetic polymers, drugs and some microbiological materials [53, 54]. The thermal dissociation and fragmentation of the sample produces a chromatogram which is a 'fingerprint' for that sample. The small molecules produced in the pyrolysis reaction are frequently identified using a GC–MS system (see Chapter 7). The pyrogram can give information on molecular structure for identification of the fragments and also information on chemical composition of the sample. Thus for a plastic sample, information on the type of monomer(s), plasticisers, extenders and other additives would be obtained. PGC, especially using WCOT columns, has proved very useful in forensic analysis for characterising plastics, rubbers and paint samples. Frequently only small fragments of sample are available from contact traces of debris at the scene of the crime. PGC requires only small samples and since in many forensic applications a comparison between test sample and reference samples is the main objective, the pyrogram fingerprint can provide the comparative evidence required.

5.12.5 Metal chelates and inorganic materials

Although inorganic compounds are generally non-volatile, GC analysis can be achieved by converting the metal species into volatile derivatives. Only some metal hydrides and chlorides have sufficient volatility for GC. Organometallics other than chelates, which can be analysed directly, include boranes, silanes, germanes, organotin and lead compounds. Several classes of derivatives, such as metal carbonyls and alkoxides, can be formed, but only metal chelates have been used extensively in analytical applications [115–117]. The chelating ligands contain oxygen, sulphur, selenium, phosphorus and nitrogen as donor atoms, but β-diketonates have proved to be the most suitable, forming thermally stable, soluble, volatile metal chelates. Suitable β-diketonates are shown below:

$$R{-}CO{-}CH_2{-}CO{-}R' \quad \text{where R and R' are } CH_3, CF_3 \text{ or } C_3F_7$$

The β-diketonates of di-, tri- and tetravalent metals have proved the most suitable and the ligands acetylacetone trifluoroacetylacetone and hexafluoroacetone the most used. In general, the more fluorinated the β-diketone the more volatile the chelate; also greater sensitivity can be achieved with the ECD.

5.12.6 Dual detector applications

A number of specific detectors or detectors with enhanced responses to certain classes of compounds are available and their specificity when compared to an FID, for example, can be used for analysis of complex mixtures. Generally, the column effluent is split 50:50 between an FID and a second detector, such as an NPD or ECD. Dual detector amplifiers and dual channel recording systems are required. An example of this application is the identification and chemical classification of drugs based on the relative response of the nitrogen selective detector NPD and FID [111–113, 118, 119]. The drugs are analysed using caffeine as reference internal standard. In addition to characterising the drug in terms of a retention time relative to caffeine, the drug is also characterised by the ratio of the NPD response to FID response determined by the caffeine standard. At least 70 drugs can be characterised by this system and with different columns other groups of drugs can be analysed. A similar approach can be used with an ECD/FID system for the analysis of polychlorinated plasticisers, pesticides and herbicides, since the ECD has considerably greater response to chlorine containing compounds [61]. A TCD connected in parallel or in series (since the TCD is a non-destructive detector) with an FID, enables samples containing both permanent gases, which are not detected by an FID, and hydrocarbon materials, to be completely analysed with a single sample injection. An example of an application is the analysis of natural gases which invariably contain C_1–C_4 hydrocarbons, nitrogen, oxygen and carbon dioxide [118, 119].

5.12.7 *Dual column applications*

Dual column analysis involves the simultaneous analysis of a sample using two columns in parallel having stationary phases with widely differing polarities such as a dimethylsilicone and Carbowax (Figure 5.1(c)). The log–log plot of retention time on the two columns for the components to be identified are constructed, thus enabling complex mixtures to be analysed when there are problems of achieving good separation. Experiment 20, Chapter 9 illustrates this application.

5.12.8 *Environmental analysis*

Environmental pollution is an age-old trademark of man and in recent years as technology has progressed, populations have increased and standards of living have improved, so the demands on the environment have increased, with all the attendant problems for the earth's ecosystems. Combustion of fossil fuel, disposal of waste materials and products, treatment of crops with pesticides and herbicides have all contributed to the problem. Technological developments have enabled man to study these problems and realise that even trace quantities of pollutants can have detrimental effects on health and on the stability of the environment. There is a vast amount of literature on the use of GC for studying a wide variety of these problems [120–122]. A variety of GC sampling techniques are used, for example, solvent vapour in the atmosphere may be analysed by adsorption on granular active charcoal followed by desorption with carbon disulphide or collected on an adsorbent then thermally desorbed for separation by WCOT columns. The sensitivity of the technique allows sub-occupational exposure limit levels of a wide range of solvents and atmosphere pollutants to be monitored (see section 5.9.8). Polynuclear aromatic hydrocarbons (PNAs or PAHs) in the atmosphere and in polluted waters have also received considerable attention due to their associated carcinogenic properties. PAH analysis is therefore important in assessing water quality. Analytes are present in the ppb ($ng\,ml^{-1}$) range, consequently samples are extracted with cyclohexane and concentrated prior to analysis by FID or ECD using WCOT columns. Over 100 PNAs/PAHs have been identified, most thought to originate from the combustion of wood and coal and oil, biodegradation processes and crude oil. Pesticides and related compounds also present a growing environmental problem. These include the polychlorinated biphenyls, chlorinated pesticides (DDT and BHC) and organophosphorus and sulphur compounds.

The Environmental Protection Agency (EPA) in the USA and the European Union have produced standard methods for analysis of pollutants in waters and waste water [123]. These include methods of analysis for about 40 chlorinated, 40 organophosphorus and seven carbamate pollutant species.

Analytes are extracted into dichloromethane, concentrated and then analysed on BP17, OV17 columns with an ECD detector for the chlorocarbons and FPD for the organophosphorus analytes. Carbamates are converted to their 2,4-diphenylether derivatives prior to analysis. GC is also valuable for fingerprinting oil spills or contamination of ground waters and for analysis of *N*-nitrosamines as their acetyl derivatives [124]. Phenols, amines and organic acids form strong hydrogen bonds and although they can be analysed on thick film WCOT columns, derivatives such as methyl esters are frequently prepared (see section 5.9). Organometallic and inorganic pollutants are frequently encountered, some of which may be determined by GC, generally after extraction into dichloromethane or cyclohexane. These include organotin biocides which are analysed as the methyl esters after extraction or as hydrides after treatment with $LiAlH_4$. Selenium in natural waters may be analysed as the 4,6-benzodiazo-dibromoselenol formed from Se(IV) and 1,2-diamino-3,5-dibromobenzene. Se(II) and Se(0) can be analysed similarly [125, 126].

References

[1] Rooney, T.A., Altmayer, L.H., Freemann, R.R. and Zerenner, E.H. *Rapid GC Separations Using Short Glass Capillary Columns*, Technical Paper Number 73. Hewlett Packard, Avondale, 1978.
[2] Sapina, W.R. *The Packed Column in GC*. Supelco, Poole, Dorset, 1974.
[3] Berezkin, V.G., Pakhomov, V.P. and Sakodinski, K.I. *Solid Supports in Gas Chromatography*. Supelco, Poole, Dorset, 1980.
[4] Bradford, B.W. and Harvey, D. *J. Inst. Petrol.*, **41** (1955) 80.
[5] Scott, R.P.W. *Gas Chromatography 1960*, p. 284. Butterworth, London, 1960.
[6] Zlatkis, A., Chang, R.C. and Schurig, V. *Chromatographia*, **6** (1973) 223.
[7] Scott, R.W.P. *Gas Chromatography 1958* (Desty, D.H., ed.), p. 1. Butterworth, London, 1958.
[8] Kovats, E. *Helv. Chim. Acta*, **41** (1958) 1915.
[9] Kovats, E. *Advances in Chromatography*, Vol. 1, p. 229. Marcel Dekker, New York, 1965.
[10] Pacakova, V. and Feltl, L. *Chromatographic Retention Indices, An Aid To Identification of Organic Compounds*. Ellis Horwood, London, 1992.
[11] Rohschneider, L. *J. Chromatogr.*, **22** (1966) 6.
[12] McReynolds, W. *J. Chromatogr. Sci.*, **8** (1970) 685.
[13] Zweig, G. and Sherma, J. *Handbook of Chromatography General Data and Principles*, Vol. 1. CRC Press, Boca Raton, FL, 1985.
[14] Moffat, A.C. *Clarke's Isolation and Identification of Drugs*, p. 193. The Pharmaceutical Press, London, 1986.
[15] Golay, M.J.E. *Gas Chromatography 1958* (Desty, D.H., ed.), p. 36. Butterworth, Academic Press, London, 1958.
[16] Ettre, L.S. *Open Tubular Columns in Gas Chromatography*. Plenum Press, New York, 1965.
[17] Jennings, W. *Gas Chromatography with Glass Capillary Columns*. Academic Press, London, 1980.
[18] Grob, R.L. *Modern Practice of Gas Chromatography*. John Wiley, New York, 1985.
[19] Jennings, W. and Nikelly, J. *Capillary Chromatography, The Applications*. Huthig, Heidelberg, 1991.
[20] Hyver, K.J. and Sandra, P. *High Resolution Gas Chromatography*. Hewlett Packard, Cheadle Heath, Cheshire, UK (1989).

[21] Horvath, C., Ettre, L.S. and Purcell, J.E. *Am. Lab.*, **6**(8) (1974) 75.
[22] de Nijs, R.C.M. *Chrompack News*, **11**(3E) (1984) 6, Chrompack International BV, Milharbour, London.
[23] Sie, S.T. and Bleumer, J.P.A. *Gas Chromatography 1968* (Harbourn, C.L.A., ed.), p. 235. Institute of Petroleum/Academic Press, New York, 1968.
[24] Novotony, M. *et al. Anal. Chem.*, **53** (1981) 407A.
[25] Jennings, W. *Comparison of Fused Silica and Other Glass Columns in GC*. Huthig, Heidelberg, 1981.
[26] Konig, W.A. *Entiomer Separations by Capillary Gas Chromatography*. Huthig, Heidelberg, 1987.
[27] Grob, K., Grob, G. and Grob, Jr., K. *J. Chromatogr.*, **156** (1978) 1.
[28] de Nijs, R.C.M., Franken, J.J., Dooper, R.P.M. and Rijks, J.A. *J. Chromatogr.*, **167** (1978) 231.
[29] Kirkland, J.J. *Anal. Chem.*, **35** (1963) 1295.
[30] Petitjean, D.L. and Lefault, C.J. *J. Gas Chromatogr.*, **1** (1963) 18.
[31] de Nijs, R.J. *J. High Res. Chromatogr. Chromatogr. Comm.*, **4** (1981) 612.
[32] de Nijs, R.J. and de Zeeuw, J.J. *J. Chromatogr.*, **279** (1983) 41.
[33] Cooper, M. PhD thesis, Nottingham Trent University, 1990.
[34] Braithwaite, A. and Cooper, M. *Chromatographia*, (1995).
[35] Chrompack, *Application Note 906*. Chrompack International BV, Milharbour, London.
[36] de Zeeuw, J., de Nijs, R.C.M., Buyten, J.C. and Peene, J.A. *J. High Res. Chromatogr. Chromatogr. Comm.*, **11** (1988) 162.
[37] Scott, C.G. *J. Inst. Petrol.*, **45** (1959) 118.
[38] Scott, R.P.W. *Gas Chromatography 1960*, p. 137. Butterworths, London, 1960.
[39] Little, J.N. *J. Chromatogr. Sci.*, **8** (1970) 647.
[40] McCreery, R.L. and Sawyer, D.T. *Anal. Chem.*, **40** (1968) 106.
[41] McCreery, R.L. and Sawyer, D.T. *J. Chromatogr. Sci.*, **8** (1970) 122.
[42] Dave, S. *J. Chromatogr. Sci.*, **7** (1969) 389.
[43] Russell, L.W. *J. Forens. Sci. Soc.*, **21** (1981) 317.
[44] Novotony, M. and Hayes, J.M. *Science*, **189** (1975) 215; **194** (1976) 72.
[45] Grob, K. *Classical Split and Splitless Injection in Capillary GC*. Huthig, Heidelberg, 1988.
[46] Grob, K. *On Column Injection in Capillary Column Gas Chromatography*. Huthig, Heidelberg, 1987.
[47] Blau, K. and Halket, J.M. *Handbook of Derivatives for Chromatography*. John Wiley, New York, 1993.
[48] Knopp, D.R. *Handbook of Analytical Derivatisation Reactions*. John Wiley, New York, 1979.
[49] Christie, W.W. *Gas Chromatography and Lipids, A Practical Guide*. The Oily Press, 1989.
[50] Kolb, B. *Applied Headspace Gas Chromatography*. Heyden, London, 1980.
[51] Hachenberg, H. and Schmidt, A.P. *Gas Chromatographic Headspace Analysis*. Heyden, London, 1977.
[52] Loffe, B.V. and Vitenberg, A.G. *Headspace and Related Methods in Gas Chromatography*. John Wiley, London, 1984.
[53] Wampler, T.P. *Applied Pyrolysis Handbook*. Marcell Dekker, 1995.
[54] May, R.W., Person, E.F. and Scothern, D. *Pyrolysis—Gas Chromatography*. Royal Society of Chemistry, London, 1977.
[55] Betz, W.R., Maroldo, S.G., Wachob, G.D. and Firth, C.M. *Am. Industr. Hyg. Assoc. J.*, **50** (1989) 181.
[56] Joint Committee Report on Adsorbents, *Am. Indust. Hyg. Assoc. J.*, **32** (1971) 404.
[57] Brown, R.H. and Purnell, C.J. *J. Chromatogr.*, **178** (1979) 79.
[58] Perkin Elmer, *Thermal Desorption Application Notes Nos 1–30*. Perkin Elmer Ltd, Beaconsfield.
[59] Glover, J.H. *Thermal Desorption in Industrial Hygiene and Environmental Analysis*. Spantech Publishers, South Godstone, Surrey, 1991.
[60] Deans, D.R. *Chromatographia*, **1** (1974) 187.
[61] David, D.J. *Gas Chromatography Detectors*. John Wiley, New York, 1974.
[62] Hill, H.H. and McMinn, D.G. *Detectors in Capillary Chromatography*. John Wiley, New York, 1992.

[63] Poole, C.F. *J. High Res. Chromatogr. Chromatogr. Commun.*, **5** (1982) 454.
[64] Poole, C.F. and Zlatkis, A. *Electron Capture, Theory and Practice in Chromatography*, J. Chromatogr. Lib. Vol. 20. Elsevier, Amsterdam, 1981.
[65] Lovelock, J.E., Maggs, R.J., Joynes, P.L. and Davies, A.J. *Anal. Chem.*, **43** (1971) 1966.
[66] Lovelock, J.E. *J. Chromatogr.*, **1** (1958) 35; **35** 474.
[67] Deveaux, P. and Guichon, G. *J. Gas Chromatogr.*, **5** (1967) 314.
[68] Littlewood, A.B. *Gas Chromatography*, p. 339. Academic Press, London, 1970.
[69] Kolb, B. and Rischoff, J. *J. Chromatogr. Sci.*, **12** (1974) 625.
[70] Patterson, P.L., Gatton, R.A. and Ontiveres, C. *J. Chromatogr. Sci.*, **20** (1982) 97.
[71] Patterson, P.L. *J. Chromatogr. Sci.*, **24** (1986) 41.
[72] Boeck, P. and Janak, J. *Chromatogr. Revs*, **15** (1971) 111.
[73] Brody, S.S. and Chaney, J.E. *J. Gas Chromatogr.*, **4** (1966) 42.
[74] Farwell, S.O. and Baringa, C.J. *J. Chromatogr. Sci.*, **24** (1986) 483.
[75] Cadwell, P.J. and Marriott, P.J. *J. Chromatogr. Sci.*, **20** (1982) 83.
[76] Sevcik, J. and Thao, N.P. *Chromatographia*, **8** (1975) 559.
[77] Krost, K.J., Hodgeson, J.A. and Stevens, R.K. *Anal. Chem.* **45** (1973) 1800.
[78] Driscoll, J.N. CRC Critical Revs, *Anal. Chem.*, **17** (1986) 193.
[79] Driscoll, J.N. and Spaniani, F.F. *Anal. Inst.*, **13** (1974) 111.
[80] Lovelock, J.E. *Anal. Chem.*, **33** (1961) 162.
[81] Driscoll, J.N., Wood, C. and Whelan, M. *Amer. Env. Lab.*, **3** (1991) 19.
[82] Driscoll, J.N. *J. Chromatogr. Sci.*, **20** (1982) 91.
[83] Longhurst, M.L. *J. Chromatogr. Sci.*, **19** (1981) 98.
[84] James, A.J.P. and James, A.T. *Biochem. J.*, **63** (1956) 138.
[85] Kiran, E. and Gillham, J.K. *Anal. Chem.*, **47** (1975) 983.
[86] Phillips, C.S.G. and Timms, P.L. *J. Chromatogr.*, **5** (1961) 131.
[87] Piringer, O. and Pascalau, M. *J. Chromatogr.*, **8** (1982) 410.
[88] Coulson, D.M. *J. Gas Chromatogr.*, **3** (1965) 134.
[89] Lovelock, J.E. *J. Chromatogr.*, **1** (1958) 35.
[90] Andrawes, F.F., Byers, T.B. and Gibson, E.K. *Anal. Chem.*, **53** (1981) 1544.
[91] Andrawes, F.F. and Gibson, E.K. Chromatogr. Comms., *J. High Res. Chromatogr.*, **5** (1982) 265.
[92] Firth, J.G., Jones, A. and Jones, T.A. *Ann. Occup. Hyg.*, **18** (1975) 63. European Patent 0 030 112 A1 (1981).
[93] Braithwaite, A. and Gibson, S. unpublished results; Gibson, S. PhD Thesis, Nottingham Trent University (1984).
[94] Jones, T.A. *Organic Semiconductor Gas Sensors*, Technical Information Note No 8, Health and Safety Executive, Sheffield, S3 7HQ.
[95] van Ewyck, R.L., Chadwick, A.V. and Wright, J.D. *J. Chem. Soc. Faraday*, **76**(1) (1980) 2194.
[96] Angel, J.B., Terry, S.C. and Barth, P.W. *Scientific American*, June, (1984) 12–23.
[97] Micro-GC CP-2002, *Chrompack News*, **22**(1) (1995) 14–16, Chrompack International, Millharbour, London, E14 9TN.
[98] Smith, R.M. (ed), *Supercritical Fluid Chromatography*, Royal Society of Chemistry, Cambridge, UK (1993).
[99] Rawden, M.G. and Norris, T.A. International Laboratory, June, 12–23 (1984).
[100] Kolb, B. (ed), *Applied Headspace Gas Chromatography*, Heyden, London (1980).
[101] Hachenberg, H. and Schmidt, A.P. *Gas Chromatographic Headspace Analysis*, Heyden, London (1977).
[102] Loffe, B.V. and Vitenberg, A.G. *Headspace and Related Methods in Gas Chromatography*, John Wiley, London (1984).
[103] Jennings, W. and Shibamoto, T. *Qualitative Analysis of Flavours and Fragrances by Glass Capillary GC*. Academic Press, New York, 1980.
[104] Lamparsky, D. *Analysis of Foods and Beverages by Headspace Techniques*. Academic Press, New York, 1978.
[105] Jeffs, A.R. *Applied Headspace Gas Chromatography* (Kolb, B., ed.), Heyden, London, 1980.
[106] Taylor, A.J. *J. Med. Microbiol.*, **11** (1978) 9.
[107] Larson, L., Mardth, P.A. and Oldham, G. *J. Clin. Microbiol.*, **7** (1978) 23.

[108] Gordon, M.H. *Principles and Applications of Gas Chromatography in Food Analysis*. Ellis Horwood, London, 1990.
[109] James, C. *Analytical Chemistry of Foods*. Blackie A&P, Glasgow, 1995.
[110] Baltes, W. *Recent Developments in Food Analysis*. Verlag Chemie, Basel, 1982.
[111] Moffat, A.C. *Clarke's Isolation and Identification of Drugs*. The Pharmaceutical Press, London, 1986.
[112] Gough, T.A. *The Analysis of Drugs of Abuse*. John Wiley, Chichester, 1991.
[113] Mills, T., Robertson, J.C., McCurdy, H.H. and Hall, W.H. *Instrumental Data for Drugs Analysis*. CRC Press, Boca Raton, FL, 1992.
[114] Ho, M.H. *Analytical Methods in Forensic Chemistry*. Ellis Horwood, London, 1990.
[115] Lederer, M. *Chromatography for Inorganic Analysis*. John Wiley, Chichester, 1994.
[116] Moshier, R.W. and Sievers, R.E. *Gas Chromatography of Metal Chelates*. Pergamon Press, London, 1965.
[117] Schwedt, G. *Chromatographic Methods in Inorganic Analysis*. Huthig, Heidelberg, 1981.
[118] Colenutt, B.A. and Thorburn, S. *Int. J. Environ. Studies*, **15** (1980) 25.
[119] Baker, J.K. *Anal. Chem.*, **49** (1977) 906.
[120] Grob, R.L. *Chromatographic Analysis of the Environment*. Marcel Dekker, 1983.
[121] Bloemen, H.J.Th. and Burn, J. *Chemistry and Analysis of Volatile Organic Compounds in the Environment*. Blackie A&P, Glasgow, 1993.
[122] Clarke, A.G. *Industrial Pollution Monitoring*. Chapman and Hall, London, 1995.
[123] US EPA, *Determination of Organic Compounds in Drinking Water, Method Series 500; GC Methods of Analysis for Solid Waste, Method Series 8000*. US Environmental Protection Agency, Cincinnati, OH.
[124] Thompson, J.F. *Analysis of Pesticide Residues in Human and Environmental Samples*. US Environmental Protection Agency, Cincinnati, OH.
[125] Shimoishi, U. and Toei, K. *Anal. Chim. Acta*, **100** (1978) 65.
[126] Shimoishi, U. and Uchida, U. *Environ. Sci. Technol.*, **14** (1980) 541.
[127] Konig, W.A. *Entiomer Separations by Capillary Gas Chromatography*, Huthig, Heidelberg (1987).
[128] DEX110 and DEX120 WCOT columns, Supelco Inc., Poole BH17 7NH, UK.

6 High performance liquid chromatography

6.1 Introduction

The pioneering work of Martin and Synge [1] led to the development of gas–liquid chromatography, an elegant chromatographic method where, even with packed columns, the efficiencies achieved were high and the technique could be used with equal facility for both analytical and preparative studies. The first commercial apparatus was marketed in 1955 and the scope and applications of the technique have continued to expand through the 1970s and 1980s with refinements in instrumentation and equipment, especially the development of capillary columns, microparticulate packings and bonded stationary phases, automation and microprocessor control. The essential prerequisite for the analysis of a mixture by gas chromatography (GC), is that each of the sample components has an appreciable vapour pressure at the operating temperature of the column. Though the limitations imposed by volatility can in many cases be overcome by the conversion of less tractable compounds to suitable volatile derivatives, errors in quantitation are common as it is difficult to ensure complete reaction and conversion given the diversity of sample matrices. In addition there still remains a large number of compounds, such as pharmaceuticals, polymers, proteins and dyestuffs, which cannot be volatilised, or even heated, without decomposition, neither are they suitable for derivitisation and therefore these compound classes cannot be analysed by GC. Indeed, only about 20% of known compounds are suitable for analysis by GC and it was recognised that there was a need for a complementary high resolution separation technique.

Liquid chromatography (LC), the generic name used to describe any chromatographic procedure in which the mobile phase is a liquid, and in particular the established techniques of thin layer and open column chromatography, was used with some measure of success for the analysis of mixtures of such compound classes.

In conventional open column chromatography, solvent is gravity fed onto a column of large (~150–250 μm) particles, and the components of the mixture are then carried through the packed column by the eluant, separation being achieved by differential distribution of the sample components between the stationary and mobile phases. However, open column classical liquid chromatography suffers from a number of disadvantages, for example:

- column packing procedures are tedious and the column is usually only used once, which makes the technique expensive;

- efficiency achieved by these large particle packings is relatively low and the analysis time lengthy, even for fairly simple mixtures; the practice of the technique is operator sensitive, e.g. sample application;
- detection of the solute in the eluant is achieved by the manual analysis of the individual fractions, which in itself is labour intensive and time consuming.

Thin layer chromatography offered an improvement in some of the above aspects, though it suffered from several limitations, especially in the areas of automation, reproducibility, quantitation and preparative studies.

Though the theoretical principles of modern LC were firmly established by Martin and Synge in the same paper as they laid the foundation for GC, it is unclear why the development and evolution of LC was so protracted. The principal factor limiting column performance in classical open bed LC was the decreased rate of diffusion of solute molecules between the phases due to the use of a liquid eluant. The key to achieving improved column efficiencies in LC lay in increasing the rate of mass transfer and equilibration of solute molecules between the mobile and stationary phases. The classical theoretical treatise of van Deemter [2] and Giddings [3, 4] acknowledged the need for improved interparticle diffusion rates and suggested that the above objectives could be attained by using small particle size packing. Concomitant with the use of small particle size were large pressure drops across the columns, and consequently equipment for high performance liquid chromatography (HPLC), such as pumps, columns and fittings, had to be capable of delivering and withstanding the required high inlet pressures. A further consequence of the reduction in column and particle size was that the volume of the detector cell had to be modified accordingly as the peak volumes expected would be much smaller.

It may be that, as the establishment of modern LC depended upon advances and innovations in several areas such as the preparation of packing material and the design of pumps and detectors, available resources were preferentially expended on the development of gas chromatography. Whatever the reason, it was not until the late 1960s, when active and intensive research and development culminated in the availability of commercial instruments for modern LC, that the potential of the technique was realised. Horvath *et al.* [5] constructed one of the first practical HPLC systems for use in their work on nucleotides, reported in 1967. About the same time successful procedures for packing microparticulate materials were being developed [6]. Modern high resolution LC, variously known as high pressure, high performance and high speed LC, and universally referred to as HPLC, has now become firmly established at the forefront of chromatographic techniques. HPLC is now used for a wide range of applications and offers significant advantages in the analysis of pharmaceutical formulations, biological fluids, synthetic and natural polymers, environmental residues, a variety of inorganic substances and trace element contaminants. High column efficiencies can be achieved comparable to capillary column GC due to the greater control and choice of both stationary *and* mobile phases. An added advantage is that many detectors used in HPLC are non-destructive, thus facilitating sample recovery and providing the opportunity for subsequent spectroanalytical studies. Chromatograms showing typical separations of some of these compound classes and illustrating the power and utility of HPLC are shown in Figure 6.1.

HPLC is now pre-eminent amongst chromatographic techniques as evidenced by the vast number of published scientific papers which cite the technique as the chosen method of analysis. HPLC is not limited as is GC in applicability by component volatility or thermal stability, which makes it the method of choice for polymers, polar, ionic and thermally unstable materials. To summarise, modern LC has the advantages that the columns are reusable, that sample introduction can be automated and detection and quantitation can be achieved by the use of continuous flow detectors; these features lead to improved accuracy and precision of analysis. In consequence the technique not only complements GC but is regarded as the most useful and expedient of chromatographic methods.

6.2 Modes of chromatography

Separation in HPLC can be achieved by exploiting a variety of sorption processes, of which the more important are shown in Table 6.1 together with the dominant sorption mechanisms.

The exact mode of chromatography operating in a given application is determined principally by the nature of the packing, though it must be appreciated that, while there may be one dominant mechanism, the modes are not mutually exclusive. The power and utility of HPLC compared to GC is derived partly from the greater range of sorption modes available and partly from the facility to modify radically the chemical nature and solvent strength of the eluant. The majority of packings are based on 3, 5 or 10 µm spherical particles with controlled pore diameters typically 50–150 Å to which may be bonded a hydrophobic or polar organic layer. For completeness at this stage a brief synopses of the above sorption mechanisms are included.

6.2.1 Adsorption

Adsorption chromatography is sometimes also referred to as solid–liquid or normal phase chromatography; the latter term indicates the use of a polar packing with a non-polar eluant. The lattice of the common porous

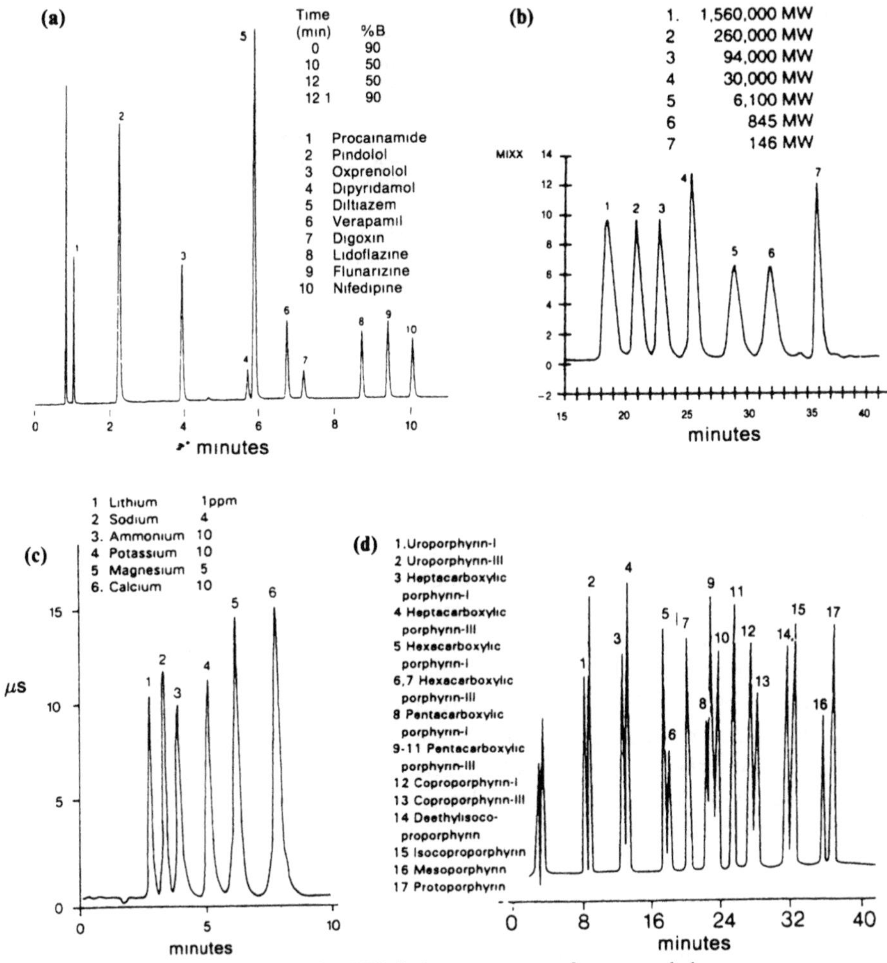

Figure 6.1 Illustrative HPLC chromatograms of compound classes.

(a) Reverse phase chromatography of cardiac drugs (permission of Supelco). Ten cardiac drugs resolved (low pH, short wavelength). Column SUPELCOSIL ABZ + Plus, 15 cm × 4.6 mm i.d., 5 µm particles, cat. no 5-9196, mobile phase acetonitrile: 25 mM KH_2PO_4 (pH 3.0), flow rate 2 ml min, detection UV, 220 nm, injection 20 µl methanol/water, 50:50 (30 µg/µl pindolol, diltiazem, 100 µg/ml dipyridamole, 50 µg/ml other analytes).

(b) Separation of styrene oligomers (permission of Machery Nagel). Polystyrenes (wide MW range). Column Phenogel 10 µ 10^5Å, 10^4Å, 10^3Å, dimensions 300 × 7.8 mm, mobile phase THF, flow rate 1.0 ml min^{-1}, detection differential refractometer, injection volume 100 µl 0.25% w/v, temperature ambient.

(c) Separation of Group I and Group II cations (permission of Dionex). Column IonPac CS12 (4 mm), eluant 20 mM HCl or methanesulphonic acid, flow rate 1 ml min^{-1}, detector suppressed conductivity, injection volume 25 µl.

(d) Separation of porphyrin isomers (permission of Phenomenex). Column Hypersil 5 µ NH_2 (APS), dimensions 250 × 4.0 mm, mobile phase A: 10% Acetonitrile in 1 M ammonium acetate pH 5.15, B: 10% Acetonitrile in Methanol, gradient 30 min linear gradient from 100% A to 65% B followed by isocratic elution at 65% B, for 10 min, flow rate 1.0 ml/min, detection UV *viz* 404 nm.

Table 6.1 Sorption modes and associated retention mechanisms

Sorption mode	Mechanism of retention
Liquid–solid adsorption	Surface adsorption on basis of polarity.
Bonded phase	Either partitioning between phases or adsorption interactions between solute and polar bonded phase.
Ion exchange	Charge interaction between solute ions and counter ionophores on packing.
Ion pair	Partitioning of neutral ion pairs between phases.
Size exclusion	Filtering effect on the basis of hydrodynamic volume.
Chiral	Diastereoisomeric interactions between solute enantiomers and chiral sites within the packing.
Affinity	Bio-specific binding of solute to immobilised ligand.

adsorbents, e.g. alumina and silica, is terminated at the surface with polar hydroxyl groups which may be free or hydrogen bonded, and it is these groups which provide the means for the surface interactions with solute molecules.

The eluant systems used in adsorption chromatography are based on non-polar solvents, commonly hexane, containing a small amount of a polar modifier, such as 2-propanol, dichloromethane or methyl-*t*-butyl ether. When the sample is applied to the column sample molecules with polar functional groups will be attracted to the active sites on the column packing; they will subsequently be displaced by the polar modifier molecules of the eluant, as the chromatogram is developed, and will pass down the column to be re-adsorbed on fresh sites. The ease of displacement of solute molecules will depend on their relative polarities; more polar molecules will be adsorbed more strongly and hence will elute more slowly from the column.

6.2.2 Bonded phase chromatography

Bonded phase chromatography (BPC) takes place either under normal phase or reverse phase conditions. In reverse phase mode the stationary phase is non-polar while the eluant is polar, e.g. methanol or acetonitrile with aqueous buffers. Bonded phase packings have superceded the classical packings where the stationary phase was distributed over the surface of the support particles and bound simply by physical forces of attraction. However, due to the problems of solvent stripping and limited hydrolytic stability, these classical systems, though developed for a few specialised applications, have been replaced by organo-bonded stationary phase materials.

6.2.2.1 Reverse phase liquid partition. The limitations of classical liquid–liquid partition chromatography (LLC) systems led to the development

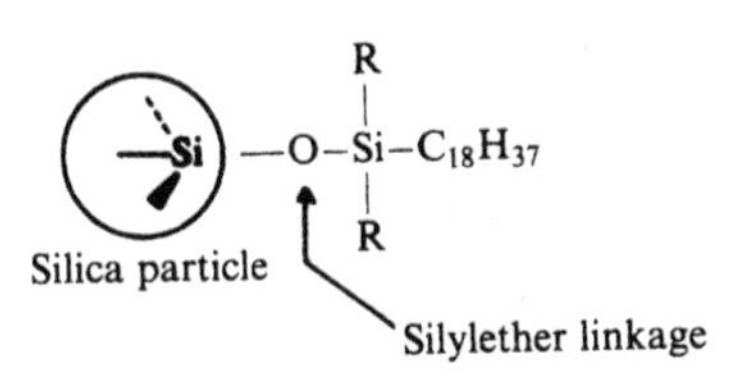

Figure 6.2 Octadecylsilane (ODS) chemically bonded stationary phases.

of packing material where the stationary phase is chemically bonded or organo-bonded to an insoluble matrix. The most common stationary phase is octadecylsilane (ODS), which is bonded to a silica support via a silyl ether (siloxane) linkage (Figure 6.2).

Such a packing material is used with a polar eluant, e.g. methanol or methanol–water; an elution technique commonly referred to as reverse-phase chromatography. Other reverse-phase packings in common use have bonded to the support one of the following: C_8 hydrocarbon chains, C_8 groupings containing aromatic moieties, phenyl groups and polar materials such as cyanopropyl and nitropropyl. The range of stationary phases, though extremely varied in character, is small in number due to the extensive control that can be exerted over selectivity by variation in the eluant composition, a feature in marked contrast to GC. These packing materials, with some restrictions on the pH of the eluant used, provide good hydrolytic stability and are resistant to solvent stripping within normal column operating pressures. Newer reverse phase packings, based on graphitised carbon and on rigid, porous microparticulate beads comprising a polystyrene/divinylbenzene matrix, are available, which though more polar than ODS can tolerate a much wider range of eluant pH (1–13).

6.2.2.2 Normal phase liquid partition. The polar characteristics of the stationary phase can be modified by incorporating ether, nitrile, nitro, diol and/or amino substituents normally at the end of a hydrocarbon chain or, on an aromatic ring, both of which are chemically bonded to the support material. The major advantage of these materials compared to the liquid–solid systems described above is the facility to undertake gradient elution. Elution is normally carried out with relatively non-polar solvents such as tetrahydrofuran, diethyl ether, chloroform and hexane.

6.2.3 Ion exchange

Ion-exchange chromatographic (IEC) techniques utilise the differing affinities of ions in solution for oppositely charged ionic groupings located on the packing. The nature of functional groups providing the sites for exchange, quaternary ammonium ($-N^+R_3$) for anions and sulphonic acid

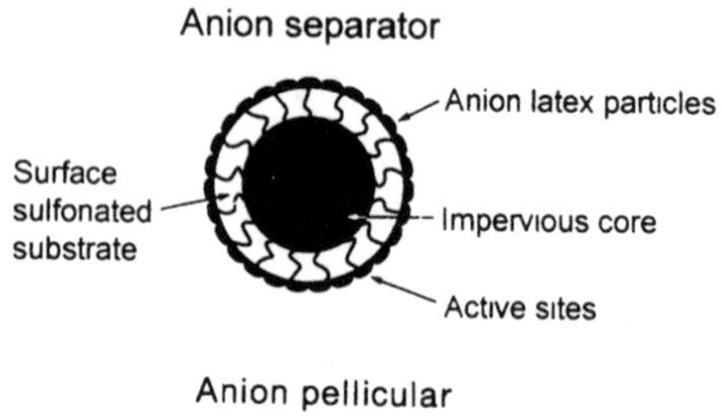

Figure 6.3 Schematic representation of a pellicular latex packing (Adapted ex Dionex, with permission).

($-SO_3H$) for cations, is the same in HPLC as in classical ion exchange chromatography. A variety of substrates ranging from cross-linked polystyrene, cross-linked polydextrans, cellulose and silica have been utilised. However, due to the problems of swelling, compressibility and mass transfer encountered with macroreticular polymeric supports and the limited pH stability of silica-based packings, the efficiencies achieved in IE–HPLC applications have been moderate.

It was not until the development of ion exchange materials based on small particle, pellicular resins layered with latex ion exchange material (Figure 6.3) that improved separation, speed, efficiency and selectivity was achieved in this mode. The characteristics of these resins can be finely tuned by altering either the degree of cross-linkage, the size or the functional group of the exchange material. A more detailed treatment of these materials and associated sorption mechanisms are presented in section 6.4.

6.2.4 *Ion pair partition*

Ion pair chromatography (IPC) is complementary to and overlaps with ion exchange and is essentially a type of reverse phase partition chromatography used increasingly for the separation of ionisable organic compounds. This mode utilises an eluant system which contains an ionic compound containing a relatively large organic counter-ion which has the opposite charge to that of the analyte ion(s), and which will form a neutral ion-pair with the ionic sample components:

$$\underset{\nearrow \atop \text{Ion sample}}{A^+} + \underset{\nwarrow \atop \text{Counter-ion}}{B^-} \rightleftharpoons \underset{\uparrow \atop \text{Ion pair}}{A^+B^-}$$

While there are different exposes of the technique, the essential prerequisite is that the counter ion contains bulky organic substituents such that the ion-pair subsequently formed will be hydrophobic in character and will be attracted to the non-polar stationary phase. This enables conventional reverse-phase packings and methanol–water based solvent systems to be used for ionic samples. A wide range of ion pair reagents are available ranging from the anionic alkyl sulphonic acids as their sodium salts to the cationic ammonium bromides, e.g. hexadecyltrimethylammonium bromide. The great utility of this technique is in its ability to be used for the analysis of mixtures which contain ionic and non-ionic analytes, and is illustrated in Chapter 9, experiment 27.

6.2.5 *Gel permeation–Gel exclusion*

Exclusion chromatography [7, 8], one of the newer of the LC methods, is variously known as gel chromatography, gel filtration and gel permeation chromatography. Exclusion chromatography is fundamentally different from the above modes and utilises the selective diffusion of solute molecules within the solvent filled pores of a three-dimensional lattice. Small molecules will permeate the pores while large bulky molecules will be excluded. Thus, separation is achieved principally on the basis of molecular weight and size, with the larger molecules being eluted from the column more quickly.

Principally two classes of material are available for exclusion chromatography either, semi-rigid cross-linked macromolecular polymers or rigid inert porous silicas or glasses—the bead size of both is in the range 5–10 μm.

The organic polymers were based on dextran or co-polymerised styrene–divinylbenzene. Size exclusion was carried out on these materials, however, due to their limited mechanical strength and a tendency to swell readily column head pressures could not exceed 300 psi (2070 kPa) leading to a concomitant reduction in eluant flow-rate. The dextrans are unsuitable for aqueous-based mobile phases since water cannot penetrate the pores due to its high surface tension though sulphonated polystyrenes are compatible. A range of styrene divinylbenzenes with a variety of controlled pore sizes and associated molecular weight exclusion ranges are available. Organic gels formed from the co-polymerisation of 2-hydroxyethylmethacrylate and ethyl dimethacrylate have hydrophilic character and can withstand pressures up to 3000 psi (20.7 MPa).

The rigid packings, porous silica and glasses, are synthesised in pore sizes ranging from 4 to 250 μm giving corresponding operating ranges of 1000–8000 and 2.5×10^5–1.5×10^6 Da. These rigid packings have numerous advantages over the organic semirigid gels, e.g.

- chemically resistant up to pH 10;
- can be used at high flow-rates, column head pressures and elevated temperatures; and
- compatible with both aqueous and organic solvents.

The deleterious surface adsorption properties can be eliminated by chemically modifying the surface.

6.2.6 *Affinity*

The chromatographic modes described so far depend for their effectiveness on differences (often small) in adsorption, partition, ionic charge or size, between analyte components of similar chemical character. Affinity chromatography is a modified gel technique [9,10] which exploits the specificity of a donor–ligand interaction, many applications involve biospecific binding between an immobilised biochemical and a protein. The technique uses packings typically based on soft gels such as agarose the ligand being immobilised by covalent bonding to the matrix. Often a short-chain alkyl group or spacer is inserted between the ligand and the matrix to eliminate steric hindrance of the binding interaction. Some applications have used wide pore silicas as the ligand support. The sample is applied to the column in a suitable buffered eluant such that the substrate becomes specifically, though not irreversibly bound to the ligand (see Chapter 4, Figure 4.14).

The composition or pH of the eluant is then altered in such a manner as to weaken the ligand–substrate bonding, thus promoting dissociation and facilitating elution of the retained compounds. Due to the nature of the technique packings are synthesised for particular applications. Thus affinity chromatography, while outside the mainstream of HPLC, has considerable potential and utility in biochemical and clinical applications, such as the purification of antigens, enzymes, proteins, viruses and hormones.

6.2.7 *Chiral chromatography*

A number of chiral stationary phase (CSP) materials are commercially available for the resolution of enantiomeric mixtures. One of the first was based on derivatives of chiral amino acids, such as phenyl glycine and leucine, which are covalently or ionically linked to a 5 μm aminopropylsilica (Figure 6.4). These phases, introduced by Pirkle [11], have been used for the separation of amino acids and hydroxy acids. Separation in this instance is dependent on at least three points of interaction between the enantiomers and the stationary phase.

Other packings have used esterified celluloses, chiral peptides and β-cyclodextrins as enantioselective media [12]. The cyclodextrins are chiral carbohydrates formed from up to 12 glucose units. The monomers are configured such that the cyclodextrin has the shape of a hollow truncated cone or barrel-like cavity within which stereospecific guest–host interactions can occur, though other features such as steric repulsion, solvent, pH, ionic strength and temperature all affect retention.

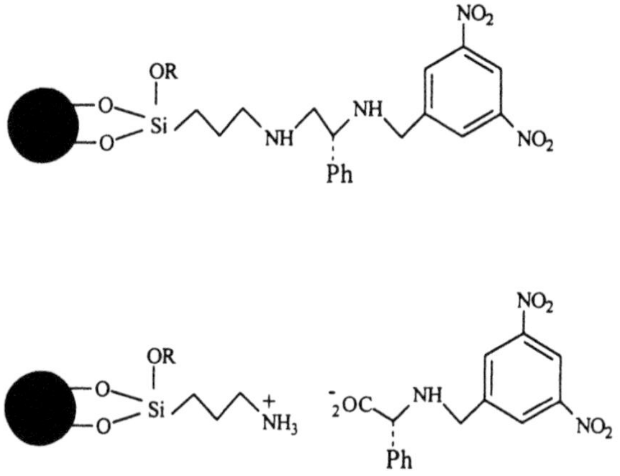

Figure 6.4 Pirkle stationary phases showing *N*-(3,5-dinitrobenzoyl)phenylglycine, ionically and covalently bonded to an aminopropyl silica.

6.3 Overview of high performance liquid chromatography instrumentation

The following is intended as no more than a brief overview of HPLC instrumentation at this stage, the individual components will be discussed in detail later in this chapter [13]. In HPLC filtered eluant is drawn from the solvent reservoirs, the eluant composition being determined by the proportion of each solvent delivered to the column via a high pressure pump and a solvent mixing system. The sample mixture is applied to the top of the column and the components of the mixture are then carried down through the column by the eluant at a rate which is inversely proportional to their attraction for the packing material. The passage of the solutes from the column is monitored by the detector and the response displayed on either a chart recorder or an integrator.

Though resolution of sample mixtures can often be effected using a solvent of constant composition (isocratic elution) gradient elution techniques, which involve changing the eluant composition in a controlled manner during the development of the chromatogram, provide a powerful tool for the optimisation of difficult separations. This has led to the development of LC systems which incorporate gradient forming and solvent mixing capability. The system illustrated (Figure 6.5) uses a low pressure mixing procedure (i.e. solvents A and B are mixed on the low pressure side of the pump); an alternative configuration of the system components would provide for solvent mixing on the high pressure side of the pump. Equipment for both approaches is modular in design and though described more fully in later sections a brief description of the essential common components follows.

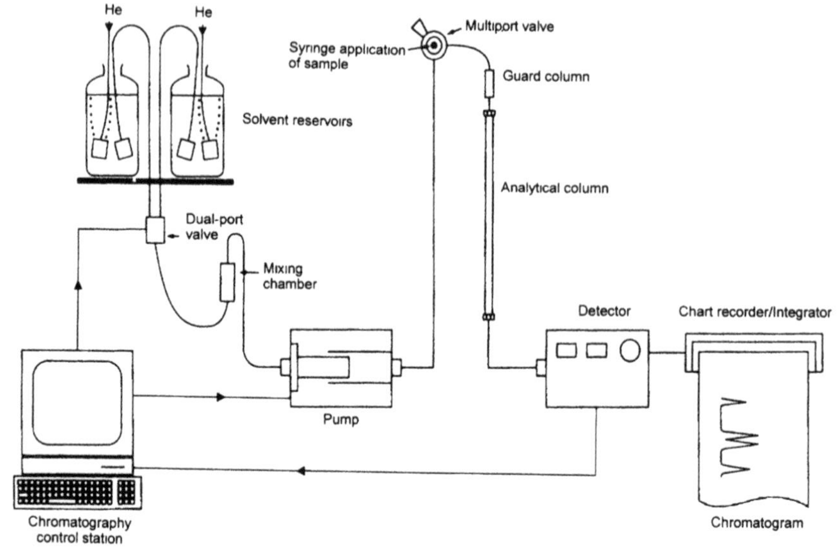

Figure 6.5 Low pressure mixing gradient HPLC system.

6.3.1 *Packings*

The fundamental problem in achieving high resolution in liquid chromatography was in attaining rapid mass transfer of solute molecules between the packing material and the mobile phase, and thus considerable effort was committed to the development of suitable support materials. These efforts culminated in the provision of microporous particles of 3, 5 and 10 μm diameter, with controlled pore size and with large unit surface areas permitting high loadings of stationary phase if required. This development allowed the theories of van Deemter and Giddings, which had postulated the potential of LC using microparticulate packings to be applied and this more than anything else engendered the rapid advancement in HPLC techniques. The nature of the stationary phases required for the different modes of HPLC has previously been indicated.

6.3.2 *Pumps*

As a consequence of the large back pressures encountered, due to the small particle size of packing used in HPLC columns, pumps must be employed to achieve acceptable eluant flow rates. The pumps may be classified as either those which provide constant inlet pressure or those which provide constant outlet flow and should be capable of delivering up to 7000 psi (48.3 MPa). They are constructed from materials which are resistant to the organic solvents and aqueous buffer solutions commonly used as eluants. The system may also have provision for microprocessor control of the pumps to provide for complex gradient elution using high or low pressure mixing.

6.3.3 *Detectors*

It is convenient to use continuous monitoring detectors located at the column exit. Various detector designs have been used and these may be classified as those which monitor a specific property of the solute or those which detect changes in a bulk property of the column effluent.

An example of the former is the ultraviolet spectrophotometer which may be of fixed wavelength (usually 254 or 280 nm) or variable wavelength design. The detector functions by monitoring the change in absorbance as the solute passes through the detector flow cell, i.e. it utilises the specific property of the solute to absorb ultraviolet radiation.

An example of the second type of detector is the refractive index monitor which functions by recording the refractive changes in the eluant as the solute passes through the detector cell. Bulk property detectors, though more versatile, are generally several orders of magnitude less sensitive than specific property detectors and the choice for a particular application is often dictated by solute characteristics.

6.3.4 *Solvents*

The eluants employed for HPLC separations may comprise water, aqueous buffer solutions, miscible organic solvents such as methanol and acetonitrile or a mixture of the above. All solvents should be of high spectroscopic purity, dust-free, and should be filtered and degassed (i.e. have dissolved gases displaced) before use. They should also, if ultraviolet detection is being employed, be transparent to the detector wavelength employed.

6.3.5 *Columns*

Columns for analytical HPLC are typically 10–25 cm long and 2.1–4.6 mm i.d. The columns are constructed of stainless steel to cope with the high back pressure and are glass lined to prevent metal catalysis of solvent–solute reactions at the high column pressures experienced. As mentioned previously, optimum packing size is $\leq 5\,\mu m$ and columns with such packing would yield $>10\,000$ theoretical plates m^{-1}. An important consequence of the small size of packing materials and columns is that the volumes of detectors and column couplings (known collectively as extra column dead volumes) must be minimised to avoid excessive loss of resolution of the chromatographic peaks. The volume of detector flowcells is commonly $\leq 10\,\mu l$.

6.3.6 *Injection*

Column performance in HPLC is critically dependent on the sample injection. Samples may be introduced in a number of ways. The earliest method was based on the GC approach and involved syringe injection of the sample onto the top of the column through an elastomer septum. The technique is operator sensitive and cannot be used at column head pressures in excess of 50 bar. In another method, referred to as stopped flow, the flow of eluant to the column is stopped and once the pressure reaches ambient the column is opened and the sample applied. The column is then sealed and eluant flow restarted. This technique is time consuming, gives irregular flow and results in poor reproducibility of retention times. Though cheap to use these two methods are seldom encountered in modern HPLC and are really only of historical interest.

The most common injection technique is via six-port valves of the Valco and Rheodyne design. These valves afford precise, accurate sample loading with minimal interruption to solvent flow. They can be used up to column pressures of 7000 psi (48.3 MPa) and the sample loading can be varied either by part filling of the loop or by changing the loop volume.

6.4 Theory

The following section is intended as no more than an overview of the theoretical principles considered relevant to a basic understanding of the factors which influence chromatographic separations in liquid column chromatography. The reader is directed to Chapter 2 and references therein for a fuller account of the principles and theories of chromatographic processes. Up to this point the discussion of the chromatographic processes in HPLC has been largely qualitative; however, many of the foregoing principles and concepts can be expressed in precise mathematical relationships.

6.4.1 *Retention*

In order to achieve separation in chromatography, the analytes must be retained in the column. The fundamental relationship between the retention volume (V_R), the quantity of stationary phase (V_S) and the distribution coefficient (K) is expressed in the following equation:

$$V_R = V_M + KV_S$$

where V_M is the volume of mobile phase within the column. The retention volume can be determined directly from the chromatogram as $V_R = F_C \times t_R$, where F_C is the flow-rate of solvent, and t_R is the retention time of the component; similarly $V_M = F_C \times t_M$, where t_M is the time taken for an unretained compound to traverse the column; however, in order for K, which measures the extent of the retention, to be evaluated, V_M and V_S must be known.

A more practical expression of solute retention is given by k, the capacity factor:

$$k = \frac{n_S}{n_M}$$

Here, n_S is the total number of moles of solute component in the stationary phase and n_M is the total number of moles of analyte in the mobile phase. It can subsequently be shown that

$$V_R = V_M(1 + k)$$

and that

$$k = \frac{t_R - t_M}{t_M}$$

Thus, the capacity factor, k, is a measure of the sample retention by the column and can be determined from the chromatogram directly.

6.4.2 *Column efficiency*

The concept of plate number and plate height in chromatography was developed by Martin and Synge who used an analogous model and theory of mass distribution as that adapted for the elucidation of processes occurring in a distillation column.

The empirical expressions derived are generally applicable to all types of column chromatography and are used universally, first as a measure of column performance and efficiency, expressed in terms of the number of theoretical plates (N) and the height equivalent of a theoretical plate (HETP) denoted by H, and secondly, to measure the resolution (R_S) attained in the chromatogram. For a fuller discussion the reader is directed to Chapter 2, though a brief resume of the expressions for these parameters is presented below.

A measure of the separating efficiency is given by the number of theoretical plates the column is equivalent to, which is defined in terms of the retention time (t_R) and the base width of the peak (w_b):

$$N = 16\left(\frac{t_R}{w_b}\right)^2 = 5.54\left(\frac{t_R}{w_h}\right)^2$$

The efficiency of any column is best measured by the HETP which is expressed as

$$H = L/N$$

where L is the length of the column. The resolution, R_S, of two compounds is defined as being equal to the peak separation divided by the mean base width of the peaks:

$$R_S = \frac{2(t_{RB} - t_{RA})}{w_{bB} + w_{bA}}$$

A more practical expression for R_S in terms of the experimental factors α, k and N, can be arrived at by substitution for t_R and w_b in terms of N. Thus, R_S as a function of α, k and N is given by

$$R_S = \underset{\text{(i)}}{\frac{\sqrt{N}}{4}} \underset{\text{(ii)}}{\left(\frac{k}{k+1}\right)} \underset{\text{(iii)}}{\left(\frac{\alpha - 1}{\alpha}\right)}$$

where the selectivity factor $\alpha = k_2/k_1$ for component bands 1 and 2 in the chromatogram. The three terms are broadly independent and thus can be individually optimised see section 6.14.

Plate theory, though useful in monitoring and comparing column performance and in giving a measure of the resolution achieved in the chromatogram, has a number of shortcomings. Some of these deficiencies arise from the simplicity of the model adopted. The main criticism, however, is that the treatment does not consider the influence of important chromatographic variables such as particle size, stationary phase loading, eluant viscosity and flow-rate of column performance; and furthermore, how these factors might be adjusted to optimise chromatographic efficiency.

6.4.3 Rate theory

The rate theory of chromatography avoids the assumptions of plate theory such as instantaneous equilibrium and isolated chromatographic contact units, and examines column efficiency with reference to the kinetic effects which operate during development of the chromatogram and which give rise to band broadening. The kinetic effects giving rise to band broadening in HPLC are similar to those considered in other modes of chromatography, though their relative contributions may be different.

Each of the rate factors contributing to the band broadening is considered in isolation and a mathematical expression evaluated for the plate height contribution in terms of the variables governing the kinetic effect under examination. These contributions are then combined to give a general expression for the overall plate height, commonly referred to as the van Deemter equation. The following terms are considered to contribute to broadening of the solute band during development of the chromatogram.

6.4.3.1 Multipath term (H_f). Broadening of a chromatographic zone as it passes through the column arises from the variable channels which the solute molecules may follow through the packing. This effect is known variously as the multipath, the eddy-diffusion or non-uniform flow term. However, in practice the solute molecules are not fixed in single channels but can diffuse laterally into other paths. These composite motions result in a decreased contribution to band broadening from this term. The contribution of this effect to the overall plate height (H) is proportional to the particle size (d_P) of the packing material within the column and is expressed as

$$H_f \sim \lambda . d_P$$

where λ is a measure of how uniformly the column is packed.

6.4.3.2 Longitudinal diffusion (H_d). Whenever a concentration gradient exists in the column then spreading, i.e. diffusion of the molecules from a region of high concentration to a region of low concentration, will occur due to random thermal processes. This interdiffusion of the two molecular species, which occurs in all directions independent of whether solvent is flowing or not, is dependent upon two factors: first, the diffusion coefficient (D_M) for interdiffusion of solute and solvent molecules; and second, the time over which this diffusion occurred, i.e. the time spent in the mobile phase (t_M). However, as t_M is inversely proportional to the mobile phase flow-rate (u), then the longitudinal diffusion contribution to the effective plate height of the column may be expressed as

$$H_{dm} \propto D_M/u$$

Furthermore, as diffusion can also take place radially, migration of solute molecules towards the column walls can occur. This could lead to appreciable band broadening on account of, first, the slower flow at the walls of the column, and second, solute molecule interaction with the walls, causing retardation of the zone. However, provided the sample is applied as a narrow concentrated band and the solvent has suitable viscosity, the rate of radial diffusion is not sufficient to become a problem, and under these conditions the column is described as operating in the infinite diameter mode.

There is also an analogous contribution to the overall plate height due to diffusive spreading in the stationary phase and using similar reasoning as above this may be expressed as

$$H_{ds} \propto D_S/u$$

where D_S is the interdiffusion coefficient of the solute and stationary phase molecules.

6.4.3.3 Slow equilibration (H_e). The assumption in the plate theory that the transfer of solute molecules between the mobile and stationary phases was instantaneous is invalid due to the finite rates of mass transfer within the mobile and stationary phase. This contribution to the effective plate height may itself be regarded as arising from two separate effects.

First, the system deviates from ideality as there is a finite rate of mass transfer of solute molecules across the chromatographic interface. The contribution to the overall HETP arising from this kinetic control of the sorption–desorption process increases with increasing mobile phase flow-rate.

The second effect, commonly labelled diffusion controlled kinetics, arises from the finite rate at which molecules can diffuse to the interface and become available for transfer, and has itself two contributions. In the mobile phase there exist velocity gradients across the channels in the packing material such that solute molecules closer to the packing will, in addition to having a shorter distance to diffuse, also flow more slowly than those in mid channel and thus the latter will have less opportunity to reach the interface and transfer. The above, collectively referred to as mobile phase mass transfer effects, may be expressed as

$$H_{\mathrm{em}} \propto d_{\mathrm{P}}^2 u / D_{\mathrm{M}}$$

Analogous arguments can be developed to take account of stationary phase mass transfer effects which can similarly be expressed as

$$H_{\mathrm{es}} \propto \frac{d_{\mathrm{S}} u}{D_{\mathrm{S}}}$$

where d_{S} is the thickness of the conceptual stationary phase and D_{S} is the diffusion coefficient of solute molecules within it.

6.4.3.4 Stagnant mobile phase ($H_{em'}$). Finally, the presence of substantial amounts of immobile solvent, which is either trapped in the interstices between the packing material or in the deep pores within the particles, is a further cause of band broadening; solute molecules entering these stagnant pools may quickly diffuse out or conversely diffuse further in, thus becoming effectively trapped.

The factors affecting the contribution of this effect to the overall plate height are as those experienced for $H_{\mathrm{em'}}$, the mobile phase mass transfer term. Thus, the mobile phase mass transfer term may be expressed as

$$H_{\mathrm{em'}} \propto \frac{d_{\mathrm{P}}^2 u}{D_{\mathrm{M}}}$$

Giddings [3, 4] has shown that the overall plate height arising from these kinetic factors can be expressed mathematically as

$$H = \frac{1}{(1/H_{\mathrm{f}} + 1/H_{\mathrm{em}})} + (H_{\mathrm{dm}} + H_{\mathrm{ds}}) + H_{\mathrm{em'}} + H_{\mathrm{es}}$$

where

$H_f = C_f\lambda . d_p$	(multipath term)
$H_{em} = C_{em}(d_P^2 u/D_M)$	(mobile phase mass transfer)
$H_{dm} = C_{dm}(D_M/u)$	(diffusion in mobile phase)
$H_{ds} = C_{ds}(D_S/u)$	(diffusion in the stationary phase)
$H_{em'} = C_{em'}\dfrac{d_P^2 u}{D_M}$	(stagnant mobile phase mass transfer)
$H_{es} = C_{ds}\dfrac{d_S u}{D_S}$	(stationary phase mass transfer)

and C_f, C_{em}, C_{dm}, C_{ds}, C_{es} and $C_{em'}$ are plate height coefficients. The equation is commonly represented in its simplified form:

$$H = Au^{0\ 33} + B/u + Cu + Du$$

The equation, though complex, shows the importance of particle size, mobile phase flow rate and diffusion coefficients and indicates that the deleterious effects on H and thus the column performance can be minimised by reducing the packing particle size, the stationary phase thickness and the solvent viscosity (thus decreasing D_M and D_S). The latter can be achieved by using elevated column temperatures.

A representation of the variation with flow rate of the individual contributions to H is shown in Figure 6.6. The main differences between the GC

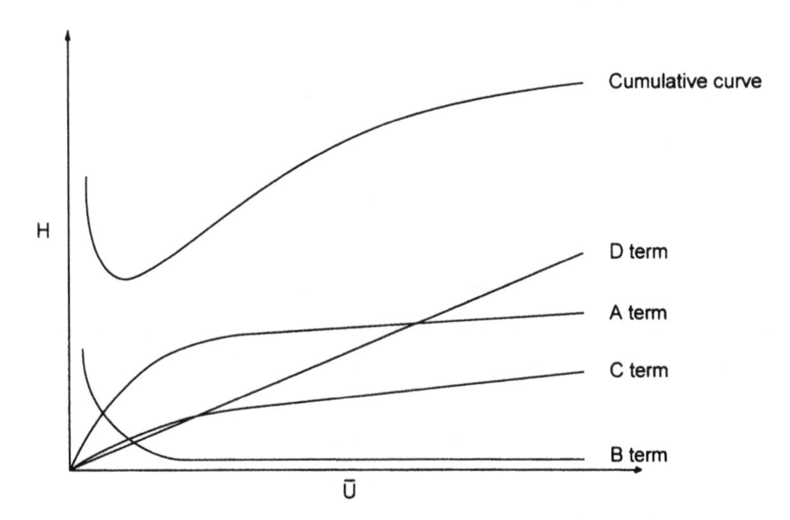

Figure 6.6 Plot of plate height (H) versus linear velocity of mobile phase ($\bar{U}$), showing contributions from the four terms detailed above.

and LC plots of H versus u arise due to the reduced rates of sample diffusion in liquids compared to gases.

D_M is 10^4–10^6 times greater in gases than in liquids. For LC columns containing large particles (>10 μm) H_{dm} is insignificant in comparison to those terms which are a function of d_P or d_P^2. However, as d_P decreases then the contribution to the overall plate height H, from H_{dm}, will become dominant at low flow velocities. In consequence the H versus u plot shows a minimum at flow velocities which are of importance for routine analytical analysis with 3 μm packings.

The other main difference between GC and LC is that the gradient of the curve is shallower for the latter and thus mobile phase velocities substantially greater than u_{OPT} can be used without seriously impairing column performance.

6.4.4 *Extra column band broadening*

The discussion of band broadening so far has been concerned with effects occurring in the column. The resolution of solutes attained can be further reduced if connecting tubing is of too large a volume or column connections contain unswept volumes. These extra-column and dead volumes are present in any chromatographic system as there is a finite volume between the point of injection and the column, between the column end and the detector, and in the detector flow cell itself. As the sample, dissolved in the mobile phase, passes through the extra-column volume broadening takes place simply as a result of the velocity gradient across the diameter of the connecting tubing and also due to the diffusional movement of the solute. The solvent and the solute in it, which is nearest the walls, will move more slowly than that in mid stream with a resultant dilution of the sample and broadening of the peak. These effects are less noticeable in LC than GC due to the lower rates of interdiffusion in liquids.

The overall peak dispersion produced is a combination of column and extra column effects, is additive and can be expressed in terms of their variances (see Chapter 2):

$$\sigma^2_{Tot} = (\sigma^2_{CInj} + \sigma^2_{Conn} + \sigma^2_{Det}) + \sigma^2_{Col}$$

It can be shown that the contribution to the overall plate height from these extra column effects is a function of a number of instrumental parameters, i.e.

$$H_{Ext\_Col} \propto \frac{\pi r^2 L}{24 D_M}$$

where D_M is the interdiffusion coefficient of analyte with the mobile phase, r is the radius of the column and L its length. It can also be shown that the width at the base of the peak is given by the following expression:

$$w_b = 4\pi^2 \epsilon_{Tot} L N^{-1/2}$$

where ϵ_{Tot} is the column porosity. By applying the above equations it can be shown that in order to limit remixing of the solutes to $\leq 1\%$, the following procedures should be adhered to:

- connecting tubing should be <0.025 mm bore and less than 20 cm in length;
- detection cells should not be larger than 10 µl and if possible smaller, especially where high efficiency, short or small particle size columns are being used; and
- there should be no voids in the system between the injector and the detector.

With narrow bore columns the flow-cell volume must be ≤ 3 µl in order that the resolution achieved on the column is not lost due to analyte spreading in the detector cell.

6.4.5 *Temperature effects and diffusion*

The variation of the capacity factor k with temperature depends upon the heat of transfer of the solute between the mobile and stationary phase and is expressed in the van't Hoft equation:

$$\frac{d \ln k}{dT} = \frac{\Delta H_{(m \rightarrow s)}}{RT^2}$$

Since the heat of transfer of solute between mobile and stationary phase is much smaller in LC than in GC, change in temperature has much less effect on the degree of retention and resolution. Nonetheless, an increase in temperature would reduce the mobile phase viscosity giving an increased rate of mass transfer and increased solute solubility. As a result, capacity factors decrease, peaks become sharper, giving better resolution and faster analysis and lower values of H. An important feature in the control of inter-diffusion of solute and solvent and of solute mass transfer is the solvent viscosity and where a choice of solvent is possible then that with the lower viscosity should be chosen, e.g. methanol in preference to ethanol since as indicated above mass transfer is optimum at lower viscosity with increase in resolution and column performance.

In GC capacity factors, retention and selectivity are controlled by adjusting the column temperature and stationary phase characteristics. In LC change in the composition of the eluant serves both purposes more effectively and thus solvent programming (otherwise known as gradient elution) is used in HPLC where temperature programming would be employed in GC.

Nonetheless, a change in temperature affects both resolution and retention characteristics and a small but significant number of applications use temperature as an operating variable. However, though the column temperature itself can be readily controlled there are still problems with regards to the thermostating of the solvent and detector.

6.5 Detailed discussion of high performance liquid chromatography instrumentation

6.5.1 Solvent delivery system

Solvent delivery modules are required to deliver a pulse free flow of eluant to the column at flow rates ranging from 0.1 to 10 ml min^{-1} with a precision of 0.5% or better. The pump must be capable of operating at pressures up to 7000 psi (48.3 MPa) and the system should incorporate a degassing unit to remove dissolved air and other gases from the eluant.

The main criteria for a suitable pump can be summarised as follows:

- the materials of construction should be inert towards the solvents used;
- the pump should be capable of delivering high volumes of solvent;
- it should be capable of delivering a precise and accurate flow;
- the pump-head should have a small volume to facilitate rapid change of solvent composition;
- it should be capable of delivering high pressures up to around 7000 psi (48.3 MPa); and
- it should deliver a pulse-free flow and hence not contribute to detector noise.

There are three main designs of solvent delivery system: (a) the reciprocating piston pump which may be of the single, dual or triple head design; (b) the syringe or displacement design which delivers constant non-pulsating flow; (c) the constant pressure or pneumatic pump.

Of these types the constant flow pumps, and in particular the reciprocating design are used in the vast majority of HPLC analytical applications [14] as they ensure reproducible retention time data, are suited to LC detectors the majority of which are concentration dependent and allow gradient elution techniques to be readily undertaken.

6.5.1.1 Reciprocating pumps. Reciprocating pumps of the single piston design (Figure 6.7) function by having a slow solvent delivery cycle compared to rapid refilling of the piston chamber. The flow rate to the column is modified by changing the length or rate of the piston stroke. The use of check valves controls the direction of flow into and out of the piston chamber. The chamber has a small volume typically 35–400 μl which allows for rapid change of eluant composition. There is, however, some pulsing of the solvent flow associated with this type of solvent delivery module due to

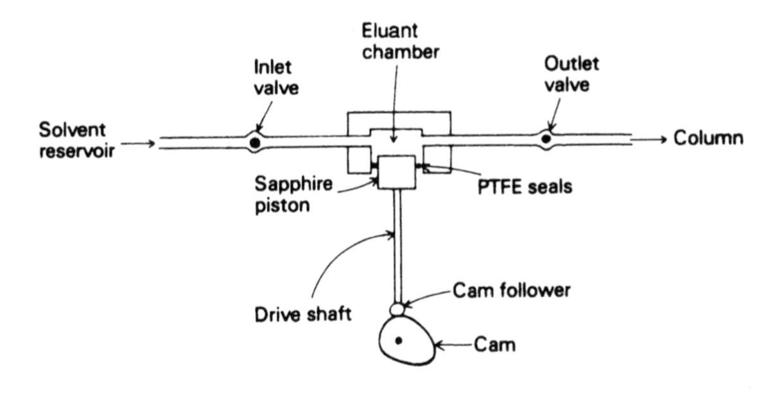

Figure 6.7 Single piston reciprocating pump.

the finite time taken to fill the piston reservoir and due to the fact that the initial part of the delivery stroke is concerned with compression of the solvent prior to pumping. Precise uniform flow is achieved by suitable design of the cam system, such that the compression portion of the delivery cycle is very small compared to the displacement segment, and by the use of microstep controlled direct drive motors.

Such drive systems largely eliminate flow rate instabilities and are capable of providing precise flow-rates to 0.01 ml min^{-1}. Further smoothing of fluctuations in flow delivered from the pump can be achieved using pulse dampners, that is Bourdon coils, which are essentially coiled tubing inserted between the piston and the column. The disadvantages of using pulse dampening coils are the attendant problems of dead volumes and unswept dead volumes introduced prior to the column, which makes change of solvent composition rather difficult and inefficient. Pulse dampners are available which have minimal dead volume and use a compressible liquid separated from the mobile phase by a diaphragm. The compressible fluid expands on the piston full cycle so maintaining system pressure and eluant flow.

Improved precision and smoothing of flow is provided by twin-piston reciprocating pumps. Here, the pistons are driven, either via a single cam (Figure 6.8) or more commonly by individual cams on a single rotating drive shaft, by a variable speed motor, approximately 180° out of phase (Figure 6.9). As piston A completes its compression stroke, piston B begins its refill cycle. These pumps provide a constant solvent flow which is easy to control, almost entirely pulse free, and with a small pump-head delivery volume, they afford rapid change of solvent composition. This makes them readily adaptable for gradient elution chromatography.

A further design uses triple headed pumps, where the pistons are effectively working 120° out of phase, with two heads in different stages of filling as the third is pumping, thus smoothing the flow considerably.

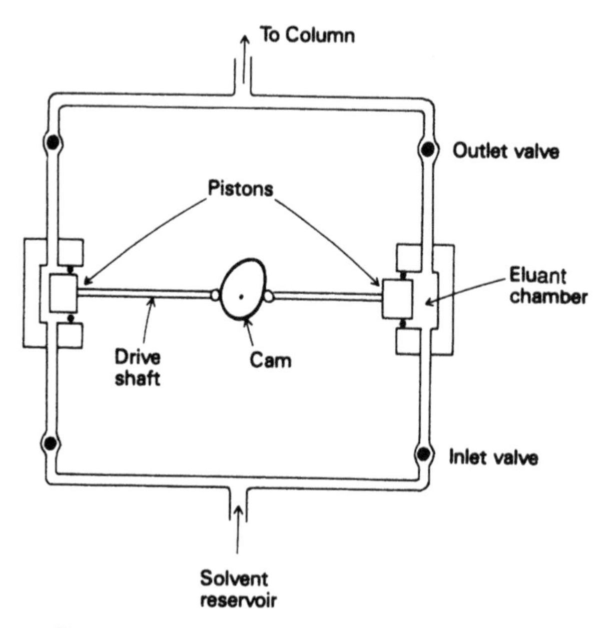

Figure 6.8 Single-cam dual piston reciprocating pump.

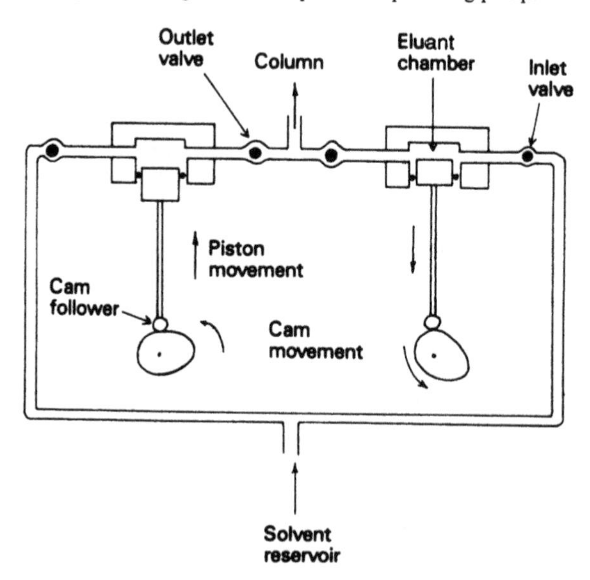

Figure 6.9 Twin-cam dual piston reciprocating pump.

The advantages of reciprocating pumps over other designs can be summarised as follows:

- they can be used continuously as theoretically they have infinite solvent reservoirs;

- they can readily be made pulse-free; and
- the low piston chamber volume facilitates rapid change in solvent composition.

6.5.1.2 Syringe pumps. The construction of this pump (Figure 6.10) is similar to the constant pressure one, except in this instance the piston delivering the solvent is driven by a digital stepping motor. The flow delivered is determined by the incremental rotational rate of the motor, the piston is therefore driven at constant speed during the delivery stroke, and hence delivers a constant flow-rate to the column. However, the flow-rate delivered though essentially pulse-free is dependent upon the solvent viscosity and the column back pressure.

This overcomes the major disadvantage of pneumatic amplifier type pumps, and makes syringe pumps ideal for reproducibility of retention time data. However, the major problem encountered with the syringe pump is the design of a suitable refill mechanism. However, it is possible to use two pumps in concert giving limited gradient elution capabilities also the finite volume of the eluant chamber (250–500 ml) is not such a disadvantage with narrow bore columns with their small flow-rate requirements.

6.5.1.3 Constant pressure pumps. Constant pressure pumps (Figure 6.11) deliver solvent via a small headed piston which is driven by a pneumatic amplifier. A gas acts on the relatively large piston area of the pneumatic actuator. This is coupled directly to a small piston which pushes the eluant through the column. Pressure amplification is achieved in direct ratio to the piston areas and thus for low inlet pressures (approximately 100 psi (690 kPa)) it is possible to obtain large outlet pressures (10 000 psi (69 MPa)).

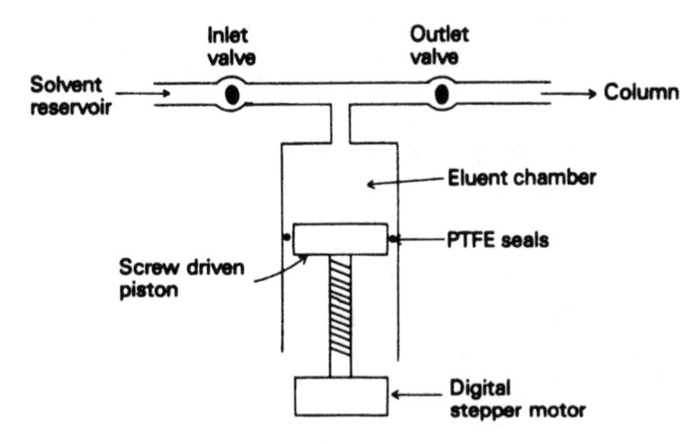

Figure 6.10 Syringe pump.

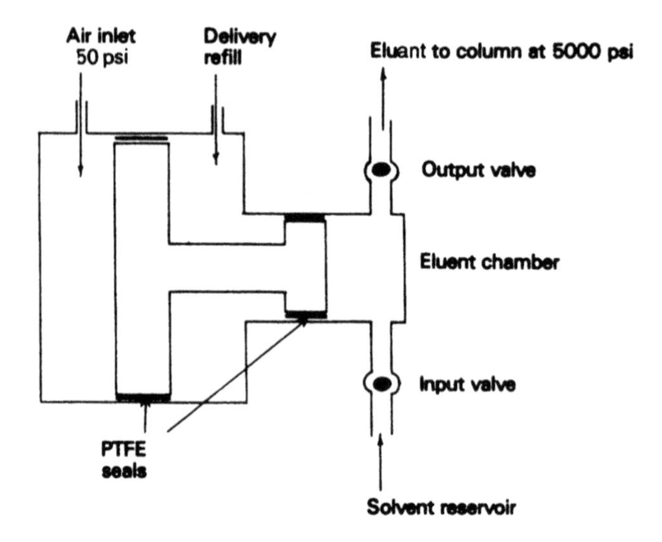

Figure 6.11 Constant pressure pump.

The main advantages of these pumps were

- low cost;
- ability to deliver high pressure; and
- stability of flow during the delivery stroke of the pump.

However, due to their mode of operation, pneumatic amplifier pumps have certain disadvantages. They are constant pressure rather than constant flow and therefore, as the elution volume is proportional to flow, fluctuations in the latter—due to, for example, partial column blockage or temperature change—can lead to poor precision and accuracy of analysis. The flow-rate is also dependent on solvent viscosity and column back pressure.

In addition, the chamber has to be refilled periodically with solvent, thus resulting in an interruption of the analysis. The design of such pumps also makes it difficult to change solvent compositions. However, though they are not ideally suited for analytical LC constant pressure pumps are still commonly used for slurry packing of columns due to their ability to deliver high flows and pressures.

6.5.2 Equipment for gradient elution

The quality of an HPLC analysis is often equated with the resolution of the peaks achieved, and as previously discussed this may be optimised by adjusting the solvent composition to obtain appropriate capacity factors for the solutes. For difficult separations, isocratic elution may be inadequate and the desired resolution may only be achieved using gradient elution techniques. Gradient elution [14] involves changing the composition of the eluant during development of the chromatogram; this change may be in the form of the solvent polarity, ionic strength, pH or combinations of these parameters. The change in solvent composition could be either stepwise, linear or curvilinear (convex or concave).

All gradient forming systems must ensure complete mixing of solvents before delivery to the column. Often the solvents employed shall have significantly different viscosities and miscibility characteristics and the mixing process shall tend to be exothermic which can result in evolution of gas if the solvent has not been degassed prior to use. Mixing chambers initially comprised a short length of column packed with ballotini beads though newer systems have a chamber packed with a proprietary fibre designed to create non-laminar flow and thus ensure mixing. In addition, these mixing chambers serve as degassing units as they are fitted with a vent which allows discharge of any liberated gases.

There are two types of system commercially available for gradient forming.

6.5.2.1 Low pressure mixing. In this system the solvent is drawn from the reservoirs at a ratio determined by the switching valve rate and then passed through a mixing chamber prior to delivery to the piston head. The switching valve is microcomputer controlled and gradient-profile data is entered via the keyboard for construction of the desired gradient. These systems are adequate down to 5% of one or other solvent. However, lower solvent compositions are difficult to obtain due to the finite actuation time of the solenoid valve as the opening time becomes comparable to the flow time in the low composition position. Further, since even with dual piston reciprocating pumps, there is a short time when no solvent is being drawn by the pump (due to the fill-cycle of the pump being shorter than the compression cycle), further inaccuracies in gradient profile can arise at low switching rates. These problems can be overcome by using a mixed solvent in one of the reservoirs.

For example, in the system shown (Figure 6.12) 99% B can be achieved with a 9 : 1 switching rate. An incidental advantage of this technique is the degassing which occurs during premixing. The main advantage of low pressure mixing systems is in cost, since only one pump is required. A disadvantage is the greater dead volumes prior to the pump which delay the effect of solvent change at the switching valve acting on the column.

6.5.2.2 High-pressure mixing. High pressure mixing uses two pumps (Figure 6.13) whose flows are controlled via a microcomputer. The total flow required for the application is selected and then the composition of A and B entered, e.g. for 75% A at 1 ml min^{-1} pump A is programmed to deliver 0.75 ml min^{-1} and pump B 0.25 ml min^{-1}. The solvent streams are

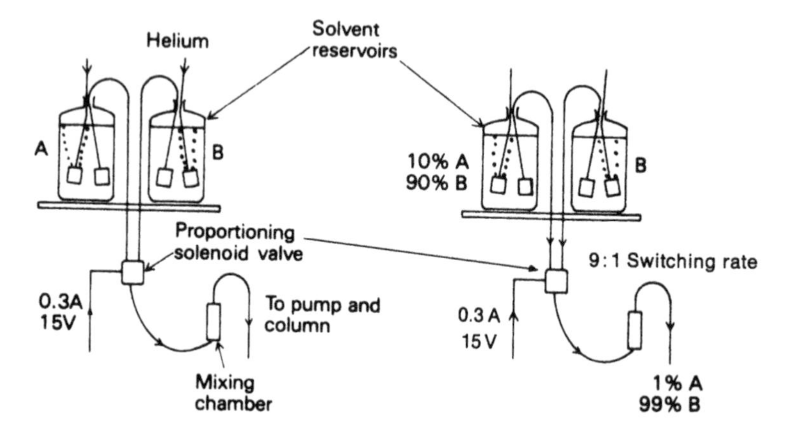

Figure 6.12 Low pressure gradient former.

then mixed between the pumps and the column, i.e. on the high pressure side of the pumps.

This system can be used at low compositions (1%) of A or B (except at low flow-rate). The main disadvantage is cost and inconvenience, as two pumps are required for this set-up and that the system is generally limited to two-solvent gradients.

6.5.2.3 Quaternary gradient systems The introduction of four solvent delivery modules has proved invaluable in developing quaternary solvent systems in the optimisation of resolution and in method development [7]. Commercially available systems such as those marketed by Perkin Elmer and Du Pont utilise low pressure solvent mixing via a four-way solenoid valve (Figure 6.14). The principal instrumental innovation is in pump design which allows for considerably extended solvent proportioning times, for example, at a flow rate of $1\,ml\,min^{-1}$, the proportioning time is 6 s; this advance provides for good accuracy and reproducibility of the complex solvent mixtures required.

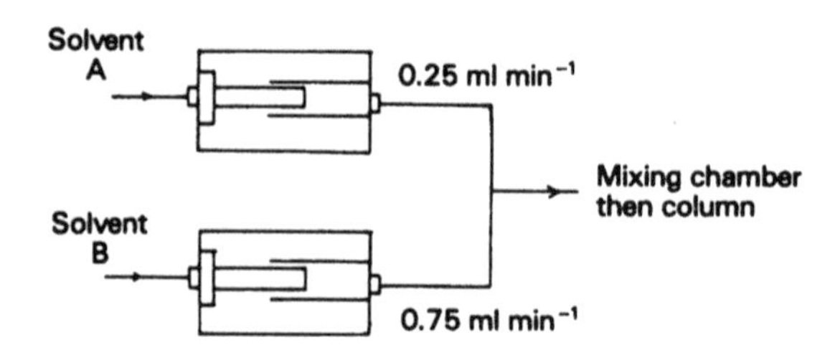

Figure 6.13 High pressure gradient former.

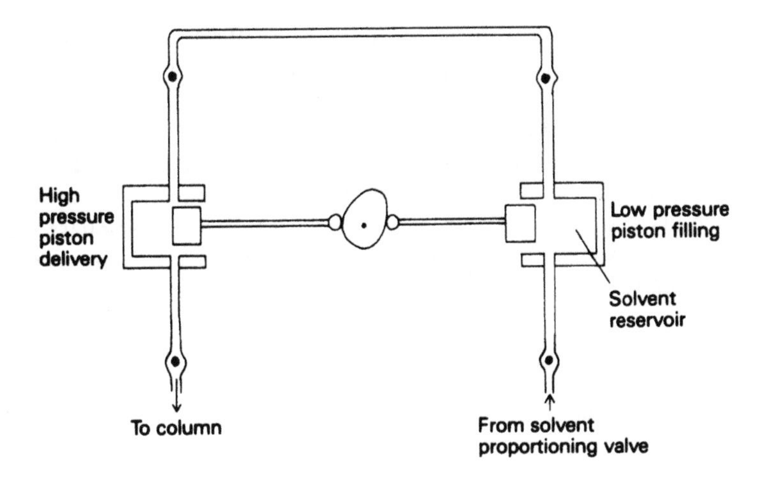

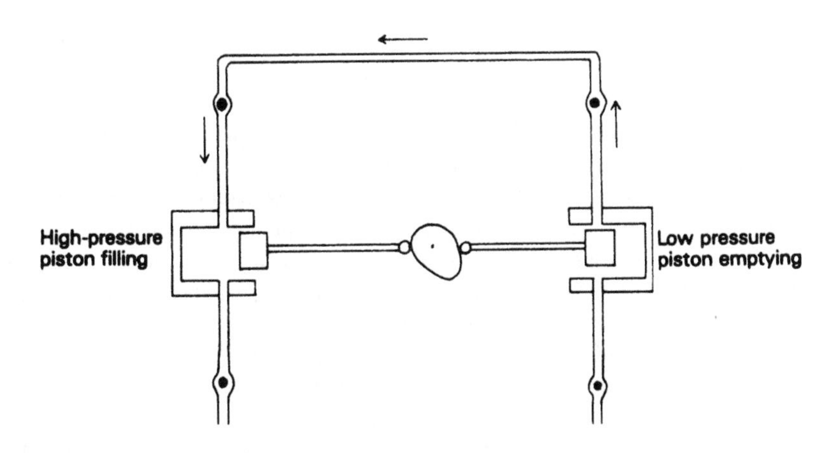

Figure 6.14 Solvent mixing pump.

The pump consists of two pistons, but solvent is delivered to the column only from one piston, the other piston chamber serves as a reservoir. As illustrated the pistons are 180° out of phase such that when piston A is delivering solvent to the column, the piston reservoir B is being replenished with solvent. The piston volumes are 100 µl and thus at a flow-rate of $1\,\mathrm{ml\,min^{-1}}$ the delivery fill cycle is 6 s. As the fill cycle takes the same time as the delivery cycle then the proportioning valves on the low pressure side have 6 s to actuate. Mixing of the component solvents occurs inside the pump and connecting tubing during the transfer of the solvent from the reservoir piston to the delivery piston.

The routines developed for determining the optimum four-solvent eluant composition are designed to yield a prechosen level of resolution for all pairs of components in a mixture by carrying out no more than seven analyses with various solvent compositions. Evaluation of these chromatograms allows selection of a set of standard conditions for the given application. The technique has a sound basis in theory and appears to have general applicability and has been used with both normal and reverse phase systems. The routines associated with the optimisation process have been reviewed by Glajch *et al.* [15] and Poile [16] and are discussed in detail in section 6.14.

6.5.3 Sample introduction

There are a number of methods for applying sample to the column which can be categorised as

- syringe injection;
- stopped flow injection; and
- valve injection.

However, syringe injection is little used today.

6.5.3.1 Valve injection. Valve injection of the sample is now the preferred and accepted technique. Sample application is rapid, the solvent flow from the pump does not have to be stopped and these systems are easy to use, readily adapted for automated injection and can operate at pressures up to 6000 psi (41.4 MPa) with reproducibility $\geq$0.2%. Six-port valves are commonly used, either fitted with an internal or an external sample loop and are an integral component of an HPLC system.

For the former, the volume of the internal loop is formed from a machined groove in the surface of the rotor. When in the load position, this channel is isolated from the solvent flow through the column and can be filled with the sample as shown (Figure 6.15a). On turning the rotor to the injection

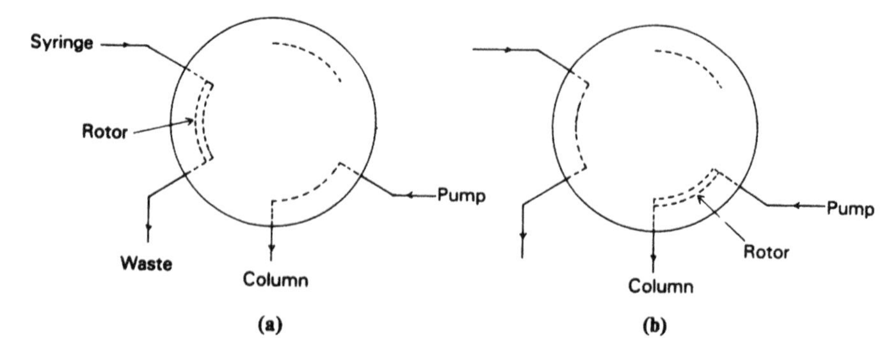

Figure 6.15 Internal loop valve: (a) rotor fill, and (b) sample inject.

position, the channel containing sample is relocated at the solvent outlet ports and the sample is flushed into the column. This system is simple and reliable to use and has the advantage that the extra-column volume due to connecting tubing is minimised.

The disadvantage of internal loops is that they are fixed and can only be altered by changing the rotor. This has led to the development of external interchangeable loop valves of the type marketed by Valco and Rheodyne. Here dead volumes are minimised and the loop can readily be exchanged for other desired sample sizes, e.g. 5 μl, 10 μl, 20 μl and 50 μl. Microsample injection valves capable of delivering 0.5–5 μl are also available. Operation of the valve is similar to those with internal loops, i.e. a load position and an inject position (Figure 6.16).

6.5.3.2 Syringe injection. This was the first type of sample application system developed and closely resembled GC and involved syringe injection of the sample through a self-sealing septum. The greatest column efficiency is achieved by application of the sample via a syringe directly onto the column bed, due to

- back-diffusion of sample being minimised; and
- application of the sample to the chromatographic bed as a small point source results in the transport of the component bands through the core of the column, well away from the walls, thus preserving the infinite diameter effect [17, 18].

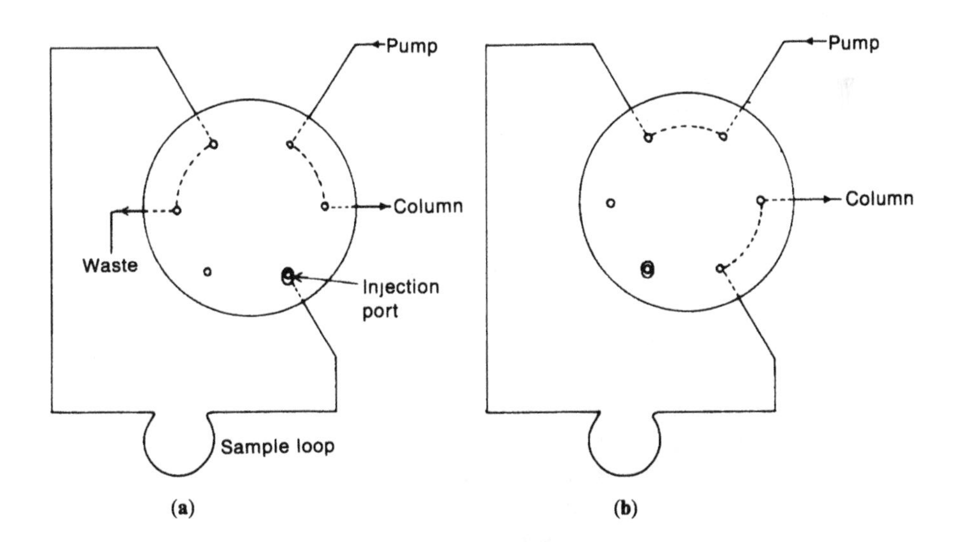

Figure 6.16 External loop injection valves: (a) load, and (b) inject.

The principal merits of such systems were: the ability to vary sample size; the efficiency of sample usage and that they were relatively inexpensive. However, on-column injection has a number of disadvantages principal amongst which is that they cannot be used at the high column back pressures associated with modern packing materials and thus this technique is simply of historical interest.

6.5.3.3 On column. One on-column sample application method involves stop-flow techniques. In this instance, the pump is stopped and isolated by a three-way valve from the column, the sample is then loaded via a syringe through an injection port which does not contain a septum. The pump is then restarted, the flow restored by switching the valve, and the sample is rapidly flushed onto the column. There is no apparent loss in efficiency but inaccuracies in retention measurement occur due to the finite time required for flow to be established and therefore this technique is redundant.

6.5.4 Detectors

The function of the detector in HPLC is to monitor the column effluent and afford a means of detecting solutes therein. Detectors function according to many different principles but all output an electrical signal which is proportional to some property of either the analyte, e.g. ultraviolet absorbance or a property of the column eluant which is modulated by the analyte, e.g. refractive index. The choice of detector is often dictated by the chemical characteristics of the analyte species and this choice may subsequently determine which eluant is used and also possibly which stationary phase and mode of chromatography.

The detector response will be related to the amount of the analyte in the column effluent though different analytes will respond to differing extents and hence the detector must be calibrated with respect to each of the analytical species of interest. Detector response factors and their determination is covered in detail in Chapters 2 and 10.

The ideal HPLC detector should have the following characteristics:

- high sensitivity;
- good stability and reproducibility;
- linear response over several orders of magnitude;
- small internal volume minimising zone broadening;
- a short response time independent of flow-rate;
- insensitive to changes in temperature and pressure;
- high reliability and ease of use;
- similar response to analytes or selective response to analyte classes; and
- non-destructive.

Unfortunately no truly universal detector which satisfies these criteria has as yet been developed for HPLC although mass spectrometry and electrochemical detection systems arguably approach this ideal.

There are continuing advances and developments in detector design and technology giving not only improvements in established detectors but also leading to new and novel systems and some excellent reviews [19, 20] have been presented which survey both existing and novel types.

The various approaches to solute detection which have been pursued have led to the development of a wide and impressive range of detectors which may be classified as follows.

- Detectors which monitor a *specific property* of the solute which is not shared with the solvent, e.g. ultraviolet absorbance and fluorescence. Possession of such a property by the solute affords its detection in the effluent.
- Detection systems which monitor a *bulk property* of the eluant, e.g. refractive index, dielectric constant and density; in this instance the solute modifies the base value of the property associated with the solvent.
- Detectors which function by *desolvating*, i.e. separating the solvent from the eluant, thus allowing subsequent detection by, e.g. flame ionisation detector (FID) or mass spectrometry (MS) of the analyte.
- Pre- or post-column *derivatisation* involving chemical reaction of the analyte.

The choice of detector is often dictated by the properties of the solutes and/or the sensitivity required from the analysis. The more important detector parameters governing the choice are listed below.

- The noise level (n) of the detector. This may be present as long term or short term noise and is caused by changes in mobile phase composition, incomplete mixing, temperature variation, voltage fluctuations in the electronics, etc.
- The response time (τ) of the detector should be no more than 0.1 of the peak width in time units. The time constant can be determined from the following expression:

$$\tau = \frac{2t_R}{5\sqrt{N}}$$

  where N is the noise level. For columns containing small microporous packings the time constant should be ≤ 0.25 s.
- Lower limit of detection, that concentration which produces a signal having twice the background noise level.
- Linearity or dynamic linear range, the concentration range over which the detector response is linear, should be of at least four orders of magnitude. If the concentration of sample is outside this range, then the sample should be diluted or an extended calibration curve constructed.

The primary functional characteristics of a range of detectors are presented in Table 6.2. The table also gives an indication of the relative use based upon literature citations [21].

6.5.5 *Specific property detectors*

6.5.5.1 Ultraviolet detectors. Ultraviolet detectors function by monitoring the light absorbed by the solute molecules from the incident beam. Ultraviolet detectors are the most commonly used type with LC systems, they are not appreciably flow or temperature sensitive, have a good dynamic linear range, but are, however, selective. The absorbance is proportional to concentration and obeys the Beer–Lambert Law which is defined as follows:

$$\text{Absorbance} = \log I_0/I_t = \epsilon cl$$

where ϵ is the absorptivity coefficient, c is the solute concentration and l is the path length. However, the Beer–Lambert Law is a limiting law and the absorbance versus concentration curve is usually only linear over low concentration ranges; also the law is strictly valid only for monochromatic radiation.

The major modification required to adapt conventional spectrometers for use in HPLC is to design suitable flow-cells. The volume of a well resolved peak in HPLC can be as little as 150 µl and considerably smaller for microbore, and thus a small volume flow-cell which will avoid excessive band broadening and which is capable of withstanding pressures of several bar is required. Typically, the flow-cell for spectrophotometric detection in HPLC has a pathlength of 10 mm and a bore of 1 mm giving a volume of 8 µl. For narrow bore columns flow-cell volume has to be reduced to 2–3 µl. Common flow-cells in use are shown in Figure 6.17.

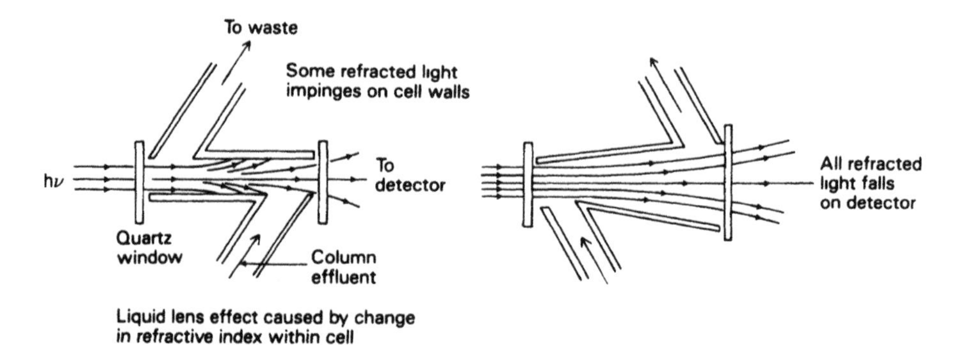

Figure 6.17 Flow profiles through conventional and taper cells.

Table 6.2 Characteristics of chromatographic detectors[a]

	Response	Detection limit for sample	Linear range	Flow sensitive	Temperature sensitive	Useful with gradient	Favourable samples	% Usage
Ultraviolet (absorbance)	S	5×10^{-10} g cm^{-3}	10^4–10^5	No	Low	Yes	Conjugated aromatics and heterocyclic compounds	78
Photo-diode (absorbance)	S	$>2 \times 10^{-10}$	10^4–10^5	No	Low	Yes		18
Fluorescence	S	$\sim 10^{-12}$	10^3–10^4	No	Low	Yes	Vitamins and steroids	31
Infrared (absorbance)	S	10^{-6}	$\sim 10^3$	No	Low	Yes	Carbonyl and aliphatic	<5
FTIR (absorbance)	S		$\sim 10^3$	No	Low	Yes		
Refractive index (RIU)	G	5×10^{-7}	10^3–10^4	No	$\pm 10^{-4}$°C	No	Universal	37
Conductometric (μmho)	S	10^{-8}	10^3–10^4	Yes	±1°C	No	Ionic substances	15
Electrochemical (μamps)	S	10^{-12}	10^4–10^5	Yes	±1°C	No	Catecholamines electroactive	21
Radioactivity	S	50 cpm ^{14}C cm^{-3}	$\sim 10^3$	No	Negligible	Yes	Labelled compounds	<5
Mass spectrometry	S	10^{-10}	10^4	No	No	Yes	Universal	5
Transport FID (A)	G	5×10^{-7}	10^4–10^5	No	No	Yes	Oxidisable hydrocarbons	<5

[a] Adapted from Snyder and Kirkland (1979) *Introduction to Modern Liquid Chromatography* (2nd edn). Wiley-Science, New York.
[b] *LC–GC*, **4** (1986) 526.

The various flow-cell designs have been developed to minimise flow disturbances within the cell and subsequent distortion of the absorbance signal. These fluctuations arise from changes in the refractive index (RI) of the eluate which distorts the incident beam and causes variance in the intensity of the light falling on the detector. These effects are referred to as the liquid or dynamic lens phenomena [22], are responsible for distorting the light beam and arise from changes in temperature, flow-rate mobile phase composition and change in the RI of the solvent due to the presence of solute. The taper flow-cells minimise the amount of refracted light falling on the cell walls and collimating lens can then be used to refocus the emerging light on the detector.

6.5.5.1.1 Fixed wavelength. The first source in common use was a mercury lamp which emits intense radiation at 254 nm which is easily isolated using simple narrow band pass filters. The addition of phosphor converters can be used to provide a line at 280 nm of adequate intensity for absorption measurements. Other filters allow measurement at 220, 313, 334, 365, 436 and 546 nm. This restricts the use of these detectors to compounds or compound classes whose absorbance profile overlaps significantly with one of these wavelengths. Many organic compounds, such as those containing conjugated chromophores, aromatic or heteroaromatic rings, can be adequately detected using the Hg lamp as a source. In addition, due to the intensity of the Hg 254 nm line, very narrow band pass filters can be used giving essentially a monochromatic source, an essential prerequisite for adherence to the Beer–Lambert Law. Fixed wavelength detectors typically have a maximum sensitivity of 0.0005 absorbance units full scale (AUFS) with a noise level better than 0.00005. Improvement on these properties would require thermostating of the detector cell to ±0.01°C and the use of high energy fixed wavelength lamps.

6.5.5.1.2 Variable wavelength. There are many classes of compound which do not absorb significantly at the above wavelengths, e.g. barbiturates and pesticides, and consequently sensitivity is very poor. To increase sensitivity and specificity the Hg lamp has subsequently been replaced by the deuterium (190–400 nm) and the tungsten (400–700 nm) lamps coupled with a manually adjustable diffraction grating monochromators. These are referred to as variable wavelength detectors and allow the wavelength for optimum sensitivity in a given application to be chosen.

A consequence of the high sensitivities demanded by LC down to 0.0005 AUFS deflection is the need to use greater band widths with variable wavelength detectors (+5 nm being typical), otherwise the resulting noise would render the detector unusable. However, as deviations from Beer–Lambert's Law can arise due to the use of non-monochromatic radiation, it is important to check the linearity of response of the sample components

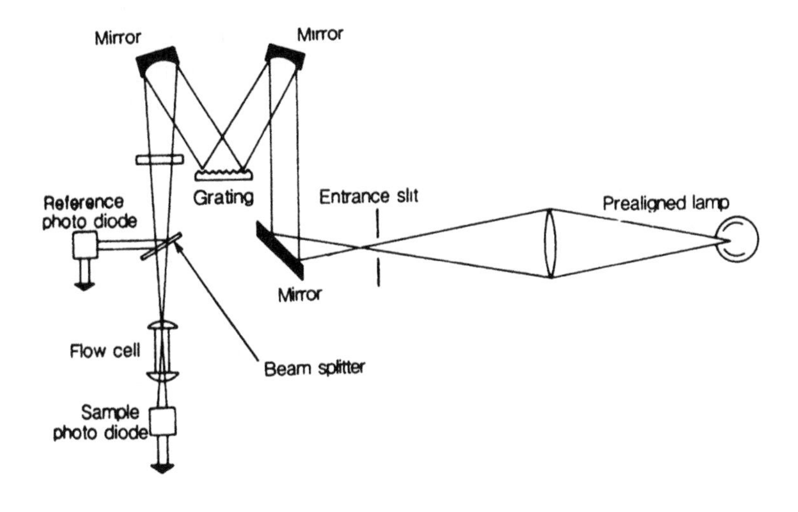

Figure 6.18 Schematic diagram of a variable wavelength detector.

over the concentration range of interest; with photometric sensitivity of this order it is possible to monitor solutes having only moderate absorptivity.

A schematic diagram of the optical layout of a typical LC spectrophotometer is shown in Figure 6.18. The instrument is described as a double-beam photometer, although the reference beam in fact passes directly onto a reference photocell.

Reference cells filled with the eluant being used for analysis were incorporated into early instrumentation. However, this served no useful purpose and in fact caused additional noise due to the difficulty in exactly matching the solvent flows, a problem especially in gradient work.

Detectors are commercially available which allow programmable wavelength switching during analysis, thus optimising sensitivity and selectivity. As many as 12 wavelengths can be selected. The systems use reversed-optics geometry, that is, the light from the source is dispersed after passing through the sample by a holographic grating onto an array of photodiode detectors. One system uses a xenon strobe lamp, thus giving a wavelength range of 190–800 nm. Deuterium lamps have also been used.

6.5.5.1.3 Absorbance ratioing techniques. Spectrophotometers are available for the simultaneous monitoring of light at two wavelengths; for example, suitable orientation of a mercury lamp and phosphor allows light of 254 and 280 nm to be directed through the flow-cell and to subsequently impinge on individual detector cells. This is a technique which provides for direct absorbance ratioing, an accurate method for determining the homogeneity of a chromatographic zone [23]. In general, the ratio of two absorbances at any two wavelengths is a constant, the ratio of the absorptivity coefficients at wavelength 1 and wavelength 2, respectively, and is diagnostic for a pure compound. If the ratio of absorbances across the chromatographic zone is not constant then this is indicative of peak overlap and incomplete resolution.

Variable wavelength detectors extend the scope of the technique. Using programmable multiwavelength detectors, wavelengths for ratio absorbance monitoring can be preselected to maximise the ratio discrepancy due to peak overlap. Sophisticated microprocessor control allows the ratio gram to be evaluated and plotted in real time whilst monitoring the effluent at additional pre-programmed wavelengths (to maximise detectability and selectivity and plotting the chromatogram).

Whilst single wavelength detection using variable wavelength ultraviolet monitors give adequate sensitivity and selectivity for quantitative and qualitative analysis further spectroanalytical data can be obtained by stopped-flow scanning [24]. Computerised data collection and processing systems have enabled spectra to be collected, added, subtracted and scaled. Thus, when the eluant flow is stopped, the solution in the detector cell can be scanned and a background subtraction of the solvent spectrum, which has been previously recorded and stored in the data system, performed; the resultant spectra can then be displayed.

6.5.5.1.4 Photodiode array spectrophotometers. Photodiode array technology allows continuous scanning [25] of the absorbance spectrum of the chromatographic eluant, thus eliminating the need to stop the flow of the mobile phase. Diode array spectrophotometers employ reversed-optic geometry, in which all of the light from the source is focused onto the sample, rather than dispersing the light in a monochromator and passing single wavelengths through the detector cell (Figure 6.19). The

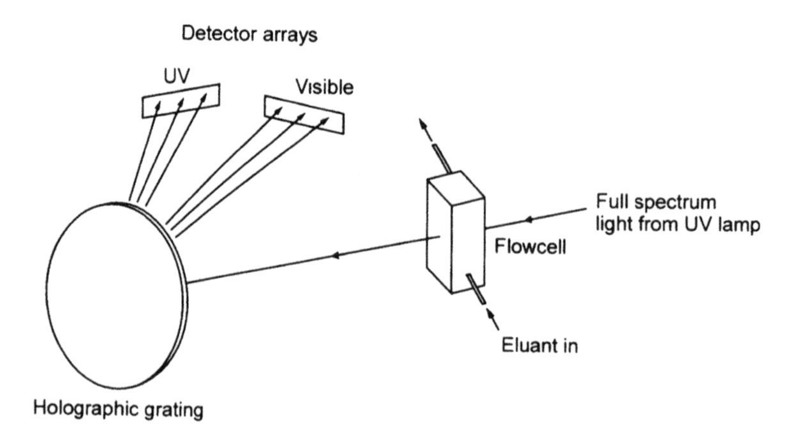

Figure 6.19 Photodiode array spectrophotometer.

full spectrum of light emerging from the flow cell is dispersed by a holographic diffraction grating into single wavelengths which are focused simultaneously onto a linear array of photodiode detectors. The photodiode array is a row of detectors up to 1056, mounted on a 1 cm silicon chip each diode receiving a different wavelength with spectral resolution of ~1.2 nm.

When transmitted radiation falls on the photodiodes, milliamp currents are generated. The electronic signal from each photodiode detector is processed to give absorbance data, which is then displayed as a spectrum across the complete spectral range of the array in as little as 0.01 s and thus many spectra can be evaluated and displayed for even the fastest eluting peak. The absorbance data can be displayed as a function of wavelength and time (Figure 6.20).

With modern computing power it is also possible to display data as isoabsorbance contours plotted in the wavelength time plane. Peak detection is also enhanced by full-screen three-dimensional displays which can be rotated to give different perspectives aiding the location of hidden peaks.

The full spectra obtained from diode array spectrophotometers are a useful aid to component identification. The examination of the many spectra, undertaken using sophisticated software and microcomputers, taken during the elution of a peak, gives information on the peak homogeneity thus aiding accurate quantitation. It is possible to optimise detectability for an analysis by averaging absorbance data either in the time axis or the wavelength axis, as random noise decreases by the square root of the number of readings taken. Furthermore the detection wavelength can be changed during the

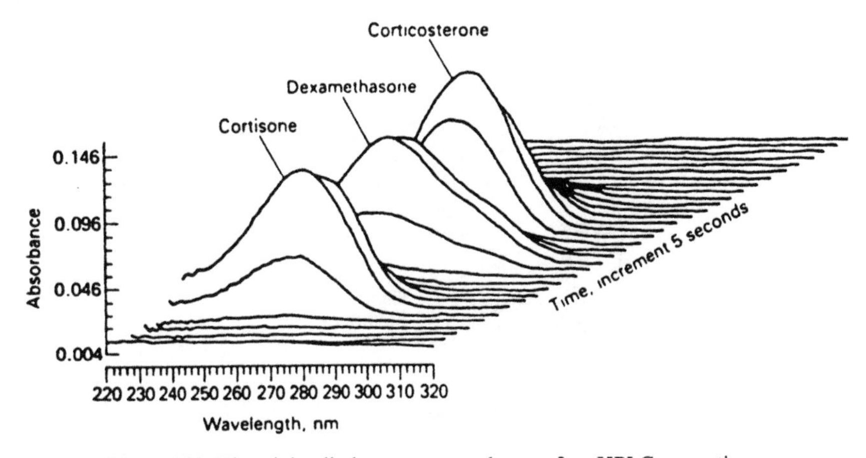

Figure 6.20 Ultraviolet diode array spectral map of an HPLC separation.

analysis as many times as necessary so that every peak is detected at its optimum wavelength.

6.5.5.2 Fluorescence detectors. A smaller number of compounds possess the ability to fluoresce, i.e. to absorb radiation of one wavelength and then subsequently to emit fluorescent radiation of longer wavelength. Fluorescing compounds typically contain highly conjugated cyclic systems, for instance, polynuclear aromatics, quinolines, steroids and alkaloids. Many compound classes which do not fluoresce can be reacted with fluorogenic reagents [26], for instance, aminoacids reacted with *o*-phthaldehyde or ninhydrin to give fluorescent derivatives. Dansyl chloride (5-(dimethylamino)-1-naphthalene sulphonic acid) has been used to good effect as a derivatising reagent as it reacts with amines, amino acids and phenols to give fluorescent derivatives.

The fundamental properties of fluorescence make this a particularly attractive basis for an HPLC detection system [27], for whereas photometers depend upon the measurement of fairly small differences between the intensity of a full and slightly attenuated beam the measurement of fluorescence starts in principle from zero intensity. At sufficiently low values of concentration (<0.05 absorbance) then the intensity of fluorescence is directly proportional to concentration with a linear range of three to four decades. Consequently, fluorescence detectors are more selective and sensitive than ultraviolet detectors in LC by a factor of 10^{-2} giving noise equivalent sensitivities of better 1 ng ml^{-1}.

The optical layout (Figure 6.21) is similar to the conventional fluorimeter in that the detector is placed at right angles to the primary incident beam which is supplied from a xenon or deuterium source. However, due to the flow-cell dimensions and orientation only a small amount of the fluorescent energy will fall on the photodetector. Increased sensitivity is provided by the

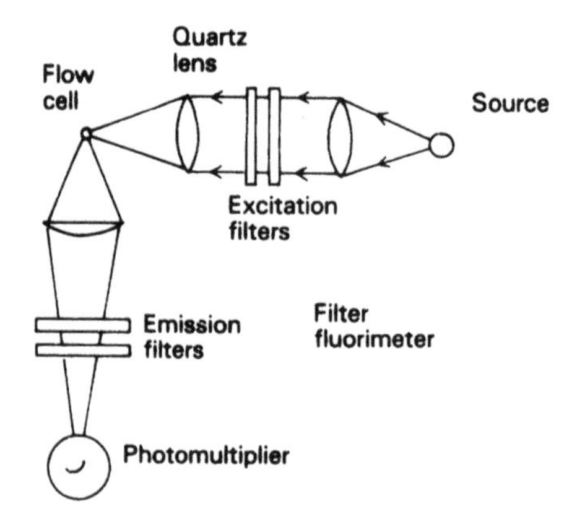

Figure 6.21 Schematic diagram of a filter fluorimeter.

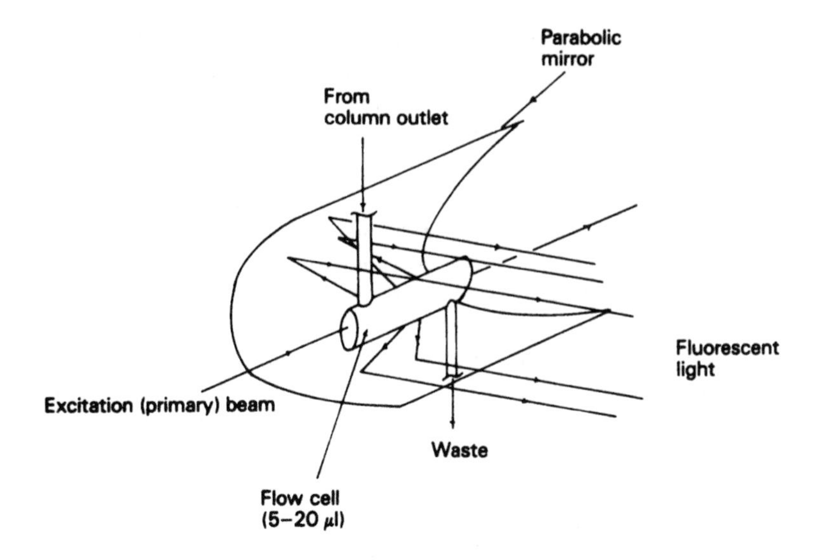

Figure 6.22 Fluorimeter flow-cell for HPLC.

use of parabolic collecting mirrors which focus the fluorescent radiation onto a photomultiplier tube (Figure 6.22).

Notwithstanding this larger flow-cells (10–25 μl) are used to give improved sensitivity at the expense of reduced resolution. As in conventional fluorimetry, scatter of the primary beam in the direction of the detector occurs and must be removed either by filters or monochrometers. Likewise, narrow band pass or cut-off filters or monochrometers are used to select the wavelength of the exiting beam.

Spectrofluorimeters are available where both excitation and fluorescent wavelengths can be chosen using variable monochromators. The excitation energy may be provided either with a deuterium lamp (190–400 nm) or a tungsten lamp (250–600 nm). Fluorescence measurements in LC are susceptible to the usual limitations such as quenching and turbidity. They are, however, relatively insensitive to fluctuations in solvent pulsing and temperature and are suitable for use with gradient systems [27].

6.5.5.3 Infrared detectors. The eluant can be monitored by observing its absorbance of infrared radiation [28, 29]. Variable wavelength detectors, fitted with low volume flow-cells (1.5–5 μl, pathlength 1.5 mm), are commercially available which cover the spectral range 2.5–14.5 μm (4000–690 cm^{-1}). The detection wavelength can be chosen to respond to a particular compound class for instance, alkenes (~6 μm), thus making it highly specific. It has also been used in the analysis of non-polar lipids [30] which have a specific absorbance at or about 5.75 μm due to the carbonyl function. Alternatively a wavelength can be chosen at which many compound classes will absorb, for instance, all compounds with aliphatic C–H bonds absorb at $\sim$3.4 μm. Quantitative studies can be readily carried out since as with ultraviolet photometers, absorbance is proportional to concentration.

The major limitation of the infrared detector in LC is that the mobile phase must have adequate energy transmission at the wavelength of interest. Many of the common HPLC solvents absorb in the infrared and one must be chosen which has a suitable spectral window (>30% transmittance) for a given application. Some useful solvents for this detector are acetonitrile, chloroform, methylene chloride and tetrahydrofuran. Solvent opaqueness can be reduced and more spectral windows provided by using flow cells with short path lengths however though reducing the high background, sensitivity will also be diminished. The sensitivity of the infrared detector is comparable to the refractive index detectors (5–15 μg ml^{-1}, $\sim$1 μg using CH band), though it has the advantage that it is relatively insensitive to fluctuations in temperature or flow and can also be used in gradient elution work.

Fourier transform infrared spectrophotometers for use with HPLC are commercially available and as with ultraviolet photodiode array spectrophotometers enable the absorbance data to be displayed as a function of wavelength and time. Further details are given in Chapter 7.

6.5.5.4 Electrochemical detectors. Electrochemical detectors are gaining in popularity as they afford high specificity, sensitivity and wide applicability and offer the potential for becoming the much desired general all purpose detector for HPLC study [31–34]. Whilst polarography and voltametry are recognised as well-established techniques for metal ion estimation, the technique is not specific to metals, but can be readily applied to the detection of a wide range of organic compounds which are electroactive.

Compounds containing a wide range of functional groups can be oxidised or reduced in solution provided the potential applied is greater than the electroactive analyte species' half-wave potential, e.g. ascorbic acid can be readily oxidised giving an anodic peak at +0.23 V (SCE) due to the oxidation of the enediol system, whereas organic disulphides are reduced at a half-wave potential of −1.0 V (SCE) (Figure 6.23). In HPLC a fixed potential is applied to the detector electrode which is adequate to oxidise or reduce the species of interest and the current flowing across the cell continually monitored. The currents generated are in the nA range and can be measured with high precision and accuracy affording detection limits commonly 10^4 better than that achieved with ultraviolet spectrophotometers.

Electrochemical detection coupled to LC (ECLC) is now used for the selective monitoring of a wide range of trace organic components (Table 6.3) in a variety of matrices, such as encountered in environmental, pharmaceutical and clinical studies. The technique is most suitable for the detection of electro-oxidisable compounds. Monitoring of electro-reducible compounds is complicated by the high background currents generated by dissolved oxygen

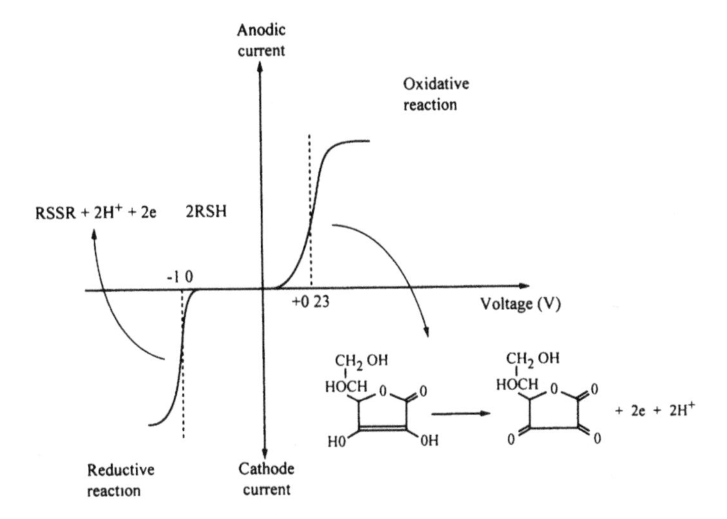

Figure 6.23 Amperiometric trace detection showing current potential waves for a typical organic disulphide and ascorbic acid.

and both eluant and sample solutions must be exhaustively degassed before use; other reducible trace impurities in the eluant such as metal ions must also be excluded.

Detectors based on amperometry and coulometry are commercially available though the former are the more often encountered. Depending on the construction of the cell and the flow characteristic of the eluant through the cell it will operate in either the amperiometric or coulometric mode. In the former the current is diffusion limited and only a small proportion of the analyte is affected (1–20%), response is proportional to concentration of the analyte as it passes through the cell. In the latter the cell geometry is modified giving a large surface area working electrode over which the eluant is passed as an extremely thin film. This ensures complete conversion of the electroactive analytes and the detector is therefore mass sensitive.

Table 6.3 Compound types (functional groups) sensed by electrochemical detectors

Oxidation	Reduction
Hydrocarbons	Olefins
Azines	Esters
Amides	Ketones
Amines	Aldehydes
Phenols	Ethers
Quinolines	Diazo compounds
Catecholamines	Nitro compounds

As with other polarographic systems ECLC is limited by the fact that the test solution must be electrically conducting and thus ECLC is restricted to reverse phase systems and to ion exchange systems where the aqueous phase contains an electrolyte. The detectors are flow sensitive and thus pulse-free constant flow reciprocating pumps must be used. In addition, the solvents must be thoroughly degassed as dissolved oxygen interferes with redox reactions occurring in the cell.

Detector cells can be made which contain only two electrodes, a working electrode and a reference electrode. A preselected potential equal to or greater than the half wave potential of interest is applied constantly across the electrodes. However, two electrode systems give a non-linear response as the voltage drops across the eluant as the current flow changes. Thus electrochemical detectors typically employ a three electrode cell. The additional electrode known as the auxiliary or counter-electrode, serves to carry any current generated in the flow cell thus enabling the reference electrode to ensure a fixed potential despite the decrease in the internal resistance of the detector cell.

EC flow-cells are relatively simple in design and construction. The electrodes are mounted in a suitable matrix (PTFE). The reference electrode is either of Ag/AgCl or a standard calomel type; the auxiliary electrode comprises either glassy carbon, platinum or stainless steel. Successful working electrodes have been formed from a variety of materials, for instance, carbon paste and gold. The former is most commonly used, it has good conductivity, low chemical reactivity, is inexpensive and has a wide range of working potentials, −0.8 to 1.3 V, versus the calomel electrode. Mercury and amalgamated gold are the choice for reductive studies.

A typical electrochemical detector is depicted in Figure 6.24. Regardless of the electrodes chosen the cells have a number of common design features. The working electrode which usually forms one face of the flow-cell is located directly opposite the column inlet with the auxiliary electrode located directly opposite and the reference electrode located downstream. This geometric arrangement produces flow-cells of very low working effective or internal volume, typically 1 μl or less making these detectors highly suitable for use with microbore columns [35].

As electrochemical reaction occurs decomposition products can be deposited on the electrode surfaces, altering the surface characteristics and causing noise drift and reduced response. Carbon paste electrodes are particularly susceptible, thus must be frequently recalibrated and may need periodic cleaning and replenishing. Alternatively the contaminants can be removed by pulsing or stepping the potential to a more positive or negative value as required. The selectivity and sensitivity of EC detectors can be further enhanced by using pulsed amperiometric techniques [36, 37] where multiple potentials are applied in a repeating sequence. In the pulsed mode, the detector measures current during a short sample interval and thus sensitivity

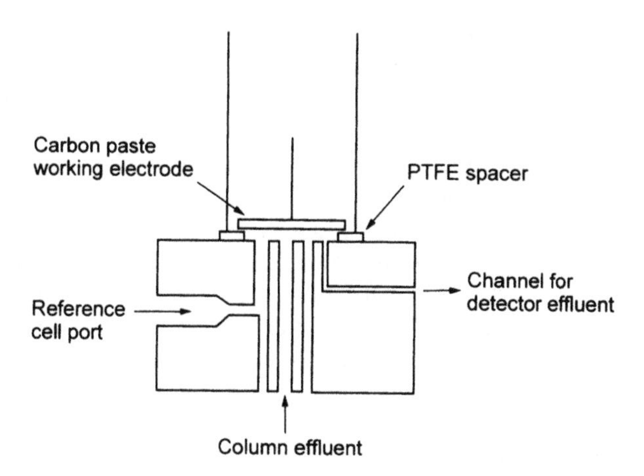

Figure 6.24 Schematic diagram of a typical electrochemical detector using a glassy carbon working electrode.

is reduced. CHOH bearing species such as carbohydrates and alcohols can be detected using pulsed amperiometric detection with a gold working electrode.

Dual working electrode systems have been developed which can be operated with the electrodes in parallel or in series. In the later one working electrode is located upstream of the other and is complementary in function. This system can be used with analytes which undergo reversible oxidation or reduction and in use enhances selectivity. In the former configuration the electrodes are mounted at 90° to one another in the flow-cell. Different fixed voltages can be applied to the electrodes and the response monitored and ratioed to check peak purity. Alternatively, one electrode can be operated as a cathode the other as an anode and oxidants and reductants determined simultaneously.

A number of other methods such as differential pulse, reverse pulse, square wave voltammetry and static mercury drop techniques have been investigated with regards to improving selectivity and sensitivity however their potential remains largely unrealised.

6.5.5.5 Conductivity detectors. Many HPLC applications involve the separation of mixtures of ionic compounds. Consequently a continuous record of the electrical conductivity of the effluent can be used to detect the eluting species. Commercial conductivity meters with flow-cells of approximately 1 μl are available for the monitoring of inorganic ions, e.g. Na^+, NH_4^+ and K^+; also anions, halides, NO_3^- and SO_4^{2-}. The HPLC cell is a scaled down version of the normal laboratory conductivity cell and consists of a pair of electrodes mounted in a low volume PTFE flow-cell.

The cell is mounted in an AC Wheatstone Bridge and thus small changes in the electrical conductivity of the eluant can be detected. For most inorganic and organic species in aqueous media the conductivity is proportional to concentration.

However, the separation of ionic species is usually achieved using aqueous based eluants containing electrolytes. The high ionic strength of the eluant thus reduces the sensitivity of the conductivity detector and makes it very susceptible to fluctuations in solvent flow and solvent composition and prevents gradient elution work. This restriction has been overcome by using suppressor columns for removal of the background electrolyte [38]. Alternative systems are being developed and marketed which use only a single column [39], thus avoiding any broadening due to dead volumes, though the reproducibility of these chromatographic units is unclear. The detectors are sensitive to temperature fluctuation and require rigorous thermostatic control.

6.5.5.6 Refractive index detectors. There are three types of commercially available RI detectors, namely, deflection, Fresnal and interference. Each measures the change in the refractive index of the base eluant due to the presence of analytes and hence in principle should provide the basis for a universal detector. Difficulties in RI detection arise due to the sensitivity of the solvent RI to fluctuations in temperature, pressure, the presence of dissolved gases and eluant composition. The RI of a solvent changes by 4×10^{-4} per °C and by 4×10^{-5} per atmosphere thus baseline noise is $\sim 1 \times 10^{-7}$ with a mixed solvent giving corresponding sensitivities of $\sim 5\,\mu g\,ml^{-1}$.

6.5.5.6.1 Principles. For dilute solutions the additivity law of refractive index is applicable. Thus, the composite RI (N_C) of a solution in the flow-cell can be expressed in terms of the RI of pure solvent and solute, N_1 and N_2, respectively, and the volume of each present, V_1 and V_2, respectively, where the sum of V_1 and V_2 equals the flow-cell volume:

$$N_C = \frac{N_1 V_1}{(V_1 + V_2)} + \frac{N_2 V_2}{(V_1 + V_2)}$$

that is

$$N_C = \frac{N_1 V_1 + N_2 V_2}{(V_1 + V_2)}$$

If only solvent is present in the flow-cell then from examination of equation (6.17) it can be seen that the RI is equal to N_1.

Refractometers, irrespective of design, are differential monitors and hence the signal (S) resulting due to presence of solute in the flow-cell may be expressed as

$$S = N_C - N_1$$

N_1 may be expressed as follows:

$$N_1 = \frac{N_1(V_1 + V_2)}{(V_1 + V_2)}$$

Substituting for N_C and N_1:

$$S = \frac{V_2}{V_1 + V_2}(N_2 - N_1)$$

Therefore

$$S \sim c(N_2 - N_1)$$

where c is the concentration of solute. Thus, for a given solvent the detector signal is proportional to concentration.

A consequence of using a bulk property for detection is that this property of the solvent must be controlled very closely; the refractive index of the eluant is sensitive to fluctuations in pressure, temperature and composition. Whilst the pressure and composition can be controlled using pulse dampners and reciprocating pumps, the limits of sensitivity and stability of the RI detector are determined by temperature. The temperature must be controlled to +0.0001 K for accepted noise levels. Fluctuations in the RI caused by temperature and noise changes are compensated for by use of a reference cell.

However, refractometers, despite the above disadvantages, most closely approach the ideal of a universal detector in that most compounds modify the refractive index to some extent. Both negative and positive changes in the solvent RI can be detected and recorded using integrators capable of processing positive and negative signals.

6.5.5.6.2 Fresnel refractometer. The refractometer shown (Figure 6.25) is based on Fresnel's Law, which states that the fraction of light reflected at a glass–liquid interface varies with the angle of incidence and the RI of the liquid. The beams produced by the source and collimating lens are focused onto the reference and sample prism–liquid interfaces. The light is then refracted through the liquid in the cell, reflected off the backing surface and then passed through the liquid–prism interface on to the detection system.

With mobile phase in both cells the beams are refracted equally and light beams of equal energy fall on the dual element photodetector. With analyte in the sample cell the light beams are refracted to differing extents and different amounts of radiation fall on the photodetector.

The disadvantages of this system are (a) two prisms are required to cover the range of RI of solvent encountered (1.31–1.55); (b) small volume cells (5 μl) help enhance sensitivity; however the limited sensitivity (×1000 less than ultraviolet) and stability are the principle limits to the universal application of this technique.

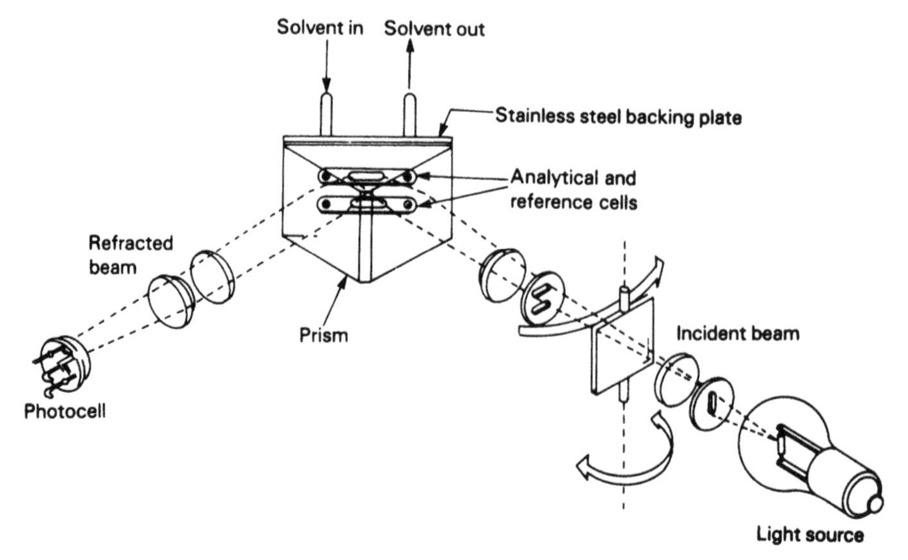

Figure 6.25 A Fresnel prism RI detector. (Reproduced by permission of Perkin Elmer Ltd.)

6.5.5.6.3 Deflection refractometer. In the deflection refractometer light from the source is collimated by the lens and falls on the detector cell, which consists of sample and reference sections that are separated by a dividing plate (Figure 6.26). As the incident light passes through the cell it is refracted, reflected and refracted again, before passing through the lens and on to the photodetector.

The signal produced is proportional to the position at which the beam strikes the detector which is dependent upon the RI within the sample

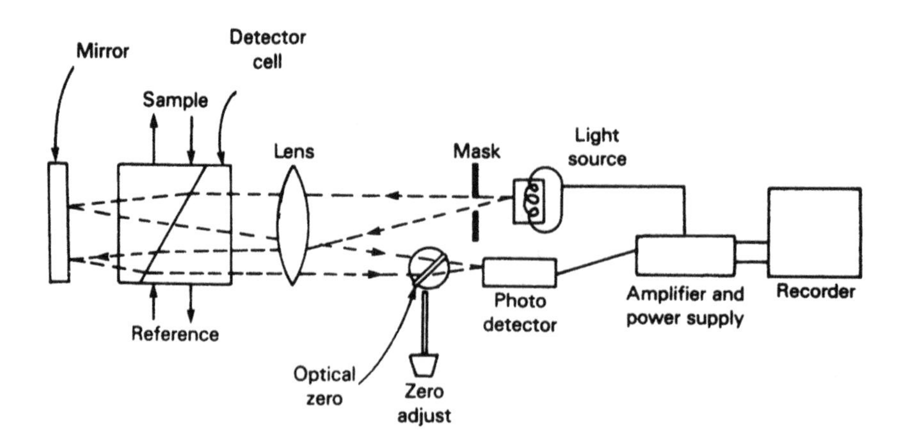

Figure 6.26 Deflection refractometer detector. (Reproduced with permission of Waters.)

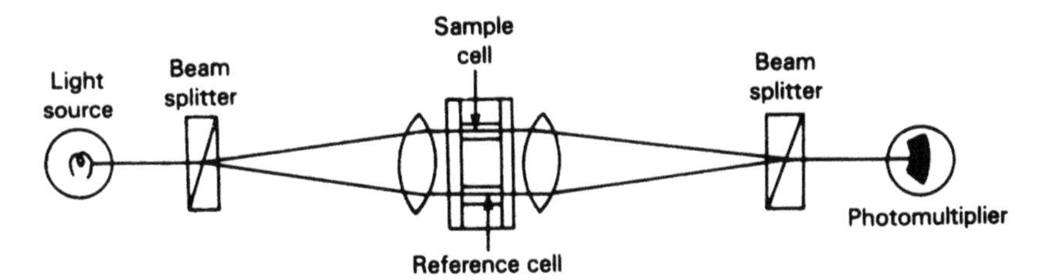

Figure 6.27 Shearing interferometric refractometer detector. (Reproduced by permission of Optilab.)

compartment. Though larger flow-cells are used a single flow-cell is adequate for the normal range of applications.

6.5.5.6.4 Shearing interferometer. In a shearing interferometer radiation of known wavelength (546 nm) is split into two beams of equal intensity by the beam splitter. These beams are then focused by the lenses and pass through the reference and sample cells (Figure 6.27). (Typical sample flow-cell volume is 5 μl with a 3.2 nm pathlength.) The beams are then focused onto a second beam splitter, B, which corrects the wavefronts so that the beams are in phase and interfere.

When solvent is present in both sample and reference cells constructive interference occurs. However, if solute is present in the sample compartment, the sample beam has a different optical path length, which can lead to destructive interference when the sample and reference beams interact after beam splitter B as the beams are now out of phase. A relatively long wavelength is chosen as this provides for a greater linear working range. This is the most sensitive of currently available refractometers though it still suffers from the same inherent sensitivity to environmental factors.

The advantages and disadvantages of RI detectors can be summed up as follows.

- Excellent versatility and detector of choice for a wide range of sample types for instance carbohydrates, sugars and gel permeation work.
- Moderate sensitivity—to optimise the sensitivity, the refractive index increment between solute and solvent should be maximised. However, even under the optimum conditions, the sensitivity of RI detectors, defined as being equal to the noise level, is 10^{-7} riu, equivalent to ~1 ppm of sample in the eluant, approximately a factor of 10^3 less than ultraviolet photometers.
- Generally not useful for trace analysis.
- Sensitive to small temperature fluctuations—the temperature coefficient of the RI is 1×10^{-4} riu per °C.
- Accurate temperature control is required; however, inclusion of heat exchangers increases the dead volume which causes band broadening and limits column efficiency.

- Difficult to use with gradient elution due to the inability to match the RIs of the sample and reference streams.
- The detectors are non-destructive.

6.5.5.7 Evaporative light scattering detectors. Recently a new detector type has become commercially available, known as the evaporative light scattering detector (ELSD), but also referred to as the evaporative analyser or mass detector [40, 41]. This detector is relatively simple in design and operation and is of considerable interest due to its apparent potential as a universal detector.

The mobile phase eluate from the chromatographic column is passed through a nebuliser, where it is converted into a fine spray or mist in air or nitrogen and the resulting droplets carried down through a controlled temperature stack (Figure 6.28). The volatile mobile phase is evaporated to leave a mist of small particles—probably droplets of high concentration of solute rather than solid particles. The cloud of analyte particles then passes through a light or laser beam and the scattered light is detected at right angles to the flow by silicon photodiodes.

The mechanism of light scattering is complex and is a result of a number of different mechanisms—Raleigh and Mie scattering, reflection and refraction [42]. The relative contribution of these processes is dependent upon the radius of the droplet compared to the wavelength of light of detector.

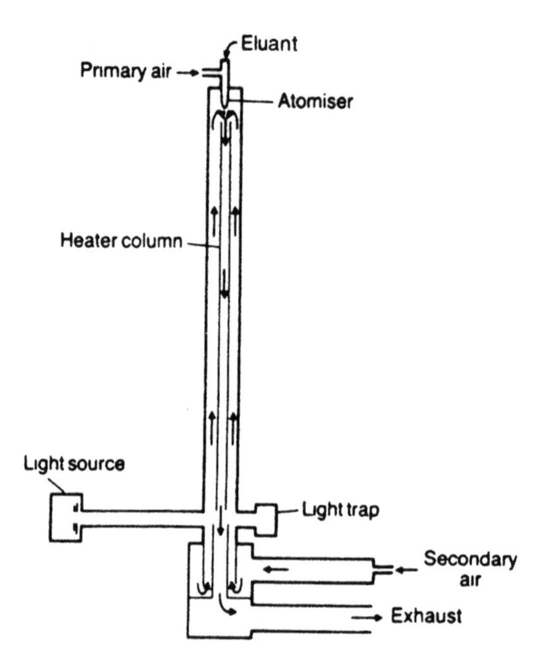

Figure 6.28 Schematic diagram of an evaporative light scattering detector.

At ratios of $\sim 5\,(r/\lambda)$ reflection and refraction dominate and give optimum sensitivity; at lower values the less effective Raleigh and Mie processes dominate, while at larger values the response declines as the surface area to volume ratio of the droplet decreases.

Thus the extent of light scattering and detector sensitivity is dependent upon the radius of the suspended droplets. Furthermore, reproducible detector response depends upon holding operating parameters, such as nebulisation and gas flow rate and stack temperature constant.

ELSDs respond to any non-volatile solute and thus approach the ideal of a universal detector. Sensitivity is enhanced using lasers as light sources with detection limits of 5 g per 25 μl reported, a considerable improvement compared to RI detection. In addition, ELSDs are not as sensitive to ambient conditions such as temperature and pressure and eluant flow rate.

The detectors have specific requirements, however, e.g. a supply of high quality air or nitrogen at $5\,l\,min^{-1}$ and efficient fume extraction of the exhaust from the detector stack is mandatory.

6.5.5.8 Desolvation/transport detectors. The principle of transport detectors, typified by the moving wire detector (Figure 6.29), was based on the concept of physically separating the solvent, which is necessarily volatile, from the involatile solute. The transport wire is passed through a coating block where eluant from the column is applied. The solvent is then evaporated, and the wire plus solute then passes to a pyrolysis or combustion

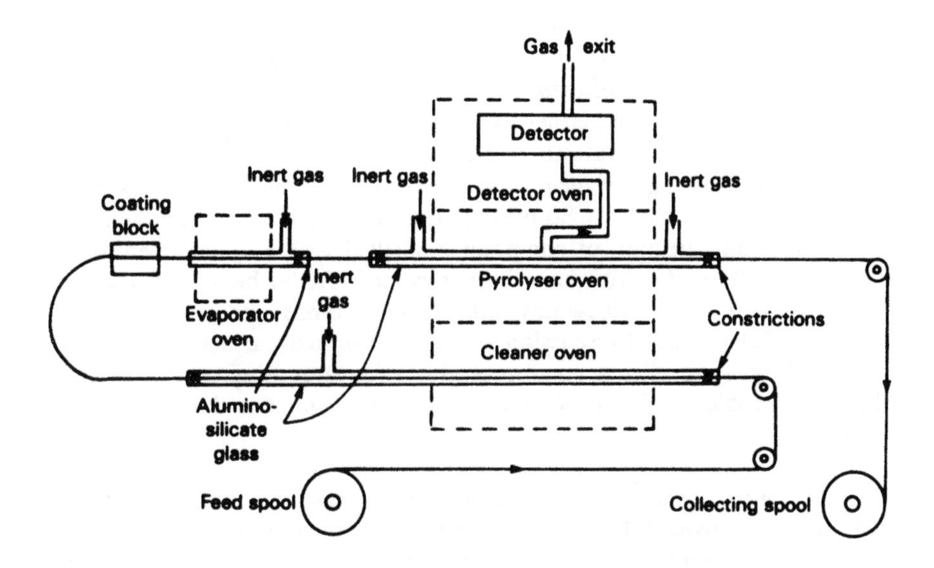

Figure 6.29 Moving belt interface coupled to an FID or ECD.

unit. The pyrolysis unit, at approximately 800°C, is purged with nitrogen gas and the gaseous products from the unit passed directly to an FID.

In the alternative combustion system, oxygen was used as the auxiliary purge gas. The sample is oxidised to carbon dioxide in the combustion chamber. The resultant gases are then passed to an auxiliary reactor, mixed with H_2 and passed over a Nickel catalyst. The carbon dioxide is thus converted to methane, which is then passed to the FID for detection and quantitation.

The transport detector is ideal for most gradient elution applications, the major limitation being those which use involatile buffers. The full potential of the FID cannot be realised due to the deficiencies of the transport system. The overall detector response is dependent on temperature stability and on the coating procedure which is related to both viscosity and surface tension of the solvent and sample. Due to these drawbacks particularly the lack of sensitivity these detectors were soon withdrawn.

However, improvements in design which increase the proportion of sample combusted and hence the detector's sensitivity have led to a resurgence of interest and detectors operating on similar principles to those elucidated above are now commercially available (Tracor Instruments) albeit at a cost. The effluent from the column is coated onto a moving quartz belt, the volatile solvent removed *in vacuo* and the involatile residue passed through a dual flame ionisation detector system. A number of interesting applications have been reported [43].

6.5.5.9 Radioactivity detectors. Radioactivity detectors are based on standard Geiger counting and scintillation systems, and specifically monitor radiolabelled compounds as they elute from the LC column [44]. Unfortunately, the common labels, ^{14}C, ^{3}H, ^{32}P and ^{35}S are low-energy β-emitters and hence the sensitivity of these detectors is low. Improvement in detector sensitivity can be achieved either by increasing the volume of the flow-cell or by reducing the flow-rate to give a longer residence time of sample in the flow-cell. This improvement is at the cost of chromatographic performance and speed of analysis. Stopped flow monitoring has also been utilised. Separation and detection of compounds labelled with stronger β-emitters (e.g. ^{131}I, ^{210}Po and ^{125}Sb) have been reported. Radiochemical detection has been used most successfully with large columns and with strongly retained solutes. Useful discussion of the operational parameters such as static and dynamic efficiency and procedures for optimising sensitivity and chromatographic efficiency in radiochemical LC monitors has been presented [45].

6.5.5.10 Mass spectrometric detection. The success in coupling GC with MS detection provided the impetus for extending this combination technique to LC and a number of LC–MS instruments are commercially available. The attractions of such a system are: (a) aids identification of compounds;

(b) single ion monitoring (SIM) can improve overall sensitivity and selectivity; (c) use of different ionising processes may provide additional information on the molecular structure of the compounds.

For LC–MS to become a reality an interface had to be designed which was capable of providing a vapour sample feed consistent with the vacuum requirements of the mass spectrometer ion source and of volatilising the sample without decomposition. Various enrichment interfaces have been developed such as the molecular jet, vacuum nebulising, the direct liquid introduction inlet and thermospray systems.

MS detection may well prove to be the most informative of all detectors for HPLC as not only does it afford quantitative analysis, but SIM procedures offer high sensitivity and selectivity and in principle it is a universal detector and can be used for all analytes. The power and utility of this approach has been extended further by the development of LC–MS–MS systems and these and other of the hyphenated techniques are discussed in detail in Chapter 7.

6.5.5.11 Other detectors. The above discussions have been concerned with those detectors most commonly employed in routine HPLC analysis and which are commercially available. Many other detectors have been developed to monitor specific solute properties in column effluents and new detection systems continue to be reported in the literature, many directed to meeting the requirements imposed by microbore and capillary separation technologies. The interested reader is directed to the reviews of Yeung [46] and Fielden [47] for a more detailed discussion of detector types employed in HPLC.

Some of these less used systems have limited applications in specific areas and combine HPLC with, for instance, chemiluminescence techniques [48], viscometry [49], optical activity measurement [50], piezoelectric crystals for mass scanning [51], atomic absorption and emission spectrometry [52–54], photoacoustic monitors [55], nuclear magnetic resonance [56], electron spin resonance [57], Raman [58] and photoconductivity measurement [59]. Details on these and other innovative detection systems are presented in the review by Bruckner [60].

6.5.5.12 Sample derivatisation. Derivatisation [61, 62] of a sample is undertaken principally for two reasons. First, there is no detector for HPLC that has universally high sensitivity for all solutes; hence, the use of a suitable chemical transformation of the solute can greatly extend the sensitivity and versatility of a selective detector. Second, sample derivatisation may be undertaken to enhance the detector response to sample bands relative to overlapping bands of no analytical interest. These reactions can be carried out either before or after the passage of solute through the column. The reagents are classed as either fluoregenic, i.e. non-fluorescent molecules which react with analytes to form fluorescent derivatives, or chromagenic

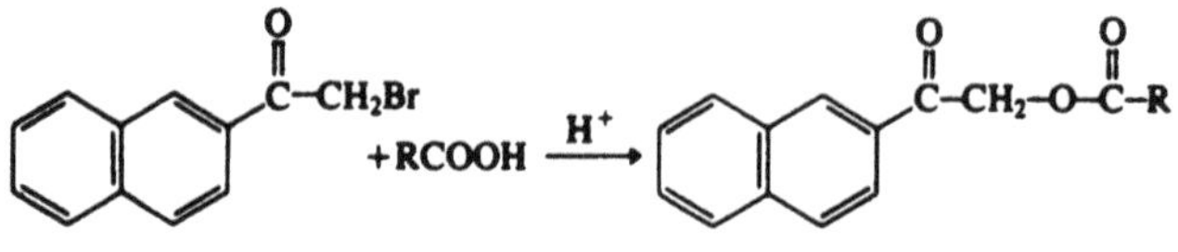

Figure 6.30 Derivatisation of carboxylic acids with naphacyl bromide.

reagents which react with the analyte to yield a derivative that strongly adsorbs ultraviolet or visible radiation.

An example of pre-column derivatisation, to enhance the spectrophotometric response of the sample, is the reaction between naphacyl bromide, the derivatising agent, and carboxylic acids (Figure 6.30). The ester product contains a chromophore unit and the sample derivatives are now able to absorb ultraviolet radiation.

An example of post-column derivatisation is the reaction of aminoacids with ninhydrin (Figure 6.31); again a chromophoric unit is present in the sample derivatives.

The replacement of ninhydrin by *ortho*-phthaldehyde [63] as the derivatising agent introduces a fluorescent group into the molecule and thus enables the highly sensitive fluorescent detector to be employed.

There are several factors to consider before choosing either pre- or post-column derivatisation. The former can be carried out manually, off-line from the system and places little restriction on the choice of reagents, solvents, length of reaction time, etc., all of which are important restrictions in post-column applications. The reaction, however, may give rise to more than one product which can subsequently complicate the chromatogram. For quantitative work, the extent of the reaction must be precise and reproducible for different sample concentrations.

Post-column derivatisation techniques, though imposing restrictions on the reagents and solvents that can be used has the advantages that it may be readily incorporated as part of a fully automated system and thus will

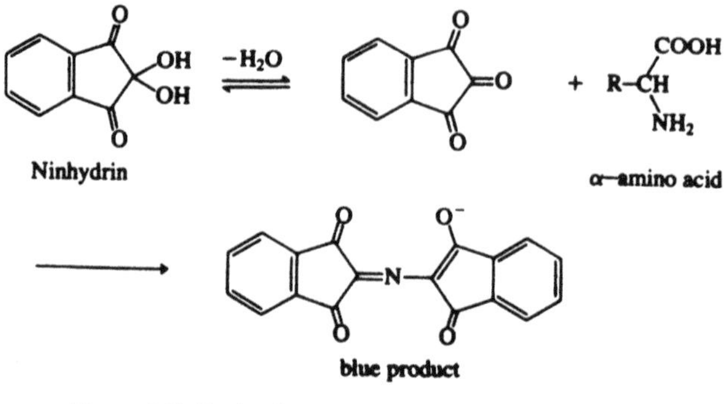

Figure 6.31 Derivatisation of amino acids with ninhydrin.

not complicate the chromatography even if the reaction yields more than one product. However, a major consideration is the extra-column effects which can impair the overall resolution of the LC system. For post-column reactors it is not essential that the derivatisation is carried out to completion, only that the extent of reaction is reproducible.

Non-chemical derivatisation has been employed for the detection of photoionisable species. The effluent is passed through a quartz tube where it is irradiated with ultraviolet light and the photoionised products passed to the conductivity detector [64, 65]. Krull has reported the detection of the active components, typically organonitro derivatives, of a large number of explosives and drugs by *in situ* LC decomposition [66].

In conclusion, it must be emphasised that each reactor must be tailored to its particular application. For every post-column reaction [67] there will be a set of optimum operating parameters such as temperature, concentration of reagent, flow-rate of reagent, reaction time, etc. These variables have to be studied in detail in order to optimise the configuration and operating conditions of the reactor. Thus, each final reactor design will tend to be specific for a particular application, and it may well be limited to a small range of eluant flow-rates.

6.6 Column packings and stationary phases for liquid chromatography

The majority of packings for modern LC are based on microporous particles of varying size, shape and porosity (Figure 6.32). The surface of these particles can subsequently be modified by either physical or chemical means to afford access to any of the classical modes of chromatography. The most common material used is silica, as it can withstand the relatively high pressures in use and is available with large surface area (200–300 $m^2 g^{-1}$) and small particle size. The importance of microporous silica is discussed in a number of reviews [68–70]. Other microporous packing materials based on alumina, zirconium and ion exchange materials are also in use.

Microporous particles (3–10 μm) give columns that are as much as 20 times as efficient as porous layer-bead or pellicular (40 μm) packings. Whilst modern LC is based almost exclusively on microporous packing materials it is informative to relate the advances in particle design with the attempts to eliminate the deleterious effects on column performance since the latter as expressed by H is related to experimental variables, such as, the particle size (d_P), the nominal stationary phase thickness (d_S) and the mobile phase velocity (u).

6.6.1 *Totally porous beads*

Initially packing materials were based on porous beads, usually silica or alumina, of 20–40 μm particle size; they gave poor column efficiencies as

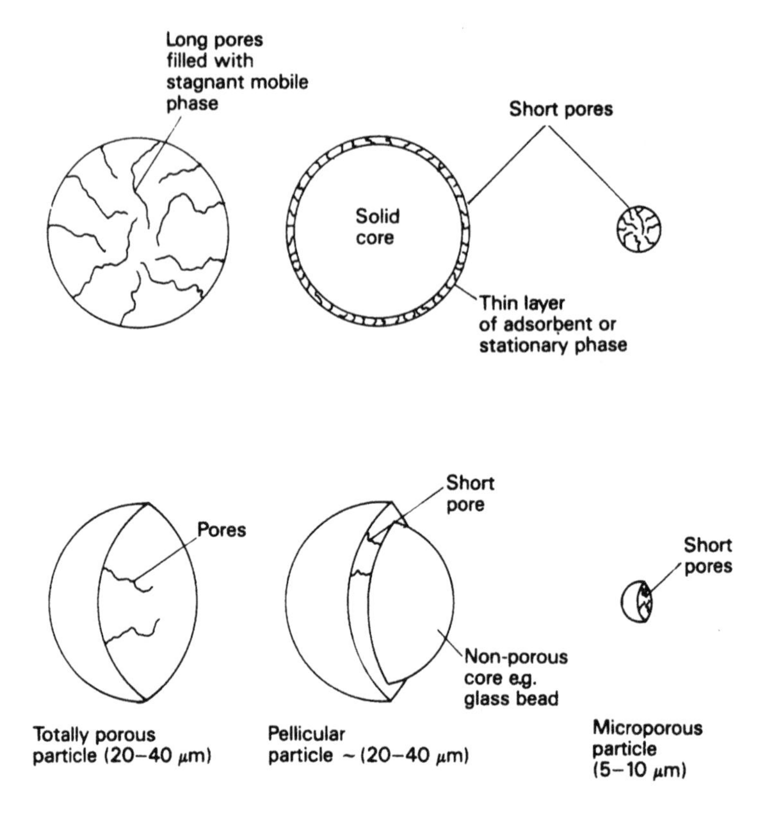

Figure 6.32 Characteristics of liquid chromatography packing particles.

the particles were totally porous and had large volumes of associated stagnant mobile phase. This led to the A and C terms being substantial and dominant in the expression for H, which effectively reduced to

$$H \approx Au^{0.33} + Cu$$

thus illustrating the importance of good column packing procedures on chromatographic performance.

6.6.2 *Pellicular packings*

Pellicular packings also known as porous layer beads or controlled surface porosity supports (that is, solid impervious beads of high mechanical strength) have the impervious core constructed from materials such as stainless steel, glass, silica, alumina or ion exchange resins. The chromatographically active layer on the surface is commonly silica, alumina, an ion exchange material or a cellulose derivative. An important advantage of the solid impervious core was that it did not swell when wetted and could withstand high inlet pressures without deformation. The shallow pore structure led to a marked improvement in the column performance due principally to the virtual elimination of the stagnant mobile phase and the associated increased rate of mass transfer between the mobile phase and porous layer, hence, the van Deemter equation reduces to

$$H \approx Au^{0.33} + B/u + Du$$

The large particle size allows H to be further approximated to

$$H \approx Au^{0.33}$$

A major disadvantage of pellicular packings is the low surface area associated with the shallow pores which limits sample capacity.

6.6.3 *Microporous particles*

The use of microporous particles of 3, 5 and 10 μm ushered in the era of genuine HPLC. Because of the small particle size stagnant mobile phase mass transfer effects are minimised as with pellicular packings whilst multipath (H_f) and mobile phase mass transfer effects ($H_{em'}$) being due to d_P and $(d_P)^2$, respectively are substantially reduced. A further advantage of these columns is afforded by the large surface area with the concomitant increase in sample capacity. The totally porous microparticle is generally a high surface area, small porosity silica gel or alumina.

Angular silicas such as Partisil and Lichrosorb are prepared by the reaction of sodium silicate with hydrochloric acid. Spherical particles, e.g. Spherisorb are synthesised by the hydrolytic polycondensation of polyethoxysilane followed by emulsion precipitation or spray-drying.

These synthetic procedures can be tailored to the production of microporous particles (3–10 μm) of controlled porosity (6–10 nm) and specific surface area (200–400 $m^2\,g^{-1}$). Packings based on these particles give a 10-fold increase in column performance thus allowing shorter columns to be used consequently giving faster separations and requiring smaller operating pressures. The surface of these particles can subsequently be modified by either physical or chemical means to afford access to any of the classical modes of chromatography.

6.7 Adsorption chromatography

The mechanism involved in adsorption chromatography is based on the selective distribution of solute molecules between solution and adsorption on the surface of the packing. This later process is competitive between solute and mobile phase molecules, especially the polar modifier. The only

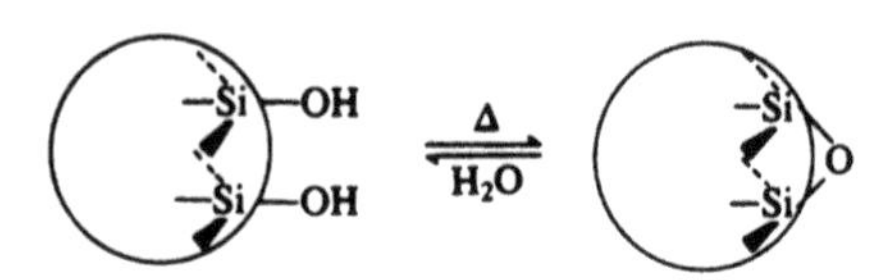

Figure 6.33 Wide pore silicas.

materials used for liquid–solid chromatography (LSC) are silica gel, alumina and graphite. Silica gel is the most extensively used though alumina is preferred in some applications, e.g. due to its basic character it has proved to be good for the separation of acidic compounds. Due to the extensive and preferred use of silica packings the following discussion will be restricted to silica gel.

Silica gel has a highly active surface layer of randomly distributed silanol groups. The mechanism of adsorption is thus the interaction of the surface silanol groups with any polar functions which may be present in the solute molecules, e.g. alcohols, amines, ketones and carboxylic acids.

As in classical LC (Chapter 3), the method of preparation and activation of the packing material must be consistent in order to obtain reproducible separations for a given application. The main method of activation involves dehydration. Associated with the silanol surface groups is water of hydration which may be classified as belonging to one of three types:

1. loosely bound water easily removed by heating or by solvent extraction;
2. strongly held water; and
3. water present as OH in adjacent silanol groups which can be removed by strong heating.

Removal of water in the first two cases is reversible; in the last irreversible.

The surfaces of narrow pore silicas are covered mainly by hydrogen bonded hydroxyl groups, while the surfaces of wide pore silicas are covered mainly with isolated hydroxyl groups. In wide pore silicas, heating at 110°C produces siloxane groups from adjacent silanols (Figure 6.33). The siloxanes are weak adsorbent sites and are not important chromatographically. The reaction is reversible by heating in the presence of water.

In narrow pore silicas water is eliminated by reaction between hydroxyl groups on adjacent sites (Figure 6.34). No reverse reaction is possible and there is an effective loss of surface area due to the cross-linking effect.

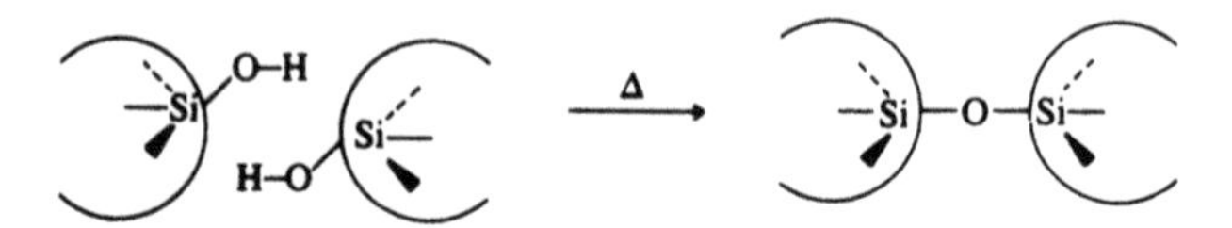

Figure 6.34 Narrow pore silicas.

The preparation of silica gel and other adsorbents and the subsequent grading—Brockmann Numbers—is presented in greater detail in Chapter 4.

The vast majority of silicas used in HPLC applications today are of the totally porous spherical and irregular type, with surface area of the order of 200–500 $m^2 g^{-1}$ and a range of pore sizes 4–400 nm. For analytical scale work 3, 5 or 10 μm particle size is used; preparative separations operated in the overload mode, use larger particle size, 40–60 μm, with larger surface areas 500–800 $m^2 g^{-1}$.

The eluant pH is restricted to between 3 and 8 pH units to avoid damage to the silica matrix. This deleterious effect can be minimised by passing the eluant through a pre-column inserted between pump and injector, thus pre-saturating the eluant with silica. Almost all the available silicas whether spherical or irregular are very similar in terms of separation efficiency and retentive properties.

6.7.1 Solvent systems

A variety of solvent systems can be used for adsorption chromatography, ranging from non-polar to polar. When a mixed solvent system is used then the polar component will be adsorbed onto the surface and will modify the adsorptive capacity of the packing material. Examples of polar modifiers are butyl chloride, dioxane and ethyl acetate. During the chromatographic process, solute molecules may displace the polar modifier and then subsequently be displaced themselves, or they may not interact with the silanol layer as such, but with the adsorbed layer. In the situation where the adsorbed layer is strongly held it is difficult to distinguish between adsorption and partition chromatography. Adsorption chromatography functions best for the separation of non-polar or moderately polar compounds using moderately polar eluants. It does not function well with high polarity eluants.

Compounds with polar substituents can interact with the active surface silanol groups. The strength of the adsorption is proportional to the polarity of the compound and so LSC is best used for the separation of mixtures into functional group classes. In addition, because of the sterically controlled interactions between adsorbent surface layer and the sample molecules LSC can also provide for the separation of structural isomers including polyfunctional isomeric compounds. LSC is not suitable for the separation of compounds only sparingly soluble in the eluant, such as polyaromatics, fats and oils, nor for the separation of the members of a homologous series. It is the preferred mode for preparative scale analysis because of the higher sample capacity and greater pH stability. Due to the difficulty in achieving rapid equilibration with solvents containing polar modifiers, gradient elution work is to be avoided with adsorption chromatography.

Table 6.4 Solvent strength parameter, polarity indices and physical properties for a range of solvents

Solvent	Viscosity at 25°C (cP)	Solubility parameter, δ	Eluant strength, ϵ^{ϕ}	Polarity index, P′
n-Hexane	0.30	14.9	0.01	0.1
Methylbenzene	0.55	18.2	0.29	2.3
Dichloromethane	0.43	19.8	0.32	3.4
1,2-Dichloroethane		20.0		3.7
Trichloromethane	0.57	19.0	0.26	4.4
Carbontetrachloride	0.90	8.60	0.18	1.6
Tetrahydrofuran	0.46	18.6	0.57	4.0
Ethanol	1.08	25.9	0.88	4.3
Methanol	1.326	29.4	0.95	6.6
Water	1.333	47.8	Large	10.0
1,4-Dioxane	1.2	20.4	0.56	4.8
Acetonitrile	0.34	23.9	0.65	6.2

6.7.2 *Solvent optimisation*

Control of sample retention in adsorption chromatography is achieved almost exclusively by modifying the composition of the mobile phase. Minor changes in the eluant strength have dramatic effects on k and α values. The elution power of a solvent is measured by its solvent strength parameter, ϵ^0, which is in effect the solvent adsorption energy per unit surface area. However, often a single solvent is unable to effect resolution of the sample components and binary and ternary solvent mixtures must be employed. A measure of the solvent strength of these mixtures may be determined from the solvent polarity indices, a term coined by Snyder [71]. A range of solvent properties are presented in Table 6.4.

The polarity index, P', is a numerical measure of the relative polarity of various solvents as determined from their solubility in some specific solvents. The polarity index P_{AB} for a mixture can then be readily determined from the polarity indices of the pure components and their respective volume fractions (ϕ_A, ϕ_B), thus

$$P_{AB} = \phi_A P_A + \phi P_B$$

Any desired polarity index can be obtained by mixing the appropriate amounts of solvents. An increase in the polarity of the solvent mixture means a stronger eluant and hence smaller k values. This is expressed in the following relationship:

$$\frac{k_2}{k_1} = 10^{(P_1 - P_2)/2}$$

Thus, a two unit change in polarity index results in a 10-fold change in k. Often satisfactory resolution can be achieved by adjusting the capacity factor by modifying the solvent strength. If, however, separation is still incomplete due to $\alpha \Rightarrow 1$, then α can be increased by changing the chemical nature of the solvent while maintaining its polarity. A fuller discussion of solvent optimisation procedures is presented later in section 6.14.

6.8 Liquid–liquid partition chromatography

Classical liquid–liquid chromatography (LLC) originated from the Nobel Prize winning work of Martin and Synge in 1941. LLC systems consist of a mobile phase and a stationary phase which is held on the support by physical forces of adsorption. The column materials are commonly prepared using the solvent evaporation technique developed for GC. In order to avoid solvent stripping of the stationary phase, it is necessary for the eluant to be saturated with the stationary phase component. The mobile and stationary phases should have contrasting polarities and be immiscible.

Though many of the applications of LLC are now carried out more conveniently using bonded phase techniques the method has some advantages which make it a useful adjunct to the other modes of column chromatography. The major benefits derive from the ease of renewal of the stationary phase, the degree of selectivity that can be achieved (as a wide range of liquids can be selected for the stationary phase) and the reproducibility of the chromatography.

LLC systems can be classified as either normal or reverse phase. The former are more stable systems due to the strong bonding that exists between the stationary phase and the support material. However, immiscible pair systems, such as isopropanol/hexane, suffer the disadvantage that sample components tend to have partition coefficients which are either very small or very large with little differentiation.

Reverse phase systems, such as squalane/water/alcohol, require the eluant to be saturated with the squalane stationary phase to avoid solvent stripping; however the binding forces are so weak that shearing occurs at high pressures. The other main disadvantages associated with LLC are that as the eluant must be saturated with the stationary phase, gradient elution techniques cannot be used and to maintain the required stationary phase loading the temperature must be controlled to $\pm 0.5°$. Some sample types with their respective mobile and stationary phases are shown in Table 6.5.

Table 6.5 Samples, stationary and mobile phases

Stationary phase	Mobile phase	Sample
Cyanoethylsilicone	Water	Coumarins
Squalane	Water/AcN	Hydrocarbons
Polyethylene glycol	Pentane	Alcohols, esters
β',β-Oxydiproponitrile	Hexane	Phenols

6.9 Chemically bonded stationary phases for high performance liquid chromatography

The severe restrictions on classical LLC with regard to solvent stripping of the stationary phase from the analytical column and the incompatibility with gradient elution techniques, led to the development of a range of chemically bonded stationary phases almost all of which are based on 3, 5 and 10 μm silicas. The silica supports are totally porous and may be spherical or irregular in shape but have a very narrow particle size distribution. Bonded-phase chromatography (BPC) though it continues to evolve is a mature technique and as practised in its various forms is the most widely used for the following reasons:

- polar, non-ionic, ionic and ionisable molecules can be effectively separated using a single column and mobile phase;
- stationary phases of a wide range of polarities are available;
- it permits a wider choice of eluant;
- the predominant eluant is water-based with organic modifiers such as methanol and acetonitrile which are all readily available and inexpensive;
- gradient elution techniques can be used without fear of stripping the stationary phase;
- the systems come to equilibrium much faster with little evidence of irreversible sorption and tailing; and
- there is no restriction on solvent inlet pressure.

The principal limitation is in the range of pH of the eluant used, restricted to 2.0–8.5, as cleavage or hydrolysis of the stationary phase can occur; likewise with certain oxidising solutions. The preparation of the bonded materials involves the derivatisation of the surface silanol groups. The bonded phases available may be classified as follows:

- hydrocarbon groups, such as octadecyl ($C_{18}H_{37}$), but also groups with shorter chain lengths such as C_1, C_2, C_8 and aryl;
- polar groups such as amino and cyanopropyl, ethers and diols; and
- ion exchange groups such as sulphonic acid, amino and quaternary ammonium.

6.9.1 Synthesis of bonded phase materials

The synthetic methods used for the preparation of bonded phase materials are illustrated in Figure 6.35. One of the first reported bonded phases, the alkoxy silanes (1) also referred to as silicate esters, was prepared by the direct esterification of silanol groups with alcohols. The major disadvantage of this packing material was its limited hydrolytic stability, as it is readily hydrolysed by aqueous alcohol eluants.

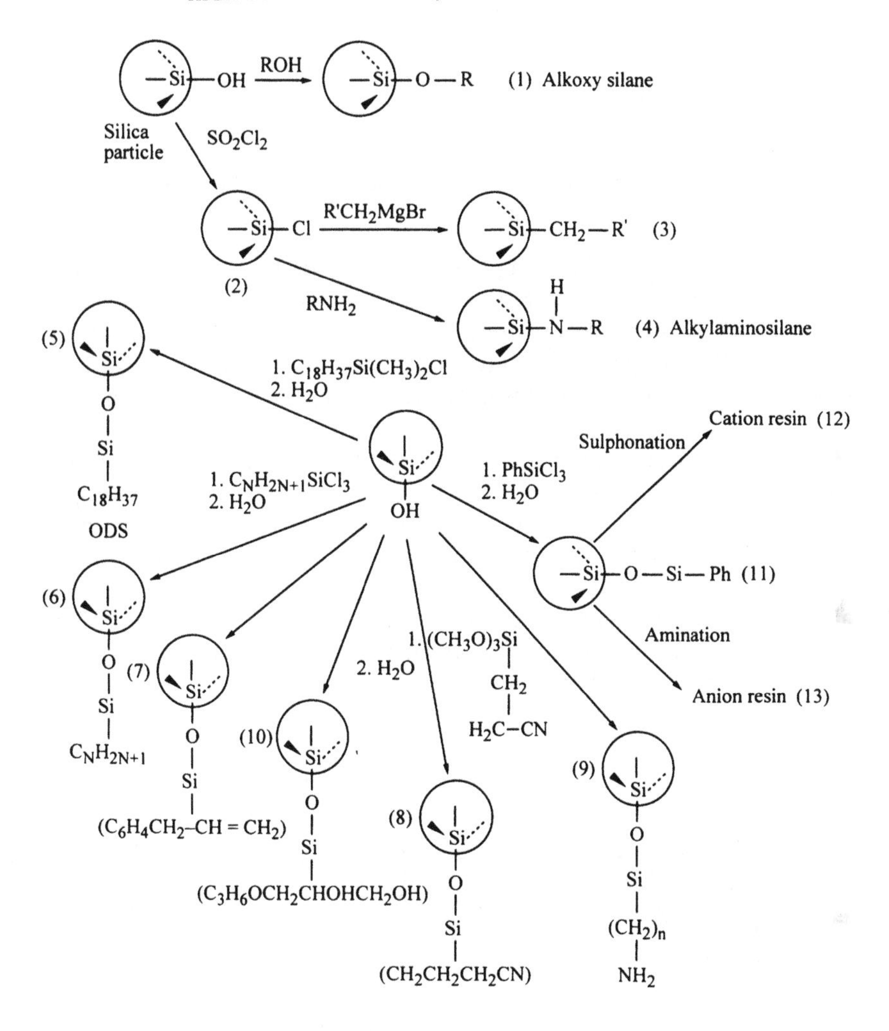

Figure 6.35 Reaction schemes for preparation of chemically bonded stationary phases.

Other syntheses were developed, leading to the linkage of stationary phases via Si–C and Si–N bonds. The common step in these reaction sequences was the chlorination of the silanol groups using thionyl chloride (2). Organic groups can then be bonded directly to the surface on reaction with Grignard or organolithium derivatives (3). Alternatively, reaction of the chlorosilane with amines gives alkylaminosilane (4) linked bonded phases. While Si–C bonds have good hydrolytic stability, the reaction and handling of organometallic reagents with silica gel is difficult and leads to incomplete coverage and deposition of hydrolysed reagent. The Si–N bonds have limited hydrolytic stability and are restricted to aqueous eluants in the pH 4–8 range.

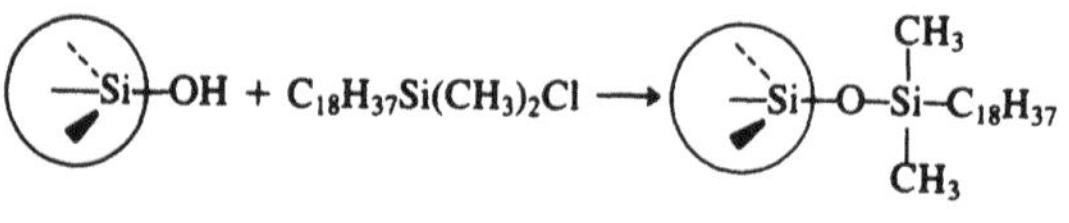

$$\mathrm{Si{-}OH} + C_{18}H_{37}Si(CH_3)_2Cl \longrightarrow \mathrm{Si{-}O{-}Si(CH_3)_2{-}C_{18}H_{37}}$$

Figure 6.36 Reaction of silica with monofunctional dimethylsilanes.

The most widely used bonded phase materials are those derived from the reaction of the surface silanol groups with organochlorosilanes, which leads to linkage of the stationary phase to the support via a siloxane (Si–O–Si) bond. For ODS ($C_{18}H_{37}Si$–) bonded phases the reagent used is octadecylchlorosilane (5). If the silane has only one chlorine then the reaction yields a well defined highly regular monofunctional stationary phase (Figure 6.36) with a surface loading of some 8–12% carbon. Residual silanol groups cause tailing and these groups can be blocked or 'end-capped' using trimethylsilylchloride (Figure 6.37). If, however, di- and trichloro silanes are used then some cross-linking can occur, giving an undefined ligate on the surface. This reaction is followed by hydrolysis of the residual chlorines and reaction with trimethylchlorosilane (Figure 6.37). The resulting material is a silica base with a denser, thicker ligate layer on the surface of the silica about 15–25% carbon loading.

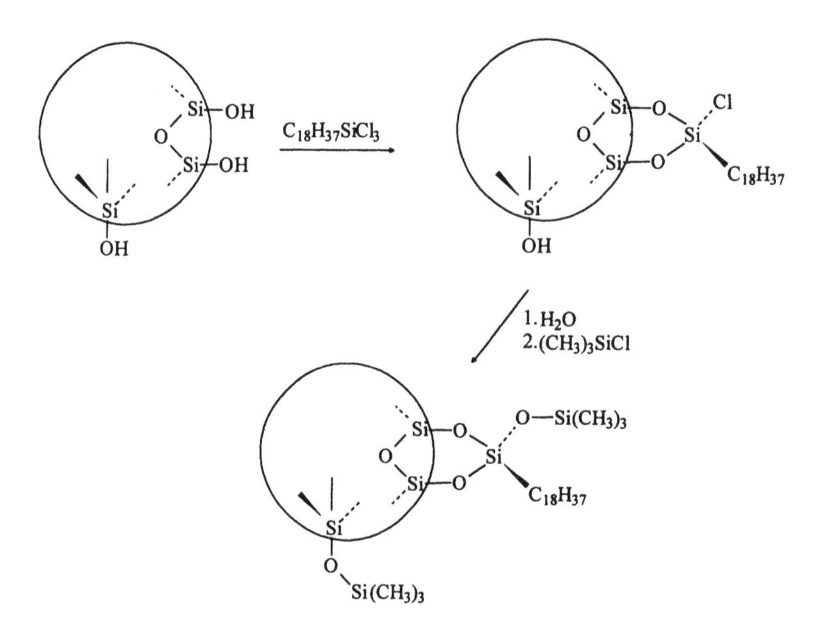

Figure 6.37 Reaction of silica with octadecyltrichlorosilane and subsequent end-capping yielding a 'polymeric' stationary phase.

In practice, there are significant differences in the performance and selectivity of bonded phase packings which are ostensibly of the same specification. This is in part due to the different chemistries employed by manufacturers to produce the bonded phases, which greatly affects the attainable coverage, but more importantly to the efficiency and extent of end-capping procedures. A fully end-capped material will have a 1–2% increase in carbon loading. Materials which are not fully end-capped shall have residual silanol groups with concomitant adsorption properties. There is a trend towards using dense monolayers of monofunctional dimethylsilanes.

A typical commercial material will have 40–60% of the available silanol sites derivatised. The advantage of these supports is the stability of the siloxane linkages to column inlet pressures and hydrolysis, thus allowing use of solvents at 6000 psi (41.4 MPa) in the pH range 2.0–8.5. Using other trichlorosilane derivatives as reactants, a variety of functional groups can be bonded to the silica support (Figure 6.35), for instance, alkyl $C_8(-O-Si-C_7H_{14}CH_3)$, C_2 and C_1 (6), alkenephenyl $(-O-Si-C_6H_4CH_2-CH=CH_2)$ (7), cyanopropyl $(-O-Si-CH_2-CH_2-CH_2-CN)$ (8), aminoalkyl $(-O-Si-(CH_2)_nNH_2)$ (9), diols $(-O-Si-C_3H_6-OCH_2-CHOHCH_2OH)$ (10) and phenyl (–O–Si–Ph) (11). The preparation and characterisation of bonded phases has been thoroughly reviewed by Sander and Wise [68].

Sulphonation and amination of phenylalkyl groups can be carried out, so giving materials with ion exchange capability (12) and (13). Ionogenic groups can also be introduced into the alkene phenyl bonded phase by suitable chemical modification. The above are the common chemically bonded stationary phases, which are marketed under a variety of trade names. The phases commercially available and their differences in alkyl chain length, percentage carbon loading and particle shape are indicated in Tables 6.6–6.8.

6.9.1.1 Stationary phase considerations. The retention mechanism(s) by which bonded phases retain solutes is complex and not fully understood [71–73]. It may be that it is comparable in behaviour and function to conventional partitioning in liquid–liquid phase systems or there may be competition between eluant and solute molecules for 'position' on the stationary phase. It is likely that there are several mechanisms operating however it is adequate to regard the stationary phase as a conventional, physically retained liquid with which analytes interact by the conventional sorption processes.

In reverse phase chromatography, retention times increase with increase in chain length of the bonded phase. Longer solute retention generally provides for enhanced resolution and thus varying chain length of bonded phase is a further aid to optimising the resolution.

The percentage carbon loading is not so important a consideration as is the percentage coverage of the silica. The degree of surface coverage greatly

Table 6.6 Commercially available reverse phase packing materials

Packing material	Phase	Particle shape and size (μm)	Pore size (Å)	Pore volume (ml g^{-1})	Surface area (m^2 g^{-1})	Carbon load (%)	Bonded phase coverage (μmol m^{-2})	End capping
Hypersil SAS	C_1	3,5,10 spherical	120	0.7	170	3.0 Monomeric	5.29	No
Spherisorb	C_1	3,5,10 spherical	80	0.5	220	4, Monomeric	1.08	Partial
Phenomenex IB Sil	C_1	3,5,10 spherical	125	0.75	165	2.5 Monomeric	4.48	No
Ultremex	C_1	3,5,10 spherical	80	0.8	200	3, Monomeric	4.50	No
Ultremex	C_2	5 spherical	80	0.8	200	3, Monomeric	4.49	Yes
Hypersil	C_4	5,10 spherical	300	0.6	50	2.0 Monomeric	4.8	Yes
Nucleosil	C_4	5 spherical	120	0.65	200	N/A, Monomeric	N/A	No
Nucleosil 300	C_4	5,10 spherical	300	0.8	100	1, Monomeric	1.41	Yes
Nucleosil 500	C_4	7 spherical	500	0.8	35	1, Monomeric	4.04	Yes
Nucleosil 1000	C_4	7 spherical	1000	0.85	25	1, Monomeric	5.66	Yes
Nucleosil 4000	C_4	7 spherical	4000	0.7	10	<1, Monomeric	N/A	Yes
Selectosil	C_4	5,10 spherical	300	0.9	110	2, Monomeric	2.61	Yes
Spherisorb	C_6	3,5,10 spherical	80	0.5	220	6, Monomeric	2.51	Yes
Ultremex	C_6	3,5,10 spherical	80	0.8	200	13, Monomeric	3.29	Yes
Lichosorb	C_8	5,7,10 irregular	100	—	250	7 Monomeric	2.8	—
Hypersil	C_8	5,10 spherical	300	0.6	50	3.0, Monomeric	5.24	Yes
Hypersil Mos 1	C_8	3,5,10 spherical	120	0.7	170	7.0 Monomeric	3.85	No
Hypersil Mos 2	C_8	3,5,10 spherical	120	0.7	170	7.0 Monomeric	3.85	—
Phenomenex IB Sil	C_8	3,5,10 spherical	125	0.75	165	7.5 Monomeric	4.29	Yes
Phenomenex IB Sil BD	C_8	3,5 spherical	125	0.75	165	8.0 Monomeric	4.61	Yes
Nucleosil 100	C_8	3,5,10 spherical	100	1.0	350	9, Monomeric	2.49	No
Nucleosil 120	C_8	3,5,10 spherical	120	0.65	200	7, Monomeric	3.27	No
Nucleosil 300	C_8	5,10 spherical	300	0.8	100	2, Monomeric	1.72	Yes
Nucleosil 500	C_8	7 spherical	500	0.8	35	1, Monomeric	2.42	No
Partisil	C_8	5,10 irregular	85	—	350	8.5 —	2.33	Yes
Selectosil	C_8	3,5,10 spherical	110	11	330	8, Monomeric	2.38	No
Selectosil 300	C_8	5,10 spherical	300	0.9	110	3, Monomeric	2.38	No
Spherisorb	C_8	3,5,10 spherical	80	0.8	220	6, Monomeric	2.51	Yes
Ultremex	C_8	3,5,10 spherical	80	0.8	200	8, Monomeric	3.81	Yes
Hypersil	C_{18}	3,5,10 spherical	120	0.7	170	10.0 Monomeric	2.84	Yes

Kromasil	C_{18}	5,7,10,13,16 spherical	100	0.9	340	19, N/A	3.1	Yes
Lichosorb	C_{18}	5,7,10 irregular	60	N/A	500	17 Monomeric	2.2	No
Lichosorb	C_{18}	5,7,10	100	N/A	300	11 Monomeric	2.2	No
Nucleosil 100	C_{18}	3,5,10 spherical	100	1.0	350	14, Monomeric	2.06	Yes
Nucleosil 100AB	C_{18}	5 spherical	100	1.0	350	25, Polymeric	4.51	No
Nucleosil 120	C_{18}	3,5,7,10 spherical	120	0.65	200	11 Monomeric	2.69	Yes
Nucleosil 300	C_{18}	5,10 spherical	300	0.8	100	6 Monomeric	2.72	Yes
Nucleosil 500	C_{18}	7 spherical	500	0.8	35	2 Monomeric	2.45	Yes
Nucleosil 1000	C_{18}	7 spherical	1000	0.85	25	1, Monomeric	1.69	Yes
Partisil ODS (1)	C_{18}	10 irregular	85	N/A	350	5 N/A	0.63	No
Partisil ODS (2)	C_{18}	10 irregular	85	N/A	350	16 N/A	2.43	No
Partisil ODS (3)	C_{18}	5,10 irregular	85	N/A	350	10.5 N/A	1.45	Yes
Phenomenex IB Sil	C_{18}	3,5,10 spherical	125	0.75	165	11.0 Monomeric	2.74	Yes
Phenomenex IB Sil	C_{18}	3,5 spherical	125	0.75	165	12.0 Monomeric	3.62	Yes
Selectosil	C_{18}	3,5,10 spherical	110	1 1	330	13.0 Monomeric	1.99	Yes
Selectosil 300	C_{18}	5,10 spherical	300	0.9	110	7, Monomeric	2.93	Yes
Spherisorb (1)	C_{18}	3,5,10 spherical	80	0.5	220	7, Monomeric	1.67	Partial
Spherisorb (2)	C_{18}	3,5,10 spherical	80	0.5	220	12, Monomeric	2.72	Yes
Uttremox	C_{18}	3,5,10 spherical	80	0.8	200	13.0 Monomeric	3.29	Yes
Zorbax	C_{18}	3,5,7 spherical	70	N/A	330	20, Monomeric	3.37	Yes

Table 6.7 Commercially available normal phase packings

Packing material	Phase	Particle shape and size (μm)	Pore size (Å)	Pore volume (ml g^{-1})	Surface area (m^2 g^{-1})	Carbon load (%)	Bonded phase coverage (μmol m^{-2})	End capping
Nucleosil 100	Phenyl	7 spherical	100	1.0	350	8, mono	1.96	No
Nucleosil 120	Phenyl	7 spherical	120	0.65	200	6, mono	2.49	No
Nucleosil 300	Phenyl	2 spherical	300	0.8	100	2, mono	1.56	No
Selectosil	Phenyl	5,10 spherical	110	1.1	330	8, mono	2.08	No
Spherisorb	Phenyl	3,5,10 spherical	80	0.5	220	3, mono	1.08	Partial
Ultremex	Phenyl	3,5,10 spherical	80	0.8	200	3, mono	1.19	Yes
Hypersil CPS1	CN	3,5,10 spherical	120	0.7	170	4.0 mono	3.55	No
Hypersil CPS2	CN	5 spherical	120	0.7	170	4.0 mono	3.55	Yes
Phenomenex IB Sil	CN	3,5,10 spherical	125	0.75	165	4.5 mono	4.15	No
Phenomenex IB Sil	CN	3,5 spherical	125	0.75	165	4.0 mono	3.68	Yes
Spherex	CN	3,5,10 spherical	100	0.8	180	2.5 mono	2.03	No
Spherisorb	CN	3,5,10 spherical	80	0.5	220	3.5 mono	2.37	No
Iltremex	CN	3,5,10 spherical	80	0.8	200	2.5 mono	1.83	No
Zorbax	CN	3,5,7 spherical	70	—	330	5 mono	2.41	Yes
Hypersil	NH_2	3,5,10 spherical	120	0.7	170	2.0 mono	2.05	No
Nucleosil 100	NH_2	5,10 spherical	100	1.0	350	N/A mono	N/A	No
Phenomenex IB Sil	NH_2	3,5,10 spherical	125	0.75	165	2.2 mono	2.33	No
Selectosil	NH_2	5,10 spherical	110	1.1	330	3.3 mono	1.8	No
Spherex	NH_2	3,5,10 spherical	100	0.8	180	2.0 mono	1.94	No
Spherisorb	NH_2	3,5,10 spherical	80	0.5	220	2.0 mono	1.58	No
Ultremex	NH_2	3,5,10 spherical	80	0.8	200	2.0 mono	1.74	No
Nucleosil 100	diol	7 spherical	100	1.0	350	0	—	No
Spherex	diol	5,10 spherical	100	0.8	180	2.0 mono	1.37	No

Table 6.8 Commercially available ion exchange materials

Packing material	Phase	Particle shape and size (μm)	Pore size (Å)	Pore volume (ml g^{-1})	Surface area (m^2 g^{-1})	Carbon load (%)	Bonded phase coverage (μmol m^{-2})	End capping
Hypersil SAX	Anion	5 spherical	120	0.7	170	2.7 mono	1.56	Yes
	Anion	5,10 spherical	300	0.6	50	1.4 mono	2.28	Yes
Partisil	Anion	10 irregular	85	—	350	0.85 (N^+R_3)	—	No
Phenomenex IB Sil	Anion	3,5,10 spherical	125	0.75	165	7.5 mono	4.29	Yes
Selectosil	Anion	5,10 spherical	110	1.1	330	1 meq/g	N/A	No
Spherex	Anion	5,10 spherical	100	0.8	180	0.6 meq/g	2.25	No
Spherisorb	Anion	5,10 spherical	80	0.5	220	4 mono	0.4 mm/g	N/A
Partisil SCX	Cation	10 irregular	85	—	350	0.40 (SO_3H)	—	No
Selectoxil	Cation	5,10 irregular	110	1.1	330	1 meq/g	N/A	No
Spherex	Cation	5,10 spherical	100	0.8	180	0.4 meq/g	2.88	No
Spherisorb	Cation	5,10 spherical	80	0.5	220	6 mono	—	No
Ultremex	Cation	5 spherical	80	0.8	200	0.4 meq/g	2.88	No

affects the retention, chemical selectivity and pH stability of the resulting bonded phase. Residual silanol groups and their adsorption effects can cause tailing of the peak. It is thus preferable to use a reverse phase packing with a high percentage coverage, or one in which the residual silanol groups have been end-capped. As to the choice of long or short alkyl chain generally polar solutes are better separated or short-chain alkyl phases, and non-polar on long-chain alkyl phases. Variation in selectivity between the same packings of different manufacturers is generally due to differences in percentage carbon loading and the degree of end-capping.

Of the other stationary phases, the amino phases have been used extensively to separate sugars and peptides; the nitrile phase has found application in the separation of porphyrins. An important consideration in the use of polar bonded-phase materials is an awareness of the reactivity of the terminal functional group, for example, aminoalkyl bonded phase should not be used for the chromatography of carbonyl compounds due to possible condensation reactions and formation of Schiff's bases.

6.9.1.2 Choice of eluant. The number of stationary phases encountered in HPLC is small as the selectivity can be readily adjusted by variation in the nature and strength of the eluant. This contrasts with the situation in GC, where the selectivity can only be adjusted either by altering the stationary phase, the support or the column temperature. The plate efficiency of bonded-phase materials is at least equal to that of the support materials, while there is evidence to suggest that they have substantially higher load capacities. The mechanism of solute retention on bonded phase materials is not fully understood, however it is satisfactory to consider the organo-bonded layer as a thin liquid film as this allows reasonable prediction of retention behaviour. The solvents commonly employed in normal BPC, i.e. with polar bonded stationary phases, such as cyanopropyl, are hydrocarbons containing small amounts of a polar solvent. Typical polar eluants in order of decreasing solvent strength are methanol, chloroform, methylene chloride and isopropylether. The resolution is systematically investigated by varying the solvent strength.

In reverse phase BPC water is used as the base solvent; the eluant strength is adjusted by using organic modifiers most commonly methanol and acetonitrile. The resolution is optimised as before by modifying the capacity factors through changes in the solvent strength. Due to the similarities in the proton donor and acceptor properties between methanol and water, substantial quantities of the former are required before it modifies the behaviour of water. Acetonitrile, on the other hand, has a much more pronounced effect at lower concentrations due to the marked difference in solvent properties, such that a change in organic modifier from methanol to acetonitrile can lead to a variation in the order of elution of sample components. Highly polar and ionisable molecules such as acids have such a high affinity for aqueous eluants that they wash through the column without being retained at all by the packing material. Modifications, however, can be made to the technique such that even highly dissociated molecules can be successfully chromatographed by BPC. This has been achieved by using ODS-silica materials with aqueous buffer solutions containing no organic modifier. Solute retention is influenced by eluant pH, which controls the degree of dissociation of the solute and hence its partition between the bonded organo-phase and the aqueous mobile phase. For effective chromatography the eluant pH should be $>\mathrm{p}K_a$ of the ionisable component. Non-ionisable compounds show little change in retention with variation of pH.

The ion suppression technique can be used to great effect for the analysis of weak acids or bases. For the analysis of acidic compounds the technique consists of the addition of a small amount of acetic or phosphoric acid to the mobile phase. By reducing the eluant pH dissociation of the sample molecules is suppressed. They thus have decreased affinity for the eluant and are retained to a greater extent by the ODS phase. The range of BPC is considerably extended using techniques such as ionic suppression and this mode of LC using ODS bonded phases finds wide application.

New stationary phases and packings continue to be developed if not to improve on existing products to cater for new demand, e.g. a number of companies now market columns designed specifically for the separation of fullerenes and fullerene derivatives. These and other new packing materials are discussed in a review by Majors [74].

6.10 Chiral chromatography

The biological and pharmacological activity of chiral compounds depends upon their stereochemistry [75] with many showing enantioselective differences in their pharmacokinetics and pharmacodynamics, e.g. the D-isomer of penicillamine has useful therapeutic properties while the L-isomer is toxic. This growing awareness of the relationship between chirality and activity provided the impetus to develop chromatographic techniques for both the analytical and preparative scale separations of racemates.

Until the mid-1980s the available packings and stationary phases in HPLC did not enable separation of racemates since enantiomers possess identical physical and chemical properties and chromatographically cannot be distinguished as $\alpha \Rightarrow 1$. Thus, chiral separations involved pre-column derivatisation of the D- and L-isomers of a racemate by reacting with an optically pure asymmetrical reagent so yielding a mixture of diastereoisomers which have different physical properties and hence could be separated by normal or reverse phase chromatography. In HPLC, for example, Marfey's reagent, 1-fluoro-2,4-dinitrophenyl-5-L-alanine amide, is used to form

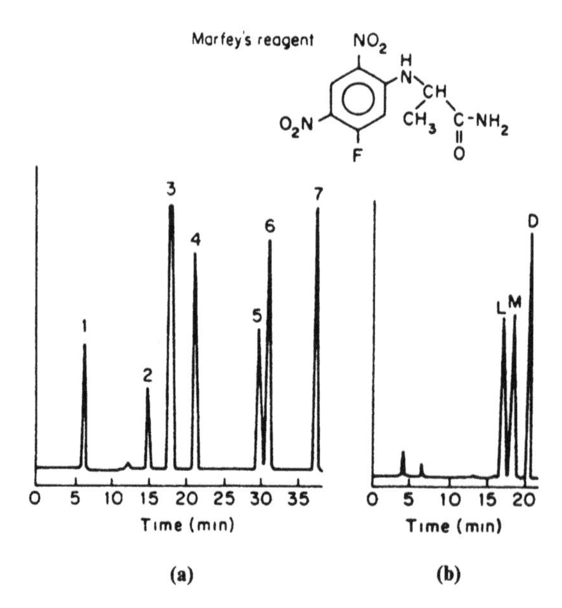

Figure 6.38 The separation of (a) D- and L-amino acids, and (b) D- and L-dopa enantiomers after reaction with Marfey's reagent. Peaks (a) 1, L-Gen; 2, D-Gm; 3, Marfey-OH; 4, L-Met; 5, D-Met; 6, L-Phe. (b) L, L-dopa; D, D-dopa; M, Morfey-OH. (Reproduced by permission of Pierce Chemical Co.)

diastereomeric mixtures from optically active amines and amino acids (Figure 6.38).

However, this procedure relies on high purity of the derivatising agents and is further complicated by the fact that the rates of reaction of the enantiomers are often different which results in formation of diastereoisomers in differing proportions to the enantiomers present in the racemate.

An alternative approach for the direct resolution of enantiomers by HPLC is to use a chiral stationary phase. This technique relies on the formation of transient/temporary diastereoisomers between the sample enantiomers and the CSP. Differences in stability between the diastereoisomers is reflected in differences in retention times, the enantiomer forming the less stable complex being eluted first.

6.10.1 Chiral stationary phases

A wide variety of CSPs are now commercially available for HPLC. These materials are commonly classified under five subgroupings (Table 6.9).

The use of CSPs is the most popular approach towards the resolution of enantiomers because of convenience, the wide range of different CSPs available, that there are few practical problems as separation depends almost exclusively upon the chiral interactions between analyte and stationary phase and is little influenced by mobile phase composition.

Table 6.9 Classification of chiral stationary phases

Descriptor	Mechanism
Pirkle or brush type bonded phases	Three-point interaction
Helical chiral polymers (polysaccharides)	Attractive hydrophobic bonding
Cyclodextrins and crown ethers	Host guest interaction within chiral cavity
Immobilised enzymes	Chiral affinity
Amino acid metal complexes	Diastereomeric complexation

6.10.1.1 Pirkle or brush type phases. The development of CSPs was pioneered by the innovative work of Pirkle [76, 77] and in recognition of his contribution the whole range of materials functioning by π-donor, π-acceptor mechanisms are named after him. These types of stationary phase are formed in general by covalently linking an optically pure amino acid derivative to a γ-aminopropyl silanised silica gel. One of the most successful classes of Pirkle's phases was the alkyl or aryl substituted *N*-(3,5-dinitrobenzoyl)phenylglycines (Figure 6.39).

The chiral recognition processes upon which the resolution of the enantiomers depends requires at least three points of interaction between the solute and the CSP, of which one must be stereochemically dependent. These points of interaction are provided by π-donor, π-acceptor aromatic fragments, the facility to hydrogen bond and the dipole stacking inducing structure in addition to the chiral centre within the stationary phase.

When three points of interaction occur a transient diastereomeric complex is formed between the enantiomer and CSP. This complexing is rapid and reversible and the degree of separation achieved is related to the energy difference between the diastereomeric pairs the more stable of which will be retained the longest. It is possible to reverse the order of elution of the enantiomers by simply changing the chirality of the stationary phase. Due to the range of interaction sites available within these stationary phases they are not limited to a single compound class and have been used for the separation of enantiomers of alcohols, polyaromatic diols, β-hydroxysulphides, diacylglycerols and a wide range of heterocyclic compounds of pharmaceutical interest.

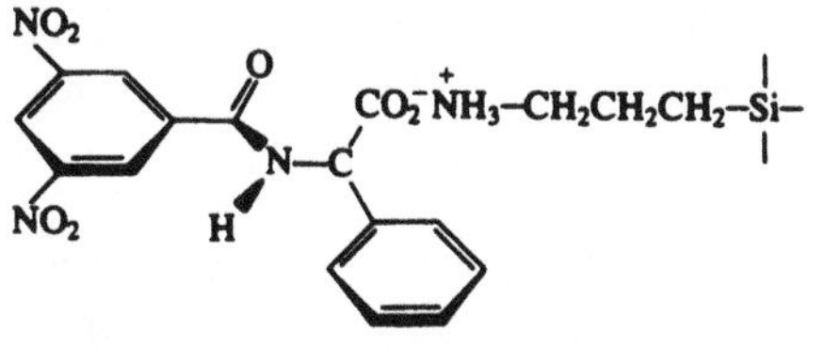

NO_2 O $CO_2^- \overset{+}{N}H_3-CH_2CH_2CH_2-Si-$ N—C NO_2 H

Figure 6.39 'Pirkle'-type stationary phase—ionically bound (*R*)-*N*-(dinitrobenzoyl)-phenylglycine.

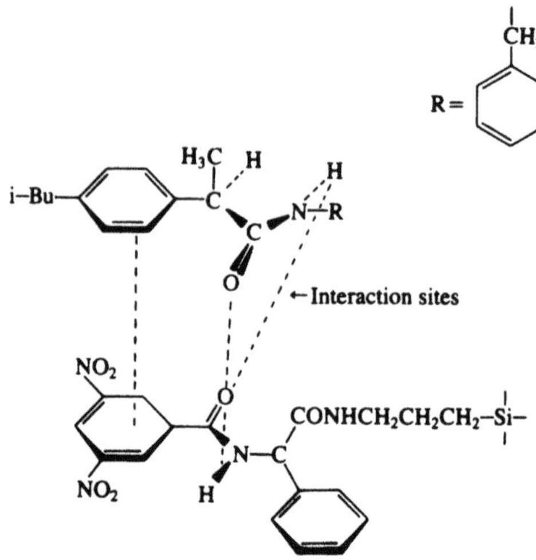

Figure 6.40 Interaction between chiral stationary phase and amide derivative of (*R*)-ibuprofen.

An interesting example of the application of this stationary phase was the separation of the anti-inflammatory drug, ibupofren, after derivatisation to its amide [78] (Figure 6.40).

6.10.1.2 Helical chiral polymers based on polysaccharides. A number of stationary phases based on naturally occurring polysaccharides such as cellulose and amylose [79] have been used for the separation of racemates. These polysaccharides comprise a linear chain of glucopyranose units linked to one another via a 1,4-glycosidic bond. The individual molecules of cellulose associate or aggregate with one another in regular structures which have crystalline properties and a highly ordered helical structure.

The derivatised glucose can act as a chiral site and result in diastereomeric interactions with enantiomers which together with the steric fit requirements within the cavity and the different interactions with the cellulose strands provides the basis for enantioselective interaction and subsequent resolution. The acetate, benzoate and phenylcarbamate glucose ester give superior resolution and selectivity compared with the parent material. Hydrophobic mobile phases are most commonly encountered though aqueous based eluants can be used with many versions of these materials. These stationary phase packings have been used to separate a wide range of pharmaceutical compounds [80].

6.10.1.3 Cyclodextrin bonded phases. Cyclodextrins are cyclic chiral carbohydrates composed of six, seven or eight glucopyranose units: the α-, β- and γ-cyclodextrins, respectively. Only the β-cyclodextrins have been found to be of use in chiral chromatography. The monomers are configured

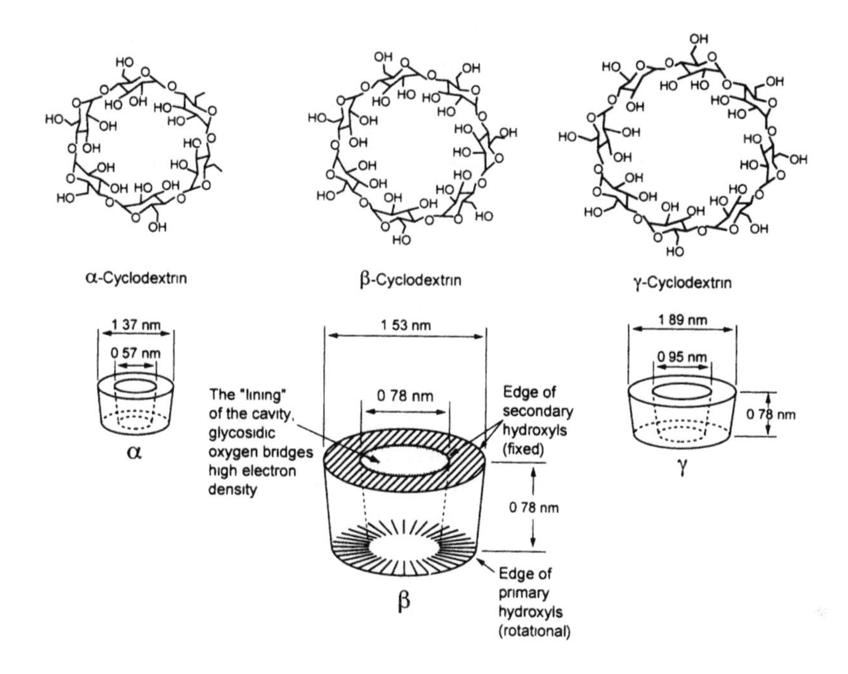

Figure 6.41 Schematic three-dimensional presentation of the geometries of cyclodextrins

such that the cyclodextrin resembles a hollow truncated toroid. The internal lining of the cyclodextrin is essentially hydrophilic composed of methylene and oxygens of the 1,4-glycosidic bonds. The external surface is hydrophilic and is surrounded by a sheath of hydroxyl groups. The upper and lower rims of the cyclodextrin torus carries a number of secondary hydroxyl groups (Figure 6.41). The internal surface of the cavity is chiral with some thirty five asymmetric centres.

When cyclodextrins are used with reverse phase solvents then the mechanism of retention functions primarily by inclusion complexation. It is necessary when working under reverse phase conditions to have an aromatic substituent within the solute. This ensures stereoselective binding of the analyte through interaction with the glycosidic oxygens, otherwise solutes occupy random positions within the cavity with a consequent loss of enantioselectivity. It also appears that the solute must have a functional group in the correct position to hydrogen bond with the hydroxyl groups on the upper edge of the cavity. These CSP have been used successfully for the separation of structural and positional isomers as well as enantiomers.

6.10.1.4 Immobilised protein columns. A further group of chiral stationary phases has been developed by immobilising naturally occurring asymmetrical peptides, e.g. bovine serum albumin, α_1-acid glycoproteins, ovomucoid and chymotypsin on a silica support. These materials show stereoselective affinity for a variety of small chiral biomolecules as they have numerous different binding sites which offer specificity for a wide range of compound classes. Immobilisation of these proteins on silica gel involves very complex chemistry [81] and uses spacers and anchoring agents and only the first of the above named is commercially available.

The mechanism(s) underlying enantioselectivity and chiral separation are complex owing to the diversity of protein structural features, conformations and the number of chiral sites. Though these materials have some inherent advantages in that the protein is capable of recognising various polar and ionic groupings, hydrophobic areas and the three dimensional structure of the analyte species, the capacity of these packings is low due to the small number of binding sites and due to the strength of binding of analyte to protein broad tail poorly resolved chromatographic bands result.

The growing need to resolve enantiomers in both biological and pharmaceutical studies provides a continuing impetus for the design of chiral stationary phases, a number of which are now commercially available, e.g. Chrompack, J.T. Baker Chemicals and Phase Separations. A number of review articles have been presented [82, 83] most recently by Levin and Abu-Lafi [84].

6.11 Ion exchange chromatography

The primary process of IEC involves adsorption and desorption of ionic species from ionogenic groups located in the packing. IEC was the first of the traditional column techniques to be exploited for modern LC, due principally to the need for the fast routine analysis of amino acids and protein mixtures.

Modern IEC may be used with one of a number of types of resin/packing:

- polymeric porous particles;
- pellicular and superficially porous particles; and
- totally microporous particles with bonded phases.

Polymeric porous particles, also referred to as macroporous or macroreticular, are formed from the co-polymerisation of styrene–divinylbenzene. The porosity can be modified by altering the degree of cross-linkage, though this consequently has an adverse effect on the mechanical strength, the degree of wetting (and hence swelling) and the resin capacity (see Chapter 4). A compromise packing is used which has an intermediate degree of cross-linkage and has small particle size ($\sim 5\,\mu m$). However, these are the least efficient of the ion exchange materials available due to the slow diffusion of sample species into the lattice. They have limited capacity and are restricted to use with small ions.

Pellicular packings are formed by coating the ion exchange resin (~2 μm) onto an impervious, inert core (30–40 μm). Also falling within this classification are superficially porous particles. These are constructed by first coating the core material with a thin layer of silica microspheres (0.2 μm) and this layer is then coated with the ion exchanger. The principle disadvantage with these materials is the low exchange capacity. Other polymeric materials used to coat the core material and support the ionogenic groups are cellulose and polydextrans.

Finally, ion exchange capability can be conferred on packing materials by chemical modification of the existing BP. This involves using in the initial silanisation reaction a monofunctional organochlorosilane where one of the hydrocarbon moieties contains an alkene functional group. The vinylated silica then undergoes an addition reaction with styrene. Ionogenic groups, quaternary ammonium for anion exchange and sulphonic acid for cation exchange, may then be introduced as discussed previously. These packings are available as 5–10 μm spherical particles. They have good mass transfer properties, can be used at high flow-rates and have much improved ion exchange capacity (0.5–2 meq g^{-1}).

Selectivity and retention in ion exchange analysis is affected by a number of parameters:

- the size and charge of the solvated sample ion;
- the pH of the mobile phase;
- the total concentration and type of ionic species in the mobile phase;
- the addition of the organic solvents to the eluant; and
- the column temperature.

These features are similar in nature to those affecting separation in traditional open column IE and have been discussed in detail elsewhere.

Initial application of ion exchange to modern LC depends on the analyte having a specific property such as ultraviolet absorbance, fluorescence or radioactivity. As many ion exchange methods require the presence of complexing agents (EDTA, citrate) and various electrolyte additions to achieve the required resolution, conductivity detectors could not be used without modification of the technique, since this parameter is a universal property of ionic species in solution.

6.11.1 Latex-based ion exchange materials

A further class of ion exchange materials have been developed [85] which employ a highly sulphonated impervious core formed from an extensively cross-linked styrene–divinylbenzene (PS–DVB) polymer which has a particle diameter of some 10–25 μm. For anion exchangers these beads are then layered with completely aminated latex particles (Figure 6.42).

Selectivity is determined principally by the characteristics of the latex particles and can be readily modified by changing the extent of cross-linkage,

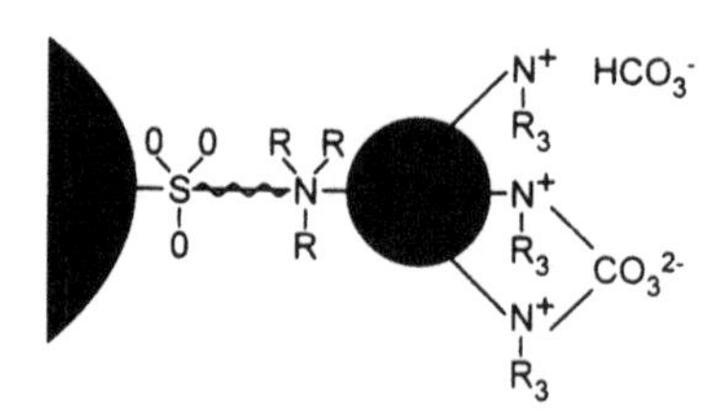

Figure 6.42 Schematic representation of a latex-based anion exchanger. (Reproduced by permission of Dionex (UK) Ltd.)

the size or the functionality of the ionogenic group. Cation exchangers are based on a similar composite structure, though now the impervious core is layered with quaternary ammonium groups and the latex particles carry sulphonic acid ionogenic groups. These materials though complex are superior to and have a number of advantages over silica based bonded phase ion exchangers and the sulphonated or aminated PS–DVB resins.

6.11.2 Ion exchange with conductivity suppression

The high background conductivity associated with eluant buffers precluded the use of conductivity detection and was a major obstacle in the development of ion exchange chromatography. This restriction was overcome through the work of Small *et al.* [86], who developed a general technique for the removal of background electrolytes. The technique developed utilises a second scrubber column also referred to as a stripper or suppressor column (Figure 6.43).

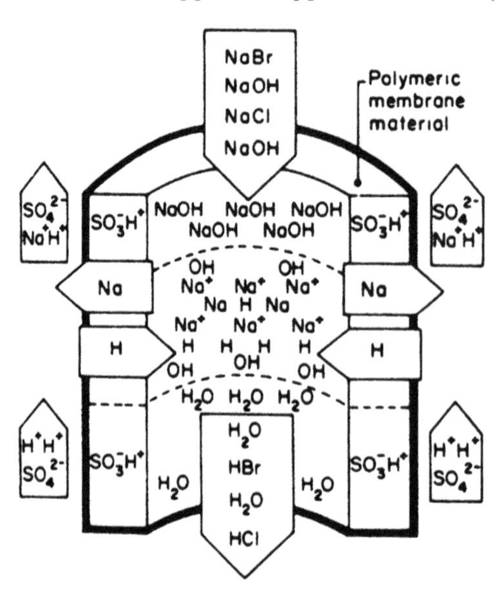

Figure 6.43 Hollow fibre suppressor column for ion chromatography which exchanges hydrogen ions for sodium ions during an anion separation. The halide ions remain as mineral acids. (Reproduced with permission of Dionex Corporation, Sunnyvale, CA.)

This secondary column effectively converts ions due to the background electrolyte to a molecular species of limited ionisation leaving only the species of analytical interest as the major conducting species in deionised water.

For the separation of a mixture of cations the eluant commonly used is a dilute solution of HCl. The analytical column achieves the required separation while the scrubber column, in this instance a quaternary ammonium hydroxide resin, combines with the H^+ ions in the eluant and at the same time converts analyte ions to their hydroxides. The configuration of the instrumentation is depicted in Figure 6.44.

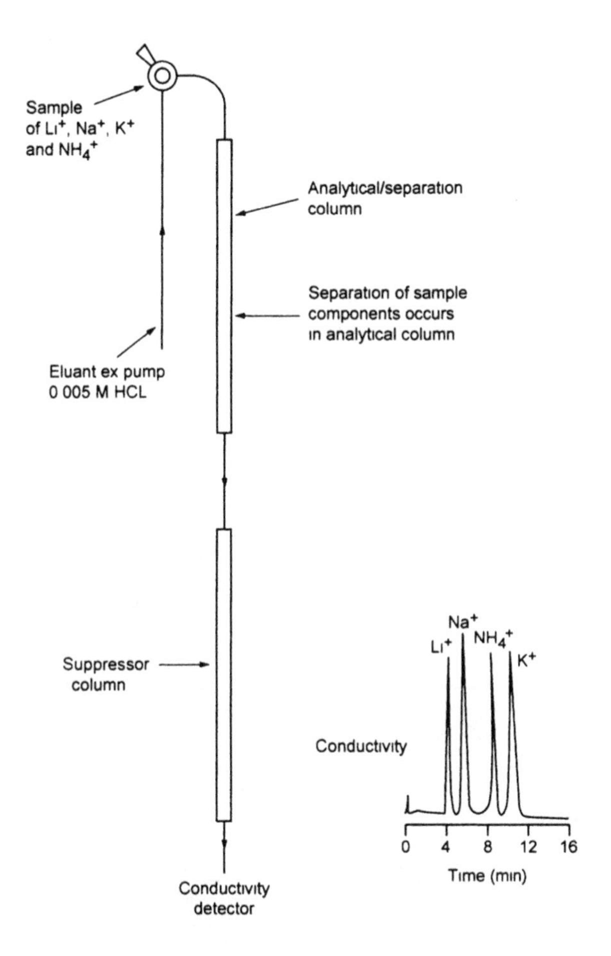

Suppressor column reactions

$$\text{Resin } \overset{\oplus}{N}R_3OH + HCl + K^{\oplus} + Na^{\oplus} + Li^{\oplus} \rightarrow \text{Resin } \overset{\oplus}{N}R_3Cl^{\ominus} + H_2O + K^{\oplus} + Na^{\oplus} + Li^{\oplus}$$

Figure 6.44 Ion chromatograph featuring suppressed conductivity.

The eluant then passes directly to a conductivity cell, where the analyte ions can be monitored in a background of deionised water, thus allowing quantitative determination from measurement of peak height or area.

For anion analysis the suppressor is in the form of an acid ion exchanger. The bicarbonate or carbonate anions typical in eluants for anion analysis are converted to carbonic acid.

$$Na^+(aq) + HCO_3^-(aq) + Resin^- H^+(s) \rightarrow Resin^- Na^+ + H_2CO_3$$

As above, the eluant then passes to the detector and the anion analytes are monitored in a weak aqueous solution of carbonic acid. A typical anion separation is shown in Chapter 9, experiment 32. These columns, however, have some attendant problems, e.g. the resin eventually becomes exhausted and must be regenerated which is both time consuming and inconvenient and in anion analysis the carbonic acid generated via the suppressor gives an intense negative peak which can interfere with the detection and quantitation of early eluting peaks. An alternative form of suppressor is based upon a sandwich membrane design (Figure 6.45). The eluant and regenerant flow in opposite directions, the latter flows on both sides of the membranes while the eluant flows lengthways between the membranes. For the analysis of anions the membranes are cation exchangers and would be used with an acid

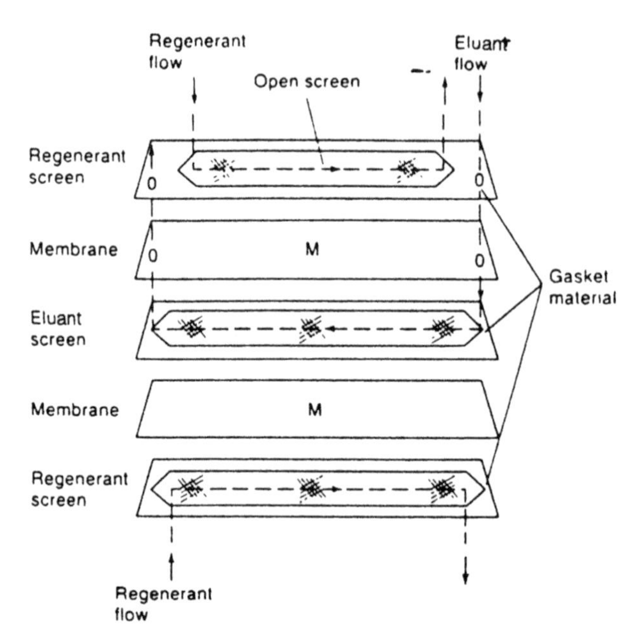

Figure 6.45 Micromembrane suppressor column for ion chromatography. (Reprinted by permission of Dionex Corporation, Sunnyvale, CA.)

regenerant which passes through the screens. For cations an anion exchanger is used with an alkali regenerant. These units can be used continuously and with their increased capacity allow a wider choice of eluants thus expanding the range of applications.

A recent development in suppressor technology utilises an electrolytic cell incorporated within the suppressor cartridge. This unit which is self-regenerating and maintenance free comprises three chambers separated by either a pair of anion exchange or cation exchange membranes. The eluant flows through the central chamber while water or spent eluant is passed in the opposite direction through the anode and cathode compartments of the electrolysis cell. For cation suppression anion exchange membranes would be used in the electrolysis cell and vice versa. Hydroxide produced by electrolysis of water at the cathode diffuses through the membrane and enters the eluant stream where it combines with and removes hydronium ions. Eluant anions pass through the anode anion exchange membrane so maintaining electronic neutrality. Though modified eluants have to be used with this unit it achieves further reduction in background noise and the signal, sensitivity and dynamic range of the system are enhanced.

6.12 Ion pairing

The utility of ion suppression techniques in the analysis of ionisable molecules by reverse phase chromatography is limited to samples of weakly basic or acidic compounds. The analysis of stronger acids ($pK_a < 3$) or stronger bases ($pK_b > 8$) would require eluant of pH, <2 or >8, respectively. Reverse phase chromatography with chemically bonded stationary phases is, however, restricted to eluant pH >2 and <8 for reasons previously discussed.

An attractive alternative to ion exchange and ion suppression analysis of ionic samples is the technique commonly referred to as IPC [87]. The pH of the eluant is adjusted in order to encourage ionisation of the sample, for acids pH 7.5 is used and for bases pH 3.5. The chromatographic retention is altered by including in the mobile phase an ion pair reagent (IPR) containing a large bulky ion of opposite charge—the counter-ion. The reagents used to provide counter-ions are similar to those exploited in liquid–liquid extraction procedures. Some commonly encountered IPRs are shown in Table 6.10.

There are three basic models proposed to describe the ion pair mechanism: ion pair, ion exchange and ion interaction.

Ion pair. This hypothesis postulates formation of a tightly bound ion pair of zero charge. The eluant pH is adjusted as previously described and then an IPR containing a counter-ion (A^-), which has the opposite charge to that of the compounds to be determined, is added to the sample and

Table 6.10 Typical reagents employed in ion pair chromatography

Negative counter-ion donors	Positive counter-ion donors
Methanesulphonic acid (Na salt)	Tetrabutylammonium hydroxide
Pentanesulphonic acid (Na salt)	Tetraethylammonium hydroxide
Hexanesulphonic acid (Na salt)	Tetrabutylammonium phosphate
Heptanesulphonic acid (Na salt)	Hexadecyltrimethylammonium bromide
Oxtanesulphonic acid (Na salt)	Trihexylamine
2-Naphthalenesulphonic acid (Na salt)	Triheptylamine
Dodecylsulphonic acid (Na salt)	Trioctylamine

subsequently forms an uncharged ion pair.

$$\underset{\text{IPR}}{\text{HA}} \Longleftrightarrow \text{H}^+ + \text{A}^- + \text{B}^+ \Longleftrightarrow \underset{\text{ion pair}}{\{\text{A} - \text{B}\}^0}$$

Ion exchange. This postulates that the lipophilic tail of the counter-ions locates onto the bonded stationary phase, effectively causing the column to behave as an ion exchanger.

Ion interaction. This suggestion is based neither on ion pair or ion exchange phenomena, though the lipophilic ions are adsorbed onto the surface but are associated with a primary ion giving an electrical double layer. The analyte then interact dynamically with this double layer by both electrostatic and van der Waal's type forces.

Whatever the model used to describe the ion pair phenomenon, the technique has gained wide acceptance and application because of the inability of ion exchange to separate samples containing both ionic and neutral materials and of the limitation of ion suppression to the analysis of weak bases and acids and the latter's inability to cope with ionic materials.

The retention of analytes in IPC can be controlled in a number of ways:

- by varying the alkyl chain length of the counter-ion [88];
- by varying the concentration of the IPR;
- by combining with ion suppression;
- by modifying the solvent strength.

The influence of these parameters on the retention of vitamins is illustrated in Chapter 9, experiment 27.

6.13 Size exclusion chromatography

Size exclusion chromatography (SEC), though an established technique for the separation of macromolecules using open column systems, met with limited success when applied to modern LC, as many of the commercially available packings did not meet the constraints and instrumental demands of HPLC. A wide variety of packings have now been developed which have led to a spectacular growth in the HPLC analysis of macromolecular samples, such as polydisperse polymer samples and biological materials, such as proteins and carbohydrates (Figure 6.46(a)) and also relatively low-molecular weight materials (Figure 6.46(b)).

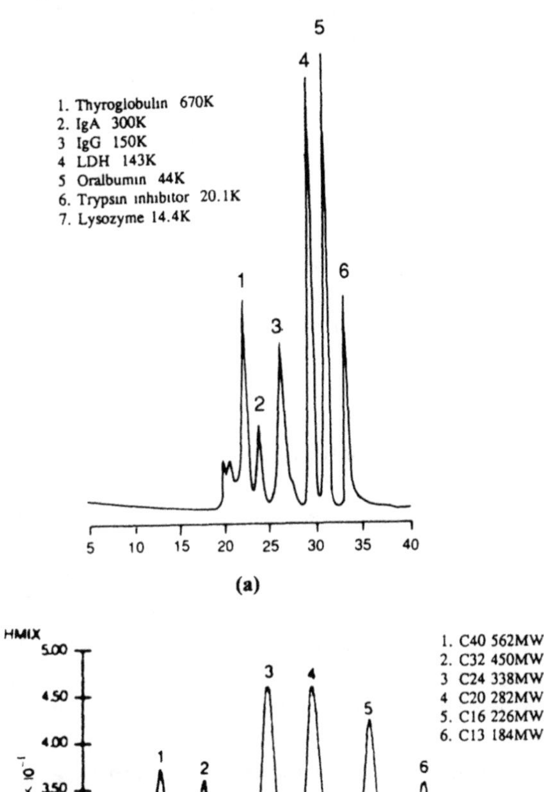

Figure 6.46 Illustrative separations achieved by SEC. (a) SEC of proteins on BIOSEP-SEC-S3000, 600 × 7.8 mm, mobile phase 50 mM NaH_2PO_4 buffer, pH 6.8, flow rate 0.5 ml min^{-1}, detection UV at 280 nm, sample protein mixture, 20 μl injected. (b) SEC of closely related hydrocarbons. Column, phenogel 5 μ 50 Å, 100 Å, 500 Å; dimensions 300 × 7.8 mm, solvent THF, flow rate 1.0 ml min^{-1}, detection differential refractometer, injection volume 100 μl 025% w/v, temperature ambient.

SEC differs from other LC modes in that separation takes place exclusively due to differences in molecular size, consequently solvent selection is simpler, as the requirements of the solvent are simply that of sample solubility and packing compatibility. Bearing in mind the rationale of the separation process, all the sample components should elute between the interparticulate volume and the pore volume. The exclusion limit defines the molecular weight–hydrodynamic volume above which an analyte cannot gain access to the pore structure of the packing and above which no retention shall occur. The permeation limit on the other hand is the molecular weight below which analytes have unrestricted access to the pore structure. It should be recognised that SEC is also now considered as an effective means of separating small molecules [89].

The principles causing retention behaviour, separation variables, molecular weight calibration and associated terminology such as interparticle and intraparticle volume, selective permeation, fractionation range and molecular hydrodynamic radius are as for open column size exclusion (Chapter 4).

6.13.1 Column packings

Support materials may be subdivided into two classes both of which are available in 5–10 μm particle size: (a) semirigid cross-linked polymer gels, and (b) inorganic materials with controlled pore size, such as microporous silicas.

6.13.1.1 Semirigid gels. A number of polymers of this general classification are available, though many are restricted in use due to solvent incompatibility and pressure restriction. Polyacrylamide gels are compatible with aqueous systems and can be used for the analysis of water-soluble compounds. Such gels are available with a molecular weight exclusion limit of 5×10^5, though it is only the smaller pore size material which has the required mechanical strength for modern LC (even here the upper pressure limit is 200 psi (1380 kPa)).

Cross-linked polydextrans have also been used for the analysis of aqueous samples in the molecular weight range 100–10^8. The utility of these materials has been extended by the synthesis of hydroxypropylated derivatives which allows use with polar organic solvents. These packings suffer the restrictions incumbent with large particle size (30 μm) and low mechanical strength.

Cross-linked polymers of styrene and DVB are the most popular packings of the above class. Extensive development has led to the synthesis of 10 μm spherical packings of controlled pore size and pore volume. A high degree of cross-linkage confers excellent stability allowing the packings to be used at elevated temperature, flow and pressure (6000 psi (41.4 MPa)). They are available in a range of pore sizes, 50–10^6 Å as well as mixed pore size corresponding to a fractionation range of 100–(5×10^8) Da, and find application in the analysis of both polymer and low-molecular-weight materials.

Swelling and shrinkage is minimal and the packings are compatible with most organic solvents though not acetone or methanol. Suspension co-polymerisation of 2-hydroxyethylmethacrylate and ethylene dimethyl-acrylate gives hydrophilic porous packings which can be used with aqueous and polar organic eluants. The packings can be used at elevated pressures (3000 psi (20.7 MPa)), though they have a restricted molecular weight operating range.

6.13.1.2 Rigid packings. Glasses and silica-based particles have found increasing application in SEC. Some of these materials have chemically modified surfaces in order to reduce the adsorptive properties of the packing. The 10 μm packings are available in a range of pore size diameters (40–2500 Å) corresponding to a mass range of 10^2–(5×10^7) Da, are stable at pH < 10 and can be used with aqueous and polar solvents.

The packings are stable at elevated temperatures, and have high mechanical strength with good 'mass transfer' properties allowing rapid equilibration with fresh solvent. As the volume remains constant and there is no possibility of biodegradation the columns, especially those silylised, can be used routinely and indefinitely after calibration. Silica-based materials have some disadvantages principally their tendency to retain solute by adsorption and also their propensity to catalyse degradative reactions of solutes.

The techniques of polymer characterisation and of the application of SEC to biological studies is a broad and detailed subject and consequently this section can only provide a brief overview of the subject material. The interested reader is referred to the monographs by Hunt and Holding [90] and Dubin [91], and references therein.

6.14 Liquid chromatography method development

The first step in developing a new LC separation is to review the available literature for an overview, at the very least, of the experimental conditions which might be appropriate. In a purely practical sense the development of an LC procedure begins with the choice of the most promising sorption mode and solvent system. A column and solvent selection guide (Figure 6.47), based on molecular weight, solubility and ionic characteristics of the sample can be used as an empirical aid in this procedure. This procedure is merely a guide, not a substitute for experience.

Following the choice of an LC method, a column and column packing appropriate to the analysis must be selected. Established packing materials are available in 3, 5 and 10 μm size, the 3 μm give the greatest resolution (×5) compared to the 10 μm packings though the latter are adequate for all but the most demanding of separations. The relationship between

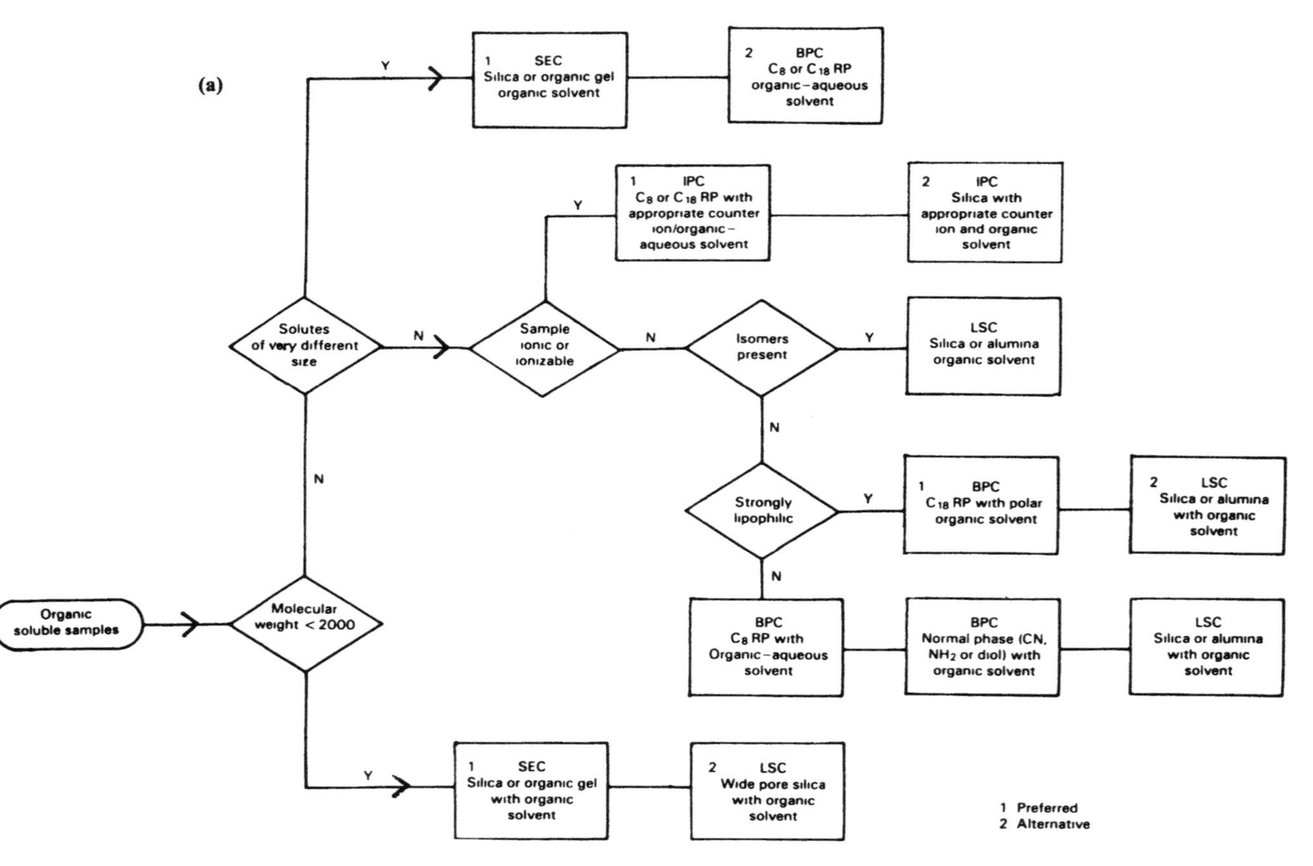

Figure 6.47 Simplified guide to column selection: (a) for organic-soluble samples; (b) for water-soluble samples.

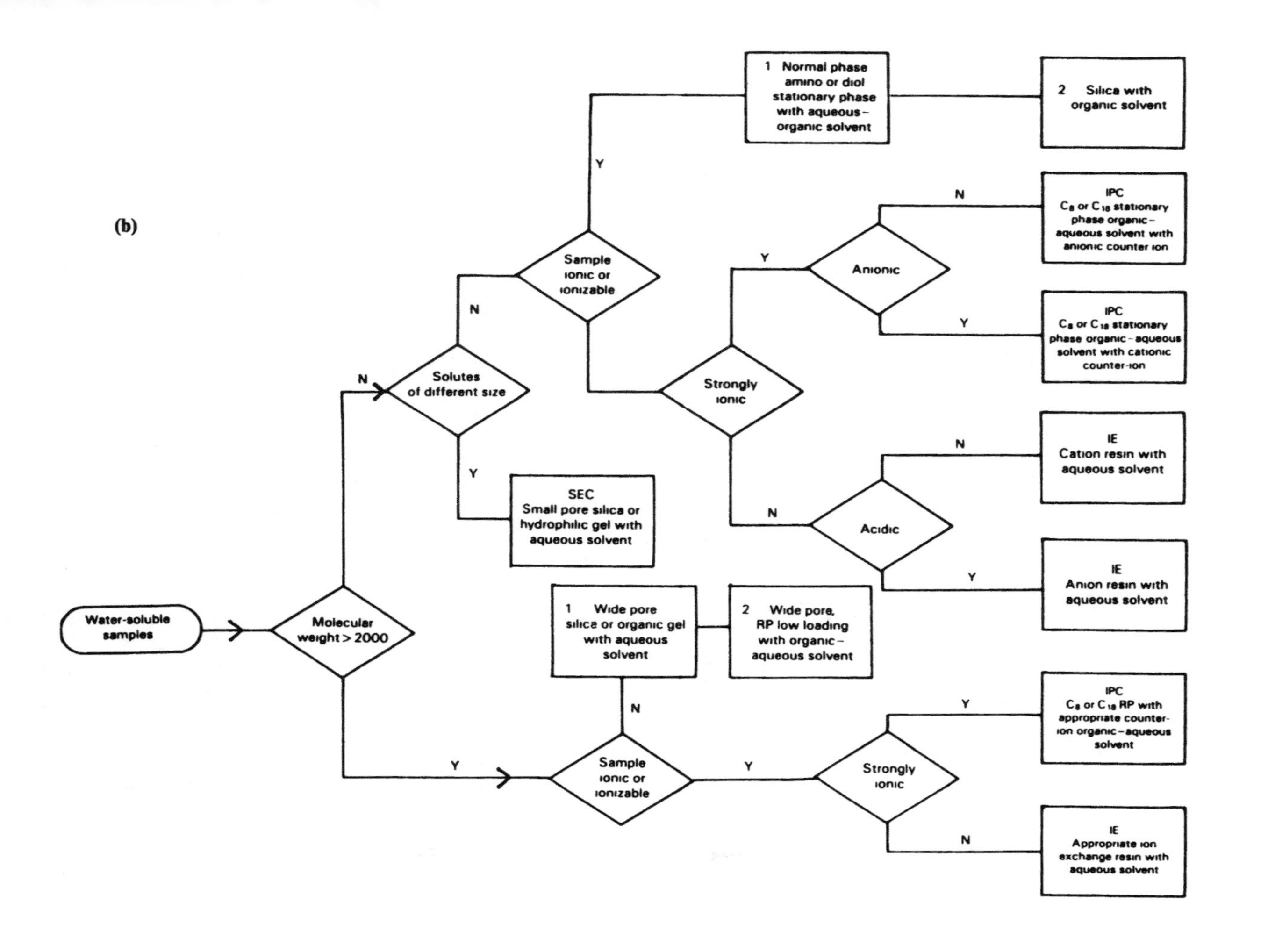
(b)
Water-soluble samples
Molecular weight > 2000
Solutes of different size
SEC Small pore silica or hydrophilic gel with aqueous solvent
Sample ionic or ionizable
1 Normal phase amino or diol stationary phase with aqueous–organic solvent
2 Silica with organic solvent
Strongly ionic
Anionic
IPC C_8 or C_{18} stationary phase organic–aqueous solvent with anionic counter ion
IPC C_8 or C_{18} stationary phase organic–aqueous solvent with cationic counter-ion
Acidic
IE Cation resin with aqueous solvent
IE Anion resin with aqueous solvent
1 Wide pore silica or organic gel with aqueous solvent
2 Wide pore, RP low loading with organic–aqueous solvent
Sample ionic or ionizable
Strongly ionic
IPC C_8 or C_{18} RP with appropriate counter-ion organic–aqueous solvent
IE Appropriate ion exchange resin with aqueous solvent

column back pressure (Δp) and particle size (d_P) is defined by

$$\Delta p \propto 1/(d_P)^2$$

and will rise by a factor of ~10 on moving from 10 to 3 μm packings. It is advised therefore that lower rates ($\leq$0.5 ml min^{-1}) are used with the latter; flow-rates for 10 μm columns are commonly 2 ml min^{-1}. Column dimensions vary, diameters ranging from 2 to 8 mm and length from 5 to 50 cm. For analytical separations a diameter of $\leq$4 mm is adequate, while the length of column required is dictated by the complexity of the sample, but is typically 10–25 cm.

6.14.1 *Solvent optimisation techniques*

When the chromatographic mode, column type, packing and dimensions have been chosen, the final stage of method development involves solvent optimisation and a choice between isocratic or gradient elution. Many separations can be achieved perfectly satisfactorily under isocratic conditions and are preferred to gradient elution techniques, as these are inconvenient due to the time required to re-equilibrate the column. A measure of the quality of separation is given by the resolution factor which can be expressed as follows:

$$R_S = \underset{\text{(i)}}{\frac{\sqrt{N}}{4}} \underset{\text{(ii)}}{\left(\frac{k}{k+1}\right)} \underset{\text{(iii)}}{\left(\frac{\alpha - 1}{\alpha}\right)}$$

where the selectivity factor $\alpha = k_2/k_1$ for component bands 1 and 2 in the chromatogram. The three terms are broadly independent and thus can be individually optimised.

The first term is a measure of the column efficiency, and N can be varied by changing the column length, the particle size of packing or the mobile phase velocity.

The expression in term (ii) of the above equation is a measure of the retention times of the peaks being considered and is only significant at small values of the capacity factor ($k < 2$). The value of the capacity factor is most readily adjusted by varying the solvent strength. Typically a unit change in the polarity index which corresponds to 10% change in the organic component shall result in approximately a threefold change in k. For reverse phase systems solvent strength increases with the proportion of the organic component in the eluant.

A measure of the selectivity of the method is given by term (iii). The selectivity factor, α, may be varied by changing the temperature, or the stationary or mobile phases. However, if $\alpha \Rightarrow 1$, then no matter how long the components stay on the column, or how many theoretical plates the column is equivalent to, there will be no separation. Thus, the greater the difference between k_1 and k_2, then the better the resolution as term (iii) becomes more significant.

Thus the resolution achieved in a chromatogram is largely dependent upon values of α and k, both of which are strongly dependent on the nature of the eluant in terms of solvent strength, organic solvent, mobile phase pH (ion suppression techniques) and eluant additives (IPRs).

In general terms, an $R_S = 1.5$ is adequate for most purposes. Values in excess of this result in over-long analysis times. In practice for isocratic elution, the following stepwise procedure is often adopted.

1. The column of highest efficiency is selected.
2. The capacity factor is optimised by adjusting the solvent strength such that the capacity factor lies within the range $5 < k < 10$. Values outside this range lead to poor chromatograms (due to reduced peak height) and overlong analysis time. To increase k values a weaker solvent is used and vice versa.
3. If, however, the resolution is still inadequate due to $\alpha \Rightarrow 1$, then the selectivity ratio must be modified by altering the nature of the mobile and/or stationary phases.

A largely empirical approach to optimising the eluant strength can be employed by obtaining a series of chromatograms of the sample with binary mixtures, stepping down the organic component successively by 10% until adequate resolution has been achieved. If necessary intermediate eluant compositions can be run to determine the optimum. The separation of a mixture of analgesics using this approach is illustrated in Figure 6.48; refer also to Chapter 9, experiment 20.

This approach can be extended to ternary solvent systems employing two organic solvents and water. Change in eluant composition in this case modifies both capacity and selectivity factors/characteristics. An example of this approach is illustrated in Figure 6.49. In chromatogram A components 1 and 2 co-elute. A solvent of the same polarity using tetrahydrofuran as the organic component is then used as the eluant. There is now a marked change in selectivity, components 1 and 2 are well resolved. However 2 and 3 now co-elute. By subsequent blending of these two base eluants the desired resolution can be obtained.

6.14.1.1 Mapping optimisation methods (MOM). The simplest approach using mapping optimisation methods is based on the use of the Snyder ternary solvent optimisation triangle (Figure 6.50). This approach towards a systematic evaluation of the best mobile phase is equally applicable to normal phase, reverse phase and ion pair chromatography and quaternary solvent systems. The underlying concepts and methodology will be discussed with reference to a reverse phase ternary solvent system.

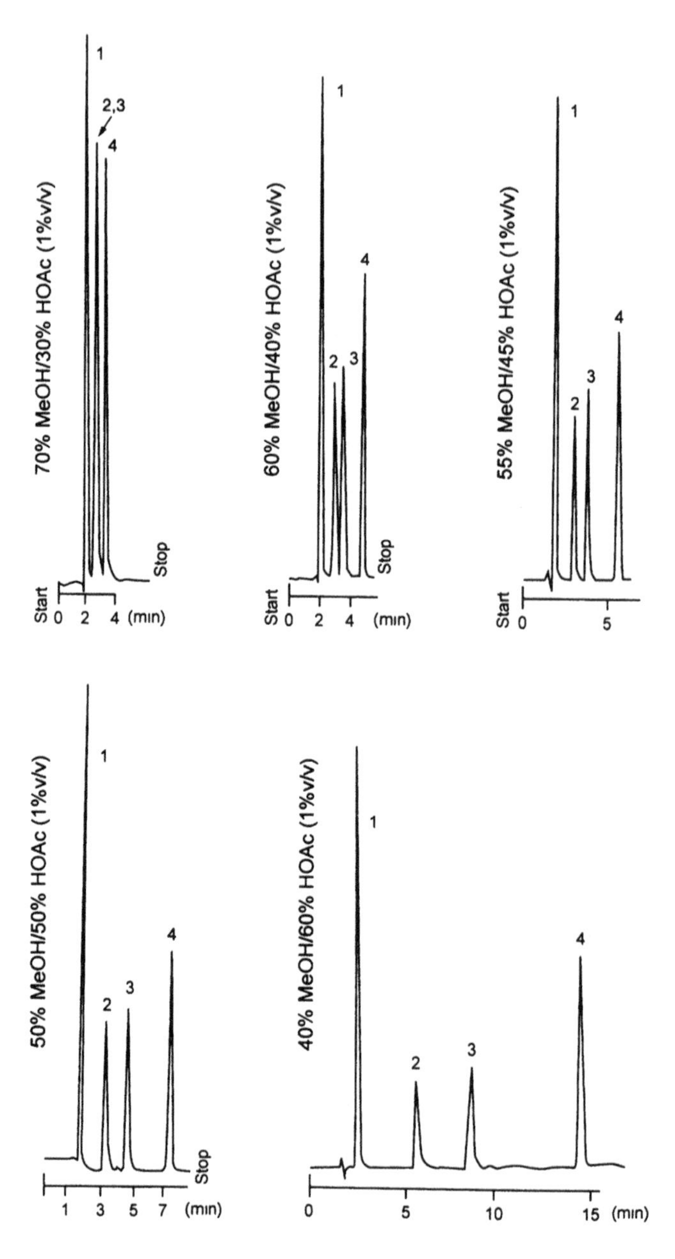

Figure 6.48 Optimising an HPLC separation using a binary mobile phase. Peaks (1) paracetamol, (2) caffeine, (3) aspirin and (4) phenacetin. Column—5 μm, C_{18}. Detection 235 nm. (a) 50% methanol/water, (b) 32% THF/water and (c) 25% THF/17% methanol/water.

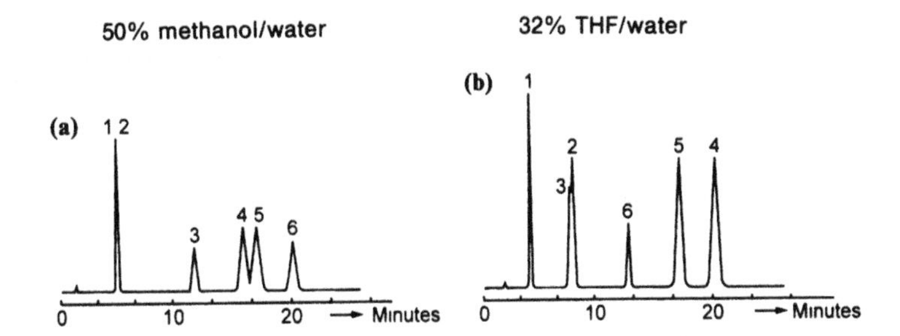

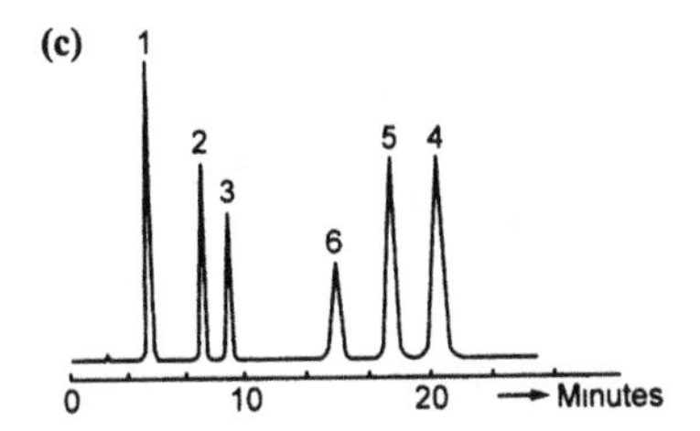

Figure 6.49 Optimisation of an HPLC separation using a ternary mobile phase (methanol/tetrahydrofuran/water). Peaks (1) benzyl alcohol; (2) phenol; (3) 3-phenylpropanol; (4) 2,4-dimethylphenol; (5) benzene; (6) diethylphthalate.

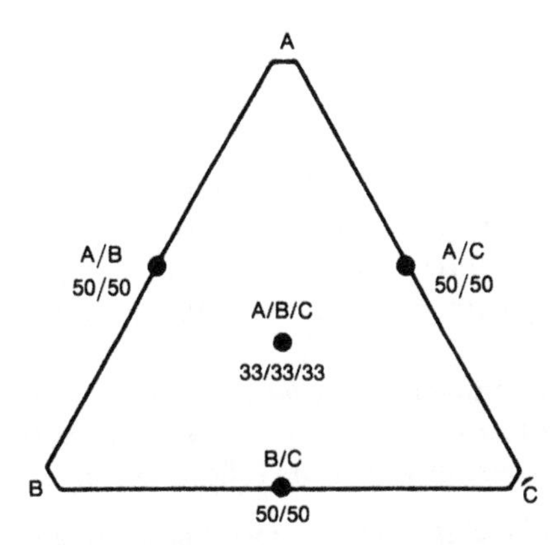

Figure 6.50 Mobile phase optimisation/selectivity triangle. The corners are isoelutropic mobile phases chosen to provide different selectivity. Intermediate points are mixtures of these binary eluants in the indicated proportions.

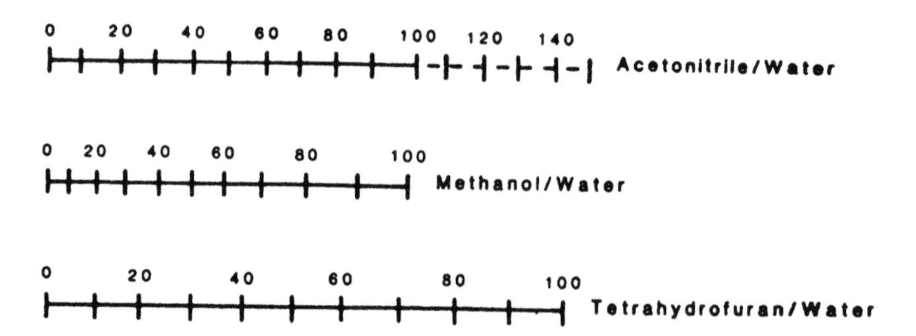

Figure 6.51 Nomograph for estimating isoelutropic mobile phases in reverse phase HPLC.

Certain general conditions are preset such as column type, temperature, flow-rate and the maximum time of analysis. Then an initial binary eluant comprising a miscible organic solvent with water is determined by running a linear gradient from which an estimate can be made of solvent composition, which would give an acceptable range of capacity factors. This mobile phase is assigned to apex A in the optimisation triangle.

Isoelutropic eluants using two other organic solvents (usually tetrahydrofuran and acetonitrile) with water are determined, which will maintain a roughly equivalent capacity factor range, though ensuring differences in selectivity. The isoelutropic compositions can be determined from polarity index calculations or from a solvent nomograph (Figure 6.51 [92]).

These solvent compositions are then assigned to apices B and C. Four other chromatograms are subsequently obtained using mixtures of A/B, B/C, A/C and A/B/C proportions as indicated in Figure 6.50. The chromatograms are then examined to identify the optimum solvent conditions. If the optimum lies between two points then it can be determined by interpolation.

The above approach can be followed as a matter of routine. However, the final choice is still somewhat subjective and may require more trial and error studies. Other disadvantages are that all chromatograms must be run before any conclusions can be drawn and if the order of elution varies at all then it is not possible to interpolate to find the optimum eluant composition.

6.14.1.2 Overlapping resolution mapping (ORM). ORM is a development of the optimisation strategy detailed above and uses mixture design statistical techniques. The approach involves the following steps:

- select three binary isoelutropic compositions as detailed above;
- chromatograph the sample using these eluants;
- evaluate the chromatograms in terms of peak overlap for each pair of adjacent peaks;

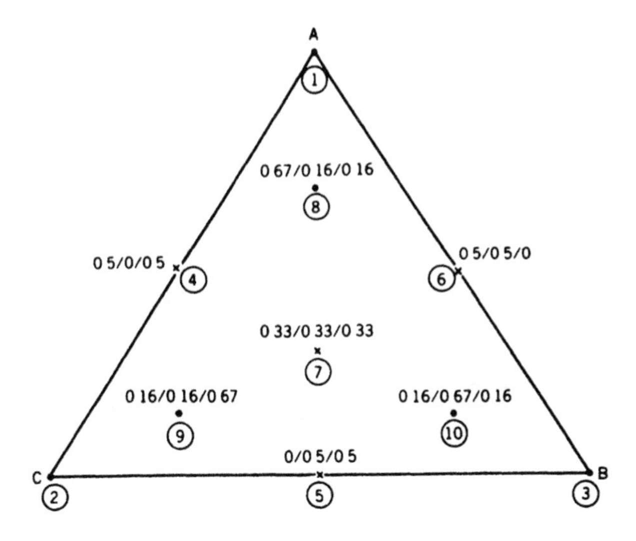

Figure 6.52 Statistical design of mixtures of ternary mobile phases.

- additional solvent compositions, up to 10, are then chosen statistically (Figure 6.52); and
- computer-generated contour maps are then plotted for each pair of adjacent analytes (Figure 6.53(a) and (b) [93]).

Superimposition of the individual contour maps produces the overlapping resolution map (Figure 6.54).

Once the computer-generated map is available then the portion identifying solvent compositions which yield $R_S \geq 1.5$ is clearly marked. Though a more exhaustive and objective approach than the MOM, the other disadvantages remain.

6.14.1.3 Simplex and directed techniques. Directed techniques embody an evaluation step after each eluant and chromatogram has been obtained. The evaluation is based on a determination of the chromatographic resolution factor (CRF), which is a numerical value giving a measure of overall quality of the chromatogram based on the extent of resolution for each pair of adjacent components and the analysis time required.

In iterative approaches only the three apex isoelutropic solvents are run before data evaluation. On the basis of the CRF values an eluant system is predicted which offers potentially the optimum separation. This approach reduces the number of eluant compositions studied to about four or five.

Simplex optimisation procedures are similar in their initial approach. However, on the basis of the CRF determinations the solvents are ranked and the poorest performing eliminated. The sample is now rechromatographed

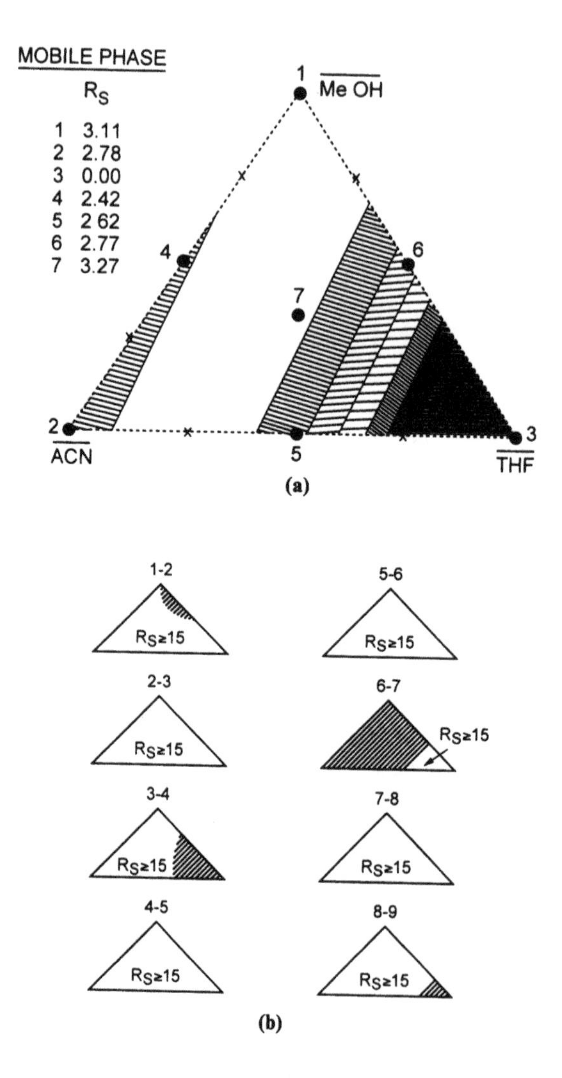

Figure 6.53 (a) Complete contour map for peaks 8 and 9. (b) Simplified resolution maps for all eight pairs of peaks. (Adapted with permission from Ref. 93.)

using a replacement eluant which is a complementary reflection of that rejected. The procedure is repeated until there is a convergence of CRF values and thus the optimum solvent composition is identified. These approaches require efficient microcomputer systems and can be readily extended to optimise automatically a range of other parameters such as pH, ion pair concentration, temperature and flow-rate [94].

6.14.1.4 Gradient elution. The factors influencing chromatographic resolution and various approaches for establishing optimum conditions using

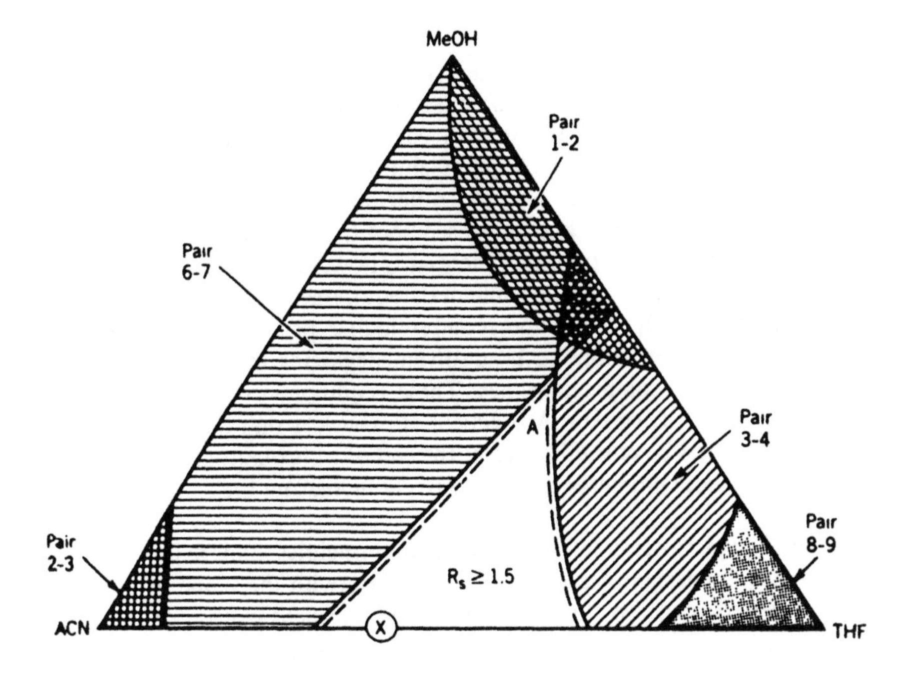

Figure 6.54 Overlapping resolution map combining data in Figure 6.53(b).

isocratic elution techniques have been discussed above. However, for difficult separations isocratic elution may not afford adequate resolution as the capacity factors (k) for some bands may be small while others may be large. Thus, for samples where k values differ widely it is not possible to adjust the solvent strength such that all bands elute in the optimum range $1 < k < 10$, using a constant strength eluant. Therefore, for difficult separations the desired resolution may be achieved using gradient elution techniques. This involves changing in a controlled manner the solvent strength during the development of the chromatogram, thus allowing for optimisation of the capacity factors for the individual bands.

A general approach to the elution problem described above is to use the technique of gradient elution. For example, a reverse phase gradient elution would use a linear profile from 10 to 100% of organic modifier in water, thus allowing a measure of the range of the capacity factors of the sample components. If the retention times of the sample components are close, this indicates a relatively narrow band of capacity factors. Therefore, in this instance isocratic elution using a solvent system containing slightly less of the organic modifier at the elution time of the bands of interest, should provide adequate resolution. The stepwise procedure for optimisation can then be utilised. If the selectivity proves inadequate then it may be that either an alternative solvent or stationary phase or alternative LC mode should be investigated.

If the initial evaluation run indicates the sample components vary markedly in capacity factor value, then a gradient elution system may prove more effective. The design of a gradient separation and programme requires consideration of not only the solvent components, but also the gradient range, shape and steepness. Gradient elution is effective in dealing with the general elution problem and has other advantages, such as improved resolution of early bands and increased sensitivity for components which elute at higher solvent strength. There are, however, associated disadvantages both in terms of the instrumental requirements and in the practice of the technique. The technique requires higher performance pumps and additional instrumentation, such as solvent degassing modules and mixing chambers and solvent programmes. These requirements add considerably to the cost of an LC system. A practical drawback is that the initial solvent composition must be re-established at the end of a chromatogram before the next analysis can commence, thus effectively increasing the sample analysis time.

A detailed discussion of the theoretical principles underlying gradient elution and the design of such separations is outwith the scope of this text. The interested reader is directed towards an excellent expose of the topic by Snyder *et al.* [95].

A number of alternative techniques have been developed to tackle the general elution problem, such as flow and temperature programming, and especially column switching techniques. Coupled column chromatography involves the use of three and sometimes four different stationary phases. The sample is introduced to the primary column for a preliminary separation and then the various polar fractions are switched to the appropriate secondary columns and then to the detector. Coupled column chromatography provides enhanced resolution, and as isocratic elution is used the technique is particularly useful for the routine assay of samples with a wide range of capacity factor values. Further details on these various techniques may be found in a review by Koenigbauer and Majors [96].

6.15 Quantitative analysis

The methods of quantitation and the criteria for precise and accurate determination for LC are similar to those used in GC, though there are a number of important differences. External standard calibration—i.e. where the detector response to a solution of known concentration is measured and then a calibration curve is constructed—is the recommended method for quantitation in LC. It is imperative that the linearity of detector response is confirmed over the concentration range of interest with standards prepared in a matrix similar to the sample. Table 6.2 details detector characteristics. The increased precision obtained compared to GC is attributable to the high degree of accuracy and reproducibility of sample loading afforded by sample loop microvalve injection.

The internal standardisation technique actually increases the analytical error due to the measurement of two peak areas and should be reserved for samples undergoing pretreatment of pre- or post-column derivatisation to account for variable sample recovery or conversion. Quantitative analysis when applied to gradient elution systems affords reduced accuracy and precision due to the practical disadvantages of constancy of flow, reproducibility of gradient formation and solvent mixing–demixing.

6.16 Preparative liquid chromatography (PLC)

In addition to the analytical-scale assay of trace impurities modern LC can be employed in the preparative mode (PLC) for the isolation of appreciable amounts of pure component [97, 98]. Where only a few milligrams (<5 mg) of a compound is required then a few repetitive injections on an analytical scale column (~4.6 mm i.d.) may provide the requisite amount. The analytical scale equipment required has been described in detail earlier in this chapter.

However, there are many other instances when larger quantities (50 mg–50 g) of pure material may be required. Separations of this scale are normally accomplished using larger bore columns which are designed for use with proprietary LC analytical-scale equipment: The principal features of the methods are as follows.

- High resolution preparative procedures use a large bore column (≥10 mm) packed with analytical grade packing material (≤10 μm particle size), thus giving a comparable degree of resolution and speed of separation as the parent analytical procedure. The flow-rate required (u) for the preparative procedure is given by

$$u = u_1 \left(\frac{d_1}{d_2} \right)^2$$

  where u_1 is the solvent flow-rate in the analytical procedure, d_1 is the i.d. of the preparative column and d_2 is the i.d. of the analytical column. The maximum sample size is obviously dictated by the separation factor, α, but should be in the region 10–50 mg of compound [99].
- The second mode is referred to as the 'sample overload' technique and sacrifices column efficiency and resolution in favour of sample throughput. The use of wide diameter columns (>10 mm) packed with large microporous particles (>50 μm) allows isolation of gram quantities of material in relatively short elution times as the column can be operated at high flow rates without seriously degrading resolution. By its nature this technique

is best used with samples containing one principle component or sample mixtures where the solutes are well resolved. The technique of peak shaving or cutting, that is, collecting a portion of a chromatographic zone, followed by recycling, can give a sample of the desired purity.

Another advance has been the development of radial compression column technology by Waters, both for preparative and analytical procedures. In the preparative mode the packing material is constrained in a flexible plastic tube which in turn is housed in a metal cylinder. The cylinder is pressurised with nitrogen, which causes the flexible walls of the tube to be compressed against the packing, thus eliminating column voids and wall effects. The efficiencies obtained are noticeably improved and the large columns used up to 300 mm in length by 50 mm diameter have excellent sample capacity and allow the separation of ~10 g quantities of component [100].

Regardless of the procedure employed preparative HPLC has most commonly been used in the adsorption mode with silica packings due to their sample versatility and loading capacity; however large bore preparative columns are now commercially available, packed with any of the proprietary reverse-phase and ion exchange media.

6.16.1 *Practical aspects*

An essential preliminary to preparative scale HPLC is to establish the solvent system for the separation. This can be determined either on an analytical-scale HPLC system or through TLC studies. The solvent system determined by the former method may be applied directly to the preparative HPLC. However, while TLC provides an inexpensive alternative and allows many solvent systems to be examined, at little cost and effort, the results must be treated with caution, when the chromatographic layer is silica. First, samples are more strongly retained in TLC due to the greater surface area of TLC silica compared to column materials and hence a solvent system should be chosen such that the $R_f \leq 0.3$. Second, where the components of mixed solvent systems have widely different affinity for silica, for instance, ethanol and chloroform, then an effective solvent gradient exists on a TLC plate, the solvent front being richer in the less polar component. On a PLC system fully conditioned with this mixed solvent this effect is not present and to take account, an eluant of lower solvent strength should be used. If the solvents are of similar strength then the mixed solvent system may be transferred directly from TLC to PLC.

In order to obtain reproducible chromatography whether in analytical or preparative scale work the column must be conditioned prior to use. As a guide the amount of solvent required for equilibration is ~15–20 column volumes. If equilibration is incomplete this can lead to poor reproducibility and separation.

Ideally, the sample should be dissolved in the mobile phase and applied to the column in as small a volume as possible. The maximum injection volume without a loss in resolution is given by

$$V_{inj} = V_M\left[(k_B - k_A) - \frac{2}{\sqrt{N}}(2 + k_A + k_B)\right]$$

where k_A and k_B are the capacity factors of the peaks of interest and V_{inj} may be calculated from the data from the analytical chromatogram.

Solvents of greater eluting power should not be used to dissolve the sample as this disturbs the system equilibrium. The sample solution can be applied via a six-port valve of the Rheodyne type fitted with an appropriate sized loop. Loops up to 10 ml are commercially available. Alternatively the sample can be applied to the column via a small volume secondary pump, though this has the disadvantages associated with stopped-flow techniques (discussed previously in this chapter). This complex subject has been reviewed by Guichon [101].

In PLC it is advisable to use both protector and guard columns [102]. Where any polar, ionic or basic mobile phase that could dissolve the column packing is being used as an eluant, then a precolumn of ~40 μm silica should be fitted between the injector and the pump. The column should be of similar length but approximately half the diameter of the preparative column. This ensures that the eluant is saturated with silica. As in analytical work a guard column should be inserted after the injector to retain undesirable sample impurities and to act as a final filter. The dimensions are somewhat smaller than the guard column both in length and i.d. in order to maintain efficiency.

The instrumental modifications required are minimal. Commercially available analytical reciprocating pumps can be readily modified (at cost) with preparative head assemblies which provide solvent delivery capacity of up to 100 ml min^{-1}. If ultraviolet detection is being used then due to its sensitivity a stream splitter is located at the column outlet, normally with a 5–10% split ratio. The detector flow can be recombined with the major flow stream before passage to the fraction collector. Pure samples can be obtained by collection of suitable fraction cuts, which can be checked by analytical LC before bulking. RI detection is popular in preparative work as the detector has adequate sensitivity and is universal in application.

Preparative systems are used routinely in industrial processes [103] allowing isolation of 100–1000 g quantities of material though flow-rates of up to 10 litres min^{-1} are required.

6.17 Microcolumns in liquid chromatography

Throughout the development of HPLC workers have strived to improve chromatographic performance by either increasing the efficiency of separation, reducing the analysis time, or a combination of both. One approach, using the conventional 4.6 mm i.d. columns, is the reduction of the packing material particle size and further improvement of the packing technique. This increases efficiency and facilitates the shortening of columns to reduce analysis time. Columns with 3 μm packings give high-speed separations with efficiencies approaching the theoretical limit where $H = 2d_P$ (i.e. HETP is equal to twice the particle diameter). However, due to the high pressure gradients encountered with decrease in particle size, it is unlikely further advances can be achieved at the moment with particles below 3 μm.

The quest for improved efficiency provides the continuing impetus to the study of reduced diameter columns, and though still in their early stage of development, these new column technologies are having considerable influence on the practice of HPLC [104,105]. The incentive for the development of microcolumns for HPLC lies in the various practical advantages they have over standard analytical columns [106]:

- due to the narrow bore, high linear flow-rates can be achieved with substantially lower total solvent volumes and thus solvent flow-rates of 50 μl min^{-1} or less are used (*cf.* 1 ml min^{-1} with conventional columns). This provides for substantial savings in solvent consumption and is of importance where solvent costs are a consideration for instance, use of ion pairing reagents and deuterated or other exotic solvents;
- the increased linear velocity of eluant allows in theory high speed separations; furthermore the reduced flow-rates and the concomitant reduction in the column backpressure allow columns to be readily connected in series, enabling efficiencies of 10^5 theoretical plates to be realised. The small peak volumes result in increased mass sensitivity and can provide a 20-fold enhancement in detector response [107];
- the reduced solvent flow has enabled the direct interfacing to mass spectrometers [108] and flame based detectors [109].

Four main categories of microcolumn types can be distinguished [110]: open tubular [111]; packed open tubular [112]; microbore and narrow microbore [113]. The principal features of these column types are outlined in Table 6.11.

Table 6.11 Microbore column types

Column type	Material of construction	Column i.d. (μm)	Particle size (μm)	Column length (m)	Flow-rate (μl min^{-1})
Microbore	Stainless steel	500–2000	5	0.1–1	1–50
Narrow microbore	Fused silica Stainless steel	100–200	3	0.1–20	0.1–20
Open tubular	Fused silica	10–50	—	1–100	0.01–1
Open tubular packed	Pyrex glass	40–80	10–30	1–100	0.05–2.0

6.17.1 *Microbore packed columns*

These are similar to conventional HPLC columns but have an i.d. of about 0.5–2.0 mm. They are packed with materials in the range 3–10 μm using established high pressure (~25 000 psi (172.5 MPa)) slurry packing techniques. Theoretical plate counts of 30 000 m^{-1} using 10 μm particles in 1 mm i.d. columns have been achieved. Higher resolution can be achieved by connecting columns in series and avoiding introduction of any dead volume; for columns of 5 μm silica columns 50 000 plates m^{-1} and total efficiencies of 10^6 plates have been reported [114]. A drawback of small bore columns packed with large particles (~10 μm) is that the column efficiency/unit length is low and increased efficiency has to be paid for at the cost of long analysis times. High speed, high efficiency small bore column separations can only be achieved with efficient packing of small particles ~3 μm.

6.17.2 *Narrow bore packed columns*

Narrow bore packed columns are characterised by having i.d. in the range 0.1–0.2 mm and particle size of 3–30 μm. Ishii and co-workers [115] evaluated several types of column materials, for instance, stainless steel, Pyrex glass, PTFE and fused silica glass, and concluded that the latter gave the best column efficiency.

Using proprietary packing techniques detailed by Verzele [116] microparticulate columns of up to 2 m can be prepared. For columns of ~200 μm i.d. packed with 3 μm packings typical efficiencies are of the order of 10^5 plates m^{-1} [106]. Narrow bore packed columns have a number of advantages additional to those mentioned for other microcolumns, for instance, fused silica has extremely high mechanical strength allowing the use of inlet pressures up to 12 000 psi (82.8 MPa). Thus long length columns can be used; silica glass has good optical transparency and thus on-column ultraviolet and fluorescence detection can be employed; finally good flexibility allows easy coupling of column to detector and pump, and also allows columns to be coiled thus saving space and simplifying oven design.

6.17.3 *Open tubular columns*

As in capillary GC these are columns made of narrow tubing on the inside of which is coated or bonded the stationary phase. Theory by Yang [117] shows that for capillary columns to rival packed columns then

$$(h\nu)dc^2 = (h\nu)d_P^2$$

where h is the reduced plate height, ν is the reduced velocity, dc is the capillary diameter and d_P is the particle diameter, which means for an open tubular column (OTC) the bore must lie in the 1–10 μm range. However, this suffers from the concomitant reduction in sample capacity. Tijssen has shown that this performance may be achieved by larger bores of 10–30 μm if very tightly coiled columns and high flow velocities are used. The practical limitation of capillary columns is caused by dispersion in the detector/injector systems producing peak broadening. If detector volumes can be reduced to around 1 nl [118] capillary HPLC would become a highly efficient and faster technique than packed columns, when the number of theoretical plates is above 30 000. Knox [119] states that capillary columns would be 27 times faster than packed columns with plate numbers greater than 100 000. Difficulties arising from their low sample capacity may be overcome by stream splitting of the sample and etching of the interior walls to give higher stationary phase loadings. These columns are still at the experimental stage though capillary internal diameters and detector volumes as low as 2 μl and 50 pl have been reported.

6.17.4 *Microcapillary packed columns* [116]

These have mainly been developed by McGuffin and Novotny [120] and co-workers and are characterised by low column diameter to particle size ratios of 2 to 5. This is much less than small bore packed columns (50–200) or conventional columns (500–2000). Below ratios of ~2, it has been reported [101] that the packing structure collapses under the viscous flow and causes clogging of the column. The microcapillary columns are prepared by extruding a heavy walled glass tube, 0.5–2 mm i.d., packed with 10–50 μm particle size high temperatures resistant silica or alumina. For reverse phase work the stationary phases have then to be bonded *in situ*.

Chromatographic performance in terms of speed of analysis and resolving power has been found to be poor relative to conventional small particle packed columns. Although packed capillary columns have larger capacities their permeability is reduced.

6.17.4.1 Summary. Knox and Gilbert [119] and Halasz [121, 122] have considered the theoretical limits on the separation performance of OTC. They conclude that in order to match the performance and speed of analysis of conventional HPLC systems ultra low flow-rates and dead volumes are required and that it is unlikely that OTCs will offer any significant advantage. However, until the very stringent instrumental requirements are satisfied the theoretical potential of OTCs cannot be satisfactorily examined. Many workers, on the other hand, have already demonstrated the advantages of microbore and narrow bore columns in terms of low solvent consumption, increased sensitivity and in the latter, dramatically increased performance.

6.17.5 *Instrumentation*

Despite the numerous advantages the instrumental demands of microcolumn LC are considerable, and these demands are further accentuated as the requirements vary from one column type to another. A consequence of the reduced flow rates is that the detector flow-cell volume should be reduced to $\leq$10 nl for OTCs, 0.1 µl for packed microcapillaries and 1 µl for microbore columns. An additional demand of the detector is that it should have a rapid response, <0.5 s. Development of suitable detectors is paramount if the potential of micro-LC is to be realised. Study of detector systems has focused in two areas; firstly, the miniaturisation of ultraviolet, fluorescence and electrochemical systems, using in the former two systems LASERS as excitation sources and ultraviolet fibre optic and on-line cells to reduce band broadening and increase sensitivity [123, 124]; secondly, the direct interfacing with systems which previously required transport and/or concentration of the eluant. Interfacing of HPLC with mass spectroscopy has been undertaken by Barefoot *et al.* [125] and Lisek *et al.* [126] and flame systems (FPD and TSD) have been reviewed by Kientz *et al.* [127]. Jinno has reviewed the interfacing of micro-LC with ICP [128].

6.17.6 *Solvent delivery*

The reduced column diameters necessitate accurate and precise flow-rates in the region of 1–100 µl min^{-1}. A number of pumps are commercially available which meet these requirements under isocratic conditions. In HPLC, solvent gradients are normally generated by controlling the flow of the component solvents, with microcolumns this requires sub-µl flow-rates. A number of systems have been developed for both linear and exponential gradient generation based on reciprocating pumps. These systems give comparable accuracy and precision of retention time and peak area as obtained in conventional HPLC, and have been reviewed by Yang [129].

6.17.7 *Injection systems*

Extra column dispersion, as would result from large injection volumes >1 µl and dead volumes introduced by frits and fittings must be reduced to obtain optimum performance. A variety of injection systems have been reported, including those of split injection, internal loop-rotary valve, 'microfeeder' and pneumatic microsyringes [130]. Considerable advances have been made in quantitative reproducibility. Dead volumes have been minimised by locating the columns directly into the injection and detector systems.

6.17.8 *Microcolumn applications*

Higher column efficiencies are required in order to investigate the complex mixtures encountered in petrochemical, biochemical and clinical studies,

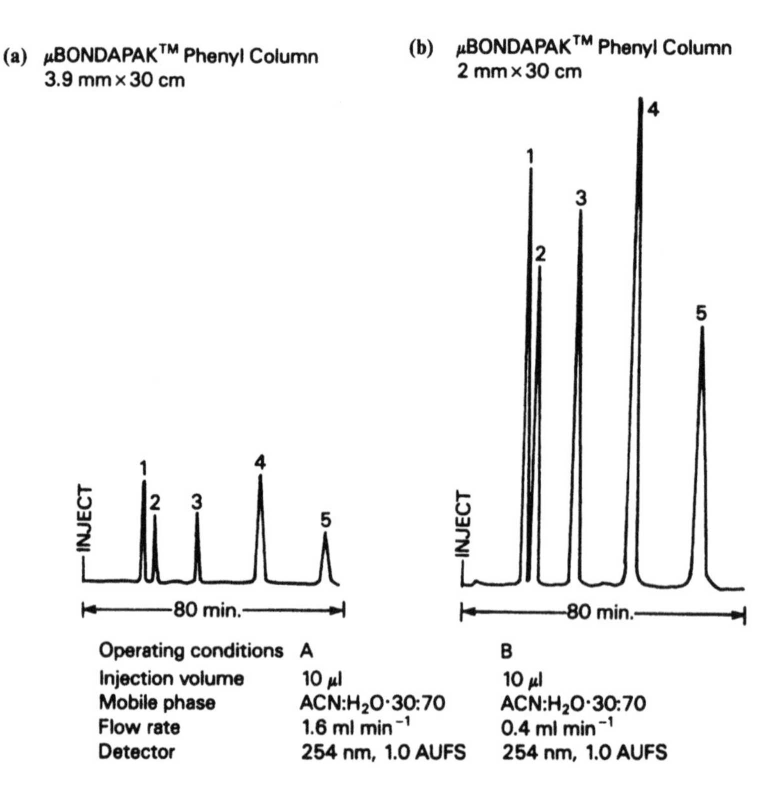

Figure 6.55 Steroid analysis using a 2 mm i.d. microbore column provides a fourfold increase in sensitivity over the 3.9 mm i.d. column. In addition, the microbore column consumed only 32 ml of solvent compared to the 128 ml for the 3.9 mm i.d. column. (Reproduced by permission from Waters.)

fermentation processes and pollution monitoring. At present only microbore (0.5–2 mm i.d.) packed columns are commercially available and considerable applications literature is available from manufacturers and suppliers. For such applications where amounts of analyte may be limited, e.g. environmental, food analysis, biological applications and forensic science, the use of microbore columns could be extremely advantageous. Figure 6.55 demonstrates the increased sensitivity achieved using a column of smaller diameter.

Where solvent costs are an important consideration the reduced solvent consumption of microbore columns will allow the use of more exotic solvents. For the other column technologies there is no substantial applications literature as the investigative work has been carried out on individually prepared columns. The following chromatograms illustrate the potential of small bore packed and microcapillary packed columns, respectively. Shown below, Figure 6.56 is an example of a highly efficient separation of

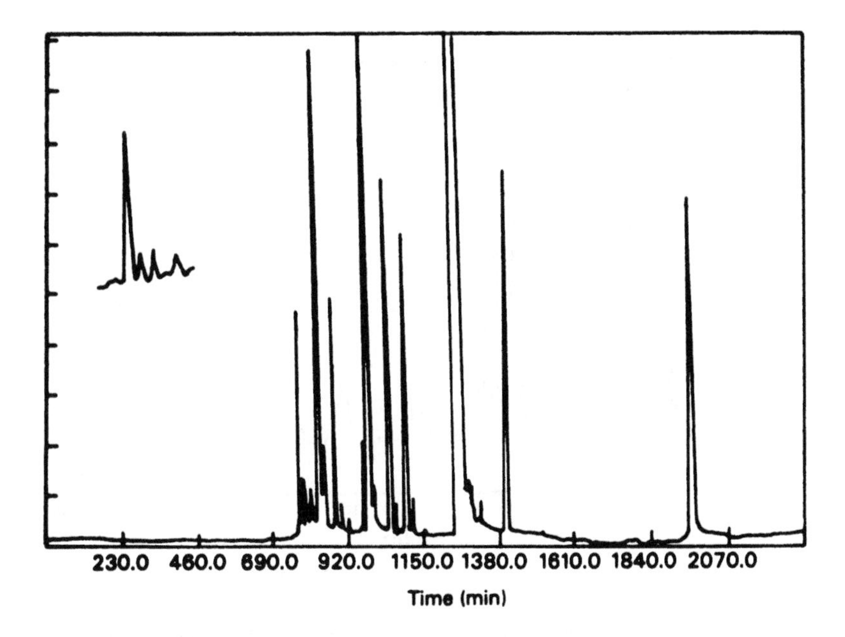

Figure 6.56 Chromatogram of an essential oil obtained with a 14 m × 1 mm i.d. microbore silica column (510 000 theoretical plates); *n*-hexane/ethyl acetate (95:5) solvent. (Reproduced with permission of Elsevier Publishing Company.)

a complex natural sample using a small bore packed column 14 m in length coupled in 1 m segments.

An example of the separation obtained of an aromatic fraction of coal tar using a microcapillary packed column is shown below (Figure 6.57 [131]); the chromatogram was obtained using a step-wise gradient in the reversed phase mode.

6.18 Applications

An indication of the versatility of HPLC, as evidenced by the variety of compound classes which can be examined, has been indicated in Figures 6.36(a) and (b). The technique is of major importance to the analyst in the pharmaceutical, food and fine chemical industries and is playing an increasingly significant role in forensic, environmental, chemical and biochemical studies [132].

For the practising chromatographer, often a search of the relevant literature will identify a similar application which with some minor modification will prove suitable for a particular need. The applications literature available is formidable; a number of technical abstracts are published [133], technical information published by instrument manufacturers and chromatography

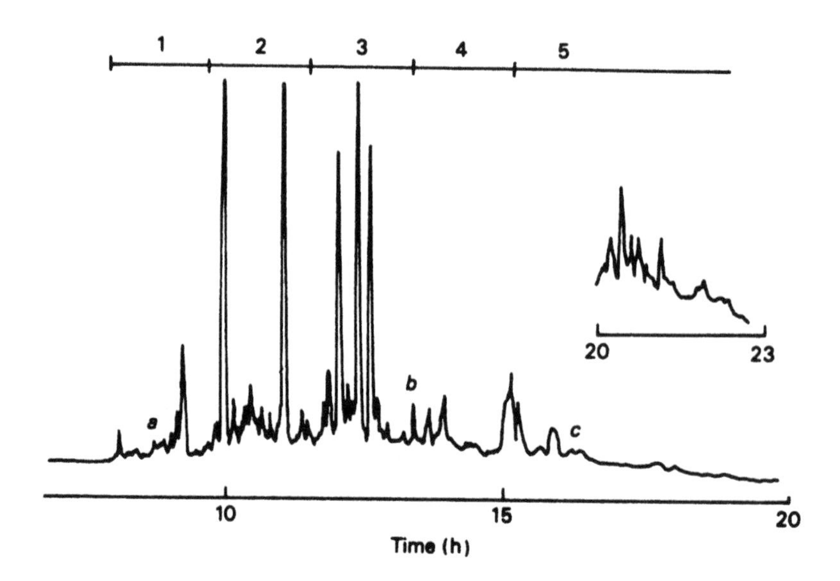

Figure 6.57 Chromatogram of the aromatic fraction of coal tar. Mobile phase, stepwise gradient: (1) methanol/water (80:20); (2) methanol/water (90:10); (3) methanol; (4) 1% methylene chloride in methanol; (5) 3% methylene chloride in methanol. Column: 55 m × 70 μm i.d., basic alumina (30 μm)/octadecyl silane. After 20 h (inset), inlet pressure and temperature were increased. Solute a is fluorene, b is dibenzo [*g*, *h*, *i*] perylene, and c is coronene. (Reproduced from Ref. 131 by permission of Elsevier Publishing Company.)

suppliers is an increasingly valuable source of reference and there are a number of excellent monographs [134, 135]. Available space does not allow a fuller discussion of the various application areas and readers requiring more detailed information are directed to the references cited above.

References

[1] Martin, A.J.P. and Synge, R.L.M. *J. Biochem.*, **50** (1950) 679.
[2] van Deemter, J.J., Zinderweg, F.J. and Klinkenberg, A. *Chem. Engng Sci.*, **5** (1956) 271.
[3] Giddings, J.C. *Anal. Chem.*, **35** (1965) 2215.
[4] Giddings, J.C. *Dynamics of Chromatography*. Dekker, New York, 1965.
[5] Horvath, C., Priess, B. and Lipsky, S.R. *Anal. Chim. Acta*, **38** (1967) 305.
[6] Majors, R.E. *Anal. Chem.*, **44** (1972) 1722.
[7] Borman, S.A. *Anal. Chem.*, **55** (1983) 384A.
[8] Dublin, P.L. *Aqueous Size Exclusion Chromatography*. Elsevier, Amsterdam, 1988.
[9] Walters, R.R. *Anal. Chem.*, **57** (1985) 1099A.
[10] Parikh, I. and Cuatrecasas, P. *Chem. Engng News*, (1985) 17.
[11] Pirkle, W.H., Hyun, M.H. and Bank, B. *J. Chromatogr.*, **316** (1984) 585.
[12] Dappen, R., Arm, H. and Meyer, V.R. *J. Chromatogr.*, **373** (1986) 1.
[13] Katrinsky, A. and Offerman, R.J. *Crit. Rev. Anal. Chem.*, **83** (1989) 21.
[14] Jandera, P. and Chuacek, J. *Gradient Elution in Column Liquid Chromatography*. Elsevier, Amsterdam, 1986.

[15] Glajch, J.L., Kirkland, J.J., Squire, K.M. and Minor, J.M. *J. Chromatogr.*, **199** (1980) 57.
[16] Poile, A.F. *33rd Pittsburg Conference on Anal. Chem. and Applied Spectroscopy*, 1982.
[17] Kirkland, J.J. *J. Chromatogr. Sci.*, **7** (1969) 7.
[18] Kirkland, J.J. *J. Chromatogr. Sci.*, **9** (1971) 206.
[19] Synovec, R.E., Johnston, E.L., Moore, L.K. and Renn, C.N. *Anal. Chem.*, **62** (1990) 357.
[20] Dorschel, C.A. *et al. Anal. Chem.*, **61** (1989) 951A.
[21] *LC–GC*, **4** (1986) 526.
[22] Little, J.N. and Fallik, G.J. *J. Chromatogr.*, **112** (1975) 389.
[23] White, P.C. *J. Chromatogr.*, **200** (1980) 271.
[24] Readman, J.W., Brown, L. and Rhead, M.M. *Analyst*, **106** (1981) 122.
[25] Jones, D.G. *Anal. Chem.*, **58** (1985) 1057A, 1207A.
[26] Lawrence, J.H. and Frei, R.W. *Chemical Derivitization in LC*. Elsevier, Oxford, 1976.
[27] Lingeman, H., Underberg, W.J.M., Takade, A. and Hilshoff, A. *J. Liquid Chromatogr.*, **8** (1985) 789.
[28] Fujimoto, C. and Jinno, K. *Anal. Chem.*, **64** (1992) 476A.
[29] Kalasinsky, V.F. and Kalasinsky, K.S. *HPLC Detection*, p. 127. VCH, New York, 1992.
[30] Christie, W.W. *HPLC and Lipids—A Practical Guide*, p. 22. Pergamon Press, Oxford, 1987.
[31] Kissinger, P.I. *Anal. Chem.*, **49** (1977) 477.
[32] Bratin, K., Blank, C.L., Lunte, C.E. and Shoup, R.E. *Int. Lab.*, (1984) 24.
[33] Rocklin, R.D., Henshall, A. and Rubin, R.B. *Am. Lab.*, **3** (1990) 34.
[34] Johnston, D.C. and Lacourse, W.R. *Anal. Chem.*, **62** (1990) 589A.
[35] Ewing, A.G., Mesaros, J.M. and Gavin, P.F. *Anal. Chem.*, **66** (1994) 527A.
[36] Lacourse, W.R. *Analysis*, **21** (1993) 181.
[37] Johnston, D.C. and Lacourse, W.R. *Electroanalysis*, **4** (1992) 367.
[38] Johnston, E.L. *Chromatogr. Sci.*, **37** (1987) 1.
[39] Jupille, T. *Chromatogr. Sci.*, **37** (1987) 23.
[40] Dreux, M. and Lafosse, M. *Analysis*, **20** (1992) 587.
[41] Christie, W.W. *Advances in Lipid Methodology*, p. 239. Oily Press, Ayr, 1992.
[42] Charlesworth, J.M. *Anal. Chem.*, **50** (1978) 1414.
[43] Dixon, J.B. *Chimia*, **38** (1986) 82.
[44] Rapkin, E. *J. Liquid Chromatogr.*, **16** (1993) 1769.
[45] Frey, M.B. and Frey, F.J. *Clin. Chem.*, **28** (1982) 689.
[46] Yeung, E.S. *J. Chromatogr. Sci.*, **45** (1989) 117.
[47] Fielden, P.R. *J. Chromatogr. Sci.*, **30** (1992) 45.
[48] Stanley, P.E. *J. Biolumin. Chemilumin.*, **7** (1992) 77.
[49] Hoagland, P.D., Fishman, M.L., Konja, G. and Clauss, E. *J. Agric. Food Chem.*, **41** (1993) 1274.
[50] Zukowski, J., Tang, Y., Berthod, A. and Armstrong, D.W. *Anal. Chim. Acta.*, **258** (1992) 83.
[51] Konash, P.L. and Bastanus, G.J. *Anal. Chem.*, **52** (1980) 1929.
[52] Duneman, L. and Fresenius', J. *Anal. Chem.*, **342** (1992) 802.
[53] Donard, O.F.X. and Martin, F.M. *Trends Anal. Chem.*, **11** (1992) 17.
[54] Hill, S.J., Bloxham, M.J. and Worsfold, P.J. *Anal. At. Spectrom.*, **8** (1993) 499.
[55] Shohei, O. and Tsugo, S. *Anal. Chem.*, **53** (1981) 471.
[56] Albert, L. and Beyer, E. *HPLC Detection* (Patoney, G., ed.), p. 197. VCH, New York, 1992.
[57] Sugata, R., Iwahashi, H., Ishii, T. and Kido, R. *J. Chromatogr.*, **487** (1989) 9.
[58] Chong, C.K., Mann, C.K. and Vickers, T.J. *J. Appl. Spectrosc.*, **46** (1992) 249.
[59] Popovich, D.J., Dizon, J.B. and Ehrlich, B.J. *J. Chromatogr. Sci.*, **18** (1980) 442.
[60] Bruckner, C.A., Foster, M.D., Lima, L.R., Synovec, R.E., Berman, R.J., Renn, C.N. and Johnson, E.L. *Anal. Chem.*, **66** (1994) 1R.
[61] Krull, I.S., Colgan, S.T. and Selavka, C.M. *Chem. Anal.*, **98** (1989) 393.
[62] Danilelson, N.D., Targove, M.A. and Miller, B.E. *J. Chromatogr. Sci.*, **26,** (1988) 362.
[63] Griffin, M., Price, S.J. and Palmer, T. *Clin. Chim. Acta*, **125** (1982) 89.
[64] Popovich, D.J., Dixon, J.B. and Ehrlich, B.T. *J. Chromatogr. Sci.*, **17** (1979) 643.
[65] Krull, I.S. and Ding, X.D. *J. Agric. Food Chem.*, (1984).
[66] Krull, I.S., Ding, X.D., Selavka, C., Bratin, K. and Forcier, G.J. *Forens. Sci.*, **2** (1984) 29.

[67] Blau, K. and Halket, J.M. *Handbook of Derivatives for Chromatography*. John Wiley, Chichester, 1993.
[68] Sander, L.C. and Wise, S.A. *CRC Crit. Rev., Anal. Chem.*, **18** (1987) 299.
[69] Nawrocki, J. *Chromatographia*, **31** (1991) 177.
[70] Nawrocki, J. *Chromatographia*, **31** (1991) 193.
[71] Snyder, L.R. *J. Chromatogr. Sci.*, **16** (1978) 223.
[72] Sander, L.C and Wise, S.A. *J. Chromatogr.*, (1993) 656.
[73] Dorsey, J.G. and Cooper, W.T. *Anal. Chem.*, **66** (1994) 857A.
[74] Majors, R.E. *LC–GC*, **6** (1993) 276.
[75] Levy, R.H. and Boddy, A.V. *Pharm. Res.*, **8** (1991) 551.
[76] Pirkle, W.H., Hyun, M.H. and Bank, B. *J. Chromatogr.*, **316** (1984) 585.
[77] Pirkle, W. H. and Popchapsky, T.C. *Adv. Chromatogr.*, **27** (1987) 73.
[78] Nicoll, G.D. *J. Chromatogr.*, **402** (1987) 179.
[79] Okamoto, Y. and Kaida, Y. *J. High Res. Chromatogr.*, **13** (1990) 709.
[80] Aboul-Einen, H.Y. and Islam, M.R. *J. Liquid Chromatogr.*, **13** (1990) 485.
[81] Andersson, S., Thompson, R.A. and Allenmark, S.A. *J. Chromatogr.*, **591** (1992) 65.
[82] Gubitz, K. *Chromatographia*, **30** (1990) 555.
[83] Taylor, D.R. and Mather, K. *J. Chromatogr. Sci.*, **30** (1992) 67.
[84] Levin, S. and Abu-Lafi, S. *Advances in Chromatography* (Brown, P.R. and Grushka, E., eds), p. 233. Marcel Dekker, New York, 1993.
[85] Pohl, C.A. and Johnson, E.L. *J. Chrom. Science*, **18** (1980) 442.
[86] Small, H., Stevens, T.S. and Bauman, W.C. *Anal. Chem.*, **47** (1975) 1801.
[87] Hearn, M.T.W. *Ion Pair Chromatography*. Dekker, New York, 1985.
[88] Bidlingmeyer, B.A. *J. Chromatogr. Sci.*, **18** (1980) 525.
[89] Groths, R.A., Warren, F.V. and Bridlingmeyer, B.A. *J. Liquid Chromatogr.*, **14** (1991) 327.
[90] Hunt, B.J. and Holding, S.R. *Size Exclusion Chromatography*. Chapman and Hall, New York, 1989.
[91] Dubin, P.L. *Aqueous Size Exclusion Chromatography*. Elsevier, Amsterdam, 1988.
[92] Schoenmakers, Billiet, H.A.H. and De Galan, L. *J. Chromatogr.*, **218** (1981) 261.
[93] Glajch, J.L., Kirkland, J.J., Squire, K.M. and Minor, J.M. *J. Chromatogr.*, **199** (1980) 57.
[94] Berridge, J.C. *J. Chromatogr.*, **485** (1985) 3.
[95] Snyder, L.R., Glajch, J.L. and Kirkland, J.L. *Practical HPLC Method Development*. Wiley, New York, 1988.
[96] Koenigbauer, M.J. and Majors, R.E. *LC–GC*, **3** (1990) 9.
[97] Guiochon, G. and Katti, A. *Chromatographia*, **24** (1987) 165.
[98] Cox, G. B. and Snyder, L.R. *LC–GC*, **6** (1988) 894.
[99] Verzele, M. and Geeraert, E. *J. Chromatogr. Sci.*, **18** (1980) 559.
[100] Jones, K. *Chromatographia*, **25** (1988) 547.
[101] Katti, A. and Guiochon, G. *Anal. Chem.*, **61** (1989) 982.
[102] Huang, J.-X. and Guiochon, G. *J. Chromatogr.*, **492** (1989) 431.
[103] Skea, W.M. *Anal. (NY), High Perform. Liquid Chromatogr.*, **98** (1989) 479.
[104] Ishii, D. *Introduction to Microscale HPLC*. VCH, New York, 1988.
[105] Takeuchi, T. *Trends Anal. Chem.*, **9** (1990) 152.
[106] Verzele, M., Dewaele, C. and De Weerdt, M. *LC–GC*, **6** (1988) 966.
[107] Simpson, R.C. and Brown, P.R. *J. Chromatogr.*, **400** (1987) 297.
[108] Ardrey, R.E. *Liquid Chromatography–Mass Spectrometry*. VCH, Weinheim, 1994.
[109] Quimby, B.D. and Sullivan, J.J. *Anal. Chem.*, **62** (1990) 1027.
[110] Jinno, K. *Chromatographia*, **25** (1988) 1004.
[111] Liu, G., Djordjevic, N.M. and Erni, F. *J. Chromatogr.*, **592** (1992) 239.
[112] Hirata, Y. *J. Microcol. Sep.*, **2** (1990) 214.
[113] Novotny, M. *Anal. Chem.*, **60** (1988) 500A.
[114] Menet, H.G., Gariel, P.C. and Rosset, R.H. *Anal. Chem.*, **56** (1984) 1770.
[115] Takeuchi, T. and Ishii, D. *J. Chromatogr.*, **238** (1982) 409.
[116] Verzele, M. and Dewaele, C. *Chromatogr. Sci.*, **45** (1989) 37.
[117] Yang, F.J. *J. Chromatogr. Sci.*, **29** (1982) 241.
[118] Scott, R.P.W. and Kucera, P. *J. Chromatogr.*, **185** (1979) 27.
[119] Knox, J.H. and Gilbert, M.T. *J. Chromatogr.*, **186** (1979) 405.
[120] McGuffin, V.L. and Novotny, M. *J. Chromatogr.*, **218** (1981) 179.

[121] Halasz, I. *J. Chromatogr.*, **173** (1979) 229.
[122] Hoffmann, K. and Halasz, I. *J. Chromatogr.*, **173** (1979) 211.
[123] Janek, M., Kahle, V. and Krechi, M. *J. Chromatogr.*, **438** (1988) 409.
[124] Edkins, T.J. and Shelly, D.C. *HPLC Detection* (Patonay, G., ed.), p. 1. VCH, New York, 1992.
[125] Barefoot, A.C., Reiser, R.W. and Cousins, S.A. *J. Chromatogr.*, **474** (1988) 39.
[126] Lisek, C.A., Bailey, J.E., Benson, L.M., Yaksh, T.L. and Jardine, I. *Rapid Comm. Mass Spectrom.*, **3** (1989) 43.
[127] Kientz, Ch.E., Verweij, A., De Jong, G.J. and Brinkman, U.A.T. *J. Chromatogr.*, **626** (1992) 59.
[128] Jinno, K. *Chromatogr. Sci.*, **45** (1989) 175.
[129] Yang, F.J. *J. High Res. Chromatogr. Chromatogr. Comm.*, **3** (1983) 348.
[130] Oates, M.D. and Jorgenson, J.W. *Anal. Chem.*, **61** (1989) 1977.
[131] Jinno, K., Fujimoto, C. and Ishii, D. *J. Chromatogr.*, **239** (1982) 625.
[132] Ahuja, S. *Trace and Ultratrace Analysis by HPLC*. Wiley Interscience, New York, 1992.
[133] (i) *Gas and Liquid Chromatography Abstracts*; (ii) *Liquid Chromatography Abstracts*; (iii) Selects, C.A. *High Performance Liquid Chromatography*; (iv) *Journal of Chromatography*; (v) *Chromatographia*.
[134] *J. Chromatogr.* e.g. Parris, N.A. *Instrumental Liquid Chromatography*, **27** (1984), Elsevier, Oxford.
[135] Brown, P.R. and Grushka, E. *Advances in Chromatography*, Marcel Dekker, New York.

7 Chromatography and spectroscopic techniques

7.1 Introduction

During the past decade considerable advances have taken place in digital and analogue electronics, personal computer systems and instrument design which has enabled dedicated gas chromatography (GC) and high performance liquid chromatography (HPLC) instruments to be designed incorporating compact integrated spectroscopic detectors, particularly mass spectrometric (MS) detectors. Control software and databases incorporating libraries of reference spectra run on fast personal computer (PC) workstations, this allows sample mixtures containing common organic analytes to be separated and identified on a routine basis using a single bench-top instrument.

Chromatography is primarily a separation technique recording the amount of analyte eluted from the chromatographic system. Although analytes separated by capillary column GC can be identified using two columns (see experiment 20, Chapter 9), and retention indices the method is limited either to a specific application or to a group of compounds. Data from HPLC ultraviolet detectors also provides limited information on analyte peaks and is only useful for simple sample mixtures. Identification of separated components is only possible if no co-elution occurs and sufficient unique information on the structure or physical properties of the compound can be obtained. This may be achieved by obtaining spectroscopic data on the eluted molecules and if a peak is sampled several times during elution then peak purity can also be established. In order to obtain the spectral data the spectrometer must be capable of carrying out a scan very rapidly and certainly several times during the elution time for each peak of a chromatogram. Developments in microelectronics, instrument design and microcomputers have enabled mass spectrometer, infrared and ultraviolet–visible spectrometers and atomic absorption instruments to be used successfully as chromatography detectors.

The chromatographic methods most suitable for spectroscopic identification techniques are GC and HPLC although TLC and recently CZE and SFC methods have received attention. GC was the first technique to be interfaced to a mass spectrometer since it proved relatively easy to transfer the gaseous effluent and GC–MS instruments were capable of sufficiently rapid scan times even in the 1970s to acquire undiscriminated spectral data [1]. Interfacing HPLC systems proved more difficult due to the problem of selectively removing a liquid mobile phase whilst transferring the analytes to the mass spectrometer. After considerable development reliable, direct and indirect interfaces have been produced. GC capillary columns with 0.15–0.18 mm i.d. and narrow bore HPLC columns have been specifically developed for mass spectrometers to reduce the mobile phase volume transferred to the ion source enabling direct column–ion source interfacing to be achieved. A quadrupole MS scans sufficiently rapidly for 'on the fly' analysis, i.e. the spectrum is obtained in a fraction of a second as the eluant emerges from the chromatograph. Development in Fourier-transform infrared instruments and ultraviolet–visible diode array detectors have enabled a range of spectroscopic detectors to be designed which provide the analytical chemist with a powerful, sensitive, and almost ideal range of analytical techniques. The power of such systems is assessed by the unique information obtained to characterise each component. This amounts to the detail or number of data points in a single spectrum with over 1000 spectra being recorded during a chromatographic run.

- UV: number of items of data per spectrum 2–10
- IR: number of items of data per spectrum 10–50
- MS: number of items of data per spectrum 20–500

Such large amounts of data can only be sensibly and rapidly analysed and compared with reference spectra using microprocessors such as the fast 32 bit processors in PCs. The main systems in use today are discussed below, and in addition to the above mentioned techniques the microwave induced plasma (MIP) detector, a helium microwave plasma emission source coupled to a GC and an optical emission spectrometer are reviewed.

7.2 Chromatographic requirements

The essential requirements of a GC or HPLC chromatographic system linked to a spectrometer follow normal practice. However, great care needs to be exercised to ensure that good separations and reproducible chromatograms are obtained. The main requirements may be summarised as follows:

- reproducible chromatograms, stable mobile phase flow-rate and pressure, achieved by using accurate control systems;
- stable stationary phase, with virtually no column bleed (GC) or dissolution (HPLC); achieved by using bonded stationary phases and operating the column at temperatures well below the upper limits (GC) and with specific solvents (HPLC);
- sharp, well resolved peaks with minimal tailing, since sharp narrow peaks produce the greatest concentrations of solute in the mobile phase;

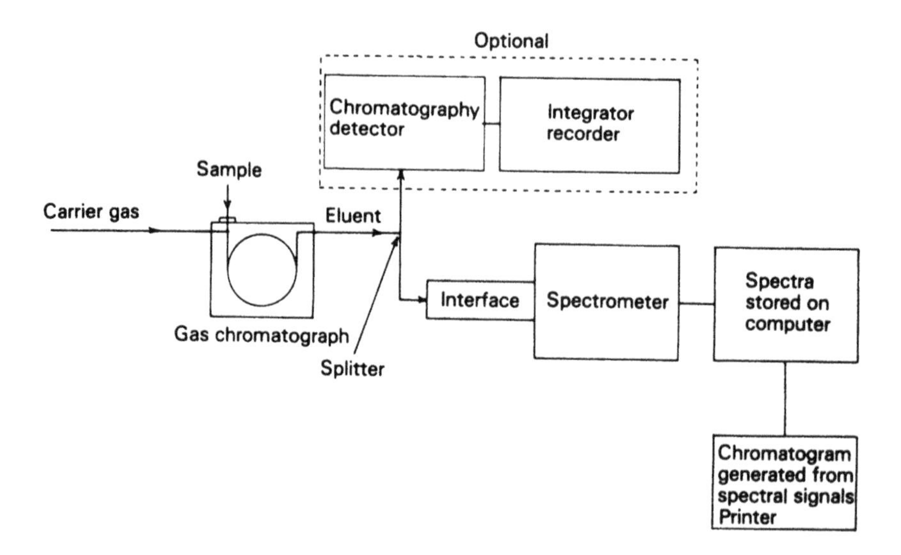

Figure 7.1 GC–spectrometer system.

- peak widths as uniform as possible throughout the chromatogram, using column temperature programming (GC) or solvent programming (HPLC);
- using a mobile phase that can, if necessary, be efficiently removed in an interface, has minimal spectral interference with the spectra of the eluted components, is thermally stable and does not react with the components at the elevated temperatures of an interface.

No modifications are required to the conventional sample introduction techniques used in GC and HPLC. The normal mobile phase flow and pressure control systems are usually suitable providing their specification meets the stability requirements of the interfaces. Normal GC and HPLC detectors can still be used to follow the progress of the chromatography and a stream splitter employed to divert most of the eluant to the interface. Split ratios are typically in the range 1 : 100 to 1 : 1000. A chromatography detector is not usually included in instruments specifically designed as integrated GC/HPLC–spectroscopic systems (Figure 7.1). The chromatogram is constructed from the total spectrum signal, that is, the total detector signal for each scan. A scan rate of at least 1 scan s^{-1} is used so a plot of total spectrum signal against time produces a well defined chromatogram. For example, in GC–MS systems, the total ion current for each scan is plotted against scan number as shown in Figure 7.2, the scan number being representative of the retention time. All the spectral data is stored in the computer and the chromatogram is available after each run has been completed to carry out library searches, identify the analytes, quantitative calculations and for output to a printer.

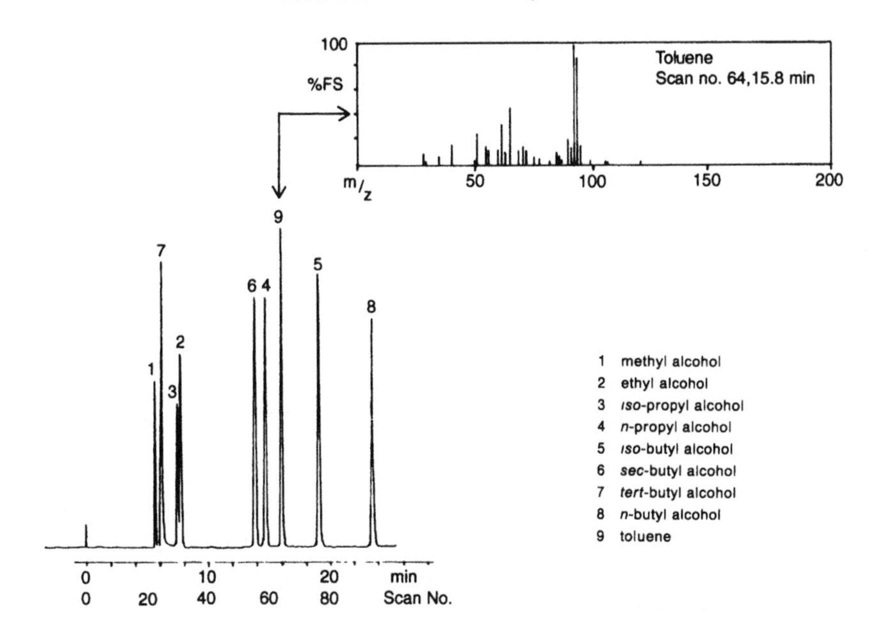

Figure 7.2 Total ion chromatogram for GC–MS analysis of alcohols with toluene as the marker and the mass spectrum of scan number 64 corresponding to peak 9, toluene. 25 m × 0.25 mm Carbowax 20M column; carrier gas: He at 1 ml min^{-1}; temperature programmed 50–200°C at 5°C min^{-1}.

7.3 Chromatographic and mass spectrometry techniques

Technological developments over the past decade, particularly to optimise ion source and quadrupole design, have produced reliable instruments. Parallel developments of narrow bore capillary GC and HPLC columns have reduced the problem of interfacing a high volume gaseous or liquid effluent with the vacuum requirements of the MS ion source. Before discussing the interfacing requirements a brief outline of MS instrumentation is presented [2–6] (Figure 7.3).

7.3.1 Ion source and inlet system

The MS ion source where electron impact is utilised is maintained at a pressure of $<10^{-3}\,\tau$ (torr). The sample inlet system is designed to release the analytes into the central region of the source at a carefully controlled rate. This is dependent on the concentration of the analyte and its physical properties. In chromatography interfaces a short length of heated silanised silica or stainless-steel capillary tubing is used to transfer the eluant directly

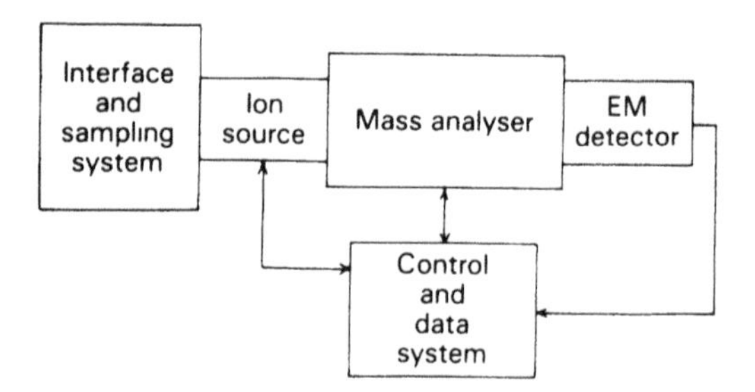

Figure 7.3 Schematic diagram of a mass spectrometer system.

to the ion source or to link an interface to the source after removal of most of the mobile phase.

There are many methods used to produce ions in the MS ion source, with the electron impact (EI) process being the most widely used. Databases containing standard 70 eV EI spectra for over 140 000 compounds are available. Chemical ionisation (CI) and field ionisation (FI) have attracted a specialised interest.

7.3.1.1 Electron impact ion source. The electron impact ion source is efficient, simple to construct, stable and produces ions with a narrow kinetic energy spread (Figures 7.4 and 7.5). The spectra obtained are specific and characteristic of the molecular structure of the analyte since the fragmentation processes produce ions in a relative abundance ratio which is characteristic of the molecular structure of the compound. The ions and their relative intensities provide a fingerprint for the compound.

Electrons emitted at a hot filament are accelerated and the resulting electron beam traverses the ion chamber to the collector anode. Interaction between the electron beam and the organic molecules (M) results in an energy transfer of over 20 eV which is sufficient to ionise most molecules; in many cases the molecular ion is unstable and subsequently undergoes

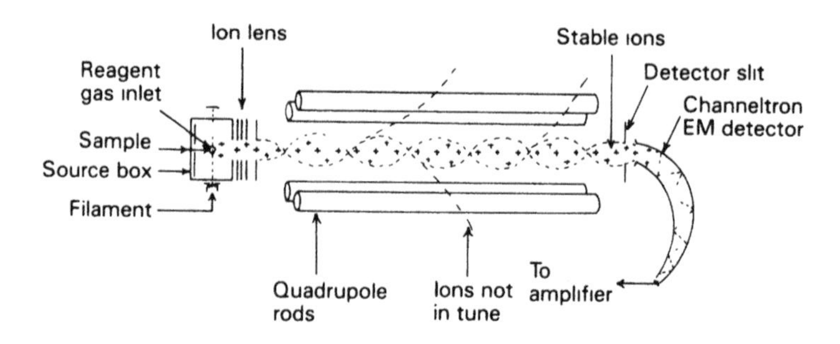

Figure 7.4 Quadrupole mass analyser with a chemical ionisation source.

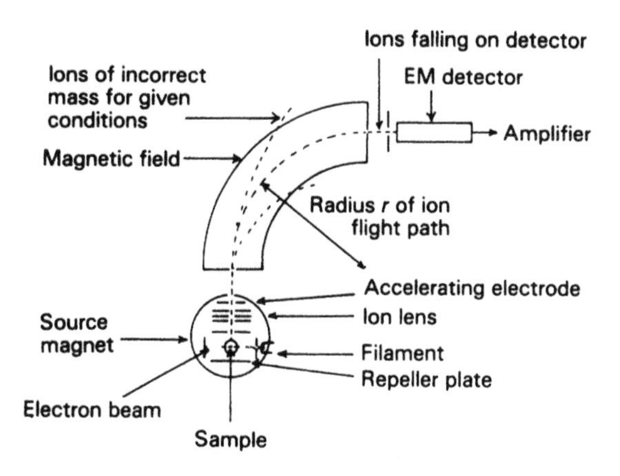

Figure 7.5 Single focusing magnetic analyser with an electron impact source.

fragmentation to form smaller ions:

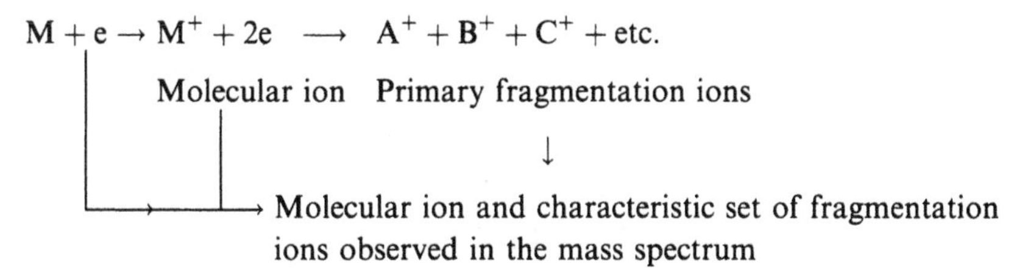

The resulting positive ions move out of the ionisation area through the slits and an ion lens system into the mass analyser, under the influence of a small positive repelling potential in the source. The degree of fragmentation, and hence the spectral fingerprint or pattern, depends on the energy of the bombarding electron beam. Ion currents are a maximum for electron energies in the range 50–80 eV, with most reference spectra being obtained at 70 eV. An EI source requires a pressure of less than $10^{-3}\,\tau$, to avoid unwanted ion molecule collisions, and a pumping system with vacuum conductance of 30–100 litres s^{-1} permitting 5–10 ml min^{-1} of vapour to be introduced into the source. Typical mobile phase flow rates and gas/vapour volumes are shown in Table 7.1.

7.3.1.2 Chemical ionisation. CI is a variation on the EI source (Figure 7.4). EI spectra frequently contain weak molecular ions or none at all due to the molecules gaining energy in excess of their ionisation potential causing extensive fragmentation. In the CI process a much lower transfer of energy occurs and a quasi-molecular ion formed by the loss $(M-1)^+$ or addition $(M+1)^+$ of hydrogen ions is produced with little attendant fragmentation. Thus, CI spectra provide molecular weight information and frequently

Table 7.1 Typical mobile phase flow rates and gas/vapour volumes

Technique	Column i.d. (mm)	Mobile phase	Gas/vapour flow-rate ($ml\,min^{-1}$)
GC packed column	6.3	Helium	30
GC capillary column	0.32	Helium	1.5
GC capillary column	0.18	Helium	0.5
HPLC normal column	4.6	Water/ methanol	$1.5 \equiv 1860\,ml\,min^{-1}$ vapour $1.5 \equiv 825\,ml\,min^{-1}$ vapour
HPLC narrow bore column	2.1	Water/ methanol	$0.3 \equiv 372\,ml\,min^{-1}$ vapour $0.3 \equiv 165\,ml\,min^{-1}$ vapour

structural features not revealed by EI spectra. The ionisation chamber in a CI source is a gas-tight design allowing pressures in the order of 1 torr to be maintained. The sample molecules and a reagent gas (usually methane or isobutane) up to 10^6 times in excess is introduced into the ionisation region where it is ionised by the electrons. A secondary ion–molecule reaction then occurs between the sample molecules in low abundance and the primary reagent ions, the resulting secondary ions proceed to the mass analyser.

Formation of principal reagent ions from electron bombardment of methane at 1 torr are CH_5^+ and $C_2H_5^+$:

$$CH_4 + e \rightarrow CH_4^+ + CH_3^+ + CH_2^+$$

$$CH_4^+ + CH_4 \rightarrow CH_5^+ + CH_3^+$$

$$CH_3^+ + CH_4 \rightarrow C_2H_5^+ + H_2$$

Subsequent reactions of analyte molecules to form secondary ions:

$CH_5^+ + M \rightarrow MH^+ + CH_4$	Proton transfer to form $m/z = M + 1$
$C_2H_5^+ + M \rightarrow MH^+ + C_2H_4$	Proton transfer to form $m/z = M + 1$
$C_2H_5^+ + M \rightarrow MC_2H_5^+$	Electrophylic addition to form $m/z = M + 29$
$C_2H_5^+ + M \rightarrow (M{-}H)^+$	Hydride transfer to form $m/z = M - 1$

Because large volumes of reagent gases are used, it is essential to have a source vacuum system with a high pumping speed in addition to the analyser vacuum pump.

7.3.1.3 Field ionisation. FI sources have been used in GC–MS instruments. Ions are produced in an FI source by the high positive electric field (typically $10^8\,V\,cm^{-1}$) produced between a wire and a knife edge. The energy available is generally 12–13 eV which is sufficient to ionise most organic molecules (ionisation potential typically 7–13 eV), but since there is less excess energy than with the EI source, less fragmentation occurs and the parent ion can usually be observed. The main disadvantage concerns sensitivity which is of the order of 100 times lower than the EI process.

7.3.2 Mass analysers

The role of the mass analyser is to separate ions emerging from the ion source according to their mass (m/z ratio). The separating capability or mass resolution, R, of the analyser is measured in terms of its ability to resolve ions to less than a 10% valley between peaks:

$$R = \frac{M}{\Delta M} \quad \text{or} \quad R_{\text{PPM}} = \frac{\Delta M}{M} \times 10^6$$

where M is the mass of the first peak of an adjacent pair and ΔM the difference in mass of the two peaks. Ideally, the peaks should be of approximately equal intensity, separated by a valley between them of 10% peak height, corresponding to a 5% peak overlap. A second parameter which influences GC–MS performance is the length of the ion flight path. A good vacuum must be maintained in the analyser region to ensure that the mean free path of the ions will be significantly greater than the analyser flight path. To avoid vacuum problems differential pumping between the source region and the analyser section is used so that a pressure difference between the source, at $\sim 10^{-3}\,\tau$, and the analyser $\sim 10^{-5}\,\tau$ is maintained.

There are two main groups of analysers, non-magnetic analysers typified by quadrupoles and ion trap analysers and magnetic analysers which use permanent or electromagnets to separate the ions under the influence of a magnetic field.

7.3.3 The quadrupole mass analyser

The quadrupole mass analyser is the most commonly encountered non-magnetic analyser (Figure 7.4) [6]. It consists of a set of four rods of circular or hyperbolic cross section in a quadrant formation mounted parallel to the z-axis. Opposite rods are electrically connected together, a voltage is applied which consists of a d.c. and r.f. (1–2 MHz) component creating an oscillating field between the rods. When an ion moves into the quadrupole field it will start to oscillate in the x, y-directions between the rods. If the mass of the ion is such that these oscillations are stable then the ion will move through the analyser to the electron multiplier. Ions of other m/z values will undergo unstable oscillations of increasing amplitude until they move out of the quadrupole field. Since there is no force along the z-axis of the rods an ion accelerating potential of only 20–30 V is required. Scanning is achieved by varying the magnitudes of the d.c. and r.f. voltages whilst maintaining a constant r.f.:d.c. ratio to produce a linear mass spectrum. A wide range of practical quadrupole spectrometers are manufactured with resolution varying from 100 for analysis of gaseous mixtures to organic quadrupoles with a resolution of 2000 and mass range to 1000. Compact quadrupole analysers with a 70 eV EI source and very fast scan times, $<10\,\text{ms scan}^{-1}$, have been specifically designed as GC/HPLC detectors. Linear spectra and easy interfacing for electronic control and data acquisition have contributed to their popularity, commercial integrated GC/HPLC–MS instruments almost all use quadrupoles.

7.3.4 Magnetic analysers

7.3.4.1 Single focusing magnetic analysers. Single focusing magnetic analysers with a resolution of over 3000 are commonly used in organic analysis (Figure 7.5). The ions formed are accelerated through the source slit into a homogeneous magnetic field and then follow a curved path, the radius (r) of which is determined by the accelerating potential (V) and magnet field strength (B). The mass (m/z) of an ion which follows a path radius r through the analyser is given by

$$m/z = \frac{B^2 r^2}{2V}$$

Since the radius is fixed in the design of the instrument, varying either B or V will result in ions of varying m/z values falling on the detector. Controlling the electromagnet and varying B at a fixed accelerating potential is commonly used to scan through the mass range. GC–MS techniques require a fast reproducible scan which may be readily controlled electronically for reliable interfacing to data acquisition systems. Scanning by varying the magnetic field is a limiting factor although newer laminated magnets are able to generate stable magnetic fields at fast scan rates. The resolution of magnetic analysers is principally determined by the radius or the ion path through the magnet and by the width of the source and detector slits, which can be pre-set to selected values.

7.3.4.2 Double focusing instruments. Double focusing analysers are employed for high resolution instruments, $R > 10\,000$. An additional ion focusing system, an electrostatic analyser, is used to generate a radial electric field to counteract the velocity dispersion and hence peak broadening that occurs in a magnetic field (Figure 7.6). The limitations imposed on scan speed, require special design and interfacing techniques, the discussion of which is beyond the scope of this text. Double beam techniques, in which the sample and a reference (e.g. perfluorokerosene, PFK) are run together in a tandem pathway through the magnetic and electric fields, are used to

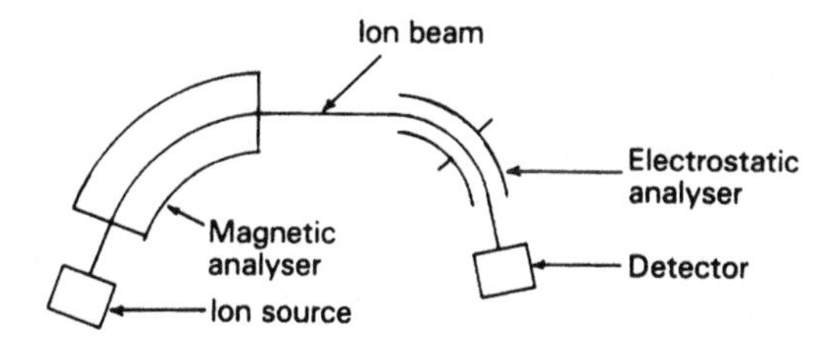

Figure 7.6 Double focusing mass analyser, Nier–Jordan geometry.

accurately calibrate the mass range to at least four decimal places, the reference ions acting as mass markers. Double focusing instruments are used where separation of ions with a mass that varies in the second or third decimal place is required, for instance in metabolic studies. Since high resolution spectra are complex it is preferable in GC/HPLC–MS to precede an attempt to obtain accurate data by a low resolution run. Thus the relevant peaks and mass range may be determined and the problem of handling excess amounts of data is reduced.

7.4 Gas chromatography interfacing techniques

The separated components emerging from a GC column are present in picogram amounts in the carrier gas stream. If the column eluant is to be coupled to a mass spectrometer then the volume flow-rate of carrier gas should be minimised for a given separation to achieve high sensitivities in the ion source and reduce pumping requirements. GC parameters are selected to obtain symmetrical sharp peaks which are eluted in the minimum carrier gas volume and with the best peak height to width ratio that can be achieved. Clearly, the components need to be resolved, as co-eluting peaks would not produce pure mass spectra which could then be compared to a library of reference spectra. A number of factors need to be considered to optimise the GC system:

- the carrier gas needs to be inert and easily removed in an interface;
- the carrier gas should not interfere with the spectra nor the total ion current;
- there should be almost zero column stationary phase bleed, the stationary phase being used within its operating temperature range;
- injection technique and split ratio should be optimised to deliver the sample onto the first few plates of the column without overloading;
- peaks should be symmetrical, sharp and well resolved ($R_S > 1.0$) to obtain the highest concentrations of component molecules in the carrier gas;
- temperature programming is used to maintain a satisfactory peak profile throughout the run so that later eluting peaks are not broad; and

- the GC system should be chosen and set up for minimum peak volume of the emerging peaks.

GC systems which meet these requirements use WCOT capillary columns with bonded stationary phases and helium carrier gas. Helium is an inert gas, with a high ionisation potential of 24.6 eV. WCOT columns are capable of separating complex samples containing over 100 components (see Figure 5.9) having high column efficiencies (N_{eff}), minimum band broadening and low helium flow-rates. Although a wide range of stationary phases are available only four or five are required to successfully analyse over 90% of samples. Capillary columns can be directly connected into the ion source of the mass spectrometer or alternatively an interface can be used which selectively removes most of the helium and transfers the components to the ion source. Integrated GC–MS instruments almost exclusively use capillary columns and direct interfacing which maintains chromatographic resolution and transfers all the analytes to the ion source for maximum sensitivity. Packed columns have largely been replaced by capillary columns but interfacing techniques are available that can cope with the high carrier gas volumes of these columns. Wide bore capillary and micropacked columns are less demanding. They have a lower volume flow-rate of carrier gas than packed columns and larger sample volumes can be used than with narrow bore capillary columns. Gases such as carbon dioxide, nitrogen, argon and methane have been used as carrier gas in specialised applications, particularly methane which also forms the reactant gas in a CI source.

The EI ion source pressure is typically less than $10^{-3}\,\tau$. The main reasons for maintaining these pressures are to

- prevent the filament burning out;
- reduce background spectra;
- prevent ion molecule reactions occurring and thus altering the fragmentation pattern;
- maintain the mean free path of the ions; and
- avoid high voltage discharge.

Since the source pressure has to be maintained within the pumping capacity or conductance of the vacuum system, low carrier gas flow rates are necessary or various types of interface are used to reduce the carrier gas component in the GC effluent. Interfaces and transfer lines have to be maintained at or above the maximum column temperature used.

7.4.1 Direct interfacing

Direct interfacing is used in bench-top GC–MS instruments which are equipped with narrow bore WCOT capillary columns and the carrier gas flow-rate is less than 2 ml min^{-1}. Recently introduced 0.15 and 0.18 mm i.d. columns with optimum flow-rates of 0.3 and 0.5 ml min^{-1}, respectively, have high column efficiencies, rapid analysis times and produce narrow symmetrical peaks over the full temperature programmed run (see the chromatograms in Figure 5.9). The end of these columns is fitted directly into the ion source. No heated transfer line is required since the quadrupole mass analyser is located on the side of the column oven (Figure 7.7a).

7.4.2 Indirect interfaces

An alternative interface used with capillary columns, particularly where a range of columns might be used in the GC, is the open split interface (Figure 7.7b) [1, 7]. The split ratio is controlled by the ion source pressure, column flow-rate, flow-rate of the purge gas and the predetermined length of the restrictor. The only variable is the purge gas flow-rate which is readily adjusted via a needle valve to give the desired flow of carrier gas and component into the ion source. By increasing the purge flow-rate, solvent dumping and transfer of selected peaks can be achieved. The main disadvantage of the open split interface is that only a proportion of the sample is transferred to the ion source.

One of the most commonly used interfaces is the jet separator (Figure 7.7c). The GC carrier gas passes through a jet and expands into a partial vacuum, only the heavier sample molecules pass through into the collecting jet, the lighter helium being pumped away. Frequently, two stages are used to give the desired performance and careful design is necessary to obtain an efficient separation. A separate additional pumping system is required.

Alternative separators are mainly based on effusion or diffusion principles. The effusion separator generally consists of a tube through which the carrier gas flows at slightly reduced pressure (Figure 7.7d). A range of porous materials are used, for example, glass, stainless steel, silver, or PTFE. The rate at which molecules diffuse through to the low pressure side is inversely proportional to the square root of the molecular weight. Thus, the carrier gas will be preferentially removed from the effluent. Again additional pumping is required.

Semipermeable membrane separators use a variety of polymeric materials, the most popular being a silicone rubber membrane (Figure 7.7e). The carrier gas passes over a thin membrane supported on a fine metal mesh or glass scinter. The inorganic carrier gas is insoluble in the silicone polymer so most of it passes on to the exit. However, the organic material is attracted to the silicone membrane by a process akin to attraction to the stationary phase in a GC column and then diffuses through into the lower pressure ion source. The rate of diffusion of the component molecules through a polymer membrane is a function of specific solubility, the diffusion characteristics

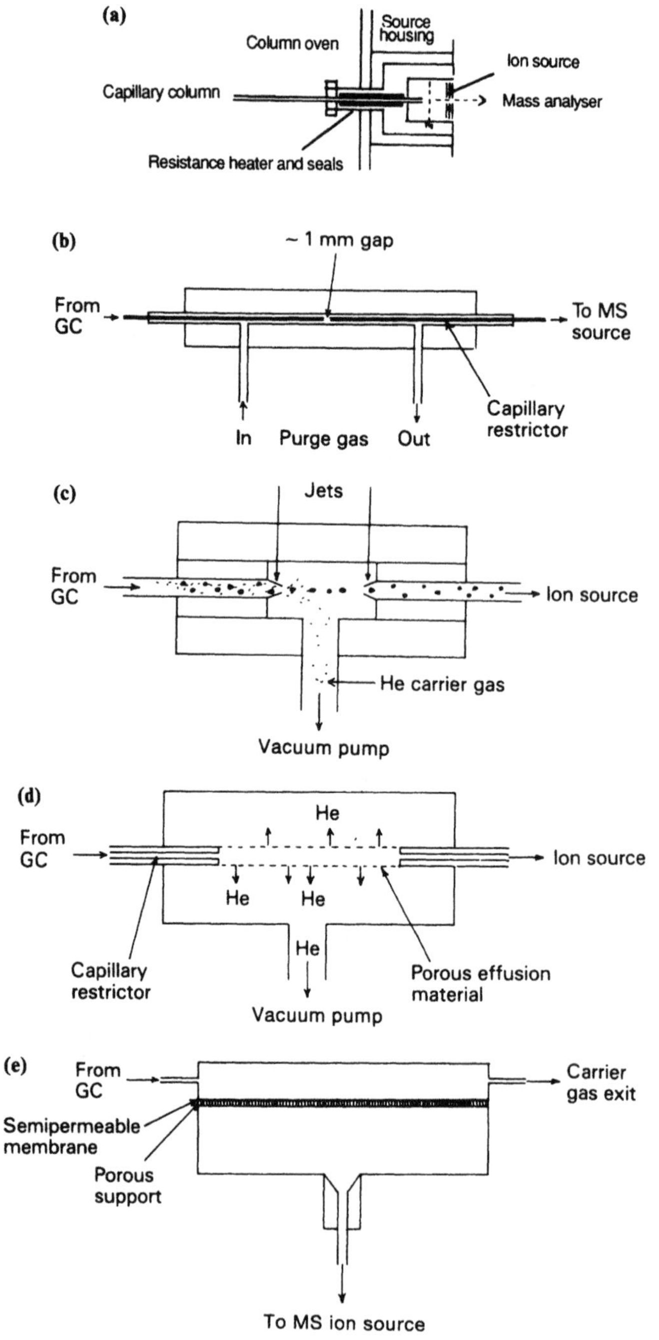

Figure 7.7 GC–MS interfaces. (a) Direct injection from capillary columns. (b) Open slit interface. (c) Molecular jet separator. (d) Effusion separator. (e) Diffusion separator.

of the molecules and the physical dimensions of the membrane. In practice, one or two stage membrane separators may be used. The main advantage of membranes is that no additional pumping system is required. However, loss of chromatographic resolution due to the diffusion process occurs limiting their use to the analysis of less complex mixtures. Membrane interfaces have been used in integrated GC–MS instruments because of their simple design which results in a compact overall system.

7.5 High performance liquid chromatography interfacing

HPLC–MS interfaces were developed after GC interfaces and considerably more work was required to produce efficient interfaces capable of dealing with the large vapour volumes from a liquid eluant [8, 9]. Possible solutions to the problem include

- removing the mobile phase solvent by vaporisation without loss of the analytes before the sample is introduced into the ion source;
- splitting the eluant from the HPLC column so that only a fraction is transferred to the ion source, such an interface would have poor sensitivity;
- using an interface that incorporates a large pump with the capacity to remove large volumes of solvent vapour;
- using a narrow bore column to obtain small solvent flow-rates; and
- ionisation of the eluant directly at atmospheric pressure as in chemical ionisation and electrospray interfaces.

Early interfaces used liquid nitrogen or helium cryogenic techniques to remove the solvent vapour, but these were rather cumbersome and not too efficient. Moving belt transport systems were also one of the first interfaces to be developed incorporating a flash vaporiser to remove the solvent before the sample reached the ion source. The main approaches used today are based on thermospray, atmospheric pressure and particle beam interfacing techniques [10].

7.5.1 Moving belt interface

The moving belt system overcomes the major interfacing problems of removing the solvent, vaporising the components without decomposition yet maintaining a high solute yield. The main limitation is the volatility of the analytes. Eluant from the HPLC system is introduced directly onto a continuously moving polyimide or stainless-steel belt (Figure 7.8). An infrared heater/evaporator removes much of the aqueous or polar solvents, removal is completed as the sample moves through two differentially pumped vacuum locks. Less than $10^{-7}\,g\,s^{-1}$ of solvent enters the ion

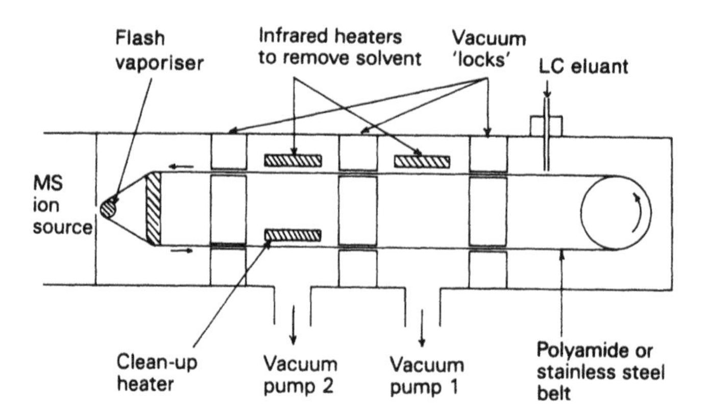

Figure 7.8 HPLC moving belt interface.

source. Inside the mass spectrometer the belt passes through a flash vaporisation chamber attached directly to the ion source. The sample is vaporised so quickly that little decomposition occurs and the molecules pass immediately into the ion source. The belt then passes over a clean-up heater to remove any residual sample that might cause interference with the next belt cycle. Up to 2.0 ml min^{-1} of volatile non-polar solvents or up to 1.0 ml min^{-1} of polar solvents may be introduced and subnanogram sensitivity can be obtained.

7.5.2 Thermospray interfaces

Thermospray interfacing techniques permit the continuous introduction of liquid eluant directly into the ion source of a mass spectrometer (Figure 7.9(a)). The method was developed by Vestal and incorporates a novel ionisation technique [11, 12]. The column eluant is transferred into the interface through a resistively heated capillary tube of approximately 0.1 mm i.d. The solvent is vaporised and forms a supersonic jet of vapour containing electrically charged mist droplets. The droplets continue to vaporise as they move across the ion source increasing the field gradient across the charged droplets. Eventually ions are expelled from the droplet and are directed by a repeller electrode to the orifice leading to the mass analyser. The ionisation efficiency is increased if the HPLC eluant contains a volatile electrolyte such as 0.1–1% of acetic acid or ammonium acetate. The system provides stable vaporisation and ionisation for flow rates up to 2 ml min^{-1} of a polar mobile phase. No external ionising source is required to achieve the CI type spectra at sub-nanogram detection levels for a wide range of non-volatile solutes. Figure 7.9(b,c) shows a mass chromatogram and mass spectra of some components [11–13].

(a)

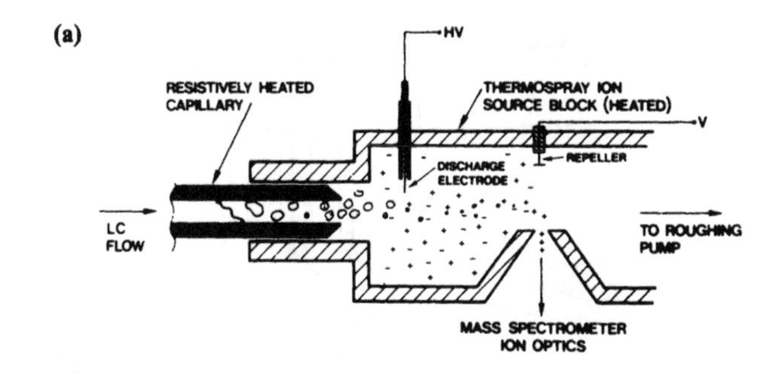

(b)

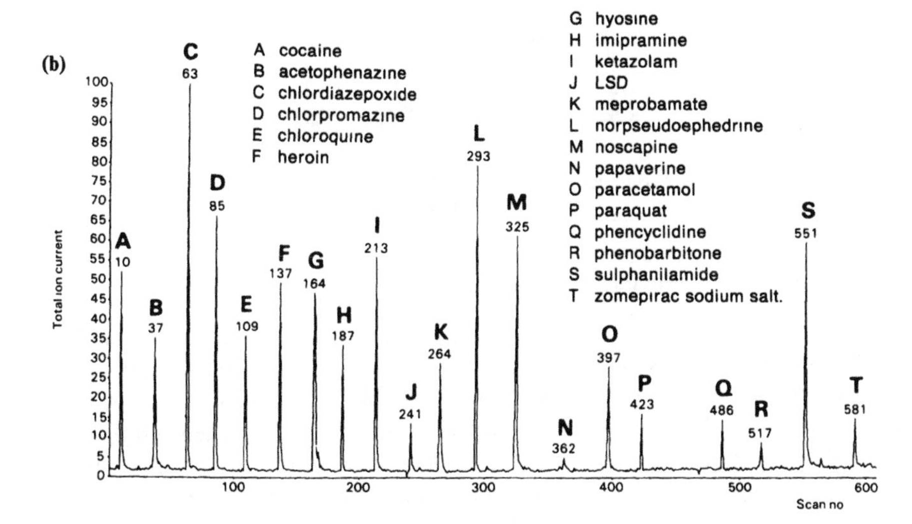

(c)

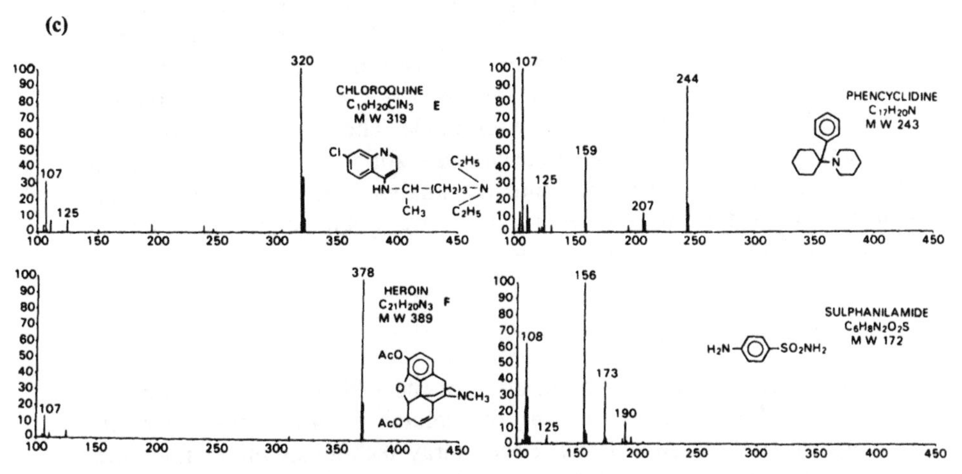

Figure 7.9 HPLC–MS thermospray system used in the analysis of a mixture of drugs. (a) Thermospray interface, (b) total ion chromatogram, and (c) mass spectra of peaks E, F, Q and S. (Reproduced by permission of VG Instruments.)

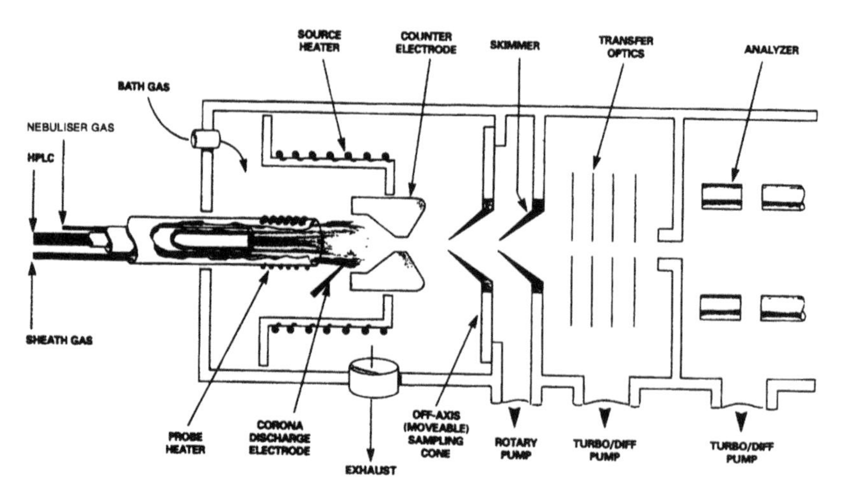

Figure 7.10 Atmospheric pressure ion source for LC–MS. (Reproduced by permission from VG Instruments.)

7.5.3 Atmospheric pressure ionisation interfaces

Atmospheric pressure ionisation techniques are able to accept the eluant directly from an HPLC column at flow-rates up to 2.0 ml min^{-1} and are suitable for a wide range of volatile and non-volatile analytes. The HPLC eluant is converted into an aerosol by a high velocity jet of nitrogen nebuliser gas and then vaporised in the heated chamber at the tip of the inlet tube. The vapour is swept by the nitrogen sheath gas into the ion source where ionisation takes place in the discharge formed between the corona discharge electrode and the counter-electrode (Figure 7.10). Primary ions are formed from the solvent which subsequently react with analyte molecules to form analyte ions by proton abstraction forming $(M+H)^+$ ions in the positive mode and $(M-H)^-$ in the negative ion mode. The ions travel through the sampling cone into an intermediate pressure zone at 1 torr before moving through the skimmer electrode into the analyser. Structural information can be obtained from fragmentation spectra obtained by adjusting the voltage applied to the sampling cone.

7.5.4 Electrospray interface

The electrospray interface is a variation of the API source where a metal-sheathed 0.1 mm capillary carrying the column eluant is directed at the primary electrode. Conventional electrospray techniques utilise flow-rates of approximately 10 μl min^{-1}, therefore eluant from HPLC columns has to be split before the interface. Recently developed high flow electrospray designs are able to accept up to 1 ml min^{-1} of eluant and can be connected

(a)

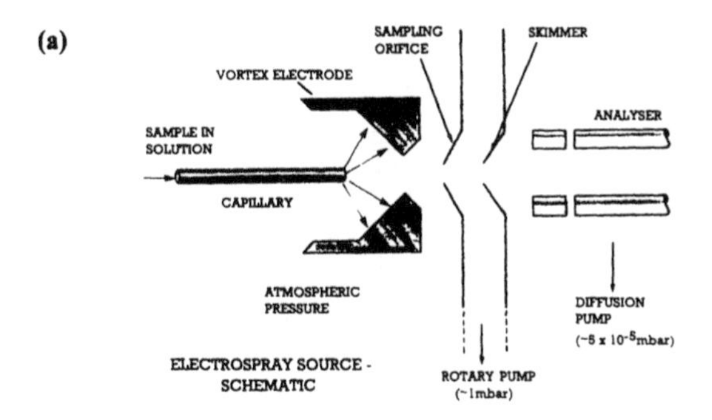

Figure 7.11 Electrospray LC–MS techniques: (a) electrospray ion source.

directly to HPLC columns without compromising sensitivity and reproducibility. Eluant emerging from the capillary tube is directed at the vortex electrode (Figure 7.11). The capillary jet is held at approximately 5 kV with respect to the sampling orifice. Coulombic repulsion forces cause the plume of charged droplets formed at the capillary tip to spread out. Solvent progressively evaporates until the electric field at the droplet surface has sufficient strength to desorb ions which are subsequently drawn through the conical nozzle into the analyser. Figure 7.11 includes an ion chromatogram of pesticides in water and an example of a positive ion electrospray mass spectrum.

Electrospray MS is suitable for direct on-line interfacing of capillary electrophoresis (CE). CE is a high resolution technique which is suitable for the separation of a wide range of samples such as peptides, surfactants, and pollutants at femtogram levels. The nl min^{-1} flow-rates used in CE are not adequate for direct analysis by electrospray ionisation so a make-up flow of solvent, for example 50:50 methanol/1% formic acid (aq) at 20 μl min^{-1} is introduced prior to the electrospray capillary inlet [13]. Figure 7.11c includes a CE electropherogram and a selected ion recording chromatogram of a sample containing 20 p mols of a mixture of herbicides.

7.5.5 Particle beam interface

Particle beam LC–MS is a rapidly developing complimentary interface to thermospray techniques and provides a method of linking conventional HPLC systems with eluant flow-rates of 0.3–1.0 ml min^{-1}, to an EI ion source to obtain the classical EI spectra which can be compared to conventional reference spectra (Figure 7.12). A capillary GC column may be connected to the same interface [10]. LC eluant enters the interface together with a stream of helium to form an aerosol of droplets which move through the desolvation chamber maintained at room temperature and pressure. The

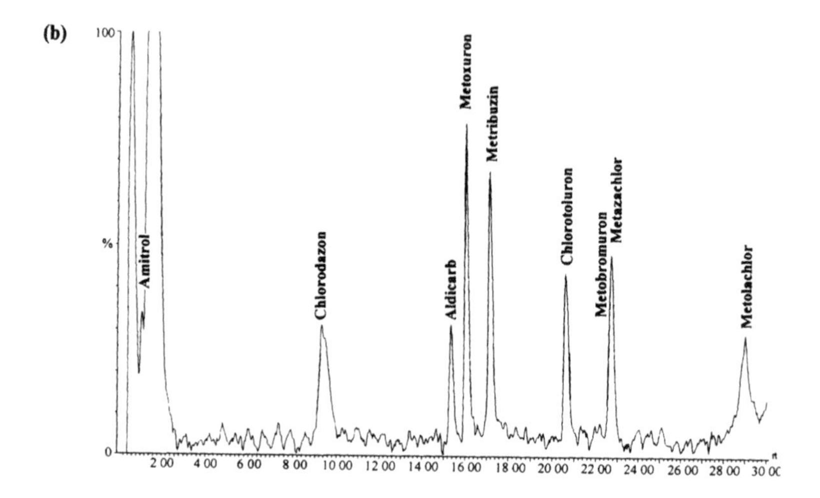

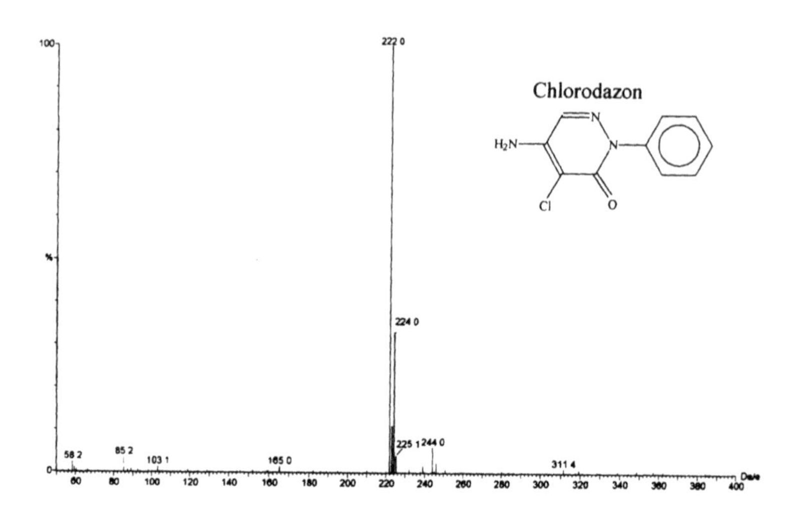

Figure 7.11 (b) Total ion chromatogram (top) and positive ion electrospray spectrum of chlorodazon from the LC–MS analysis of a pesticide mixture (10 pg μl^{-1}) using an API ion source.

solvent evaporates producing a mixture of vapour, particles containing the analyte molecules and helium. This enters the first low pressure region through a narrow nozzle, a process which accelerates the mixture to form a jet of particles directed to the skimmer nozzles. Solvent vapour and helium are skimmed by the two nozzles and pumped away leaving the heavier analyte particles to pass directly into the ion source where they strike a heated wall and are vaporised and ionised by electron impact or chemical ionisation processes.

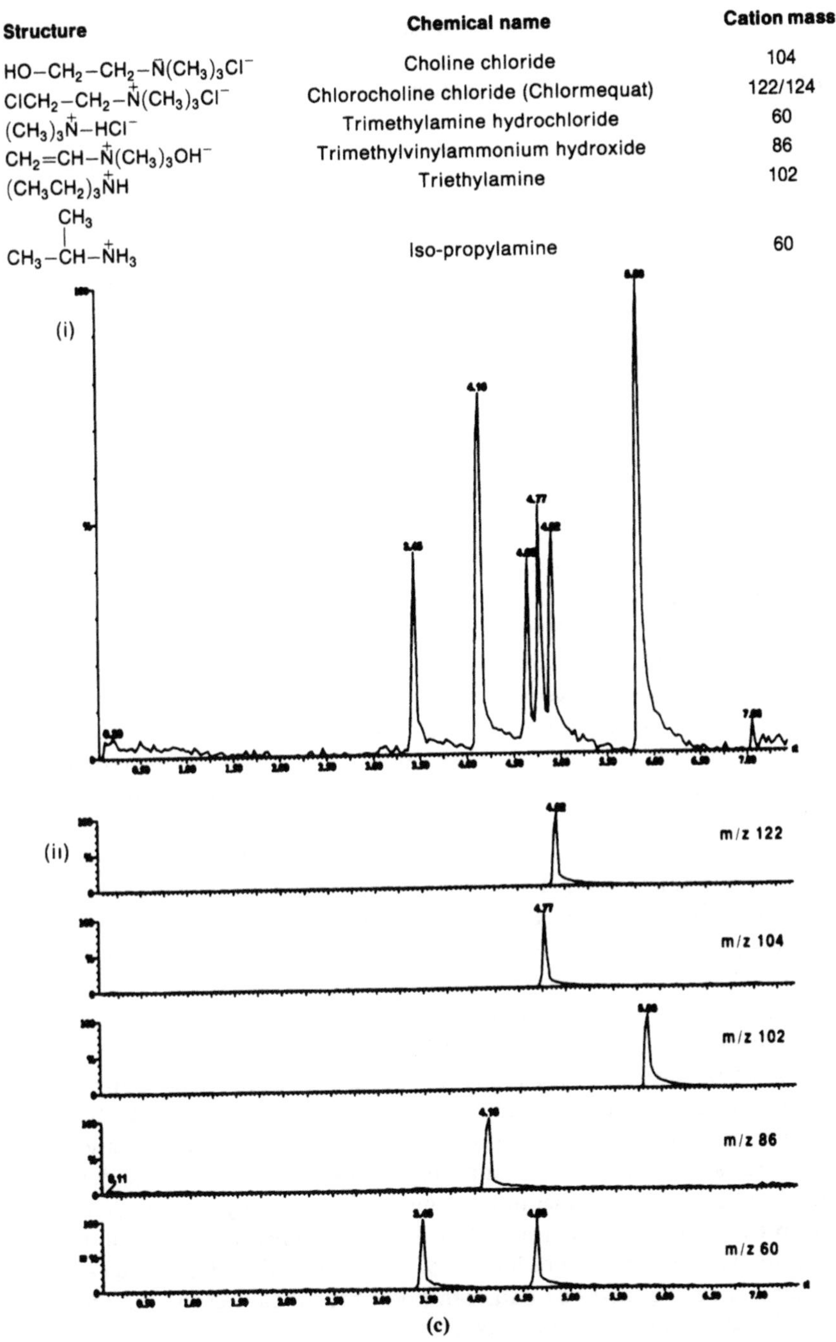

Structure	Chemical name	Cation mass
$HO{-}CH_2{-}CH_2{-}\overset{+}{N}(CH_3)_3Cl^-$	Choline chloride	104
$ClCH_2{-}CH_2{-}\overset{+}{N}(CH_3)_3Cl^-$	Chlorocholine chloride (Chlormequat)	122/124
$(CH_3)_3\overset{+}{N}{-}HCl^-$	Trimethylamine hydrochloride	60
$CH_2{=}CH{-}\overset{+}{N}(CH_3)_3OH^-$	Trimethylvinylammonium hydroxide	86
$(CH_3CH_2)_3\overset{+}{N}H$	Triethylamine	102
$CH_3{-}CH(CH_3){-}\overset{+}{N}H_3$	Iso-propylamine	60

(c)

Figure 7.11 (c) Capillary electrophoresis electrospray data from analysis of a mixture of four herbicides with triethylamine and *iso*-propylamine as internal standards, 3 pmol of each compound separated on a 20 cm capillary using indirect ultraviolet and single ion recording. (i) CE electropherogram, (ii) single ion chromatogram for each component. (Reproduced by permission of VG Instruments.)

7.6 Chromatography and mass spectrometry data systems

MS data systems form an integral part of the total chromatography–MS package. The computer system is based on a workstation frequently a PC running under MS Windows. The software comprises four integrated modules:

1. control software;
2. methods files;
3. data acquisition and processing; and
4. data analysis.

Chromatography and MS instruments contain their own specific control logic and microprocessors. However, software is required to check that all parts of the system are functioning within specification and that the mass analyser is tuned, that is, an autotune routine is run to calibrate the mass range, set the quadrupole rod voltages and the attenuation of the mass spectrometer detector. Calibration is carried out using the peaks in the spectrum of a standard reference compound such as PFK or FC43 (perfluorotributylamine). On quadrupole instruments the calibration may be retained for several weeks due to their inherent stability, although for quantitative work calibration should be carried out daily. The calibration routine memorises the spectral position and instrument parameters for the reference peaks of the calibration material and then calibrates the m/z scale over the whole mass range. An accuracy of ±0.01 mass units with quadrupoles and ±60 ppm with magnetic instruments can be achieved. The data system determines parameter settings and instrument functions required to run the method file selected for the analysis. Control algorithms interpret the run parameters in terms of the instrument control functions such as mobile phase flow rate and pressure, temperature programme or solvent programme, autosampler sequence and split ratios. Control data is then down loaded to the GC, HPLC and MS instruments.

A methods file is created for each analysis to specify the standard conditions to be used such as autosampler sequence, use of blanks and QA samples, column temperature programme, mobile phase composition and flow-rate, split ratios, mass range to be scanned, scan rate and sample identification codes. A methods menu leads the analyst through the full set-up sequence. Methods are filed for future reference and for use in routine analysis. The software can be linked to other standard packages such as word processors so the methods data can be linked to the results file and then transferred to the analytical report.

Considerable amounts of data are generated during an analysis. Data acquisition algorithms are continuously sampling the mass spectrometer detector signal at a predetermined scan rate usually about $5\,s^{-1}$. At least 10 scans are required to accurately describe a chromatographic peak. Each

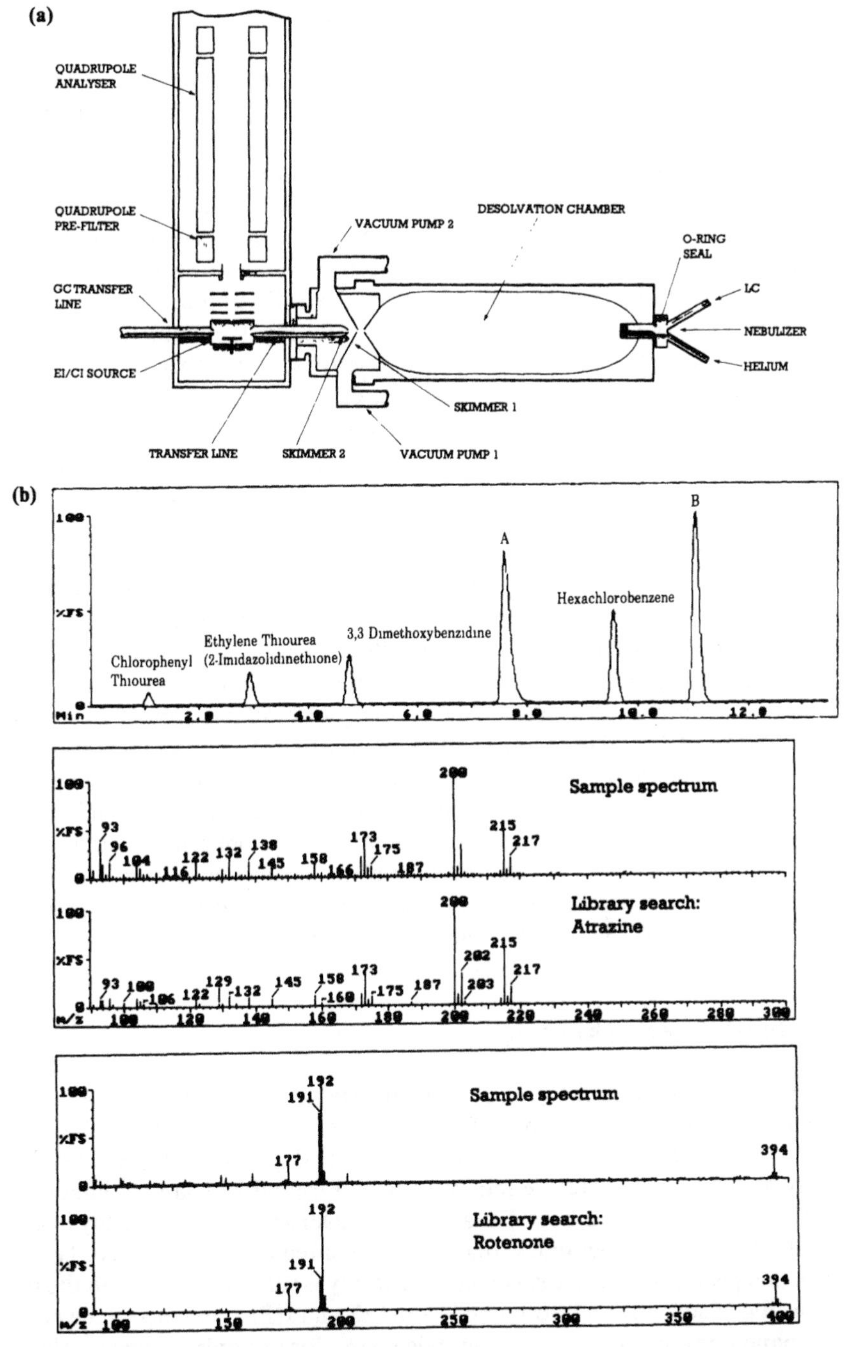

Figure 7.12 Particle beam HPLC–MS interface and data system reports: (a) VG particle beam line interface; (b) total ion chromatogram for a series of pesticides obtained using the LINC Particle Beam interface, in the EI mode. Peaks A and B are identified below as Rotenone and Atrazine by the library search.

scan will contain up to 400 m/z values and their signal intensities and the analysis time may be over 30 min as illustrated by the chromatograms in Figure 5.9. Over 8 million data points may therefore be collected which together with instrument and sample information forms an extremely large data file. Algorithms are used to reduce the data collected, options can be selected so that data is only collected when a peak has been detected or a mass fragment is recorded in the mass spectrum. High capacity hard disks are used for data storage for subsequent analysis. Data collected is used to construct the total ion chromatogram by plotting the total ion current (signal intensity) for each scan against time or scan number (Figure 7.2). Alternatively a single ion chromatogram can be constructed by plotting the signal obtained from monitoring the abundance of a characteristic ion (Figure 7.11c). Chromatography integration algorithms calculate the usual chromatography data such as retention time, peak area, peak height and width and quantitative results based on internal or external standards.

Data files are used to construct analytical reports for routine analyses which include the methods data and chromatographic results and where required confirmation of the presence or absence of components in the sample. Alternatively, the chromatogram can be manually evaluated by examining all or selected peaks by comparing the spectra at peak maximum with a database of reference spectra. A range of data bases are available from the Wiley-NBS library of 140 000 spectra to libraries of drugs or environmental data. Spectra corresponding to the sides of a peak can be examined to assess peak purity and background subtraction can be used to reduce spectral interference due for example to column bleed or contamination. Chromatograms and spectra may be printed out, compiled into an analytical report or incorporated into a LIMS data file. A detailed account of data processing and analysis is given in Chapter 8. Figures 7.11 and 7.12 include examples of data analysis.

7.7 Infrared spectrophotometry

Infrared spectrophotometry is a familiar established analytical technique which provides identification of compounds by fingerprint spectra, of which a vast library is available. Both liquid and gaseous samples may be easily analysed and therefore modifications of established sample handling techniques have enabled both GC and HPLC instruments to be readily interfaced. Ideally, scan times of less than 1 s are required to be able to record each peak and peak shoulders. Instrument sensitivity is sufficient so that on the fly recording of spectra can be obtained from GC and HPLC eluants which contain nanograms of sample per ml mobile phase, for example, 10 ng sample in 100 μl GC–IR sample cell. Fourier transform infrared (FTIR) instruments are able to meet these criteria but until recently the instrumentation and computer system have been too expensive for routine use. The new generation of

FTIR instruments are controlled by microcomputers and incorporate data handling facilities. A schematic diagram of a typical instrument is shown in Figure 7.13. Conventional spectrophotometers achieve the spectral dispersion using a diffraction grating, however, FTIR instruments operate on a completely different principle. In Fourier transform optics the source radiation passes into a Michaelson interferometer consisting of a beam splitter and moving mirror [14, 15] (Figure 7.13). The beam splitter passes approximately half the radiation to a moving mirror where it is reflected

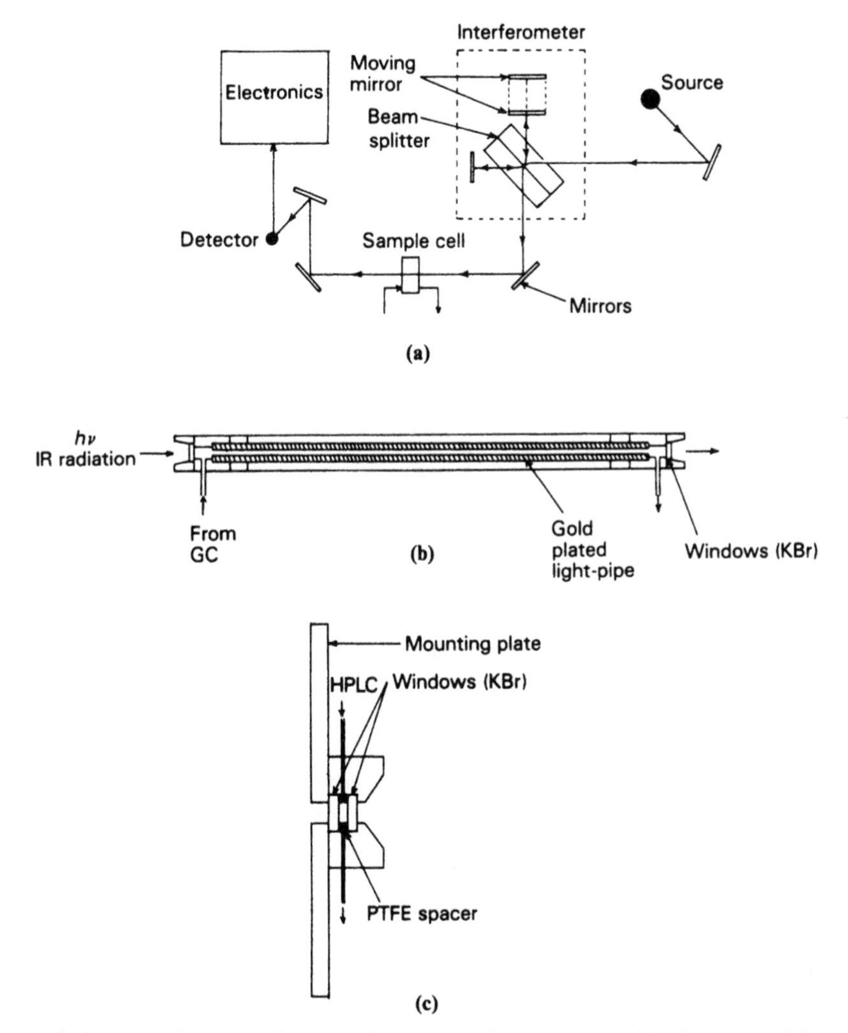

Figure 7.13 (a) Infrared interferometer Fourier transform instrument; (b) GC–infrared light-pipe sample cell; (c) HPLC–infrared microliquid sample cell.

back again. The rest of the radiation is reflected to a stationary mirror whereupon it too is reflected back. The relative path lengths of the two beams are varied by the moving mirror thus introducing a phase difference between the two beams. After recombination interference occurs between the beams and the resulting encoded radiation passes onto the sample cell and detector. The latter requires a rapid response and is usually a liquid nitrogen cooled mercury–cadmium telluride (HgCdTe) detector. The spectral range is covered by varying the difference in path length of the two beams which is achieved by translation of the moving mirror. The resulting detector signal is an interferogram consisting of the complex Fourier transform of the spectrum. The frequency domain spectrum is obtained by carrying out an inverse Fourier transform using a 'fast Fourier transform' computer program. The resolution and scan speed of FTIR instruments is sufficient for on-line interfacing to GC and HPLC instruments [16–18].

Interfacing consists mainly of transferring the GC or HPLC column eluant via a heated capillary line to a flow-through sample cell. The volume of the cell must be less than the mobile phase volume corresponding to the peak width at half height of the first eluting peaks. This ensures that resolution of the peaks is maintained and closely eluting peaks do not merge in the sample cell.

The sample cell system for GCIR consists of a micro-light-pipe cell, typically a 12 cm long silica tube with an i.d. of 1 mm and reflective gold plated inner surfaces giving a cell volume of approximately 100 μl. The ends of the cell are fitted with spring loaded potassium bromide or zinc selenide windows (Figure 7.13b). The GC transfer line and cell are heated to a preset temperature between 50 and 350°C.

HPLC sample flow-cells have internal volumes of less than 20 μl, with a typical path length of 1 mm and window area of 2–3 mm (Figure 7.13c). NaCl or KBr window materials may be used with many organic solvents and PTFE and polyethylene windows are available for both organic and aqueous based solvents. Narrow bore or microcolumns (2–3 mm i.d.) have lower flow volumes than normal columns, typically 0.3–0.5 ml min^{-1} and are therefore ideally suited for infrared detectors. Micro-HPLC-FTIR techniques may employ a direct flow cell or solvent elimination techniques [16–18]. Considerable care is required to match the chromatographic system with the sample cell to avoid loss of resolution.

7.7.1 Data collection and processing

The spectrophotometer monitors infrared absorbance in the sample cell by carrying out repetitive scans for the predetermined wavelength or frequency range at a scan rate of up to 20 scans s^{-1}. Consecutive scans can be added prior to the Fourier transformation. The fast Fourier transform process is a mathematical treatment of the recorded data that converts time domain data into frequency domain spectra [19, 20]. That is, the intensity values at specific moving mirror positions (located using a laser reference beam) are converted to intensity values at the corresponding frequencies. Infrared data is presented as real time Gram Schmidt or total response chromatograms and a display of the current infrared spectrum on the VDU screen. The Gram Schmidt technique is a mathematical process that uses the absorption information contained in the interferogram. The magnitude of the difference between absorbance values in similar parts of the fly interferogram and a pre-recorded reference interferogram corresponds to the amount of analyte in the sample cell. The Gram Schmidt plot is therefore used to construct the chromatogram in real time and for quantitative chromatographic analysis whilst the spectra are used for subsequent qualitative evaluation by comparison of the data with a library of reference spectra. Automatic baseline correction of spectra, subtraction of spectral interference such as column bleed and background correction can be carried out. Subtraction of spectra to obtain pure spectra can be performed prior to data evaluation and to view spectral differences. Stacked plots in three dimensions are used to display infrared spectra against time to show the spectral features of the peaks in the chromatogram (Figure 7.14). The chromatogram may also be plotted as a series of windows covering a specified spectral range, useful for distinguishing compound types and to assist in the identification of separated components. If a solvent is present in the sample, as in HPLC, the solvent spectrum (previously recorded) can be subtracted from the recorded data to produce the solute spectrum. Wide variations in sensitivity of GCIR systems can occur due to the variation in infrared molar absorptivity coefficients between substances, up to 50:1 between a small polar molecule and larger hydrocarbons. Databases such as the Aldrich, Sadtler and EPA libraries are available and provide an added dimension in component identification in the analysis of mixtures.

7.8 Ultraviolet–visible spectrophotometry

Chromatography–ultraviolet–visible spectrophotometry techniques are generally used with LC, the ultraviolet–visible detector being the most widely used HPLC detector. Scan times of conventional instruments which use the rotation of a diffraction grating to disperse the spectrum over a single detector element are too slow to record the spectra of rapidly eluting peaks in real time. Stopped flow techniques overcome some of the problems but adversely affect the chromatogram, thus most chromatograms are recorded at predetermined wavelengths. Developments in diode array detector technology have produced a range of instruments which can record a complete spectrum in as little as 0.01 s. It is therefore feasible to obtain multiple spectra of a single rapidly eluting peak. Multielement

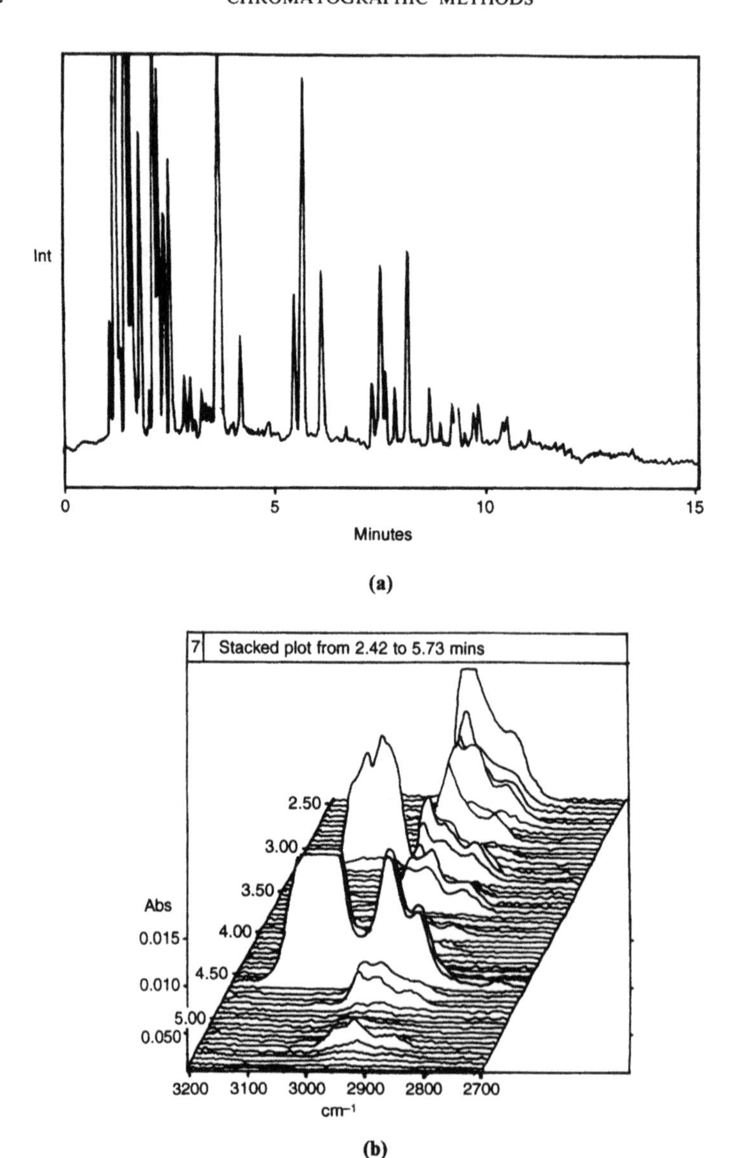

Figure 7.14 GC–FTIR analysis of petrol using a 25 m × 0.25 mm BP-1 capillary column: (a) Gram–Schmidt chromatogram; (b) stack plot. (Reproduced by permission of Perkin Elmer Ltd.)

array spectrophotometers use a detector consisting of 256, 512 or 1024 photodiode detector elements arranged in line on a semiconductor chip. The dispersed spectrum falls on to the array so that each diode records a small band within the required wavelength range thus recording all the spectrum at the same instant. Resolution of better than 1 nm is possible depending on the wavelength range scanned (Figure 7.15a). As with infrared the total intensity for a single spectral scan is used to construct a chromatogram and selected spectra are automatically or manually evaluated. Characteristic data such as λ_{max} and relative absorbance values may be referenced to a database of spectra to assist with identification of the components and used for quantitative calculations. Conventional HPLC flow-through sample cells are used and the extremely rapid response enables the cell contents to be scanned in less than 0.1 s. Sensitivities are similar to conventional HPLC systems. Chromatography is a separation technique and therefore it is often important to check for co-elution. Peak shape, asymmetry and the presence of shoulders are an unreliable indication of peak purity. At least three scans on each peak are obtained and the spectra compared (Figure 7.15b). This can indicate whether one or more components are present in the peak. Repetitive scans are stored in memory. Scan mode can be selected to automatically record spectra sequentially at predetermined time intervals, at peak maxima, and at a specified peak height (e.g. at 10%, 25% or 50% peak height), inflection points and the baseline. When these are combined with multiple peak recording a spectral map of a complete chromatogram can be obtained (Figure 7.15c). Further capabilities include signal to noise optimisation, parallel detection of up to eight wavelengths, creation of library data bases and adding specific processing requirements for report formatting, error checking and quality control/quality assurance procedures.

7.9 Atomic spectroscopy

Trace element analyses are often required for the determination of toxic metals such as chromium, mercury and lead in environmental samples and for monitoring the workplace environment. Conventional methods requiring extraction and separation procedures are time consuming. However, recent developments in GC and HPLC interfaced to atomic absorption and plasma emission spectrometers have enabled on-line analyses to be carried out. Ideally, the GC or HPLC column should be connected directly to the spectrometer sample cell or sample area to avoid dilution and loss of resolution. In practice a short heated transfer line of stainless steel or silica is used which has an internal diameter smaller than the column i.d.

GC may be interfaced to an atomic absorption spectrometer (AAS) via a heated stainless-steel tube (2 mm o.d.). The column effluent is introduced into the fuel mixture of the air–acetylene flame either via the nebuliser or just below the burner rail [21–24]. The nebuliser method presents a problem if the sample contains volatile analytes such as the alkyl lead

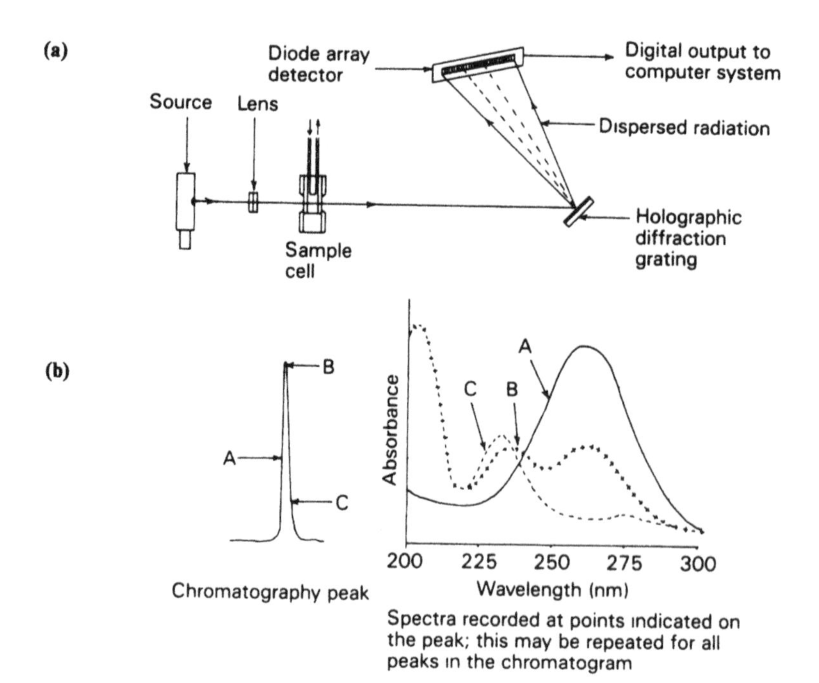

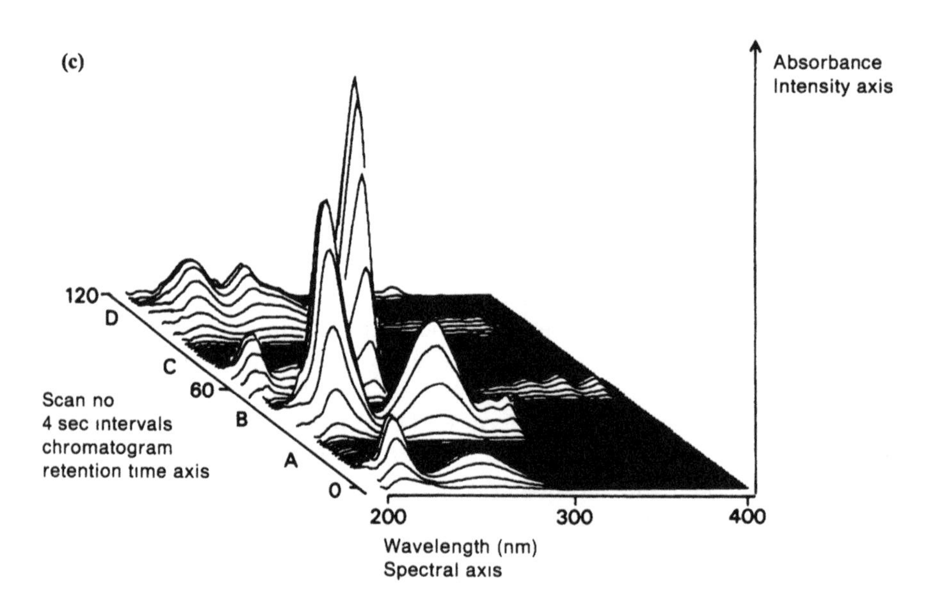

Figure 7.15 HPLC–UV diode array instrument and chromatograms. (a) HPLC diode array UV detector system; (b) UV spectra recorded at three points on an HPLC peak to enable peak purity to be determined; (c) Isometric map obtained by plotting successive spectra from an HPLC separation of polynuclear aromatics. A, naphthalene; B, fluorene; C, anthracene; D, chrysene.

and organomercury compounds and there is also considerable dilution of the sample in the mixing chamber. When the column eluant is introduced into the burner head the analytes are mixed with the fuel and oxidant gases just before the flame, reducing dilution effects to a minimum and retaining the chromatographic resolution. Eluant lead levels of 20 ng have been successively detected from the analysis of environmental samples containing tetra alkyl lead compounds. Chromium compounds have been analysed using a similar interfacing technique with a claimed detection limit of 1 ng chromium. The samples were digested with sulphuric acid/hydrogen peroxide and the solution obtained treated with trifluoroacetyl-acetone. The resulting chromium di-ketonate complex was extracted into hexane then separated on a packed column with 10% SE 30 stationary phase [25].

An alternative approach to interfacing is to use an electrically heated silica tube furnace (7 mm i.d., 6 cm long) with the column eluant passing through a side arm into the cell [26]. Capillary columns can also be connected to the silica cell using an aluminium or stainless steel transfer line [27, 28]. A graphite tube atomiser has been used instead of the silica tube furnace for detection of tetra alkyl lead. The transfer line for the capillary column consisted of a short length of tantalum or alumina tubing fitted into the side wall of the graphite tube.

The interfacing of the eluant from an HPLC column to a flame AAS has been achieved by matching the eluant flow-rate of approximately 1 ml min^{-1} to the intake rate of the nebuliser. Five tetra alkyl lead compounds were analysed on a reverse phase C_{18} column using acetonitrile/water mobile phase directly connected to the inlet capillary of the AAS nebuliser [29]. Each compound was detected at the 10 ng level.

Plasma emission spectroscopy has a distinct advantage over atomic absorption, since many elements may be monitored simultaneously. This permits on-line trace elemental analyses to be rapidly carried out on a variety of samples [30, 31]. GC interfacing is relatively straightforward. The column effluent is transferred by a heated line to the plasma source unit of the spectrometer at a carefully controlled flow rate to maintain a stable plasma. The plasma source may be one of three types, microwave induced plasma (MIP), direct coupled plasma (DCP) and inductively coupled plasma (ICP), all of which are commercially available. Figure 7.16 shows the versatility of MIP systems and the chromatogram of a test mixture with the corresponding 'elemental chromatograms' produced by parallel monitoring of preselected atomic lines, for example, carbon 193 nm, hydrogen 486 nm, oxygen 777 nm, sulphur 182 nm. A diode array detector element or charge coupled device can be substituted for the array of detectors to give a more compact instrument and the capability of monitoring a wider range of elements [32]. A conventional capillary GC instrument is used with helium as the carrier gas since excitation of the eluted compounds is by a microwave

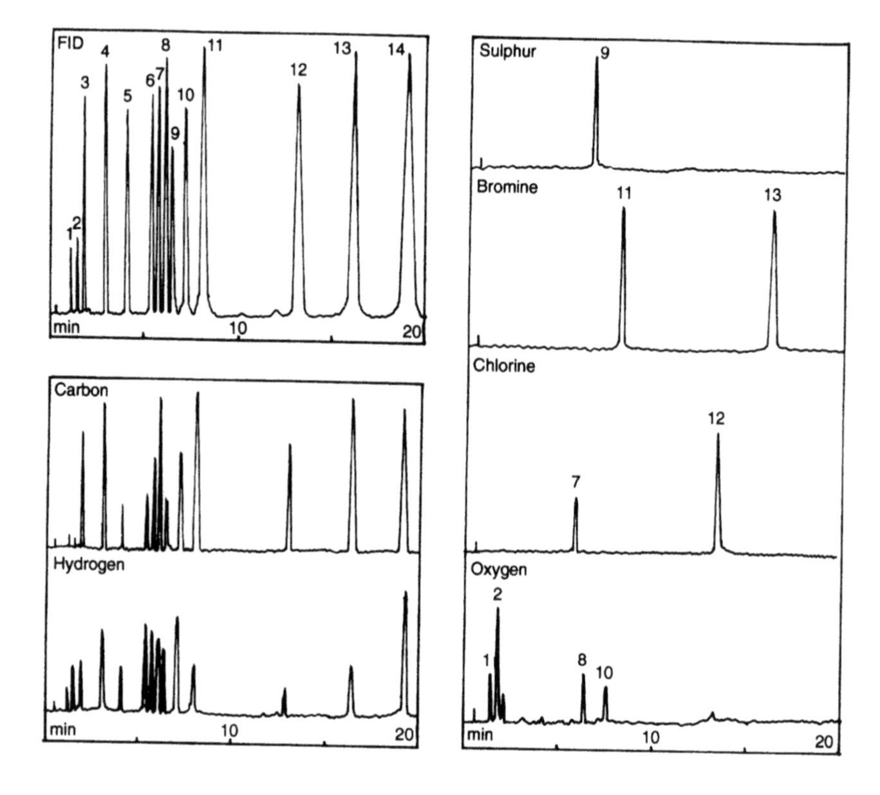

Figure 7.16 GC–MIP analysis of a test mixture showing FID and elemental chromatograms. 1, Deuteroacetone; 2, nitroethane; 3, fluorobenzene; 4, toluene; 5, *n*-butyliodide; 6, *n*-nonane; 7, chlorocyclohexane; 8, anisole; 9, diethyl disulphide; 10, octanone-2; 11, bromobenzene; 12, *o*-dichlorobenzene; 13, *o*-bromotoluene; 14, *n*-undecane.

excited helium plasma. The end of the capillary column is located just before the plasma chamber and helium make up gas is added to maintain the plasma (Figure 7.17).

HPLC is virtually incompatible with an MIP and only ICP has received much attention. The development of a suitable interface has proved a challenge with a nebuliser design finding most favour [33–35]. Initial problems with hydrocarbon or halocarbon solvents have now been overcome by modifications to the nebuliser design. Reverse phase solvents present fewer problems. Sensitivity and linearity of response vary from element to element but generally lie within the range 10 ng to 10 mg, the lower detection limits being comparable with flameless atomic absorption analysis. The lower flow-rates of microbore HPLC columns are more compatible with the nebuliser flow rate of an ICP allowing a simple T-junction interface to be used to analyse samples containing copper and zinc diketonates [36]. The potential and usefulness of the technique can be extended by using

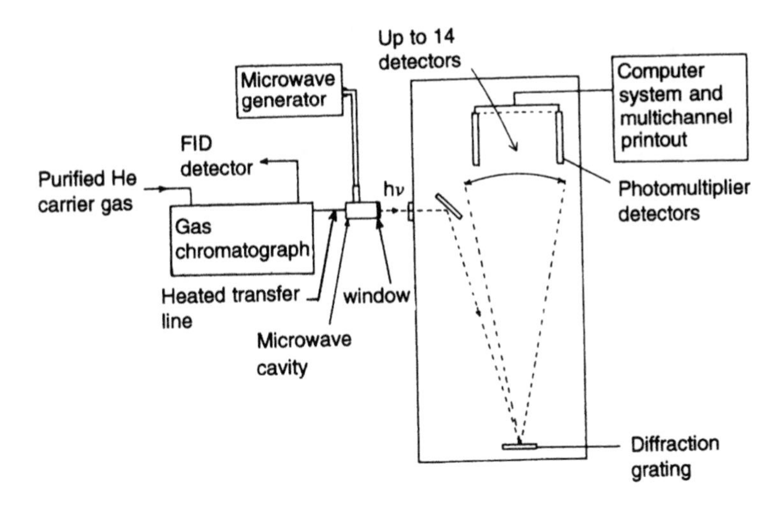

Figure 7.17 GC–microwave induced plasma system.

derivatisation reagents containing an organometallic or inorganic derivatising species [37].

References

[1] McFadden, W.H. *Techniques of Combined Gas Chromatography/Mass Spectrometry*. Wiley Interscience, London, 1973.
[2] Skoog, D.A. and Leary, J.J. *Principles of Instrumental Analysis*, p. 420. Saunders/Harcourt Brace Jovanovich, New York, 1992.
[3] Constantin, E. and Schnell, A. *Mass Spectrometry*. Ellis Horwood, London, 1990.
[4] Chapman, J.R. *Practical Organic Mass Spectrometry*. Wiley, Chichester, 1993.
[5] Prokal, L. *Field Desorption Mass Spectrometry*. Dekker, New York, 1989.
[6] Dawson, P.H. *Quadrupole Mass Spectrometry*. Elsevier, Amsterdam, 1976.
[7] Karasek, F.U. and Clement, R.E. *Basic Gas Chromatography–Mass Spectrometry, Principles and Techniques*. Elsevier, Amsterdam, 1988.
[8] Ardrey, R.E. *Liquid Chromatography–Mass Spectrometry*. VCH, Weinheim, 1994.
[9] Yergey, A.L., Edmonds, C.G., Laws, I.A.S. and Vestal, M.L. *Liquid Chromatography Mass Spectrometry*. Plenum Press, New York, 1990.
[10] Mellon, F.A. *Liquid Chromatography/Mass Spectrometry*, VG Monographs in Mass Spectrometry No. 2. Fisons–VG Instruments, Altringham, Cheshire, 1992.
[11] Blakely, C.R. and Vespal, M.L. *Anal. Chem.*, **55** (1983) 750.
[12] Arpino, P. *Mass Spectrom. Rev.*, **9** (1990) 631.
[13] Fisons, *Application Notes 209 and 211*, Fisons–VG Instruments, Altringham, Cheshire, WA14 5RZ, UK.
[14] Griffiths, P.R. *Chemical Infrared Fourier Transform Spectroscopy*. Wiley, New York, 1975.
[15] Vidrine, D.W. *Fourier Transform Spectroscopy*, Vol. 2, p. 129. Academic Press, New York, 1979.
[16] Brown, R.S., Hauster, D.W., Taylor, L.T. and Carter, R.C. *Anal. Chem.*, **53** (1981) 197.
[17] Kuehl, D.T. and Griffiths, P.R. *J. Chromatogr. Sci.*, **17** (1979) 471; Kuehl, D.T. and Griffiths, P.R. *Anal. Chem.*, **52** (1980) 1394.

[18] Griffiths, P.R. *et al. Anal. Chem.*, **58** (1986) 1349.
[19] Cooley, J.W. and Tukey, J.W. *Math. Comput.*, **19** (1966) 297.
[20] Cooper, J.W. *Transform Techniques in Chemistry* (Griffiths, P.R., ed.), Chapter 4, pp. 69–108. Heyden, 1978.
[21] Kolb, B., Kemmer, G., Schleser, F.H. and Wiedeking, E. *Z. Anal. Chem.*, **21** (1966) 166.
[22] Chau, Y.K., Radziuk, B., Thomassen, Y., Butler, L.R.P. and Van Loon, J.C. *Anal. Chim. Acta*, **108** (1979) 31.
[23] Coker, D.T. *Anal. Chem.*, **47** (1975) 386.
[24] Chau, Y.K., Wong, P.T. S. and Saitoh, J. *J. Chromatogr. Sci.*, **14** (1976) 162.
[25] Wolf, W.R. *Anal. Chem.*, **48** (1976) 1717.
[26] Chau, Y.K., Wong, P.T. S., Bengert, G.A. and Kramer, O. *Anal. Chem.*, **51** (1979) 186.
[27] Robinson, J.W., Kiesel, E.L. *et al. Anal. Chim. Acta*, **92** (1977) 321.
[28] Forthsyth, D.S. and Marshall, W.D. *Anal. Chem.*, **57** (1985) 1299.
[29] Messman, J.D. and Rains, T.C. *Anal. Chem.*, **53** (1981) 1632.
[30] Krull, I.S. and Jordan, S. *Int. Lab.*, **Nov.** (1980) 13.
[31] Barnes, R.M. *CRC Crit. Rev., Anal. Chem.*, **Sept.** (1978) 203.
[32] Quimby, B.D. and Sullivan, J.J. *Anal. Chem.*, **62** (1990) 1027.
[33] Gast, C.H., Kraak, J.C., Poppe, H. and Maessen, F.J. *J. Chromatogr.*, **185** (1979) 549.
[34] Ebdon, L., Hill, S. and Ward, R.W. *Analyst*, **112** (1987) 1.
[35] Heine, D.R., Denton, M.B. and Schlabach, T.D. *J. Chromatogr. Sci.*, **23** (1985) 454.
[36] Jinno, K., Tsuchida, H. *et al. Appl. Spectrosc.*, **37** (1983) 258.
[37] Blau, K. and Halket, J.M. *Handbook of Derivatives for Chromatography*. John Wiley, Chichester, 1993.

8 Processing chromatographic data

8.1 Introduction

Data processing in chromatography falls into three areas:

1. collecting and processing detector signals to produce chromatograms and the raw data such as retention time, peak area and peak width;
2. collating and analysing data to produce qualitative and quantitative information and to generate reports for transmission and for archiving; and
3. optimising chromatography parameters.

The current generation of chromatographic instruments are sophisticated microcomputer controlled systems. Microcomputer chips such as the Intel 80486 and Pentium and the Motorola 68000 series which are used in personal computers (PCs) have considerable computing power which can easily cope with the calculations and data processing required in chromatography. A parallel range of chips have been continuously developed as controllers (for example, the 8051 series) designed specifically to monitor the status of a system and control the various functions to maintain the pre-programmed instrument conditions, such as mobile phase flow-rate and column head pressure; temperature of the column, injector and detectors; solvent and temperature programming; control of autoinjectors and flow control valves; split ratio; detector attenuation and wavelength. Control of instrument functions is so precise that reproducible chromatograms can be obtained with a repeatability of better than $\pm 1\%$ in a well-designed instrument. Processing data from the chromatograms to derive retention times, peak areas, peak width, etc. adds a small data processing error to the overall reproducibility arising not from the numerical calculations but from the integration process and the fact that Gaussian peaks are not measured precisely.

The raw signal from a detector is in the form of a minute ion current for the gas chromatography or flame ionisation detector (GC–FID) or GC and electron capture detector (GC–ECD), or voltage for high performance liquid chromatography and ultraviolet detection (HPLC–UVD) which require high specification low noise analogue electronics to condition and amplify the signal to form the analogue output, typically within a range to 10 mV or 1 V for input to an integrator. If the output is to be connected to a computer then the analogue signal requires converting into a digital form. This can be carried out within the instrument by a dedicated signal processing and integration board or by an interface card plugged into the expansion bus of a PC. Many integration software programs are available for PCs which not only process and display the detector signal but also store methods, generate reports and archive the data.

PC software is also available to assist in method development. Optimum mobile phase composition for a selected stationary phase can be derived for HPLC analysis of specific analytes and the optimum stationary phase, column and temperature programme determined for GC analysis. A PC can also be connected to a laboratory and corporate network (Ethernet) so that electronic mail can be used to 'post' reports and receive sample information. Laboratory information management systems (LIMS) are used to manage the activities of a laboratory, for example, logging and tracking samples, scheduling of the analyses, collecting and collating analytical data, generating reports and ensuring the laboratory is managed to meet good laboratory practice (GLP) and QA/QC (quality assurance/quality control) standards. GLP and QA procedures are described in the rules and regulations of national organisations such as The Food and Drugs Administration (FDA) and Environmental Protection Agency (EPA) in the USA, the Health and Safety Executive in the UK and in the directives of the European Union (EU).

8.2 The chromatogram

Chromatograms are produced by plotting the continuously varying signal from the detector against time. The time (x) axis is representative of the retention time or retention volume. The y-axis plot is a direct representation of the detector signal after suitable electronic processing to produce an appropriate output voltage for use with chart recorders, integrators or computers (Figure 8.1). The detector signal is produced in response to the measurement of some property of the sample molecules. The magnitude of the signal at any given time is proportional to the concentration or amount of sample molecules present in the detector. In the process of

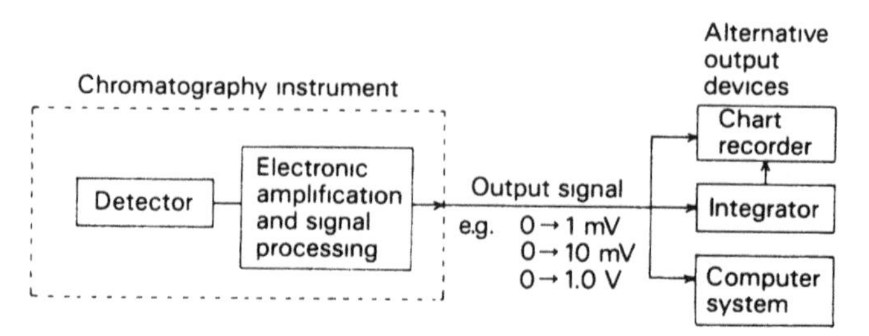

Figure 8.1 Signal processing in chromatography systems.

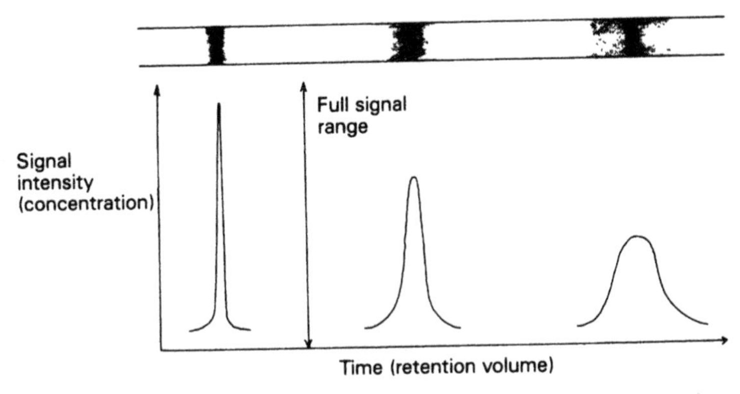

Figure 8.2 Detector response for various solute bands from chromatogram plotted on a chart recorder.

separating the components in a mixture, the chromatographic process causes band broadening. Therefore, the detector records the concentration profile of the band as a broad signal in the form of a Gaussian peak since the solute molecules will be more numerous at the centre of the band (Figure 8.2). The area under each peak is a measure of the amount of that component in the sample. Any recording device has to be able to accurately follow and record the detector signal and its response time, defined as the width of a peak at half height, $h_{0.5}$, needs to be faster to be able to respond to the rate of change of the detector output. The two techniques used for recording chromatograms

1. direct recording of a chromatogram using chart recorders; and
2. processing the detector output using computer systems, that is, integrators or personal computers with dedicated chromatography software

are discussed in the following sections.

8.2.1 Chart recorders

The essential analytical information required from a chromatogram is for analyte identification and to determine the amount present in the sample. The minimum data required is therefore retention time and peak area and an assessment of peak separation, the basic measurements needed for each peak are

- time from injecting the sample to peak maxima;
- peak height;
- peak width; and
- comparison of the baseline with the valley between peaks to assess resolution.

Chart recorders used in chromatography are y/t recorders where the y-axis is driven by the output of the instrument and therefore records detector signal intensity. The t or time axis is driven at constant speed by the chart paper drive motor, it is calibrated in seconds or minutes. The y-axis is calibrated to respond to signals over a preset range, which is set to match the output signal range of the instrument amplifier (Figure 8.1). The true detector signal is measured from the baseline, the detector background signal. The main disadvantages of a chart recorder are that electronic damping of the pen response is used to smooth the signal and reduce response to 'noise' and response times are slow. Pen response time measured for full-scale deflection is in the range 0.5–1.0 s. Peak widths measured at half height need to be slower than this otherwise the pen may not accurately follow the detector output lagging behind the true signal (Figure 8.3). Consequently, the peak height and peak maximum and hence peak area would be inaccurate. Chart recorders are most suitable for packed column GC and HPLC.

Retention time, time to peak maxima, peak height and peak width data are obtained directly from a chromatogram by taking the appropriate measurements. Resolution may be approximated by measuring the valley between peaks. Ideally baseline resolution between peaks is required for quantitative results, however, satisfactory resolution is obtained when peak overlap is less than ~2.3%, that is, when the valley is less than 10% of the mean peak height ($\bar{h}$); see section 2.6.1 and Figure 8.3.

$$h_V \leq 0.1\left(\frac{h_A + h_B}{2}\right)$$

Peak areas may be obtained by carefully cutting out the peaks and weighing each one, this produces reasonable results providing the chart

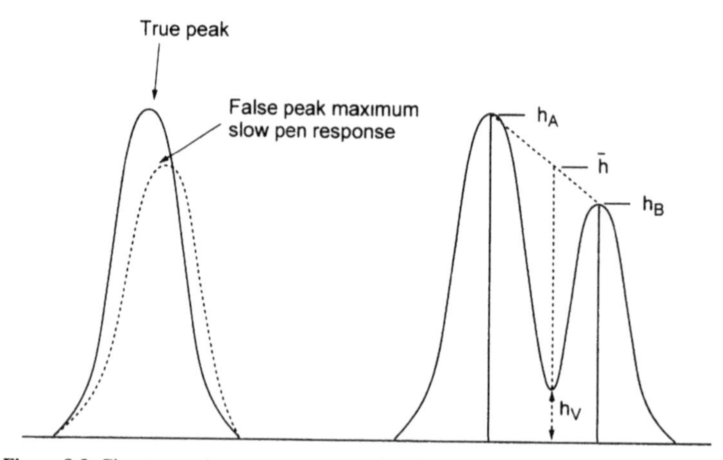

Figure 8.3 Chart recorder pen response and 10% valley measurement of resolution.

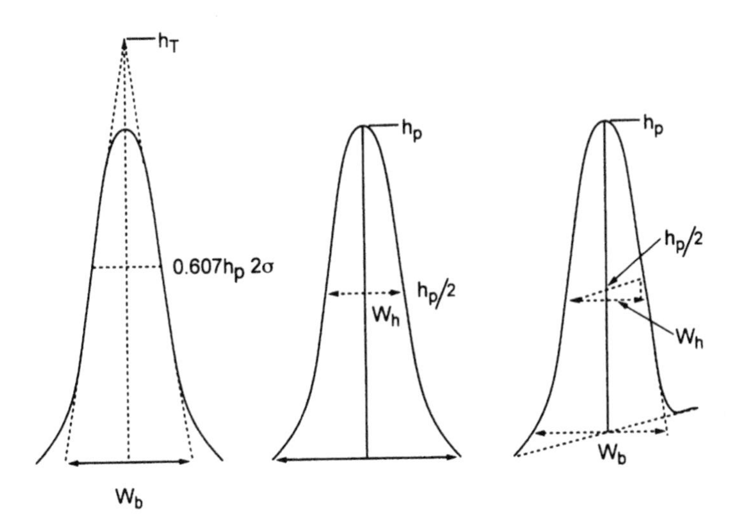

Figure 8.4 Calculation of peak areas by triangulation and direct peak measurement.

paper is of uniform thickness and moisture content. The accuracy may be improved by taking multiple photocopies of the chromatogram and determining the areas from each copy, and also preserving the original chromatogram. Relative precision can be better than ~10%. However, peak areas are usually obtained by measuring peak height and width. There are two methods:

1. by triangulation; and
2. by direct measurement of peak height and width.

Triangulation involves constructing an isosceles triangle by drawing the tangents to the inflexion points in the curve which occur at peak width of 2σ or $0.607h$ (Figure 8.4). The height of the triangle, h_T, and the base width, w_b, between the intersections with the baseline are measured and the area A_T were calculated:

$$A_T = \frac{1}{2} w_b \times h_T$$

The area obtained is a close approximation to the true peak area being 96.8% of the area under a Gaussian curve. A correction factor is therefore used to obtain the correct peak area.

$$A_{\text{correct}} = \frac{A_T}{0.968} = 1.032 \times A_T$$

The second method for obtaining peak area is to measure directly the peak height and base width of the peak and calculating the area assuming a triangle, that is, area $= \frac{1}{2}$base $\times$ height. However, Gaussian peaks 'tail' into the base line, base width is difficult to measure so peak width at half height, w_h, is used:

$$w_h = \tfrac{1}{2} w_b$$

$$A_P = w_h \times h$$

The area obtained is inaccurate being 93.9% of the true area of a Gaussian peak a correction factor is therefore required:

$$A_{correct} = \frac{A_P}{0.939} = 1.064 \times A_P$$

If baseline drift has occurred then the peak height is measured from the mid point on the line joining start and end points of the peak and peak width at half height by drawing the horizontal vector to the line drawn at $h_{0.5}$ parallel to the base line (Figure 8.4). Measurement of peak data assumes symmetrical peaks. Data from asymmetric peaks with an asymmetry ratio of greater than 1.2 will be less accurate.

8.2.2 Integrators

Early disk integrators were based on electromechanical counters, but developments in digital electronics in the late 1960s enabled integrators using voltage to frequency converters and electronic counters to be designed. No data storage facility was available so all signal processing and data print-out was immediate. All readers will be aware of the rapid developments in microprocessors and personal computers and supporting storage devices, floppy and hard disks, high resolution VDU displays and printers. Integrators are now all based on microcomputer systems, either as a dedicated chromatography integrator or as a PC running chromatography software (Figures 8.5 and 8.6). Microcomputers use digital signals for data transfer, data storage and communications so the analogue output from detectors has to be converted into the digital signals required by the computer. The analogue detector signal is converted into digital signals using an analogue to digital converter (ADC). The resulting digital data is stored and monitored so that an emerging peak is detected, the peak maximum located—to calculate retention time, and the data summed for the duration of the peak. Subsequently additional calculations such as, corrections for baseline drift and co-eluting peaks, peak width, area percentage and quantitative internal standard calculations can be carried out. A basic integrator print out is shown in Figure 8.5. Computing integrators have a wide linear dynamic range enabling them to track wide variations in peak height and width, such as solvent and solute peaks.

All the present generation of chromatographic instruments use microprocessor systems for control and therefore digital control and data signals (e.g. RS232 and IEEE488) are available for linking into computers and integrated laboratory networks (Figure 8.6). Thus, a LIMS or laboratory

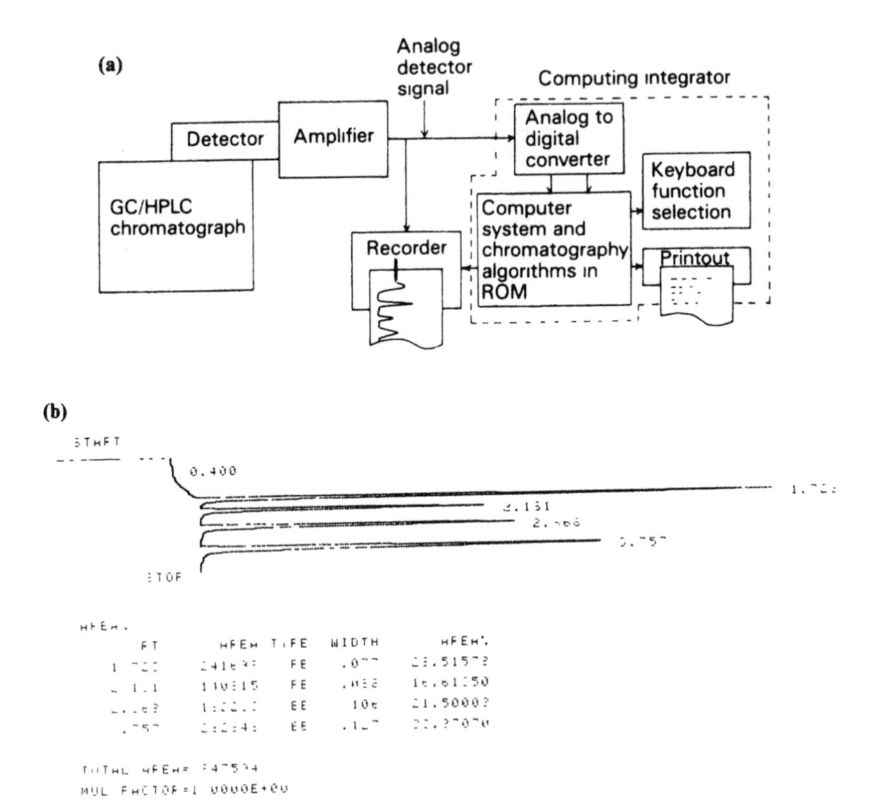

Figure 8.5 (a) Computing integrator, and (b) basic print out.

data station is able to communicate with the instruments and download sample and method information setting up the chromatograph for a particular application and analysis. When this is completed the data can be transferred from the instrument back to the computer or laboratory network for processing and reporting. It is beyond the scope of this book to discuss in detail computer systems; however, there are some excellent reviews and text books on the subject [1–3].

8.3 Data collection and processing

There are two main requirements for processing chromatographic data, accurate digitisation of the analogue detector signal and software to process the data. The software includes algorithms for detections of peaks, correction for base drift, calculation of peak areas and retention time, concentrations of components using stored detector response factors and production of the final analytical report.

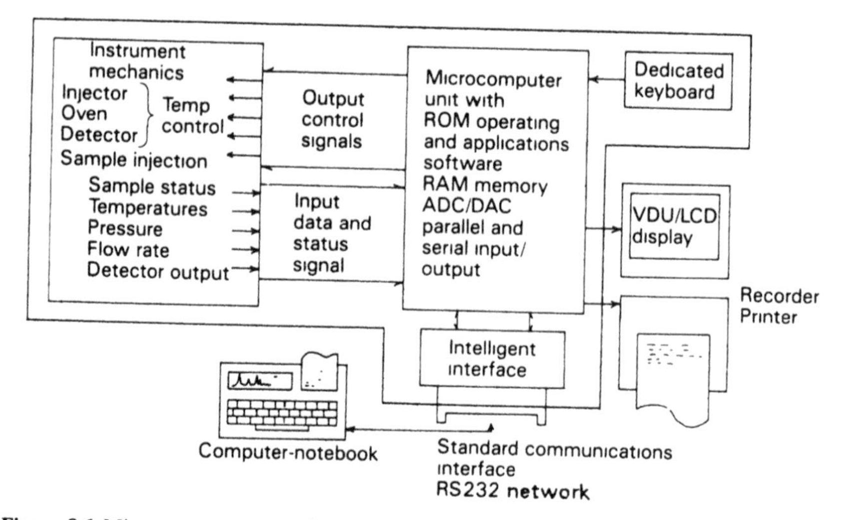

Figure 8.6 Microcomputer controlled instrument with control, ADC and communications interface.

8.3.1 Signal digitisation

The accuracy with which the analogue signal is processed is influenced by two main properties of the ADC data processing interface:

1. sampling rate; and
2. resolution of the ADC.

The analogue detector signal is converted into digital binary data required by computers via an interface unit which provides for signal buffering, amplification and attenuation, an ADC and control input/output (I/O) (Figure 8.7). It is important to use an ADC with adequate resolution since the resulting digital signals must accurately describe the analogue signal, including such details as shoulders and overlapping peaks which may be reflected in small differences in signal intensity or rapidly changing signals [4–9]. An ADC samples the analogue signal at a preselected rate, typically between 10 and 20 Hz. Recently, developed integrated ADC chips with on chip supporting electronics have enabled sampling frequencies of up to 200 Hz to be achieved. The main advantage of higher sampling rates is that more data is collected so that more effective smoothing algorithms such as the Savitsky Golay and bunching techniques may be used to smooth out signal noise [4]. The Savitsky Golay technique involves a moving window of data points (from 5 to 25 points) each window being updated from the previous one by adding a new point at the front and dropping the last point. The weighted average of each window is calculated to obtain the 'smoothed' data point.

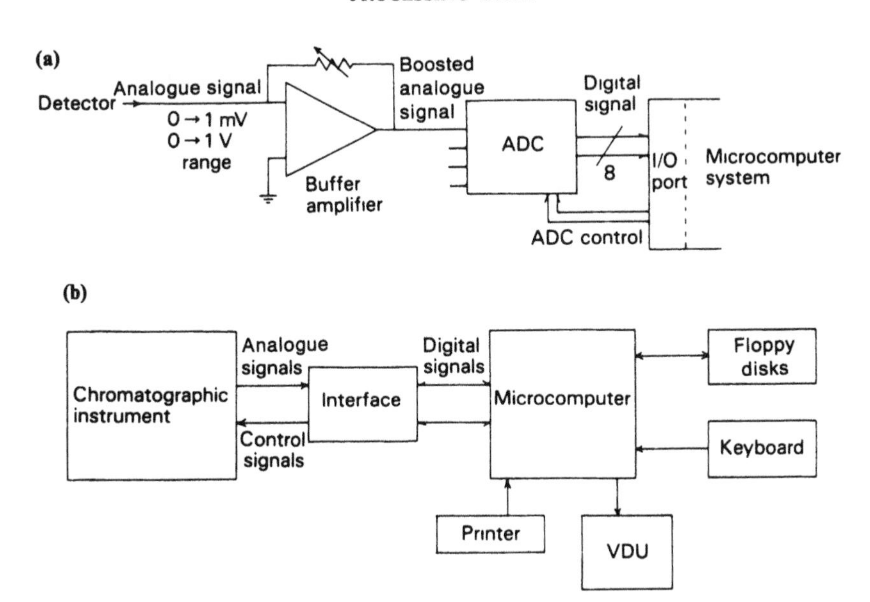

Figure 8.7 (a) Amplifier and ADC interface to produce a digital form of the detector signal. (b) Microcomputer (PC) interfaced to a chromatograph.

At least 10 points are generally required to describe a peak accurately and more if a shoulder is present, otherwise detail will be lost (Figure 8.8). Response time of a peak is defined by the time period corresponding to width at half height. For satisfactory signal processing a sampling rate of at least 10 readings per peak width at half height or at least 10 readings per second is acceptable for most chromatography.

Resolution of the detail in the detector signal is also determined by the number of binary bits (Binary digIT) used to describe the full signal range. Thus, an 8-bit ADC uses 256 (2^8) incremental steps to represent the range of an analogue signal. A 12-bit ADC uses 4096 (2^{12}) steps and therefore has a 16-fold increase in resolution (Figure 8.9). Chromatography data systems often use a 16-bit ADC with a resolution of 65 536 (2^{16}) steps. This is sufficient resolution to resolve very small baseline peaks and large analyte peaks with sufficient accuracy for quantitative calculations using one attenuation range. Autoattenuation procedures are also used, implemented in software or hardware using a multichannel ADC [10].

8.3.2 Data processing

The digital data for each ADC reading is stored in memory together with the corresponding elapsed time value, that is, time increments from the start of the run. If the sampling rate is 10 readings per second then the time increment between readings is 100 ms. At the end of the analysis the chromatogram is

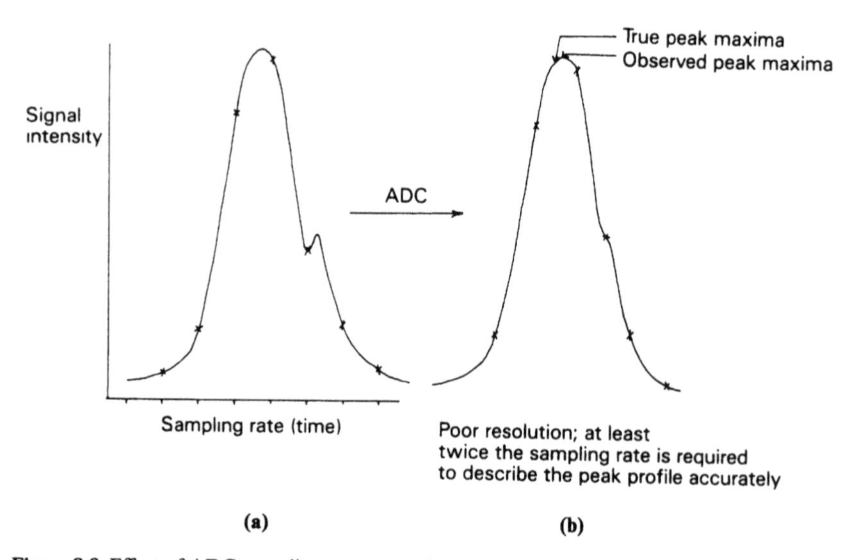

Figure 8.8 Effect of ADC sampling rate on peak representation: (a) analogue signal; (b) digital representation.

present in memory as a sequential file of ADC readings representing the y-axis signal intensity values. Since the ADC readings are taken at fixed time intervals then it is not usually necessary to hold a separate elapsed time file for retention data. This is obtained from the number of the ADC reading, N, multiplied by the time interval, Δt, between readings. Thus, when the ADC reading corresponding to a peak maxima is found the retention time,

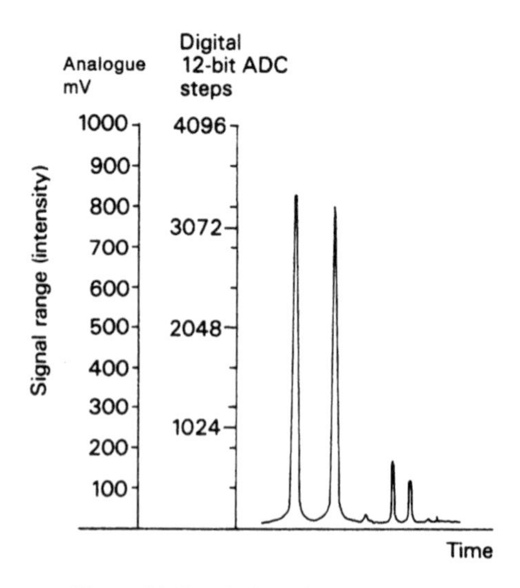

Figure 8.9 Resolution of a 12 bit ADC.

t_R, for that component is calculated by

$$t_R = N \times \Delta t \text{ (s)}$$

With a time increment of 100 ms or 0.1 s, a peak maxima occurring at the 1652 ADC reading would have a retention time of

$$t_R = 1652 \times 0.1 \text{ s} = 165.2 \text{ s, or } \approx 2.75 \text{ min}$$

The advantage of this method is that an accurate reconstruction of the whole chromatogram can be produced and subsequent inspection carried out. When all the required checks have been performed and data processing completed, the chromatographic data can be dumped to memory or a hard disk for storage and a print-out of the chromatogram and basic data such as retention time, peak area, peak width and area% produced (Figure 8.5). A simplified report may also be produced which contains essential information on the sample and results of the analysis but no chromatogram. At the completion of a run the computer will store the data and if an autosampler is being used initiate a new run. Computer operating systems and software allow multitasking of jobs so that whilst data is being collected from the new run data from the previous run is being processed and the reports generated. Alternatively, a single report can be generated at the end of all the runs in the sequence.

8.3.3 Computer hardware

A typical personal or home computer contains all the features required of a chromatography data system including high resolution display, memory and hard disks for storing and archiving data, and I/O capabilities for connection to printers, and data processing interfaces. Dedicated chromatography integrators such as the Hewlett Packard 3396A are complete microcomputer systems dedicated to signal processing and reporting. They include similar features including a microcomputer, memory, chromatography software, ADC interface and I/O lines on a single board and have a built in printer/plotter (Figures 8.5 and 8.6).

PC-based systems have two distinct advantages over dedicated integrators:

1. they have an open architecture, that is, add-on cards can be plugged into the expansion slots on the motherboard with direct access to the microcomputer system bus; and
2. they can run a wide range of standard software as well as chromatography programs.

Figure 8.10 shows the basic components of a PC system. The main component is the motherboard which contains the CPU (central processing unit) and all the support chips and software to start-up the system. These include control of communications and data transfer between all the devices

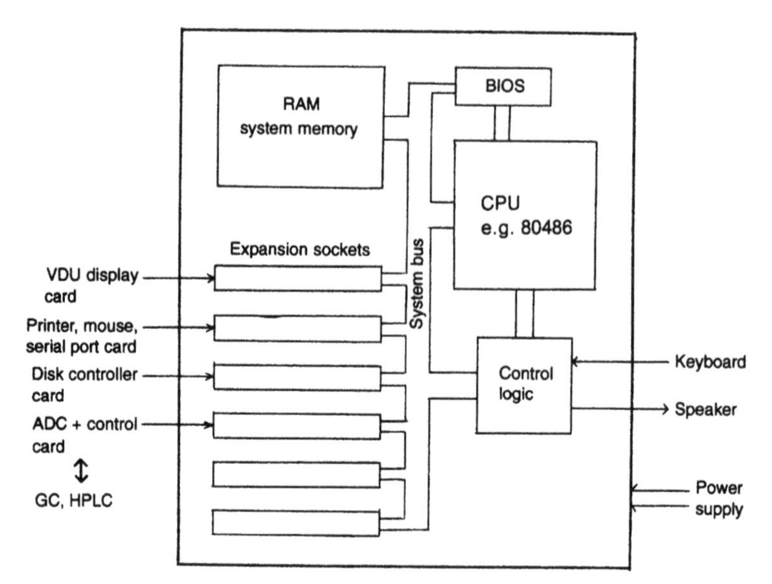

Figure 8.10 Microcomputer system.

and boards connected to the system bus. A CPU such as an Intel 80486 is the heart of the computer and processes the digital information according to the steps specified in the software program. System housekeeping, that is, control of the systems bus, timing of signals, control of data transfer between memory, the CPU and the I/O expansion slots is masterminded by a system controller chip. When a PC is first switched on the BIOS (basic input output system) software, resident in a ROM (read only memory) chip takes over to start up the system. The BIOS checks the system set-up, CPU and memory chips and which peripherals are connected such as a VDU, the mouse, hard and floppy disks and a printer. RAM (random access memory) is used to store software, program data and data transferred directly from I/O cards plugged into the expansion slots.

A basic PC system will have a video card with on board RAM for the video image, and I/O card to connect to printers, a mouse, disks (hard disks, floppy disks and CD ROMs) and serial (RS232) communications. Connection to a network such as Ethernet and to the telephone system and Internet require additional network and modem cards. ADC/control cards are commercially available from a number of manufacturers who also supply the software, including chromatography software, to operate the cards for data collection and control, and to display and printout chromatograms and results. Typical PC systems have at least 8 Mbytes of RAM, a high-resolution display such as SVGA (super-video graphics array) with an 800 × 600 pixel screen and a 340 Mbyte hard disk running under a graphical user interface such as Microsoft Windows. Such a system is capable of running the most sophisticated chromatography software and simultaneously acting as a LIMS satellite work station. LIMS are discussed later in Section 8.6. The PC can also be used to run spreadsheet, database and word processor software to analyse and archive results, generate reports and for electronic mail. Results files, tables of data, graphs, chromatograms and illustrations can be transferred between packages for producing brief analytical reports or longer project reports.

8.4 Chromatography software

Chromatography software includes the basic routines for collecting and processing detector signal data, calculating the basic results and printing a chromatogram. Additional routines can sometimes be added as BASIC-like programs for specific applications and for producing the final analytical report [11, 12]. A brief description of the main chromatography data processing routines is given below.

8.4.1 Calculation of peak area

The fundamental calculation performed on chromatographic data is the determination of the area under a peak. Although peak shapes and the base line may vary the procedure for calculation of the area is similar in all cases once the start and end of a peak has been determined. The area between these limits is integrated using a summation algorithm which can in theory be based on Simpson's Rule, simple summation of ADC readings or trapezoidal integration (Figure 8.11).

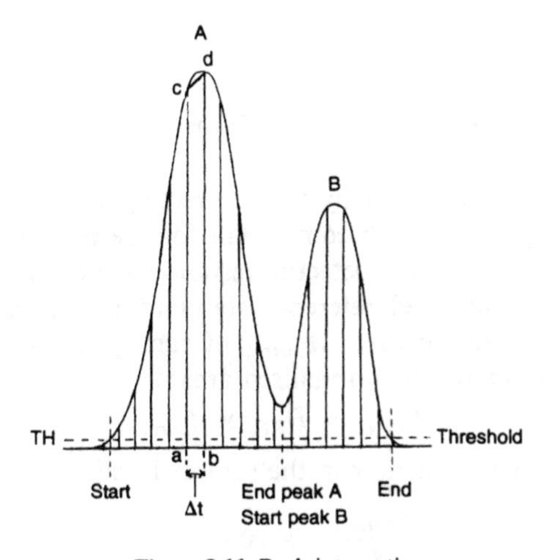

Figure 8.11 Peak integration.

Simpson's rule requires data points from the start and end of a peak so a post-peak calculation is used to find the area a distinct advantage if there is a measure of peak overlap. Trapezoidal integration and variations on this method developed by manufacturers is the preferred signal summation technique. Once the peak integration routine has been triggered by the rising signal a running count of the peak area is obtained which corresponds to the time slices for the peak or co-eluting peaks. Various peak area approximations are discussed in Section 8.4.2. The peak is divided into slices each of which is a trapezium, *abcd* (Figure 8.11), with the sloping top section, *cd*, approximating to the curve of the peak. Clearly, the more slices a peak is divided into the more closely the trapezium profile will follow the true curve and the more accurate the calculated peak area will be. The width of each slice corresponds to the time interval between the smoothed data points, thus a sampling rate giving at least 10 data points per second is required for accurate area calculations particularly for the first eluting peaks in capillary column gas chromatography. At least 10 data points should be obtained in the time corresponding to the peak width at half height, the peak time constant, for an accurate representation of the peak. The time interval is constant, therefore the total area of the integrals can be determined by summing the ADC readings corresponding to each integral between the peak limits. Since the voltage range of the detector signal is known, for example, 0–1.0 V the peak areas can be calculated in μV.

8.4.2 *Area percentage and area calculations*

Area% is frequently included in the basic set of information printed out with a chromatogram. It is obtained from the raw peak area data by calculating the total of all peak areas, the percentages contribution of each peak is then found.

$$\text{Area\% for component } A, \quad \text{A\%}_{\text{A}} = \frac{A_{\text{A}}}{A_{\text{total}}} \times 100$$

$$\text{where } A_{\text{A}} = \text{Area of } A \quad \text{and} \quad A_{\text{total}} = \sum_{i}^{n} A_i \quad \text{for } n \text{ peaks}$$

Normalised area% is obtained by first correcting the peak area from the chromatogram, A_{chrom}, of each component for the relative response of the detector to that component. Detector response factor (D_{RF}) is calculated with reference to a specified reference standard or internal standard; see Section 2.7. The corrected area, A_{correct} of each peak is then used in the equation above to obtain the normalised area%

$$A_{\text{correct}} = D_{\text{RF}_x} \times A_{\text{chrom}}$$

% of component x in the sample is calculated as

$$\text{area\%}_x = \frac{A_{\text{correct}_x}}{\sum A_{\text{correct}}} \times 100$$

Internal standard method (see Section 2.7.2) is a variation on the above, and is recommended for accurate quantitation especially in GC. It eliminates the need for accurate injections since a reference standard is included in each sample analysed. The procedure involves analysing a test sample containing known amounts of each component plus a predetermined amount of the internal standard. Since peak area is proportional to the amount of an eluted component and the detector response factor, the relative response of a component (D_{RF_x}) to the internal standard is, therefore,

$$D_{RF_x} = \frac{A_x}{A_{IS}} \times \frac{C_{IS}}{C_x}$$

where C is the amount of component x or internal standard IS.

Response factors for all components are thus calculated. A known amount of the internal standard is added to the unknown sample which is then analysed. The concentration of each component is calculated using the equation above rearranged to give

$$C_x = \frac{A_x}{A_{IS}} \times \frac{C_{IS}}{D_{RF_x}}$$

External standards are used where sufficiently accurate injection procedures are employed such as with autosamplers and in HPLC. A replicate series of one or two standard mixtures are injected and the area/unit amount of analyte calculated.

$$A_{STANDARD} \equiv x \, \text{mg litre}^{-1}$$

The unknown mixture is then analysed and the amount of the components in the sample calculated using the peak area data for the standard mixture. Therefore, if the recorded peak area for the component in a sample mixture is A_{MIX} the amount of component x is

$$\text{Amount}_x = \frac{x \times A_{MIX}}{A_{STANDARD}} \text{ mg litre}^{-1}$$

8.4.3 *Peak detection*

Peak detection algorithms use a predetermined signal threshold level above the baseline, once the threshold is exceeded peak monitoring occurs. Two methods may be used to identify the start and end of a peak (Figure 8.12).

1. Peak detection based on slope techniques monitors the smoothed ADC data points to follow the changing signal. A positive slope is detected if the difference between successive points is positive and equal to or greater than the threshold value, the start of the peak is flagged and integration begins. The end of integration is determined in a similar manner, that

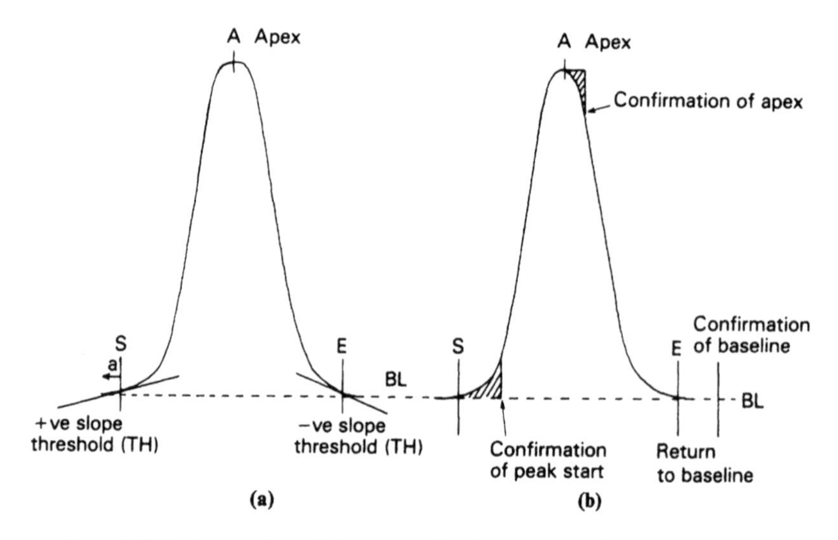

Figure 8.12 Peak detection: (a) by monitoring the slope, and (b) by area sensitivity.

is, following the negative slope of the peak tail and identifying when the difference between successive points is equal to or less than the threshold level or when a peak time threshold has been exceeded for badly tailing peaks. During integration the difference between successive data points is flagged as a positive or negative slope and is used to locate the apex of the peak at zero slope. If a rapid sampling rate is used the peak apex may be found directly as indicated otherwise the rate of change of the signal may be used to locate zero slope point.

2. A peak may also be detected by first defining an area sensitivity level. This area value is then used to identify the start of a peak by summing up the area as soon as the signal begins to increase. When the area sensitivity is exceeded integration begins and is terminated when the signal falls below the threshold level. When the signal passes through the maximum value and starts to decrease a second area sensitivity test is carried out to confirm the apex of the peak. Confirmation is flagged when the difference in area at signal maximum and the following data points exceeds the area sensitivity.

Peak height is initially calculated by the integrator directly from largest data point. The true peak apex and therefore peak height is then obtained from the curve described by the group of five or seven points of largest value. Correction for baseline drift then gives the reported peak height.
Retention time is calculated from the elapsed time from the start of the run to the time increment corresponding to the peak height. Calculations of peak height and retention time are carried out in real time so that the retention time can be printed soon after the peak maximum has occurred.

8.4.4 Baseline correction and overlapping peaks

During many chromatographic separations the baseline drifts in a linear or non linear manner in either a positive or negative direction. Correcting algorithms are used so that peak areas can be accurately calculated. If the baseline has moved upwards more than the threshold during elution of a peak then peak termination occurs when the current baseline value is equalled. A more common occurrence is when the baseline moves down and has a lower value than the original baseline. Integration is therefore continued until the slope or integral area is equal to or less than the threshold (Figure 8.13). When peak termination occurs the triangle *abc* is added to the peak area. The area of individual peaks in a set of merged peaks is most commonly calculated by detection of the valleys between peaks (zero slope between a negative and then positive slope) and dropping a perpendicular down to the baseline. A linear baseline is assumed between the start and end of the set (Figure 8.14). This free-running algorithm can also be applied to a peak on the tail of a larger peak, for example a solvent peak. Alternatively the area of a 'tail peak' can be calculated by a tangent skimming routine (Figure 8.15). The data points describing the slope profile of the larger peak are followed and peak integration begins when the valley is detected at a, providing the first baseline point, BL_1. Integration ends when the signal threshold, TH, is reached at *b*, to form the new baseline, BL_2. The area of the triangle *abc* is added to the integrated peak area *acd*. An alternative is to predict the change in baseline during the elution of a peak. This is only possible if sufficient pre-peak baseline data is available to calculate the rate of change in baseline signal and extrapolate it to forward-project the baseline (Figure 8.16).

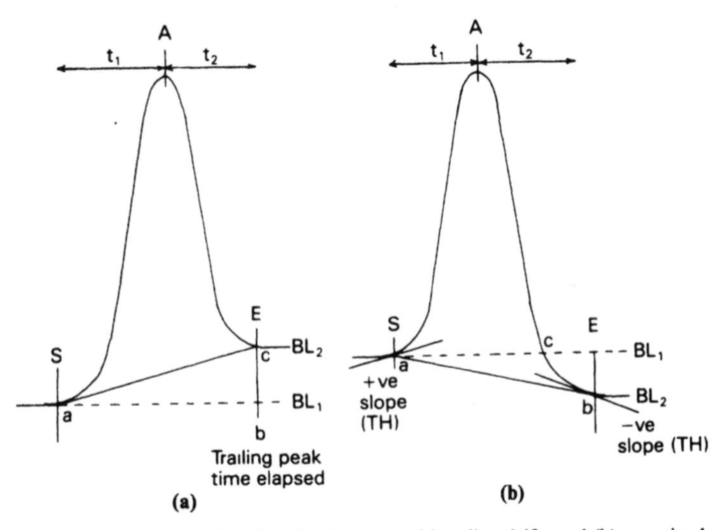

Figure 8.13 Detection of 'end of peak' after (a) upward baseline drift, and (b) negative baseline drift.

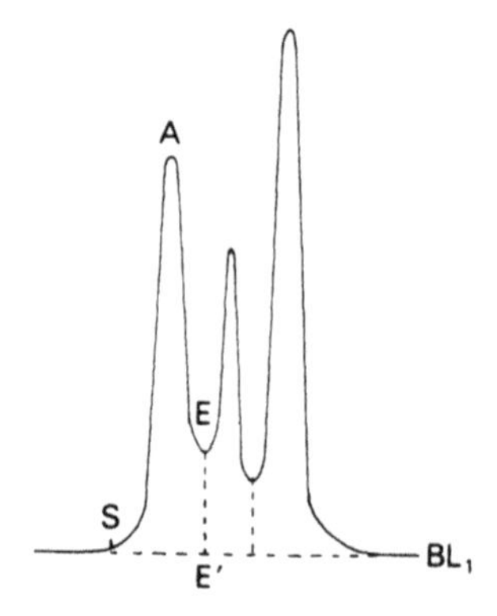

Figure 8.14 Analysis of merged peaks.

The areas of individual peaks in a cluster of peaks on a non-linear baseline is obtained using a valley to valley algorithm, integration starting and ending at the valleys between peaks. A new baseline value is established at each valley and a corrected peak area calculated (Figure 8.17). A routine to detect negative-going peaks is also used to avoid confusion with an altering baseline. Alternatively, response to all negative-going peaks can be inhibited.

Most computing integrators allow selection of the sampling rate, slope sensitivity and signal smoothing or filtering factors. All chromatographic peaks should give a positive slope value visibly greater than the baseline variation. Facilities to set the levels for rejection of noise spike peaks with a width at half height below a selected value and termination of tailing peaks after a selected time are generally included in the options menu. Reference files of chromatographic data can be accumulated either for

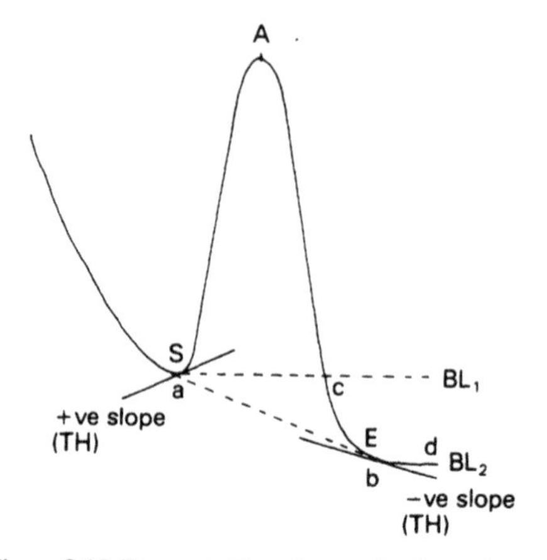

Figure 8.15 Tangent skimming to obtain peak areas.

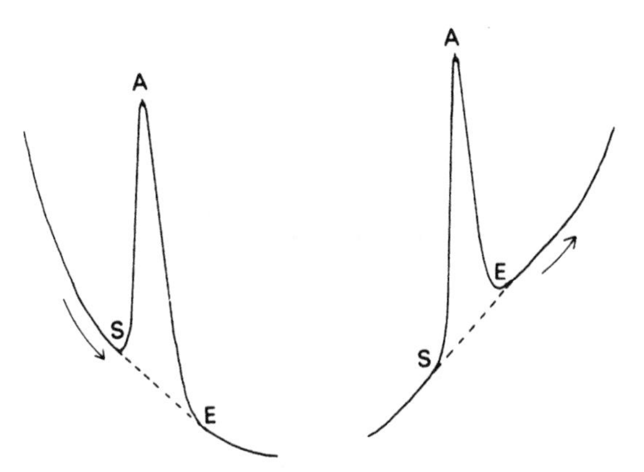

Figure 8.16 Tracking of the baseline for a side peak by predicting baseline slope.

specific analysis or group of applications. Data from a run may then be compared with the files for checking and identification purposes. A typical print-out is shown in Figure 8.18. Alternatively, the data can be exported to a LIMS, word processor, spreadsheet or report generating software.

8.5 Method development and optimisation

Personal computer systems are capable of running software to assist in selecting chromatography parameters such as column type, stationary

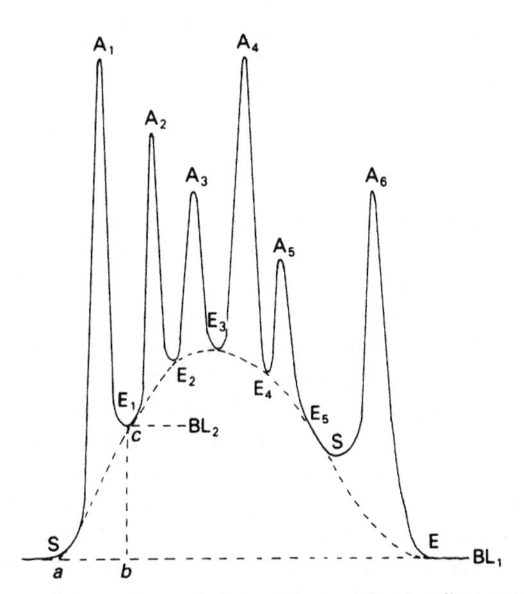

Figure 8.17 Peak cluster analysis by valley to valley baseline correction.

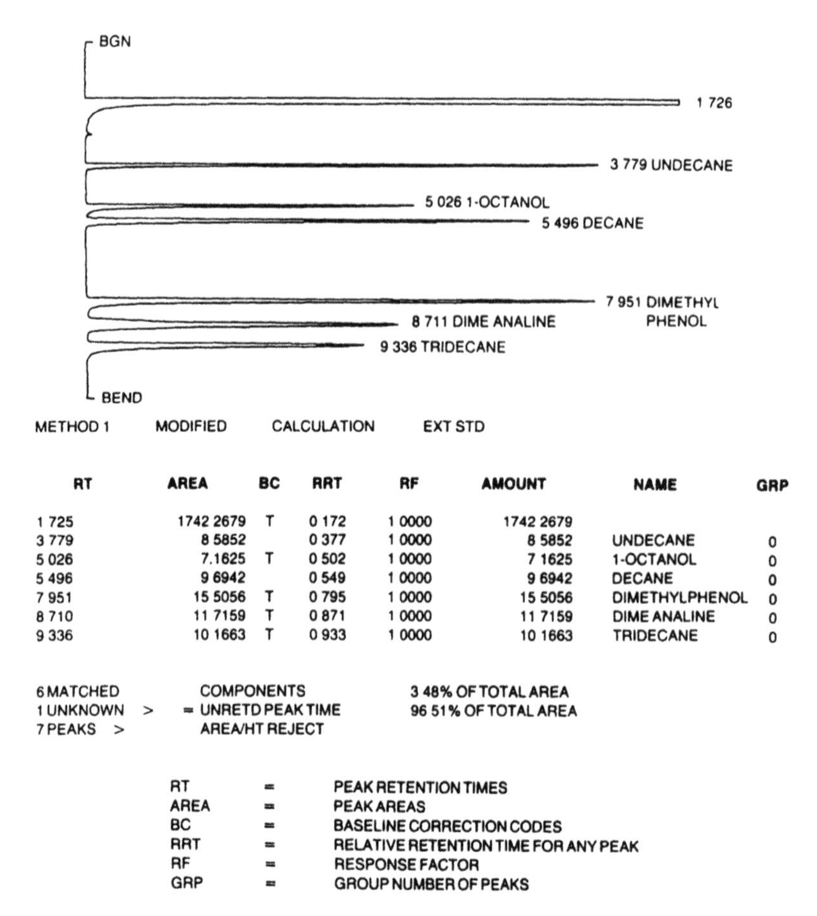

METHOD 1 MODIFIED CALCULATION EXT STD

RT	AREA	BC	RRT	RF	AMOUNT	NAME	GRP
1 725	1742 2679	T	0 172	1 0000	1742 2679		
3 779	8 5852		0 377	1 0000	8 5852	UNDECANE	0
5 026	7.1625	T	0 502	1 0000	7 1625	1-OCTANOL	0
5 496	9 6942		0 549	1 0000	9 6942	DECANE	0
7 951	15 5056	T	0 795	1 0000	15 5056	DIMETHYLPHENOL	0
8 710	11 7159	T	0 871	1 0000	11 7159	DIME ANALINE	0
9 336	10 1663	T	0 933	1 0000	10 1663	TRIDECANE	0

6 MATCHED COMPONENTS 3 48% OF TOTAL AREA
1 UNKNOWN > = UNRETD PEAK TIME 96 51% OF TOTAL AREA
7 PEAKS > AREA/HT REJECT

RT = PEAK RETENTION TIMES
AREA = PEAK AREAS
BC = BASELINE CORRECTION CODES
RRT = RELATIVE RETENTION TIME FOR ANY PEAK
RF = RESPONSE FACTOR
GRP = GROUP NUMBER OF PEAKS

Figure 8.18 Report and chromatogram from an external standard analysis. (Reproduced by permission of Perkin Elmer.)

phase and temperature programme in GC and solvent system, solvent programme and stationary phase in HPLC. Although the algorithms may involve complex iterative calculations and large files of reference data PCs with 80486 CPUs (see Section 8.3.3) are able to complete the calculations rapidly. What if scenarios can be tested and theoretical separations examined. The PC or data system can be programmed to analyse features in a chromatogram to determine whether the separation of the individual components is satisfactory. If the computer is also used to control the instrumentation variables such as mobile phase composition and flow-rate and column temperature can be reset according to criteria derived from the evaluated data and the analysis repeated until a satisfactory separation is achieved. Parameters such as total analysis time, resolution and separation factors derived from retention times and peak width are used. A fully automated system including auto injector is required as outlined in Figure 8.6. Optimisation procedures simplify the rather lengthy process of developing a new method, particularly in HPLC, and replace the rather empirical approach based on experience, guess-work and some theory [13, 14].

8.5.1 High performance liquid chromatography optimisation

HPLC optimisation has received most attention, optimisation being achieved by adjusting the mobile phase composition of a binary or ternary mixture until the required selectivity, analysis time or other criteria are met [15, 16]. There are a number of optimisation methods available which are based on linear techniques employing theoretical and semiempirical models or on a sequential or statistical search technique [17]. The most popular method uses the simplex optimisation algorithm, an efficient multidimensional sequential search technique which has been applied to a variety of analytical problems and has been adapted for HPLC [15, 16, 18, 19]. A flow diagram of a typical optimisation procedure is shown in Figure 8.19. In order to use any optimisation routine it is necessary to obtain a quantitative assessment of the separations achieved in a chromatogram. The chromatographic work function (CRF) provides a numerical description of the quality of a separation and may be used as the response input into the optimisation route [15, 16, 20]:

$$\mathrm{CRF} = \sum_{i}^{n} R + f_1(N) + f_2(t_A - t_n) + f_3(t_1 - t_0)$$

where R is the resolution between adjacent peaks, N is the number of peaks, t_A is the specified analysis time, t_1 is the retention time of the first eluted peak, t_n is the retention time of the last eluted peak, t_0 is the specified minimum elution time, f_1, f_2, f_3 are weighting factors with a value of between 0 and 3 which is selected by the operator.

A number of sequential analyses are carried out automatically with the mobile phase composition being adjusted between analyses as determined by the optimisation procedure. Typically five to 10 analyses are required for optimisation.

An alternative approach is to fit the HPLC data into well-established mathematical algorithms describing the behaviour of binary, ternary and quaternary gradient solvent systems [21]. Details of the HPLC system proposed are entered into the program. Elutropic strengths of solvent mixtures are calculated for the solvent system proposed and a solvent selectivity triangle constructed. Resolution, retention and separation factors are calculated and resolution maps and the theoretical chromatogram constructed. The analyst can then examine and optimise the chromatography.

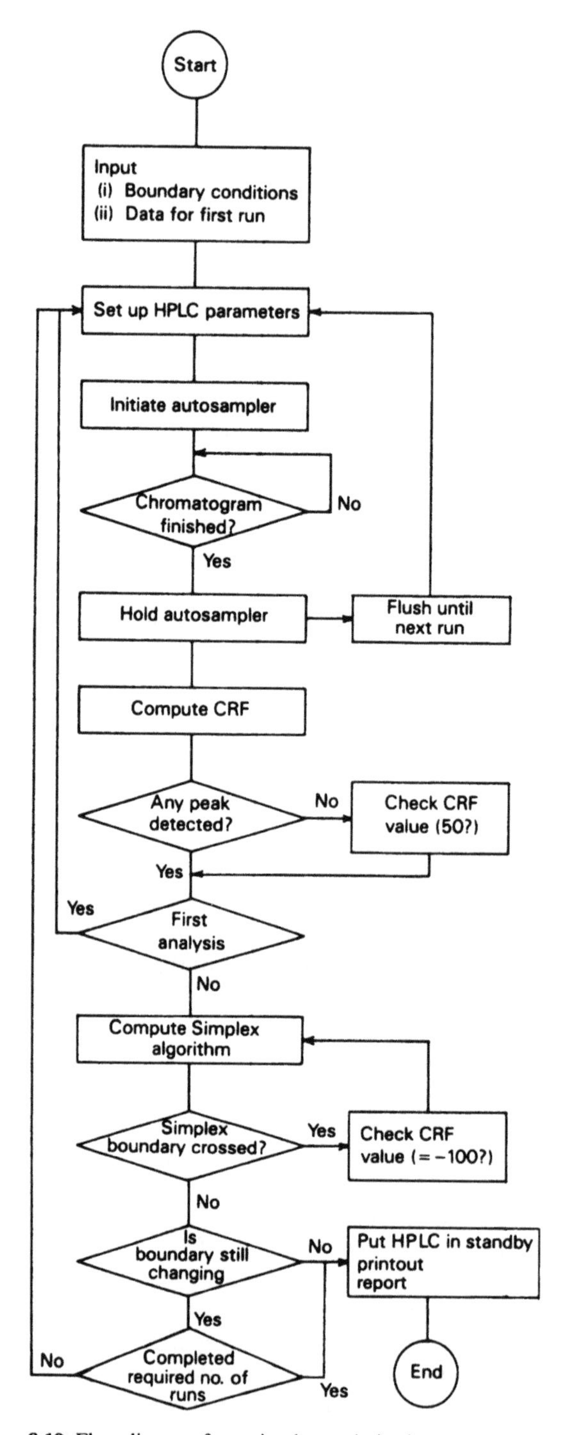

Figure 8.19 Flow diagram for a simplex optimisation computer program.

8.5.2 Gas chromatography method development

GC method development involves optimisation of temperature programme rates, carrier gas type (He, H_2, N_2, Ar) and flow rate and column type, that is, length, internal diameter, type of stationary phase and film thickness of a wall coated open tubular column. Separations can be modelled from commercial retention index (RI) libraries or the RIs can be calculated and a custom RI library created [22]. RI libraries are available for the most common analytes found in various categories of samples, for example, food and flavour volatiles, drugs and pharmaceuticals, environmental including volatile organic compounds as specified in the EPA 500, 600 and 8000 series methods, and solvents and chemicals in the C_1–C_{12} range. A typical procedure involves analysing the sample, for example, a pesticide mixture, on two columns having different stationary phases using a slow ($4°C\,min^{-1}$) and faster ($12°C\,min^{-1}$) temperature programme ramp on each column. The retention data for all four runs is then entered into the computer program to obtain the optimum operating parameters such as temperature ramp details and shortest analysis time. 'What if' chromatograms can be obtained by selecting different values for column length, film thickness and internal diameter. Figure 8.20 shows three chromatograms analysed under conditions obtained by GC optimisation, Chromatogram (a), using a 14% cyanopropylmethylsiloxane mid polar column (Rtx 1701), has the shortest analysis time but coelution of peaks 15 and 16 occurs, (b) and (c) show the optimum separations using a relatively non-polar 95% dimethyl 5% phenylsiloxane stationary phase (Rtx-5) and mid-polar 50% dimethyl 50% phenylsiloxane (Rtx-50) column.

8.5.3 Expert systems

Routine methods are developed by analysts utilising the knowledge and expertise gained through education, training and practical experience. The knowledge and expertise can be captured in an expert system and subsequently applied to routine analyses to flag problems and suggest possible solutions. Expert systems are computer programs that contain expert knowledge in the form of a database which can be used to evaluate information from analytical procedures and the data generated from analytical measurements. Informed decisions can then be made based on the criteria specified in the application and the analytical expertise programmed into the expert system. Decisions on the quality of the data, whether the procedure is working correctly and instrument performance can be carried out [23]. Expert systems contain three modules:

1. a knowledge base containing the expert information such as the details, specifications, observations and quality assurance criteria of an analytical method, the relationships between the parameters that control the

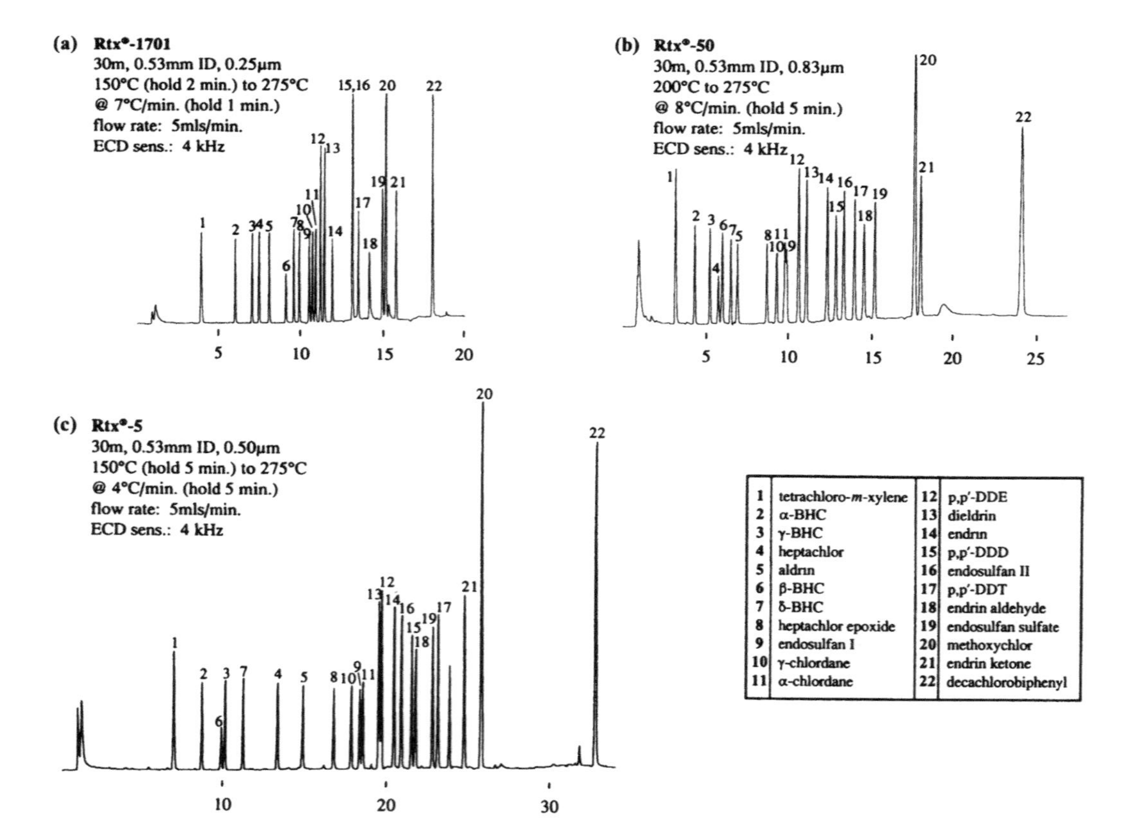

Figure 8.20 Separation of a pesticide mixture using optimised GC parameter. (Reproduced by permission of Restek Corp.)

chromatography and optimum settings and performance of the instrumentation;
2. an inference engine which contains the strategy to process the knowledge for drawing conclusions; and
3. a user interface to take care of all the communications with the user.

Such an expert system has been developed for HPLC. Initially, an 'informed guess' method is tried using suitable settings of mobile phase composition and flow-rate, stationary phase and detector wavelength and settings. The settings are selected based on the knowledge available to the analyst. However, modification of one or more parameters may be necessary to obtain the optimum chromatogram and satisfactory quality assurance. The parameters are inter-related and therefore a change in one impacts on one or more of the other variables. The expert system assists in optimisation of the method by programming in the chromatographic parameters that control the separation such as the relationships between selectivity, resolution, peak width, analysis time and solvent composition—the expert knowledge. The strategic knowledge required to evaluate problems is defined as modules in the inference engine. For example, if analysis time exceeds the specified limits the user would be referred to the solvent module for a solution to the problem, or if resolution between two peaks decreased, the selectivity module would be used. Expert systems can be built up to contain a vast array of expert knowledge for a range of analyses and applications [23].

8.6 Laboratory information management systems

LIMS reflect the progress in the application of automation and computer technology in analytical laboratories [24]. A LIMS which may be running on a PC or minicomputer, is based on a relational database and supporting software which is configured to handle the activities of an analytical laboratory. These include logging sample details, sample tracking, scheduling analyses, processing the data from analytical measurements, generating reports and archiving data and management of information required to meet regulatory requirements and good laboratory practice. LIMS are also able to handle methods management, provide productivity information and control automated instruments (Figure 8.21).

Various LIMS configurations may be implemented ranging from a dedicated multifunction laboratory data collection and reporting system to an integrated company wide system. GC, HPLC and other types of instruments generate the raw analytical data that is subsequently processed by the LIMS so direct communications between instruments, equipment and the LIMS computer is necessary. This is achieved via a network or an RS232 interface for bi-directional data transfer. Thus, control and instrument variables, as required for a specific analysis, for example, wavelength, flow-rate, solvent

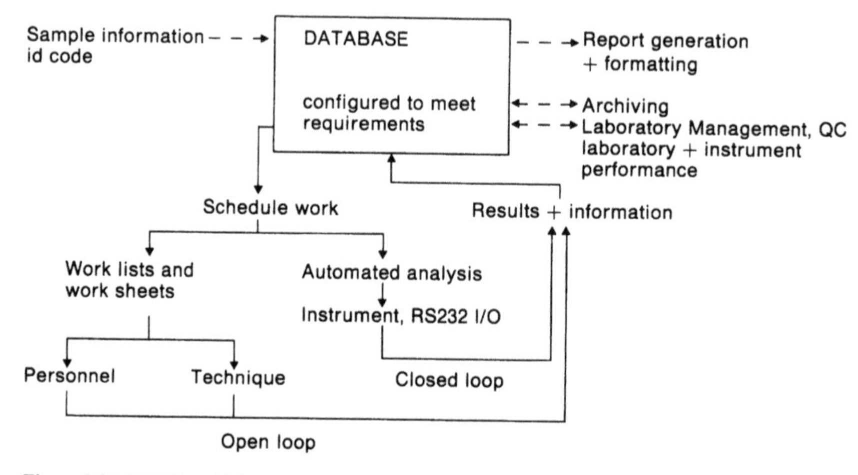

Figure 8.21 LIMS and laboratory management illustrating the main database functions and the work of a laboratory.

and temperature programmes and attenuation, can be down-loaded from the LIMS prior to sample analysis, the necessary procedural and methods software being stored on the local database. A LIMS is configured for each laboratory and instrument so dedicated software is required for the following tasks:

- sample management;
- communications, control of GC, HPLC instruments and collection of the raw data;

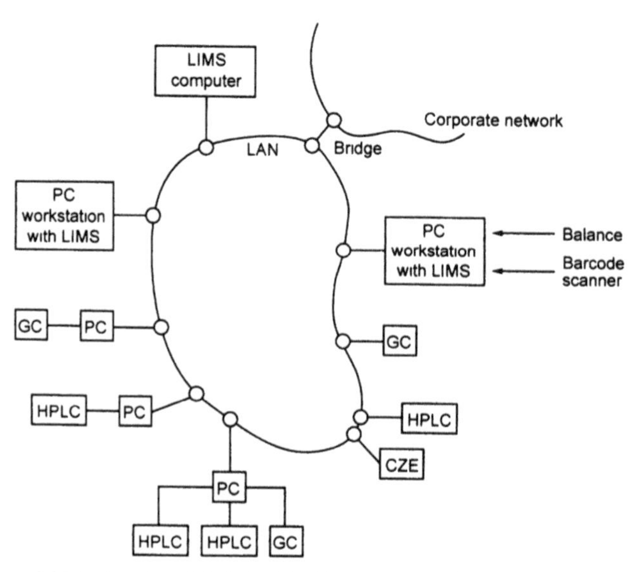

Figure 8.22 Laboratory LAN, for LIMS hardware and laboratory instruments.

- data evaluation and processing specific to the analytical technique and sample being analysed;
- reporting and archiving data; and
- communications and information transfer between computers and the central LIMS over the network.

The relevant software may be resident in the instrument, stored in a PC dedicated to the GC or HPLC or running on a laboratory LIMS computer. There are a wide range of possible computer configurations but the most flexible LIMS will employ satellite computers and distributed processing using a LAN (local area network) and the corporate network for communications (Figure 8.22).

References

[1] Carrick, A. *Computers and Instrumentation*. Hayden, London, 1979.
[2] Malcolm Lawes, D.J. *Microcomputers and Laboratory Instrumentation*. Plenum Press, London, 1984.
[3] Meadows, R. and Parsons, A.J. *Microprocessors: Essential Components and Systems*. Pitman, London, 1983.
[4] Savitsky, A. and Golay, M.J.E. *Anal. Chem.*, **36** (1964) 1627.
[5] Skoog, D.A. and Leary, J.L. *Principles of Instrumental Analysis*, Chapters 3 and 4, pp. 29–56. Saunders-Harcourt Brace Jovanovich, New York, 1992.
[6] Tarroux, P. and Rabilloud, J. *J. Chromatogr.*, **248** (1982) 249.
[7] Braithwaite, A. and Smith, B. *Comput. Appl. Lab.*, **3** (1984) 190.
[8] Lohninger, H. and Varmuza, K. *PCs for Chemists* (Zupan, J., ed.), p. 155. Elsevier, 1990.
[9] Lyne, P.M. and Scott, K.F. *J. Chromatogr. Sci.*, **19** (1981) 547, 599.
[10] Bishop, D. *Interfacing to Microprocessors and Microcomputers*. Newnes Butterworth, London, 1982.
[11] Spencer, W.A. *Anal. Chem.*, **52** (1980) 950.
[12] Kaiser, R.E. and Rackstraw, A. *Computer Chromatography*. Huthig, Heidelberg, 1983.
[13] Massart, D.L., Dijkstra, A. and Kaufman, L. *Evaluation and Optimisation of Laboratory and Analytical Procedures*. Elsevier, Amsterdam, 1978.
[14] Bunday, B.D. *Basic Optimisation Methods*. Edward Arnold, London, 1984.
[15] Ryan, P.B., Barr, R.L. and Todd, H.D. *Anal. Chem.*, **52** (1980) 1460.
[16] Berridge, J.C. *Techniques for the Automated Optimisation of HPLC Separations*. Wiley, Chichester, 1985.
[17] Grant, J.R., Dolan, J.W. and Snyder, L.R. *J. Chromatogr.*, **185** (1979) 153.
[18] Demming, S.N. and Morgan, S.L. *Experimental Design, A Chemometric Approach*. Elsevier, Amsterdam, 1991.
[19] Goupy, J.L. *Methods for Experimental Design, Principles and Applications for Physicists and Chemists*. Elsevier, Amsterdam, 1992.
[20] Carr, P.W. and Watson, M.W. *Anal. Chem.*, **50** (1979) 1835.
[21] King, B.W. *Computer Aided Optimisation for Methods Development And Validation*. Phase Separations, Deeside, Clwyd, 1993.
[22] Restek, *Overall Resolution of Pesticides in Least Amount of Time, The Restek Advantage*, Vol. 4, No. 5, p. 1. Restek Corporation, Bellafonte, PA, 1993.
[23] Maris, F. and Hindriks, R. *Intelligent Software for Chemical Analysis* (Buydens, L.M.C. and Schoenmakers, P.J., eds), Chapter 5, pp. 153–224. Elsevier, Amsterdam, 1993.
[24] Braithwaite, A. *Laboratory Information Management Systems, Concepts, Integration and Implementation* (McDowall, R.D., ed.), Chapter 3, pp. 20–33. Sigma Press, Wiley, Chichester, 1988.

9 Model or practical experiments in chromatographic techniques

9.1 Introduction

Modern chromatography boasts an impressive range of techniques and instrumentation which to the new practitioner can appear somewhat bewildering. In the past decade alone, a number of techniques have achieved maturity; for example, SFC, capillary GC, GC–MS and LC–MS while others such as instrumental TLC and CE offer quite remarkable promise and capability. Over the same period, the established techniques of GC and HPLC have been revolutionised with regards to capability and complexity.

These developments have been fuelled in part by the apparently unrelenting increase in microcomputing power, bringing with it improved control and data acquisition and processing capabilities; and partly by the improved technologies for the preparation of chromatographic media and packings in GC and LC, for example, chiral stationary phases, bonded phase capillary GC and narrow bore HPLC.

It has been the objective of the text to provide theory at a level adequate for an understanding of the underlying principles of chromatographic techniques. However, chromatography is very much an applied and practical subject and a 'hands on' approach provides much insight into the fundamentals of chromatographic methodology.

The experiments described in this chapter are representative of the myriad of applications in environmental, health and safety and food analysis, and are a few of those which have been found by the authors to be useful in teaching the fundamental techniques of chromatography to technician, diploma and undergraduate students and to members of short courses in Practical Chromatography. The experiments have been chosen to give an example of each of the main procedures, and all have been found to work well in the hands of the inexperienced chromatographer. A greater number of experiments have been included on GC and HPLC to reflect the importance of these techniques in the modern chromatography laboratory. The inclusion of experiments in capillary GC, GC–MS and CE have been included to reflect these newer developments. The experiments in these sections may be performed on the majority of commercially available chromatographs with little or no modification.

A range of more standard basic experiments is also included. In the case of column chromatography and TLC, they reflect the continued importance of these techniques to synthetic organic chemists; while a few of the others—for example, those featuring paper chromatography—enable a low-cost entry to the teaching of the fundamentals of chromatographic practice.

No particular originality is claimed, although it is not always possible to quote original sources, because several published methods may have been refined, and then further adapted, by the present authors or their colleagues. All the experimental figures and chromatograms reported have been determined by the authors (or their students) for the conditions described. Readers are referred to the Glossary, Appendix 1, for an explanation of the terms used and the chapter on the specific technique for more details on the principles.

Each experiment details the reagents and chemicals used and where possible hazards have been minimised both by the choice of experiments and by the reagents which have been selected. *Space precludes inclusion of hazard assessment data and instructors and those intending to undertake any of the above exercises should carry out the appropriate hazard and risk assessment.*

9.2 List of experiments

Section A. Paper chromatography

1. Separation of cobalt, manganese, nickel and zinc by ascending and horizontal development.
2. Quantitative separation of copper, cobalt and nickel by ascending development.
3. Ascending chromatography on ion exchange paper.

Section B. Electrophoresis and related techniques

4. Horizontal low voltage electrophoresis of amino acids.
5. Cellulose acetate and polyacrylamide gel electrophoresis of proteins.
6. Horizontal agarose gel electrophoresis of DNA.

Section C. Thin layer chromatography

7. Preparation of microplates and separation of aromatic amines.
8. Separation of simple organic compounds on fluorescent silica plates.
9. Separation of sugars on bisulphite and acetate modified silica.
10. Analysis of analgesics using normal and reverse phase TLC.

Section D. Column chromatography

11. Separation of dichromate ion and permanganate ion using an alumina column.

12. Determination of the exchange capacity and exchange efficiency of a cation exchange resin.
13. Complex elution of iron and copper using a cation exchange resin.
14. Purification of proteins on DEAE–cellulose.
15. Separation of a mixture of analgesics by flash column chromatography.

Section E. Gas chromatography

16. GC of alcohols.
17. Determination of ethanol in an aqueous solution by GC.
18. Determination of barbiturates using an internal standard.
19. Determination of whisky congeners by capillary GC.
20. Qualitative analysis by GC using retention data from two columns (polar and non-polar).
21. Study of some important parameters in GC.
22. Determination of some chlorinated insecticides by capillary GC using an EC detector.
23. Analysis of paint shop vapours.
24. Determination of water content in solvents by capillary and packed GC using thermal conductivity detection.
25. Analysis of mineral acids as their cyclohexyl derivatives by capillary GC and GC–MS.

Section F. High performance liquid chromatography

26. Analysis of barbiturates by reverse phase isocratic chromatography.
27. Ion pair chromatography of vitamins.
28. Techniques in HPLC analysis of analgesics.
29. Analysis of amino acids as their DNP derivatives.
30. Analysis of paraben preservatives by HPLC with photodiode array detection.
31. Determination of the amino acid composition of a peptide using pre-column derivatisation with *o*-phthalaldehyde and reverse phase HPLC and fluorescence detection.
32. Analysis of inorganic anions in aqueous samples.

Section G. Capillary electrophoresis

33. Determination of inorganic cations by CE.
34. Analysis of analgesics by CZE.

9.3 SECTION A. Paper chromatography

Standard tubes are available for applying accurate volumes (1–10 μl) of sample solutions to chromatographic paper and tlc plates. Alternatively, a

Pt loop 3 mm in diameter or an extruded melting point tube can be used for spotting the sample solutions.

9.3.1 **Experiment 1.** *Separation of cobalt, manganese, nickel and zinc by ascending and horizontal development*

Object

To illustrate the use of gas jars for ascending development and to compare with horizontal methods. To compare the sensitivity of three different locating reagents and different papers.

Keywords

Paper chromatography, transition metals, ascending and horizontal development.

Introduction

For basic information on development techniques and visualisation of spots in paper chromatography, see Chapter 3.

Materials and equipment

Solution: mixture of the chlorides of the four metals (about 0.05 g of each metal in 100 cm^3).
Solvent: acetone/conc. HCl/water (87:8:5, v/v/v).
Paper: No. 1, No. 3 MM (reels 3 cm wide, or cut from sheets); No. 1, No. 2, and No. 4 (12.5 cm circles), Whatman chromatography papers.
Locating reagents: (a) rubeanic acid/salicylaldoxime/alizarin (RSA); (b) diphenylcarbazide; (c) sodium pentacyanoammine ferrate(II)/rubeanic acid (PCFR).
Development apparatus: (a) gas jars 7 cm × 30 cm, with suspension hooks; (b) Petri dishes 11 cm in diameter.

Method

Ascending development. Cut three strips of No. 1 paper 24 cm long. Apply 10 μl of the solution containing about 2.5 μg of each metal to the centre of the strip about 2 cm from one end. Place the solvent in the gas jars to a depth of 3 cm and roll the jars round to wet the walls with solvent. Suspend the paper strips, sample spots to the bottom, in the sealed gas jars just clear of the solvent. Allow to stand for a few minutes, to equilibrate the liquid and vapour phases, and then lower the strips so that the ends are 1 cm below the liquid surface. When the solvent front has risen 12–15 cm remove the strips from the jars, mark the solvent front and hang to dry. Spray one strip with each of the locating reagents (a), (b) and (c). Repeat the experiment with No. 3 MM paper.

Horizontal development. Locate through the centre of the paper disc a small wick formed from a piece of the same type of paper. The wick should be long enough to reach the bottom of the Petri dish and have a diameter not more than one thickness of paper. Draw a circle of 1.5–2.0 cm diameter and apply 10 μl of the samples at 1.5 cm intervals. Allow to dry, put the paper on a Petri dish containing solvent, cover with a second dish, and develop, dry and spray as before. The horizontal methods should be tried with Nos 1, 2 and 4 papers.

Remarks

The order of elution of metals and the approximate R_f values for the ascending method are

Ni, 0.05; Mn, 0.30; Co, 0.50; Zn, 0.90

The visible solvent front is the 'dry' front. The 'wet' front has an R_f value of about 0.75, and a pale yellow band may be seen at this point (before application of the locating reagent) due to traces of iron (Figure 9.1). It should

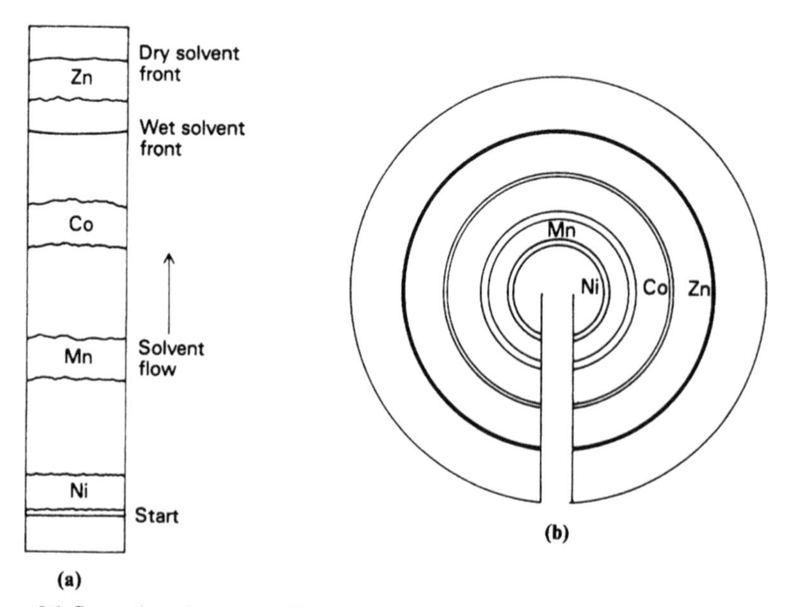

Figure 9.1 Separation of metals on Whatman no. 1 paper: (a) ascending separation; (b) horizontal separation.

be noted that traces of the iron (very soluble in the wet solvent) impurity and the zinc (very soluble in the dry solvent) give rather diffuse zones, whereas the other metals give much more compact spots.

9.3.2 **Experiment 2.** *Quantitative separation of copper, cobalt and nickel* [1] *by ascending development*

Object

To evaluate quantitatively the composition of a mixture of metal ions by visual comparison of the coloured spots with standards. An accuracy of ±5% is possible if the procedure is followed carefully, and the standards accurately made up.

Keywords

Paper chromatography, quantitative, transition metals, ascending development.

Introduction

The range of visualisation techniques and procedures for quantitations in paper chromatography have been detailed in Chapter 3.

Materials and equipment

Standard solutions: S1, 4.0 μg; S2, 2.0 μg; S3, 1.0 μg; S4, 0.50 μg; S5, 0.25 μg of each metal per 0.01 ml (solution S1 is made up of the following: $CuCl_2.2H_2O$ 282 mg, $CoCl_2.6H_2O$ 395 mg $NiCl_2.6H_2O$ 395 mg made up to 250 ml in water with the minimum amount of hydrochloric acid. S2–S5 are made by quantitative dilution of S1).
Solutions containing an unknown amount of each metal: the concentrations should fall within the limits of S1 and S5.
Solvent: butanone/conc. HCl/water (75:15:10, v/v/v).
Paper: CRL/1, No. 1, chromatography papers.
Locating reagent: rubeanic acid.
Development apparatus: 1 litre beaker and clock glass.

Method

Make up some solvent mixture and put 25 ml into the 1 litre beaker. Roll the solvent round the walls and cover the clock glass. Put about 2 ml of each standard solution and one unknown into labelled test tubes; to each add 0.5 g of potassium hydrogen sulphate, warm, and then cool to room temperature. Apply 10 μl quantitatively to each strip with a pipette, in the order S1, *u*, S2, *u*, S5 (*u* is the unknown). Label each strip. Make the sheet into a cylinder by means of a paper clip at the top, and put it in a 600 ml beaker suspended in a boiling water-bath. Leave it for 3 min to dry, and then immediately put it into the beaker containing the solvent. Replace the cover, and allow to run until the solvent front is just above the top of the slots (about 50 min). Remove the sheet, allow to dry in the air for 5 minutes, and then stand it in an atmosphere of ammonia for two minutes (a convenient way is to put it in a covered 600 ml beaker, in the bottom of which is a 25 ml beaker containing 0.880 ammonia). Immediately open out the cylinder and spray the paper evenly on both sides with rubeanic acid solution, dry and estimate the concentration of the unknown solution by visual comparison of the spots.

9.3.3 **Experiment 3.** *Ascending chromatography on ion exchange paper* [2]

Object

To illustrate the use of ion exchange papers for the separation of metals. A strong and a weak acid cation exchange modified cellulose paper are used, in the form in which they are supplied.

Keywords

Paper chromatography, ion exchange, transition metals.

Introduction

The principles underlying the separation of ions on paper have been presented in Chapter 3. The method of development is as in experiment 2; only the sorption mode differs.

Materials and equipment

Solution: iron, copper and nickel, as chlorides (aqueous solutions, about 2 mg of each metal per ml).

Eluting solution (buffer): 1.0 mol litre^{-1} $MgCl_2.6H_2O$.

Paper: (a) cellulose phosphate (p81) in the monoammonium form; (b) carboxymethylcellulose (CM82) in the sodium form (supplied as sheets in each case, cut into strips as required).

Locating reagent: sodium pentacyanoamminoferrate(II)/rubeanic acid (PCFR).
Tanks: gas jars as used in experiment 1.

Method

Prepare two tanks and one strip 3 cm wide of each paper as described for experiment 1. Apply 10 μl of the test solution as a streak; there is no need to dry the spot. Put the paper in the tank and start the run immediately. For comparison, run a strip in a third gas jar using No. 1 paper, and the solvent used in experiment 1. Allow a run of 10–15 cm, and then remove the sheets, dry and apply the locating reagent.

Remarks

Note that the strong acid paper gives a better separation than the weak acid type. The order of elution on the ion exchange paper is opposite to that in normal chromatography (Figure 9.2). Notice also that on P81 paper the iron does not move, because of the high stability of the iron-phosphate complex. The approximate R_f values are given in Table 9.1.

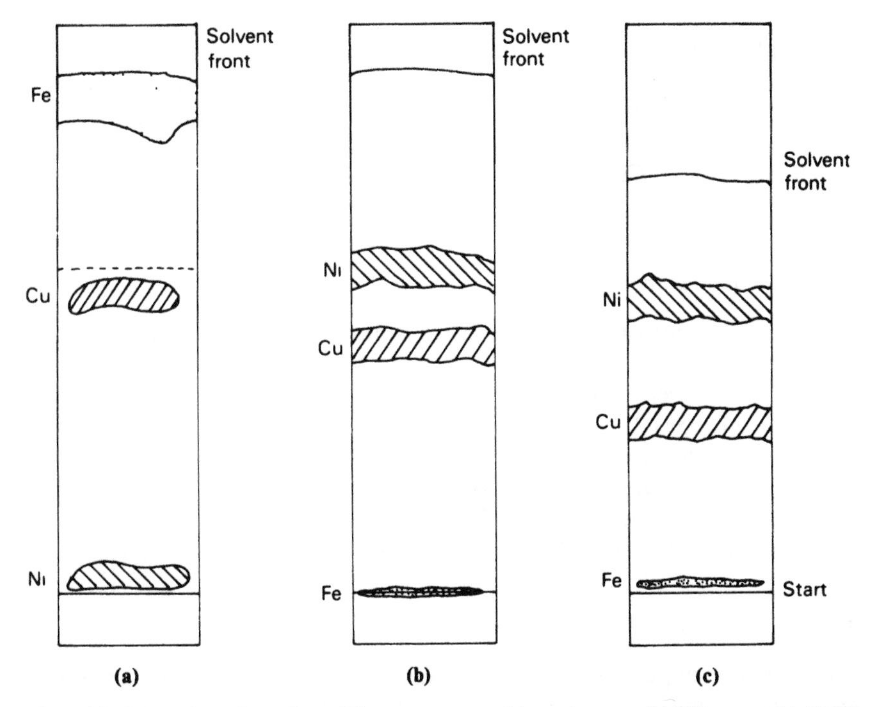

Figure 9.2 Separation of metals on Whatman papers: (a) no. 1 paper; (b) P81 paper; (c) CM82 paper.

Table 9.1 R_f values for experiment 3

Paper	P81	CM82	No. 1
Iron(III)	0.00	0.06	0.75
Copper(II)	0.60	0.27	0.65
Nickel(II)	0.75	0.75	0.05

9.4 SECTION B. Electrophoresis and related techniques

9.4.1 ***Experiment 4.*** *Horizontal low voltage electrophoresis of amino acids*

Object

To illustrate the separation of amino acids using horizontal low voltage electrophoresis.

Keywords

Electrophoresis, amino acids, ninhydrin, isatin, Sakaguchi reagents.

Introduction

Electrophoresis is more often used to separate proteins than amino acids. The degree of separation obtained for amphoteric electrolytes depends on the pH of the buffer and on the isoelectric point of the substances. This experiment illustrates the use of a horizontal low voltage paper method, which separates acidic, basic and neutral amino acids, but is not satisfactory for separating the members of one group from each other.

Materials and equipment

Solution: a mixture of arginine, aspartic acid, and glycine (0.04 mol litre^{-1} of each in water).
Electrolyte (buffer): 1 mol litre^{-1} acetic acid.
Paper: No. 1, reels 3 cm wide.
Locating reagents: (a) ninhydrin; (b) isatin; (c) Sakaguchi reagent.
Apparatus: horizontal electrophoresis tank (any horizontal electrophoresis apparatus (e.g. Shandon model U77—Figure 9.3)) could be used; in all cases manufacturers provide instructions for the use of their products.

Method

Cut four strips of paper of exactly the same length and mount them in the support frame. Number the strips and draw a line just clear of the paper support, across the end of the strips to be located in the anode compartment. Fill the

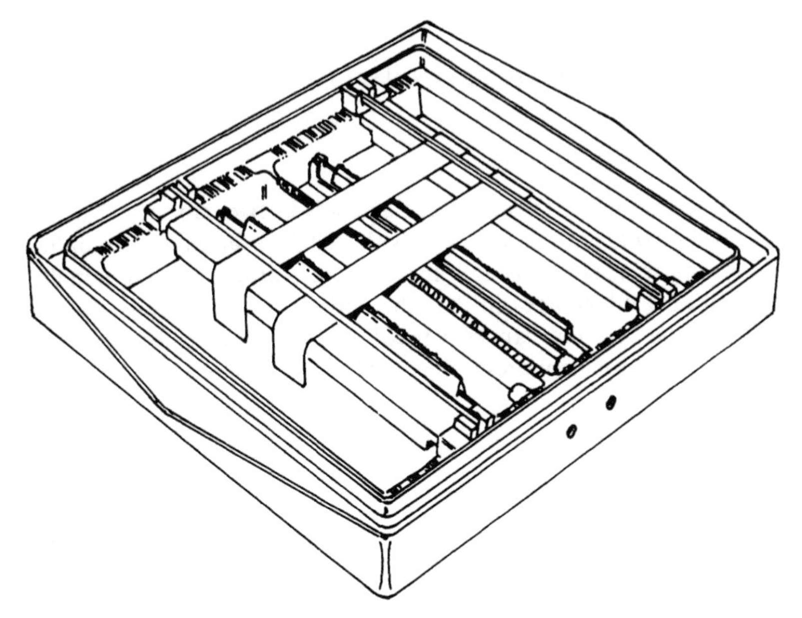

Figure 9.3 Shandon horizontal electrophoresis apparatus U77.

tank with buffer solution and put the frame in position. Moisten the strips by gently dabbing with pieces of paper dipped in the buffer solution, or with a small soft brush. Mop up any solution spilt on the tank or frame. Adjust the tension of the strips by easing into position and then apply 10 μl of the test solution with a capillary tube along the lines to within 5 mm of the edges. Replace the lid, connect up the electrodes to the power pack, and apply a potential of 200 V for 4 h; the current will be about 2 mA. Display a safety notice on the tank while the current is on, and do not touch the tank or connections without switching off. At the end of the run, switch off, lift out the frame, and dry quickly. Remove the strips from the frame, and dip two strips in ninhydrin solution and heat at 105°C for 4 min. All three acids are thus revealed, and the strips should appear identical. Dip the third strip in isatin solution and heat at 105°C for 4 min to detect aspartic acid, and treat the fourth strip with Sakaguchi reagent to detect arginine. The order of migration from the start position is aspartic acid, glycine, arginine.

9.4.2 **Experiment 5.** *Cellulose acetate and polyacrylamide gel electrophoresis of proteins*

Object

The following practicals enable two electrophoretic techniques on different solid supports for the separation of serum proteins to be studied.

Keywords

Cellulose acetate, polyacrylamide gel electrophoresis, serum proteins.

Introduction

Cellulose acetate has a lower solvent and adsorptive capacity than paper. Thus, higher voltages can be used with cellulose acetate electrophoretic media, giving sharper bands and improved resolution. Polyacrylamide gels have additional advantages in that the degree of cross-linkage and hence porosity of the medium can be varied; also, if required, the gels can be solubilised after electrophoresis thus facilitating recovery of the separated components.

(1) Cellulose acetate electrophoresis

Materials

Human serum.
Cellulose acetate strips (3 × 20 cm) CAM.
Sodium barbitone buffer (0.07 M) pH 8.6.
Lissamine green staining reagent (0.3 g dissolved in 15% (v/v) acetic acid).
1% (v/v) acetic acid (destaining reagent).
Whatman No. 1 paper.

Method

Place 150 ml of barbitone buffer in each of the two inner compartments of the electrophoresis tank and 200 ml of buffer in each of the two outer compartments. Take three CAM strips from the box and lay them with the 'Oxoid' label uppermost on a clean sheet of filter paper. Draw a faint pencil mark across each strip (leaving a 3 mm gap at each side) 5 cm from one end. Label the strips with suitable identification marks at the end of each strip. *Float* the marked strips, 'Oxoid' label down, on the surface of the buffer in the tank so that they are wetted from below by capillary action. This avoids the occlusion of air in the pores of the membrane. When the strips have been wetted completely (5–10 s) they can be submerged in the buffer. Remove the wetted strips from the tank and blot *lightly* to remove the excess buffer. Place the two curved shoulder pieces in the slots in the outer compartments so that their inner edges are 16 cm apart. Cut two 23 cm strips of Whatman filter paper from the sheets provided and moisten them in the barbitone buffer. Place the moistened strips of filter paper on the perspex shoulder pieces with one edge of the strip running along the top edge of the shoulder piece and the other edge of the strip dipping into the buffer in the outer compartment. These filter paper strips act as wicks to conduct the current from the buffer to the CAM strips. Place the moistened-blotted CAM strips in the tank at right angles to the shoulder pieces with their ends overlapping the filter paper wicks and the borders of the centre of each strip supported by the pins in the tank partitions. Pull the strips taut and secure the ends of the strips with the curved strip holders. Place the transparent cover over the tank to minimise evaporation. From this point on, do not remove the tank cover for longer than is necessary. Apply 10 μl of the serum samples to the strips in the form of a streak to within 3 mm of the edges. Replace the lid. Connect the leads to the tank so that the origins (pencil lines) are nearer to the cathode (black) than the anode (red). Connect the red and black plugs to the positive and negative terminals of the power pack, respectively. Turn on the power supply and run the electrophoresis for 60–90 min at 150 V; the meter should register a current of 4–10 mA. A current of approx. 0.4 mA cm^{-1} width is required.

Staining of serum strips. After the required time turn off the current and remove the strips. Immerse them in the Lissamine green staining reagent for 5 min. Wash the strips three times in 1% (v/v) acetic acid to remove the excess dye and blot dry. Dry the strips in a stream of warm air. Record your results both quantitatively and qualitatively. The dry strips may be rendered glass clear by immersion in Whitmore oil if required.

(2) Polyacrylamide gel electrophoresis

Materials

Human serum in 10% (w/v) sucrose.
5% (w/v) sucrose.
Solution A: gel buffer (Tris-HCl buffer pH 8.9 containing tetramethylethylenediamine (TMED) 0.17 ml/100 ml of buffer).
Solution B: 33% (w/v) solution of acrylamide in distilled water.
Solution C: 2.25% (w/v) solution of methylene–bisacrylamide in distilled water.
Solution D: an aqueous solution of ammonium persulphate (7.5 mg ml^{-1}) prepared just prior to use.
Solution E: tank buffer (Tris-glycine buffer pH 8.2).
Bromophenol blue: a 0.05% (w/v) solution in 10% (w/v) sucrose.
Staining solution: Coomassie brilliant blue.
TCA staining solution: 10% TCA in water.

Methods

Preparation of gels (7.5% acrylamide). N.B. *Acrylamide is a nerve toxin—do not ingest or allow solutions of the monomer to touch your skin.* Cap four gel tubes and stand vertically. Into a Buchner flask add 3 ml of solution A,

2.25 ml of solution B, 0.5 ml of solution C and 3.25 ml of distilled water. Cap the Buchner flask and evacuate for 30 s. Add 1 ml of the ammonium persulphate solution, mix, then add the prepared solution to the gel tubes with a Pasteur pipette to within 1 cm of the top of the gel tube. Using a hypodermic syringe gently overlay the top of the gel with distilled water, taking care not to cause any mixing, this presents oxygen (an inhibitor of polymerisation) being absorbed and also gives a gel with a flat surface. The gels will polymerise in 10–20 min. After polymerisation leave for 30 min and then remove the water from the top of the gels using a Pasteur pipette. Place the gel tubes in the apparatus and add 20, 30, 40 and 50 μl of the serum into the four gel tubes. Add 5 μl of bromophenol blue to each of the tubes and gently overlay with 5% solution of sucrose to the brim of the tube. Place solution E in the lower and upper compartments of the electrophoresis apparatus. The lower compartment is the anode compartment (connect to red leads) and the upper compartment the cathode compartment (connect to black leads). Switch on the power supply and adjust to give a current of 3–4 mA per gel tube. Follow the progress of the blue bromophenol band until it reaches 1 cm from the bottom of the tube (30–40 min).

Staining the gels. Remove the gel columns from the glass tubes by pushing water from a hypodermic syringe between the gel and the glass tube. Cut the gels at the bromophenol blue marker, place two in two separate test tubes of TCA staining solution (anode to the bottom) and heat at 60°C for 15–20 min. When the protein bands are visible rinse the gels in tap water and note your observations. The other two gels should be placed in two separate test tubes containing 'normal' Coomassie blue staining solution and left for 1 h. The gels can then be removed and destained in a mixture of methanol/acetic acid/water. The transfer of the gels to staining solution should be performed as quickly as possible in each case.

9.4.3 ***Experiment 6.*** *Horizontal agarose gel electrophoresis of DNA*

Object

To resolve a range of DNA molecules using horizontal agarose gel electrophoresis with ethidium bromide as fluorescent indicator.

Keywords

Horizontal agarose gel electrophoresis, DNA, ethidium bromide.

Introduction

DNA molecules may be separated on the basis of size. This is most often achieved by introducing the DNA into an agarose gel and then applying an electric field. The DNA which is negatively charged at neutral pH values will migrate towards the cathode. The rate of migration depends on a number of factors including the size of the DNA molecules and the agarose concentration. Linear double-stranded DNA migrates through gel matrices at a rate that is inversely proportional to the $\log_{10}$ of the number of base pairs. Furthermore, the migration rates of linear DNA fragments of a given size vary with the agarose concentration in the gel medium. Thus it is possible, by the judicious choice of different agarose concentrations, to resolve a wide molecular weight range of DNA molecules. For example, 0.3% agarose will efficiently separate DNA of size 5–60 kb, while 2% agarose will resolve fragments of size 0.1–0.2 kb. The visualisation of DNA within the agarose gels is achieved by including the fluorescent dye ethidium bromide which intercalates into the DNA in the gel on exposure of the gel to UV light.

Materials and equipment

Electrophoresis buffer: Tris borate (TBE, 50 mM, pH 8.0) with ethidium bromide (0.5 $\mu g\,ml^{-1}$).
Agarose gel: gel concentration dependent on the size of DNA to be separated.
Gel loading buffer: bromophenol blue (0.25%); xylene cyanol FF (0.25%); Ficoll (Type 400, 15%).
Apparatus: Horizontal electrophoresis tank (in all cases manufacturers provide instructions for the use of their products).

Method

Transfer sufficient agarose gel of the desired concentration into a glass bottle and add the appropriate amount of TBE. Heat the mixture in a boiling water bath or in a microwave until the agarose dissolves. Cool to 60°C then add the ethidium bromide (N.B. ethidium bromide is a carcinogen and appropriate care must be taken). Add the gel mixture to the casting tray and insert the comb to form wells for the DNA samples. After the gel has completely set, place it into the tank and completely submerge in TBE. Mix the DNA samples with the gel loading buffer and inject into the slots of the agarose gel created by removal of the comb. DNA markers of known molecular weight are included to allow the fragment size to be determined. Apply a voltage of 1–5 $V\,cm^{-1}$ across the gel. The gel should be examined under ultraviolet light and photographed after the bromophenol blue and xylene cyanol FF have migrated a significant distance into the gel. By comparison with the DNA markers determine the molecular weights of DNA in the samples.

9.5 SECTION C. Thin layer chromatography

9.5.1 *Experiment 7. Preparation of microplates and separation of aromatic amines*

Object

To illustrate a method of preparation of microchromatoplates and to use the prepared plates to identify the aromatic amines in a simple mixture by comparison with standards.

Keywords

Microchromatoplates, normal phase, aromatic amines.

Introduction

This technique for the analysis of mixtures employs an active adsorbent spread as a layer (0.05–1 mm in depth) on a planar surface, e.g. glass. Samples are applied to the base of the plate in volatile solvents ('spotting') and the chromatogram is then developed with an ascending solvent. Components are detected by spraying or exposing to the vapours of reagents which yield coloured spots. Small scale thin layer chromatography, a convenient procedure for such uses as the rapid preliminary examination of the composition of a crude product from a reaction, is based on the use of microscope slides as the support (microchromatoplates). Greater resolution can sometimes be obtained by the use of larger plates and separations on a preparative scale can be achieved with thick layer chromatography.

Materials and equipment

Solvents: methanol, chloroform and toluene.
Standard solutions: *p*-phenylene diamine, *m*-phenylene diamine, aniline, *p*-chloroaniline and *o*-nitroaniline (all 1% (w/v) in toluene).
Unknowns: two or more of the above components at the same concentration in toluene; silica gel G; calibrated solution applicators; microscope slide, beakers, watchglasses.

Methods

Preparation of microchromatoplates. Microscope slides must first be cleaned by wiping, washing with a detergent, then water and then alcohol. Do not touch the clean surfaces as this will prevent satisfactory coating. Place a mixture of 50 g silica gel G (a mixture of silicic acid and plaster of

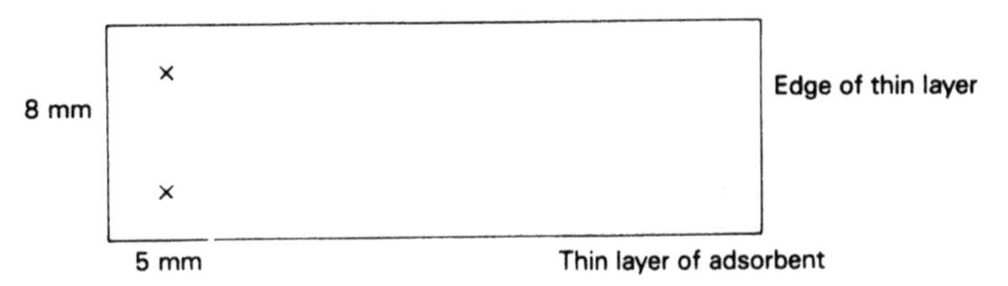

Figure 9.4 Silica chromatoplate.

Paris), 66 ml chloroform and 33 ml methanol in a 125 ml bottle, the cover of which is lined with aluminium foil. Shake for 5–10 s to bring the silica into suspension. Remove the stopper, and dip a microscope slide into the suspension to within 5 mm of the top of the slide. Withdraw the plate slowly and let any excess of solvent drain back into the jar. The dipping–withdrawal operation should normally take 1–2 s; the more quickly the plate is withdrawn, the thinner the deposit of silica. Lay the plate on a clean surface. Coat additional plates until the silica suspension starts to settle (say 30–40 s). Then suspend the silica again by shaking. The plates dry quickly (3–4 min) and are then ready for use. Carefully remove the silica adhering to the nearside of the plate by wiping with a clean tissue. Plates should be usable for several weeks (Figure 9.4).

Sample application. Apply 10 μl of test solution from a calibrated capillary by very lightly touching the plate about 5–6 mm from the bottom; the development solvent (see below) should *not* come up to the level of the spot. Each spot should be as small as possible, if possible <1 mm in diameter. Gently blowing on the capillary as it touches the plate reduces the spread of the spot.

Development. Line the inside of the beaker with a strip of filter paper, then pour in the developing solvent to moisten the paper and give a depth of 2–3 mm on the bottom of the beaker. Place the chromatoplate upright in the beaker (the end with spots on at the bottom). Cover the beaker with a watchglass. When the solvent has risen 5–6 cm up the plates, remove the plate from the beaker and immediately mark the solvent front, then allow the solvent to evaporate. In general, in choosing a developing solvent so that a separation of the components of a mixture is achieved, it is usual to start with a relatively non-polar solvent like petroleum ether. If the compounds do not move on the chromatoplate use a more polar solvent, e.g. toluene, chloroform, ethyl acetate, methanol or acetic acid, in that order. Frequently, a mixed solvent is valuable. Thus, if the compound moves only slightly with toluene, try a 9:1 mixture of toluene/chloroform.

Detection of spots. To visualise the compounds after chromatography, expose the plate to iodine vapour. Place the solvent-free plate in a sealed jar containing a few crystals of iodine. Within a few minutes reddish brown spots will appear (which are relatively permanent), warming the jar accelerates this staining process. Observe the initial colour of the spots as this often aids in the identification of unknowns.

Exercise

Make up thin layer plates as described above. Chromatograph the standards and the unknown(s) in toluene and locate the spots by iodine staining. Calculate the R_f values and hence determine which of the five standard compounds the unknown(s) contain. Confirm your conclusions by running a further chromatogram using the unknown mixture as one spot and the suspected mixture as a second spot. (The R_f values may vary with the quality of the TLC plate and hence comparisons between the unknown and the standard must finally be made on the *same* plate.)

9.5.2 **Experiment 8.** *Separation of simple organic compounds on fluorescent silica plates*

Object

This procedure describes the location of spots on adsorbent plates containing a fluorescent indicator.

Keywords

Thin layer chromatography, fluorescent indicator, normal phase.

Introduction

Plates are available commercially, which use an indicator which absorbs light at 254 nm and re-emits or fluoresces light at the green end of the spectrum, thus the plate when irradiated at 254 nm takes on a striking green colour. If a spot of compound is present which itself absorbs at 254 nm this will quench the fluorescence and the component will show up as a dark spot against the green background. Thus, the spots can be located on the developed chromatoplate by irradiating with ultraviolet light of 254 nm. This technique has the advantage that as the indicator is insoluble in the common solvents and the location is non-destructive, it allows the isolation of resolved components for subsequent spectroanalytical study. Other reagents can be added to the silica which fluoresce when irradiated at 370 nm.

Materials and equipment

Standard solutions: *m*-dinitrobenzene, fluorenone, acetone-2,4-dinitrophenyl-hydrazine, benzylidene acetone and azobenzene (all 1% (w/v) in toluene).
Unknowns: two or more of the above components at the same concentration in toluene.
UV lamp radiating at 254 nm.
Calibrated solution applicators.

Method

The procedures regarding sample application and development are as previously described (experiment 9). The component spots are located by irradiating the developed plates with short wave (254 nm) ultraviolet light. The spots can be outlined in pencil and subsequently stained with iodine if desired.

Exercise

Chromatograph the standards and the unknown(s) in toluene and locate the spots by ultraviolet detection. Calculate the R_f values of the standards and hence determine which of the five standard compounds the unknown(s) contains.

Remarks

Typical R_f values for the compounds listed are 0.46, 0.32, 0.28, 0.10 and 0.60, respectively. The absolute values may be significantly different due to the quality of the TLC plate and solvent used. However, the relative order should be the same.

9.5.3 **Experiment 9.** *Separation of sugars on bisulphite and acetate modified silica*

Object

This procedure describes two possible methods of pre-treating silica gel TLC sheets to provide a modified layer which is effective for the separation of sugars.

Keywords

Thin layer chromatography, bisulphite and acetate modified silica, sugars.

Introduction

The adsorptive properties of silica gel can be modified by incorporating substances such as bases or buffers, thus enabling gels with accurately defined pH values to be prepared. Admixtures of silica gel with sodium bisulphite and sodium acetate prove useful for the chromatography of sugars giving distinctive R_f ranges.

Materials and equipment

Sugar solutions: fructose, glucose, maltose and sucrose (all 1% (w/v) in chloroform).
Unknown(s): containing two or more of the above sugars prepared as above.
Solvents: methanol, ethanol, acetone, acetic acid, chloroform and ethyl acetate.
Chemicals: sodium bisulphite and sodium acetate.
Spray: 5% aniline hydrogen phthalate freshly prepared in glacial acetic acid.
Filter papers (Whatman No. 1).
Shallow tray or dish and oven (at 100°C).

Methods

Method A (sodium bisulphite). A solution is prepared as follows: 4 g sodium bisulphite is dissolved in 80 ml distilled water, then 120 ml ethanol are added slowly while stirring. The chromatogram sheet is submerged for 1 min in a tray containing this solution, then removed, drained and allowed to air dry on a clean piece of filter paper. It is then activated in an oven at 100°C for 30 min prior to spotting with the sugar solutions. When spotted, the sheet is eluted in the usual way and the eluant is allowed to travel approximately 8 cm up the sheet. The sheet is then removed from the apparatus and dried. The eluant consists of ethyl acetate, methanol, acetic acid and water in the ratio of 12:3:3:2.

Method B (sodium acetate). A solution is prepared as follows: 1.6 g sodium acetate is dissolved in 10 ml distilled water, then 190 ml ethanol are added slowly while stirring. The chromatogram sheet is submerged for 1 min in a tray, then removed, drained and left to air-dry on a clean piece of filter paper. It is then activated in an oven at 100°C for 30 min prior to spotting with the sugar solutions. When spotted, the sheet is eluted in the usual way and the eluant is allowed to travel approximately 8 cm up the sheet. The sheet is then removed from the apparatus and dried. The eluant consists of acetone, chloroform, methanol and water in the ratio of 16:2:2:1.

Detection. The sheet is sprayed with 5% aniline hydrogen phthalate freshly

Table 9.2

Sugars	Approximate R_f values	
	Bisulphite layer separation	Acetate layer separation
Maltose	0.22	0.11
Glucose	0.38	0.25
Sucrose	0.39	0.25
Fructose	0.44	0.30

prepared in glacial acetic acid and this is followed by gentle heating at 85°C to produce the spots. Some of these spots tend to fade rather quickly.

Exercise

Chromatograph the sugar standards and the unknown(s) on the prepared sheets using the solvent systems detailed above. Calculate the R_f values of the standards and hence determine which of the standard components are in the unknown mixture(s). Account for the different R_f values in the two systems.

Remarks

The sugars will separate accordingly to Table 9.2. By comparison with the behaviour of known sugars, an unknown one may sometimes be identified by running it in both the systems described.

9.5.4 ***Experiment 10.*** *Analysis of analgesics using normal and reverse phase TLC*

Object

To analyse an analgesics mixture by normal phase TLC using single and multiple development techniques. To analyse analgesic mixtures by reverse phase TLC.

Keywords

Thin layer chromatography, normal, reverse phase, multiple development, analgesics.

Introduction

TLC is performed on a thin layer of chromatographic stationary phase of uniform thickness supported on a sheet of inert material. Classically, this would be silica gel of about 0.25 mm thickness supported on a glass plate. Silica gel is a highly porous amorphous silicic acid, the particle size range used in TLC is 2–2.5 μm and pore size 60–124 Å. Other stationary phases are available including alumina, cellulose and more recently silica gel modified with alkylsilanes used in a reverse phase chromatographic mode, also chiral plates are available for the separation of enantiomeric compounds. Though it is unusual for retention to occur via only a single sorption mechanism the nature of the chromatographic medium will certainly dictate which is dominant. It should be remembered that the same compound mixture may be successfully chromatographed on contrasting media although this may be accompanied by changes in the degree of retention and elution order. In addition to the choice of chromatographic medium the resolution of difficult mixtures can be improved using the technique of multiple development; for example, some analgesic mixtures will separate quite well after one development, while others require several additional developments for acceptable resolution of the components of the mixture. When a number of developments are performed, care should be taken to avoid developing the sheet in areas of high humidity which will tend to deactivate it. When this type of deactivation occurs in adsorption chromatography, the R_f values of the compounds are higher and separations are not as sharp. If humid conditions cannot be avoided, then the addition of cyclohexane to the developing solvent will help to correct the deactivation of the sheet. Detection of components on the developed plate by either viewing it directly if the compounds are coloured, or if they are not by treating the developed plate with an aerosol derivatising agent; several specific spray reagents can be used to form coloured zones with certain analgesics will lower the limits of their detection. A more general method for detection of organic compounds is by fluorescence quenching, this is where a fluorescent dye is incorporated into the chromatographic medium such that the whole plate fluoresces under UV light. When an organic compound is on the plate the luminescence of the dye is inhibited by the quenching effect of the compound, and therefore appears as a dark spot. The non-destructive method of ultraviolet detection described here is more useful since the separated spots may be cut from the sheet and the pure material eluted with ethyl alcohol for subsequent spectrophotometric or other confirming analyses.

(1) Single elution normal phase TLC

Materials and equipment

Analgesics: aspirin, paracetamol, caffeine and phenacetin.
Sample: Anadin, APC or other proprietary analgesic tablet.
TLC plates: Kiesegel 10 × 5 cm 60 F_{254} layer thickness 0.2 mm.

Eluants: mobile phase 1: chloroform (7 parts), diethyl ether (2 parts), formic acid (1 part); mobile phase 2: chloroform (17 parts), methanol (3 parts), acetic acid (1 drop).
Micropipettes, TLC tanks.

Method

Prepare two TLC tanks containing solvent 1 and 2, respectively. Pour the solvent into the tank to a depth of about 3 mm and cover it with the lid. Swirl the tank gently, this is to saturate the atmosphere in the tank with solvent vapour, so as to prevent evaporation of solvent from the plate during development. Prepare solutions of the compounds aspirin, paracetamol, caffeine and phenacetin by dissolving approximately 30 mg of each in 10 ml of methanol. Prepare a mixed standard of all four components at the same concentrations. Handle the TLC plates by the edges only and try not to breathe directly onto it. Examine the surface for damage and contamination. Rule a line about 1 cm from the bottom of each plate gently with a graphite pencil. Divide this line with five small pencil marks spaced 0.8 cm apart; these are the starting points for the five 'spots'. The analgesic solutions can now be spotted onto each plate using a 5 (0.5 lambda) capillary pipette. A capillary containing the analgesic solution is touched intermittently against the plate at the starting point, so as to produce as *small* a spot as possible whilst being careful not to damage the adsorbant coating. Dry the spot with a hot air blower and view under the ultraviolet light box, if it is not clearly visible as a distinct dark spot reapply the analgesic solution in the same manner as before, and repeat this process until a dense spot is clearly visible. Aspirin in particular usually requires several applications. When all the compounds have been spotted place one plate in each tank. Lean the plate against the side of the tank such that only the uncoated side of the top edge is in contact with the tank wall, ensure that the solvent level is 3–5 mm below the starting line. Replace the lid and allow the plate to develop. The solvents will migrate up the plates by capillary action. When the solvent front is about 1 cm from the top remove the plate and before the solvent evaporates mark the line of the solvent front with a pencil. Thoroughly dry the plates with a hot air dryer, then view each plate in the ultraviolet light box and mark the outline of the spots with a pencil. Calculate the retardation factors (R_f) for each component in each mobile phase and also the corrected retardation factors for each component relative to phenacetin. From your results determine the optimum mobile phase and use it to qualitatively analyse an Anadin or an APC tablet. Extract a whole tablet into about 10 ml of methanol. Use the ultrasonic bath and a glass rod to break up the tablet and to dissolve the analytes. As some of the binding agent is insoluble allow it to settle and use the supernatant liquid for spotting the plate. Spot the sample and the mixed standard on a fresh TLC plate and elute with the optimum mobile phase. From comparison of retardation factors determine the composition of the pharmaceutical formulation.

(2) Multiple development normal phase TLC

Materials

Analgesics: salicylamide, aspirin, *o*-ethoxybenzamide, *p*-acetophenetidide, caffeine and *p*-hydroxyacetanilide.
Analgesic solutions: solutions of analgesics identified above (2% (w/v) in ethanol).
Unknown(s): powdered drug preparation, for instance a proprietary analgesic formulation, containing two or more of identified analgesics.
Eluants: system 1, comprising dichloroethane and acetic acid in the ratio of 120:1; system 2, comprising cyclohexane, dichloroethane and acetic acid in the ratio of 40:60:0.8.
Silica gel sheets UV sensitised (254 nm).
UV lamp radiating at 254 nm.
Development tanks and calibrated solution applicators.

Method
Use 10 mg of powdered drug preparation and extract the analgesics with 0.5 ml of warm ethanol. Spot 1 µl of each of the analgesic standards and the test solution on to the base line of the TLC plate, noting their positions. Develop the chromatogram to a distance of at least 10 cm using system 1. Remove the sheet, warm and completely dry off the solvent. Run the solvent up the sheet a second time to the same distance, again remove, dry and run a third time. View the fully developed chromatogram under short wavelength ultraviolet light (254 nm) and mark the separated zones. Repeat the above process using system 2. Calculate the R_f values and determine the components of the drug preparation.

(3) Reverse phase TLC

Materials and equipment

Analgesics: as part (1).
Sample: as part (1).
TLC plates: Kiesegel 10 × 5 cm RP-18, F_{254} layer thickness 0.2 mm.
Eluants: Methanol: 1% acetic acid (55:45).
Micropipettes, TLC tanks.

Method

Spot 1 μl of each of the analgesic standards and the test solution on to the baseline of the TLC plate, noting their positions. Develop the chromatogram to a distance of at least 10 cm using system 1. Remove the sheet, warm and completely dry off the solvent. If resolution is inadequate run the solvent up the sheet a second time to the same distance. View the fully developed chromatogram under short wavelength ultraviolet light (254 nm) and mark the separated zones. Calculate the retardation factors (R_f) for each component and also the corrected retardation factors for each component relative to phenacetin. Hence, determine the components of the drug preparation. Compare the results with those obtained in part (1).

9.6 SECTION D. Column chromatography

9.6.1 **Experiment 11.** *Separation of dichromate ion and permanganate ion using an alumina column*

Object

To undertake the separation of dichromate ion and permanganate ion on an alumina column and to examine the absorbance of the eluant fractions with a colorimeter.

Keywords

Column chromatography, dichromate and permanganate ion, alumina.

Introduction

As in paper chromatography, the principle of column chromatography is the separation of a mixture into its components. The mixture is distributed between a stationary phase—alumina—and a moving phase—the eluant. The solid stationary phase competes with the liquid mobile phase for the substrate and a variety of organic and inorganic liquids are used to bring about an effective separation. The eluting power of the solvent is approximately proportional to its polarity, i.e. hexane $<$ benzene $<$ ethyl acetate $<$ acetone $<$ methanol $<$ water. The two main types of elution are known as elution development and gradient elution. The former type of chromatography is used when the components of a mixture only differ slightly in polarity. The eluant is kept constant throughout the separation. In the case of gradient elution, the components usually differ widely in ease of elution. Different solvents showing increasing polarity can be used to effect a separation.

Materials and equipment

Mixture of dichromate and permanganate (~0.2 M) made in 0.05 M H_2SO_4.
Colorimeter.
Glass column, e.g. a burette.

Method

Wash the alumina with 0.5 mol litre^{-1} nitric acid and decant off any fine particles. Insert a loose glasswool plug at the bottom of the column and half-fill with dilute nitric acid. Carefully pour in the alumina in the form of a slurry so that the column is evenly packed. The column should be about 12 cm in length. Ensure that the level of the nitric acid in the column is just above the level of the alumina. It is very important to prevent the liquid level falling below the alumina level, otherwise air-pockets are formed in the alumina which greatly restricts the flow rate through the column. Using a graduated pipette, add 10 ml of the given mixture of permanganate and dichromate ion mixture to the top of the column. Adjust the flow-rate through the column to about 2 ml min^{-1} and collect the eluant in 2 ml fractions. When the level of the liquid just falls to the level of the alumina, carefully add 0.5 mol litre^{-1} nitric acid in small portions until all the first coloured band has been eluted from the column. Change the eluant to 1 mol litre^{-1} sulphuric acid, and collect 3 ml fractions. Take the most intensely coloured dichromate sample and, using the colorimeter, determine the absorbance of the sample. Use the special sample tubes provided and ensure that it is correctly located in the cell compartment. If a full-scale reading is obtained, dilute the sample by an appropriate amount to bring the reading back on scale. Note the dilution required and dilute the remaining samples by the same amount. Measure the absorbance of each sample. Likewise for the permanganate fractions. Plot the absorbance against the fraction number and comment on the shape of the absorbance curve.

9.6.2 **Experiment 12.** *Determination of the exchange capacity and exchange efficiency of a cation exchange resin*

Object

To determine the ion exchange capacity of a sulphonic acid resin in the H^+ form, and then to determine the exchange efficiency of the resin.

Keywords

Ion exchange capacity, sulphonic acid resin, exchange efficiency.

Introduction

The ion exchange capacity of a resin is a quantitative measure of its ability to take up exchangeable ions. This property and exchange efficiency reflects the accessibility of the ionogenic groups to the exchanging ions. The determination of the exchange capacity of a resin is essential for any quantitative work. The exchange capacity is defined as the number of moles of univalent cation exchanged per kilogram of resin.

Materials and equipment

ZeoKarb 225 resin.
Sodium hydroxide standard solution (0.1 M).
Hydrochloric acid (2 M).
Analar sodium chloride and copper sulphate pentahydrate.
Glass column, e.g. a burette.

Methods

Exchange capacity of resin. The resin ZeoKarb 225 is of the sulphonic acid type (strong cation exchanger) and the hydrogen (H^+) form. Take about 15 g of the resin in a large beaker, add twice the volume of distilled water, shake, allow to settle and then decant off the liquor. Repeat until the liquor is clear (usually three times). Now wash the resin by decantation into the glass column which should have a glasswool plug. From this point on the resin must always be covered with liquid to avoid air bubbles getting into the column. Leave 2 or 3 mm of water above the column. Wash the prepared resin column with a small volume of dilute hydrochloric acid to ensure that resin is in the H^+ form, then wash with distilled water until the effluent is neutral to litmus paper. Remove approximately 0.5 g resin and dry thoroughly between filter papers, weigh accurately and then place in a conical flask, add 50 ml of distilled water containing approximately 3 g of sodium chloride. The large excess of Na^+ ions ensures that the hydrogen ions are brought into solution. Shake gently for 5–10 min. Add phenolphthalein indicator (three drops), titrate slowly with standard sodium hydroxide (0.1 mol dm^{-3}). An early end-point which fades slowly may be due to the last traces of exchange still going on. Calculate the capacity of the resin.

Exchange efficiency of resin. Using the hydrogen ion form of the ZeoKarb 225 in the column, copper(II) ions are exchanged for hydrogen ion from copper sulphate in aqueous solution yielding sulphuric acid.

$$Cu^{2+} + 2R^-H^+ \rightarrow 2H^+ + (R^-)_2Cu^{2+}$$

The acid effluent is titrated with standard alkali, and thus the efficiency of analytical exchange can be evaluated. Dissolve about 0.3 g, accurately weighed, of copper sulphate pentahydrate in 25 ml of water and slowly pass this solution through the cation exchange column, collecting the effluent at the rate of about 1–2 drops s^{-1}. Then wash the column with water until the effluent is neutral to litmus paper. Titrate the total effluent against the standard alkali solution. Assume the copper sulphate to be 100% pure. Calculate the efficiency of analytical exchange for your column.

9.6.3 **Experiment 13.** *Complex elution of iron and copper using a cation exchange resin*

Object

To separate a mixture of copper(II) and iron(III) on a cation exchange resin by elution using the technique of phosphate complexation.

Keywords

Column chromatography, cation exchange, complex elution, iron, copper.

Introduction

The separation involves the stepwise elution of iron(III) by phosphoric acid and then of copper(II) by hydrochloric acid, from a strongly acidic cation exchange resin such as ZeoKarb 225, Amberlite IR-120 or Dowex 50. If hydrochloric acid alone is used the order of removal of the ions from the columns is reversed, and poorer separation is obtained. The phosphoric acid eluant modifies the activities of the ions, possibly by complex formation, giving improved resolution. The order of elution is also reversed with a phosphate eluant, iron having the smaller retention volume.

Materials and equipment

Analar copper sulphate pentahydrate and ferric nitrate.
ZeoKarb 225.
Hydrochloric acid (0.5 M).
Phosphoric acid (0.2 M).
Potassium cyanoferrate indicator solution.
Two glass columns.

Method

Make up two columns about 1.5 cm in diameter and 10–15 cm long, using ZeoKarb 225 (in the sodium form) mesh size 52–100 (8% DVB). Make a slurry of the resin in water, allow to settle, and decant the supernatant. Repeat until the washings are clear, then pack the column, and wash with water until the eluate is no longer acid. The column is then in the hydrogen form. Load each column with a solution containing about 0.5 g each of copper sulphate and ferric nitrate. Wash the column with 100 ml of water to remove the acid liberated in the exchange process. Elute one column 0.5 M HCl at a rate of approximately 100 ml h^{-1}. Collect the column effluent in 5 ml fractions. Estimate in a semiquantitative manner the amount of copper and iron present in each fraction by transferring to a test tube and comparing with a solution of known concentration. Elute the second column with 0.2 M H_3PO_4 at ~100 ml h^{-1}. Collect 2 ml fractions, and, since the phosphate complex is colourless, test each fraction with potassium cyanoferrate. When no more iron can be detected change the eluting agent to hydrochloric acid (equal volumes of concentrated acid and water); the copper is rapidly removed from the column. Collect fractions as before and test for copper with the same reagent ($K_4(Fe(CN)_6)$). Note (a) the volume eluted before iron first appears; (b) the volume which contains iron; (c) the volume collected after all the iron has been eluted, and before copper first appears; and (d) the volume which contains copper.

Remarks

Average values are (a) 15 ml, (b) 100 ml, (c) 20 ml and (d) 10 ml. Metals which readily form complexes with the eluant will tend to be eluted first. In the above experiment iron complexes strongly with phosphate whilst copper hardly complexes at all. Hence the iron is readily eluted.

9.6.4 ***Experiment 14.*** *Purification of proteins on DEAE–cellulose*

Object

The aim of the experiment is to separate glucose oxidase and catalase by ion exchange chromatography on DEAE–cellulose (Figure 9.5).

Keywords

DEAE cellulose, ion exchange, glucose oxidase and catalase.

Introduction

The first ion exchangers designed for the separation of biological molecules utilised a cellulose matrix, though cellulose, due to its hydrophilic properties, had little tendency to denature proteins. These packings suffered from low

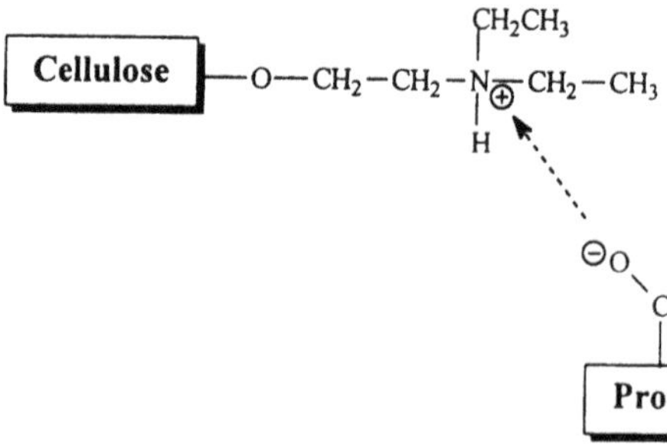

Figure 9.5 Binding of protein to DEAE-cellulose.

sample capacities and poor flow characteristics, both defects stemming from the irregular shape of the particles. It was not until the mid-1960s that cellulose gels were produced in the optimal bead form. In the production of commercial gels the polysaccharide gel is broken down and during the regeneration-bead process it is cross-linked for added strength with epichlorohydrin. The resulting macroporous bead (40–120 μm diameter) has good hydrolytic stability with an exclusion limit of $\sim 1 \times 10^6$ for proteins. The only commercially available material, DEAE (diethylaminoethyl) Sephacel can be used in the pH range 2–12 however, hydrolysis can occur in strongly acidic solutions while strongly alkaline mediums can cause breakdown of the macromolecular structure. These packings have excellent flow characteristics and increased physical strength and stability arising from the cross-linked bead structure. Re-equilibration is also facilitated as the bed volume is stable over a wide range of ionic strength and pH. The above material is used in the ion exchange separation of proteins, nucleic acids, hormones and other biopolymers. Proteins bind to this material due to electrostatic interactions between carboxylate groups on the protein and the positively changed tertiary amine.

Materials and equipment

Buffer solutions: B1 (20 mM sodium acetate and 8 mM acetic acid); B2 (40 mM sodium acetate and 40 mM acetic acid); B3 (100 mM sodium acetate and 100 mM acetic acid); B4 (105 mM sodium acetate and 105 mM acetic acid to pH 5.6).
Other solutions: glucose solution (0.7 M); hydrogen peroxide solution (3%, v/v); catalase standard solution (100 units ml^{-1}); glucose oxidase stock solution (0.03 mg ml^{-1} B4); enzyme test solution containing glucose oxidase contaminated with catalase.
DEAE Sephacell.
UV–vis scanning spectrophotometer.
Glass column.

Method

Pour a slurry of DEAE–cellulose in B1 into the column and allow it to settle. The final bed height should be about 4 cm. Wash the column with 10 ml of B1. Do not allow the column to run dry. The top of the DEAE–cellulose should be flat. Filter the solution of the enzyme if it is not clear and carefully add 5 ml of it to the column. The flow rate should be adjusted such that it is not greater than 1 ml min^{-1}. When the sample has run on to the column commence washing with 10 ml of B1 and collect the eluant in 3 ml fractions. After the B1 wash is complete wash the resin with 10 ml of buffer B2 and finally wash the column with 15 ml of buffer B3. Monitor the eluant for protein (absorbance at 280 nm). Assay those fractions which contain protein for glucose oxidase and catalase. Record your observations. Determine the recovery and purification of glucose oxidase. Select the purest fractions of glucose oxidase and catalase and determine their spectra in the range 200–500 nm. Record your observations and comment on the results.

Enzyme assays

Glucose oxidase. The basis of this assay is summarised below:

$$\beta\text{-D-glucose} + \beta\text{-FAD} \rightarrow \beta\text{-FADH}_2 + \text{D-gluconic acid}$$

$$\beta\text{-FADH}_2 + O_2 \rightarrow \beta\text{-FAD} + H_2O_2$$

$$\underset{\text{(Reduced form)}}{H_2O_2 + \text{guaiacum}} \xrightarrow{\text{Peroxidases}} \underset{\text{(Oxidised form blue } \lambda_{max} \text{ 600 nm)}}{\text{guaiacum} + H_2O}$$

Prepare solution A freshly as follows: peroxidase/guaiacum (60 ml), acetate buffer B4 (15 ml), and glucose solution (5 ml). Make 1/2, 1/4, 1/8, 1/16 and 1/32 dilutions of the stock solution of glucose oxidase stock solution using buffer B4 (1 ml of each dilution is sufficient for the assay). Add 1 ml of 1/32 enzyme dilution to 5 ml of solution A, mix thoroughly, pour into a cell and measure the absorbance at 600 nm at 3 min intervals against a control prepared by adding 1 ml of acetate buffer to 5 ml of solution A. Repeat this for each of the standard enzyme solutions. Construct a calibration graph of rate (Absorbance min^{-1}) against enzyme concentration. Assay the enzyme test solution and the protein rich fractions. Dilute these 1/500 and 1/2500 with buffer B4 before assay to take account of their high activities.

Catalase. In this assay a filter disc is first soaked in a catalase solution and then placed in a solution of H_2O_2. The oxygen produced within the filter disc causes it to float to the surface. The higher the catalase activity the more rapidly the oxygen is produced and hence the more quickly the filter floats. Set up a series of test tubes containing 5 ml of 3% H_2O. Prepare catalase

Table 9.3

Tube No.	Catalase (100 units ml^{-1})	Water
1	2.0	—
2	1.0	1.0
3	1.0	2.0
4	0.5	2.0
5	0.5	4.5

solutions from the standard provided as shown in Table 9.3. Soak the filter disc in the enzyme solution for 2 min and then transfer it to a test tube containing H_2O_2. Using a stopwatch measure the time from when the disc was placed in the test tube until it reaches the surface. Assay each dilution of the standard catalase in duplicate and construct a calibration graph transforming the data if necessary. Assay the crude enzyme solution and the protein rich fractions. It will be necessary to dilute these solutions because of high activities—try a 1/20 dilution initially.

9.6.5 *Experiment 15. Separation of a mixture of analgesics by flash column chromatography* [4]

Object

To undertake the separation of a mixture of analgesics by 'flash column chromatography' (FCC) and to examine the eluant fractions by thin layer chromatography.

Keywords

Flash column chromatography, normal phase, analgesics.

Introduction

FCC is a simple and rapid column chromatographic technique used for the separation and purification of organic compounds. It is extremely efficient, gives high recoveries and the separation of mixtures containing components with retardation factors differing by as little as 0.1 can be effected in as little as 15 min. The technique is suitable for sample loads ranging from as little as 10 mg to 10 s of grams depending on the difference in retardation factor. FCC uses a smaller and narrower range of packing than conventional preparative column chromatography and as such has to be carried out under slightly increased pressure in order to achieve optimum flow-rates. The technique is operated with a glass column which has been modified so that positive pressure, usually nitrogen or air, can be applied to the top of the column.

Materials and equipment

Analgesics: aspirin, caffeine and paracetamol.
Column packing: 40–63#, silica gel, Merck, Flash chromatography grade ex BDH.
Sand: 50–100#, white quartz ex Aldrich.
TLC plates: SIL G/UV_{254}, precoated polyester sheets ex CAMLAB.
Eluants: mobile phase 1: chloroform (7 parts), ether (2 parts), formic acid (1 part); mobile phase 2: chloroform (17 parts), methanol (3 parts), acetic acid (1 drop).
Micropipettes, TLC tanks.
FCC equipment ex Aldrich; column 25 × 584 mm.

Method

Prepare solutions of the compounds aspirin, caffeine and paracetamol by dissolving approximately 30 mg of each in 10 ml of methanol. Prepare a mixed standard containing all three components at the same concentrations. A TLC study should be undertaken, using eluants 1 and 2 as models, to identify the solvent system which gives the least retarded component a retardation factor at or about 0.25–0.30. This solvent will be used as the column eluant. The column should be packed dry. First a small plug of glasswool is inserted at the bottom of the column and covered with a fine layer of sand. The column packing is then added dry to a depth of about 18 cm and a fine layer of sand is laid on top. The chosen solvent is then poured carefully into the column without disturbing the packing. Pressure is applied to push the solvent, at a rate of 2–5 cm min^{1}, through the chromatographic bed compacting it as it passes. Care must be taken not to allow the top of the silica gel to run dry. The mixture of analgesics, in equal proportions to a total of approximately 1 g in a 20–25% (w/v) solution of eluant, should be added taking care not to disturb the top of the chromatographic bed. Approximately 300 ml of solvent should be added and the column repressurised to give the recommended eluant flow-rate. Fractions can be collected and components conveniently located by TLC.

9.7 SECTION E. Gas chromatography

The experiments described in this section were developed by the authors using a variety of chromatographs and computing integrators. The set-up parameters in the text refer to these specific instruments. The experiments, however, can readily be performed on the majority of commercially available GCs. Adjustments to the instrument parameters may be required.

9.7.1 ***Experiment 16.*** *GC of alcohols*

Object

The object of the experiment is to analyse a mixture of alcohols both qualitatively and quantitatively and also to relate the retention data to molecular structure.

Keywords

Packed column, isothermal, temperature programming, FID, qualitative, quantitative, internal standard.

Introduction

The separation of the alcohols is effected by the retention, to a varying degree, of the different alcohols by the liquid stationary phase (PEG 20M) coated onto a solid support (Chromosorb W—ratio 10:90, respectively). The small differences in the physical properties and structure of the alcohols are sufficient to cause differences in the association (H bonding, dipole–dipole interaction, etc.), between the liquid stationary phase and the alcohol (solute) molecules. Thus, the compound having least association is eluted first, followed by the others in order of the degree of the association. The retention time (the time from injection to peak maxima) is a characteristic of the compound under the conditions used.

Materials and equipment

Standards: Analar ethanol, *n*-propanol, *n*-butanol, *n*-pentanol and *t*-butanol.
Column: 10% PEG 20M on Chromosorb-W (2 m).
GC fitted with FID.
Computing integrator.

Method

For the Hewlett Packard 5890 Series II the instrument settings are as follows:

Helium carrier gas: 11 psi (note: 1 psi $\approx$ 6.9 kPa) (approx. 25 ml min^{-1}).
Hydrogen: 17 psi.
Air: 19 psi.
Column temperature: 95°C.
Injection port temperature: 200°C.
Detector temperature: 200°C.
Range/attenuation: 6/8.

Note: nitrogen may be used as an alternative carrier gas.

Qualitative experiment under isothermal conditions. Inject 1 μl samples of ethanol, *n*-propanol, *n*-butanol, *n*-pentanol and *t*-butanol in turn onto the column followed by the 3 μl sample of a mixture of the first four. Note the sequence of elution and obtain the retention times for each alcohol. Plot a graph of $\log_{10}$ retention times against

- the number of carbon atoms in the molecule; and
- the boiling points.

Note the linear response for the homologous series of *n*-alcohols and the anomalous position of *t*-butanol and suggest a reason for these results.

Quantitative experiment, conditions as above. The area of the peak for any particular compound is proportional to the amount of that compound in the mixture injected, though not necessarily with the same porportionality factor. Prepare the following mixtures: 1 ml + 2 ml and 1 ml + 3 ml of ethanol and *n*-butanol, respectively. Inject between 1 and 2 μl of each standard in turn, followed by the unknown. At the end of the chromatogram the integrator will print out retention time and peak area data from which the percentage composition of the mixture can be calculated. Note: The FID does not have the same response or sensitivity to all compounds (see Chapters 2 and 5). Some compounds produce larger detector signals than others (for a given amount) and it is therefore necessary to determine the detector response factor (D_{RF}) for each compound to be analysed. In this experiment the D_{RF} of *n*-butanol is determined relative to ethanol and can be calculated from the following:

$$D_{RF_{\text{butanol}}} = \frac{\text{Peak area butanol}}{\text{peak ethanol}} \times \frac{\text{Amount ethanol}}{\text{Amount butanol}}$$

Typically for butanol the factor is 1.67. A sample of intermediate composition can be provided as an 'unknown'.

Qualitative experiment using temperature programming. The influence of temperature programming on the resolution and analysis time of the alcohols can be investigated using temperature programming; a suggested programme for the above alcohols together with *iso*-propanol and *iso*-pentanol is as follows.

Initial column temperature: 90°C hold 1 min.
Final column temperature: 150°C.
Temperature increase: 12°C min^{-1}.

Repeat the analysis of the mixture of ethanol, *n*-propanol, *n*-butanol, *n*-pentanol and *t*-butanol then together with *iso*-propanol and *iso*-pentanol (3–5 μl injection). Compare the retention times with those from the isothermal analysis and note the improved resolution and peak shape,

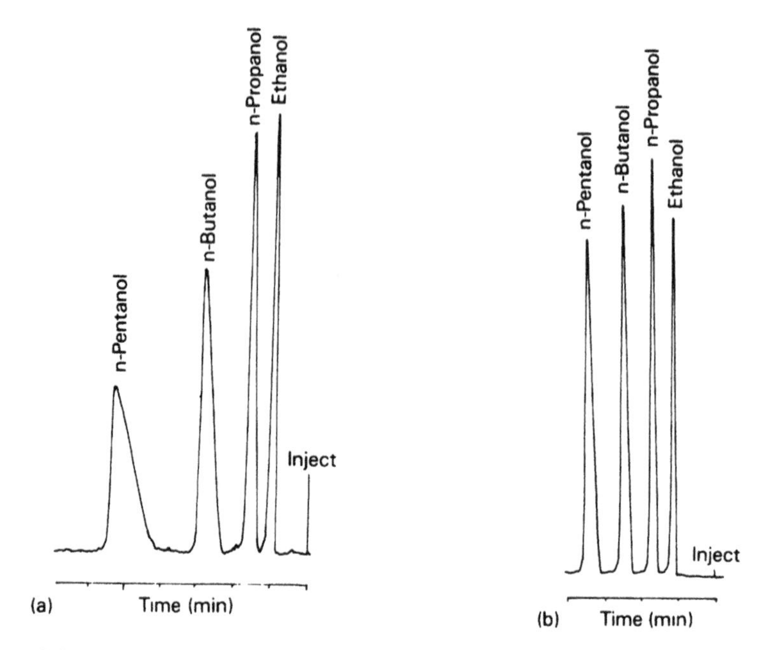

Figure 9.6 Gas chromatograms of alcohol mixture: (a) isothermal operation at 90°C; (b) temperature programmed for 90–150°C at 10°C min^{-1} with initial hold for 1 min.

particularly for *n*-pentanol. A final analysis under isothermal conditions, column temperature 150°C, can be undertaken at this point. Note the loss of resolution which results for the early eluting alcohols.

Comment

The experiment can readily be extended by having students determine the D_{RF} for all alcohols being analysed (Figure 9.6).

9.7.2 **Experiment 17.** *Determination of ethanol in an aqueous solution by GC*

Object

The object of the experiment is to determine the ethanol content of an aqueous solution using *n*-propanol as an internal standard.

Keywords

Packed column, FID, blood alcohol, internal standard.

Introduction

The analysis of alcohol solutions has become very important due to the implications of the 'Road Traffic Act (1972)' which requires the estimation of small amounts of ethanol in blood and/or urine. This experiment demonstrates the most frequently used method—GLC using an internal standard. Before the advent of GC the method most frequently used involved the chromate oxidation of ethanol. The most satisfactory method of quantitative analysis in GC involves the calibration of the detector response for the compound of analytical interest against a reference compound. This 'internal standardisation' technique involves adding a known amount of reference compound to both sample and standard solutions. The detector response factor (D_{RF}) of an analyte component relative to the internal standard (IS) can be evaluated by running a sample containing the internal standard and each of the sample components, all in accurately known concentration. Thus, the response factor for compound A can be calculated as follows:

$$D_{RF_A} = \frac{\text{Peak area } A}{\text{Peak area IS}} \times \frac{\text{Amount IS}}{\text{Amount comp } A}$$

Similarly, the response factor to the other components in the calibration mixture, relative to the internal standard, can be calculated. An accurately measured quantity of the internal standard is then added to the analysis sample and the mixture is run through the chromatograph. The amount of A can be expressed as

$$\text{Amount } A = \frac{\text{Peak area } A}{\text{Peak area IS}} \times \frac{\text{Amount IS}}{D_{RF_A}}$$

Thus, by substitution in this equation for the peak areas from the chromatogram, the relative response factors, derived from the calibration analysis, and the concentration of the internal standard added to the sample, the concentration of the components in the sample can be calculated. Since this method involves ratios of peak areas rather than absolute values, it should be noted that the precision of analysis is not dependent on the injection of an accurately known amount of sample. However, the accuracy does depend on the accurate measurement of peak area. Assay and quantitation by the internal standard is often the preferred method as it takes account of variable compound response and removes potential errors due to variation in sample injection. The 80 and 200 mg% (mg per 100 ml) standard solutions are used to confirm the linearity of response over this concentration range.

Materials and equipment

Standard solutions: 80 and 200 mg% aqueous ethanol; 25 mg% *n*-propanol (IS).

Unknown(s): simulated urine sample.
Column: 15% PEG400 on supasorb 60–80# (2 m).
Gas chromatograph fitted with FID.
Computing integrator.

Method

For the Ai Cambridge GC94 the instrument settings are as follows.

Nitrogen carrier gas: 10 psi (note 1 psi $\approx$ 6.9 kPa) (20–30 ml min^{-1}).
Hydrogen: 15 psi.
Air: 20 psi.
Column temperature: 90°C.
Injection port temperature: 175°C.
Detector temperature: 200°C.
Range/attenuation: 100.

Into a small container pipette carefully 2 ml 80 mg% ethanol and 5 ml of the *n*-propanol solution. Stopper the container and then mix thoroughly. Inject 1 µl. The chromatogram will consist of three peaks due to (i) ethanol, (ii) *n*-propanol, and (iii) water. The latter is a broad low intensity signal. Repeat the above sequence using the 200 mg% ethanol. For each analysis calculate the response factors for ethanol relative to the *n*-propanol standard solution and hence the average value. Repeat the above procedure using 2 ml of the simulated urine or blood and 5 ml of the *n*-propanol solution. Substitution of the appropriate values into the above equation will give the ethanol content of the urine or blood sample. In practice each solution (80 mg%, 200 mg% ethanol and unknown) should be analysed in triplicate to establish the accuracy and precision of the analysis (Figure 9.7).

9.7.3 ***Experiment 18.*** *Determination of barbiturates using an internal standard*

Object

To analyse a mixture of barbiturates both qualitatively and quantitatively by GC using the internal standardisation technique.

Keywords

Barbiturates, packed column, internal standardisation, silanising.

Introduction

Barbiturates are a class of drugs that act as depressants of the central nervous

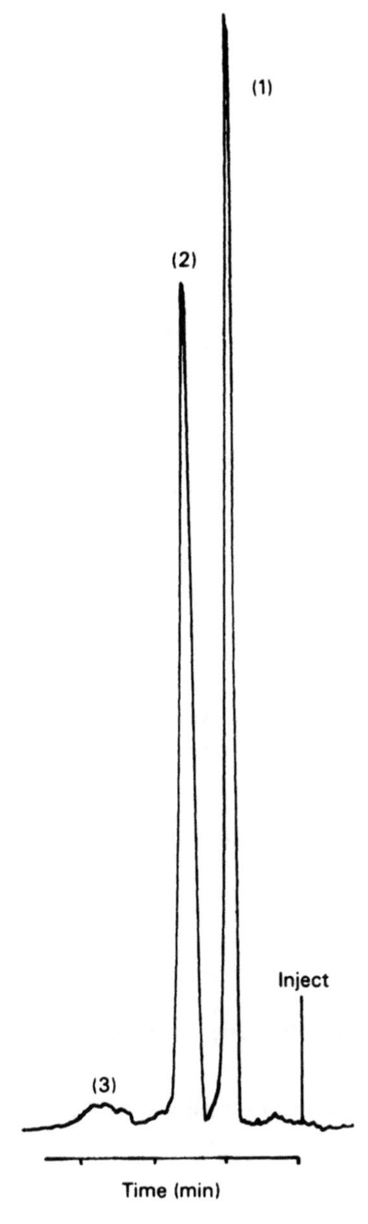

Figure 9.7 Chromatogram of standard mixture of ethanol and *n*-propanol. Conditions as detailed in text. Signals: (1) ethanol; (2) *n*-propanol; (3) water.

system. They are readily synthesised from urea and substituted malonic esters and are 5,5-disubstituted pyrimidones.

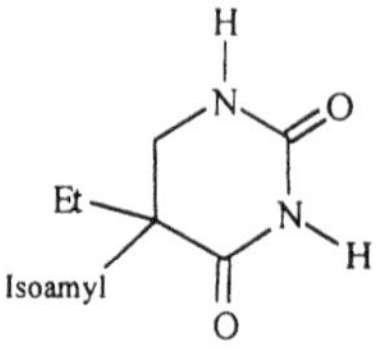

Amobarbitone

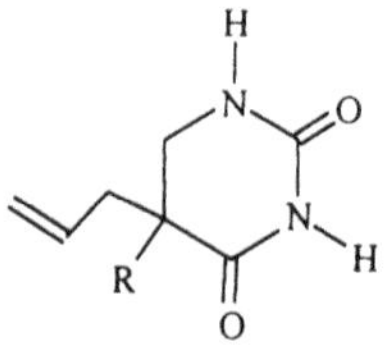

Secobarbitone (R = 1-methylbutyl)

The first barbituric acid derivative was introduced into clinical practice in 1903 and the use and misuse of this group of hypnotics has steadily increased. They are commonly encountered agents in suicide attempts and much effort has been expended in developing methods to determine barbiturates in biological fluids. Many very delicate analytical techniques involving spectroscopy, chromatography and other physicochemical methods, e.g. polarography have been established. With the introduction of partition chromatography not only could a quantity of barbiturate be determined, but also identification was quickly established too, the latter being all important if the correct treatment of an overdose is to be instigated. With the advent of GC both rapid identification and quantitation have become a reality.

Materials and equipment

Solution: 0.2% (w/v) barbitone in water (IS).
Barbiturates: butobarbitone, amobarbitone, pentobarbitone and secobarbitone.
Unknown(s): containing two or more of the above components.
Column: 10% Apiezon L on DMCS chromosorb W.
GC fitted with FID.
Computing integrator.

Method

For the Hewlett Packard 5890 Series II the instrument settings should be as follows.

Nitrogen carrier gas: 12 psi (note 1 psi ≈ 6.9 kPa).
Hydrogen: 15 psi.
Air: 20 psi.
Column temperature: 210°C.
Injection temperature: 260°C.
Detector temperature: 225°C.

Barbiturates readily adhere to glass and glass columns; liners and other glassware used in this analysis should be silanised with dimethyl-dichlorosilane prior to use. Pre-treatment of the packing by *in situ* silanisation and with successive 100 μg barbitone injections improves the column performance. If there is an initial lack of resolution then this can be overcome by following the injection of barbiturate with 2 μl of 2% dimethyldichlorosilane so that the silane chromatographs through the barbiturate. The silane will appear on the solvent front providing the injection was within ten seconds of the barbiturate injection. Prepare the following solutions.

- Solutions of amobarbitone, butobarbitone, pentobarbitone and secobarbitone with internal standard, barbitone, as follows. Dissolve 50 mg of barbiturate in ethanol, add to 50 ml of barbitone stock solution and dilute to 100 ml with ethanol (solutions A, B, C and D).
- Solution of unknown as detailed above (solution E).

Using solutions A, B, C and D determine the retention times of the barbiturates and their response factors with respect to barbitone as internal standard (see experiment 18). Using this data identify and quantify the barbiturates in the unknown. Inject 1 μl of each solution in duplicate. Wash syringe thoroughly with ether between injections (Figure 9.8).

9.7.4 **Experiment 19.** *Determination of whisky congeners by capillary GC* [5–7]

Object

To determine the concentrations of a range of congeners in a variety of proprietary whiskies.

Keywords

Whisky, congeners, capillary column, internal standard.

Introduction

Alcoholic beverages contain a wide variety of materials other than ethanol and water. The particular combination and level of these compounds derives from the materials used and the fermentation process employed and confers on each beverage its distinctive colour, aroma and taste. These impurities, which are referred to as congeners, are generally more toxic than alcohol and are often associated with the deleterious after effects of alcohol consumption. The total congener content among alcohol beverages varies from as little as 3 g per 100 litres in vodka to about 285 g per 100 litres in bourbon and comprises principally acetaldehyde, ethyl

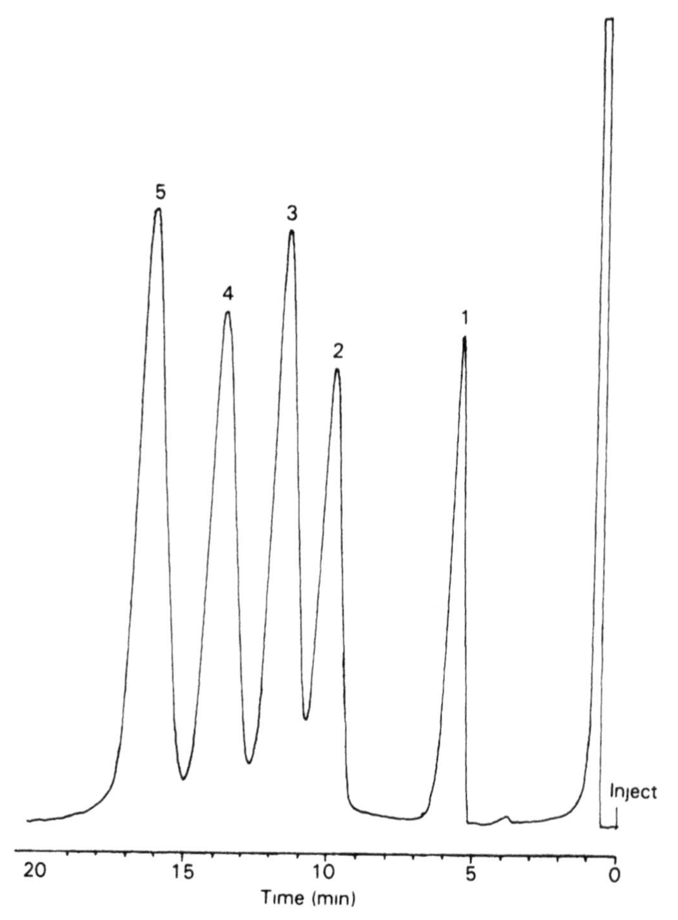

Figure 9.8 Chromatogram of a standard mixture of barbiturates, column and conditions as described in text. Peaks: (1) barbitone; (2) butobarbitone; (3) amobarbitone; (4) pentobarbitone; (5) secobarbitone

formate, ethyl acetate, methanol, *n*-propanol and *iso*-amyl alcohol [8]. A knowledge of the exact nature and amount of these impurities is essential to the monitoring of product quality in the brewing industry. GC is the preferred technique for such analysis. The most satisfactory method of quantitative analysis in GC involves the calibration of the detector response for the compound of analytical interest against a reference compound. This 'internal standardisation' technique involves adding a known amount of reference compound to both sample and standard solutions and thus it takes account of variable compound response and removes potential errors due to variation in sample injection. The detector response factor (D_{RF}) of an analyte component relative to the internal standard (IS), can be evaluated by running a sample containing the internal standard and each of the sample components, all in accurately known concentration. Thus, the response factor for compound A can be calculated as follows:

$$D_{RF_A} = \frac{\text{Peak area } A}{\text{Peak area IS}} \times \frac{\text{Amount IS}}{\text{Amount comp } A}$$

Similarly, the response factor to the other components in the calibration mixture, relative to the internal standard, can be calculated. An accurately measured quantity of the internal standard is then added to the analysis sample and the mixture is run through the chromatograph. The amount of A can be expressed as

$$\text{Amount } A = \frac{\text{Peak area } A}{\text{Peak area IS}} \times \frac{\text{Amount IS}}{D_{RF_A}}$$

Thus, by substitution in this equation of the peak areas from the chromatogram, the relative response factors derived from the calibration analysis, and the concentration of the internal standard added to the sample, the concentration of the components in the sample can be calculated. Since this method involves ratios of peak areas rather than absolute values, it should be noted that the precision of analysis is not dependent on the injection of an accurately known amount of sample.

Materials and equipment

Stock solution: a mixture comprising ethyl acetate, *n*-propanol, 2-methyl-1-propanol and 2-methyl-1-butanol (2.5% (v/v) in 40% (v/v) ethanol/water).
Internal standard solution: 2.5% (v/v) *n*-butanol in 40% (v/v) ethanol/water.
Unknown(s): proprietary whiskies.
Column: 25 m, bonded phase fused silica Carbowax, film thickness 0.25 µm, diameter 0.53 mm.
Capillary GC fitted with temperature programming unit and FID.
Computing integrator.

Method

For the Ai Cambridge GC94 the instrument settings are as follows.

Helium carrier gas: 10 psi (1 psi ≈ 6.9 kPa).
Hydrogen: 15 psi.
Air: 30 psi.
Column temperature: initial, 45°C hold for 2 min; ramp, 8°C to 150°C.
Injection port temperature: 175°C.

Detector temperature: 225°C.
Range/attenuation: 100.
Injector: splitless.

Prepare a 500 ppm analyte standard solution as follows: pipette 1 ml of stock solution and 1 ml of the internal standard solution into a 50 ml volumetric flask then dilute to volume with 40% (v/v) ethanol/water. Prepare sample solutions by diluting 1 ml of internal standard to 50 ml with proprietary whisky. Make duplicate 0.5 µl injections of the standard solution and of all sample solutions. From the peak area data of the standard solutions calculate the detector response factors for each component relative to *n*-butanol. The parts per million (% v/v) amounts of each component in the sample(s) can be determined. As an alternative to capillary GC the analysis can be carried out on the following packed column—15% PEG400 on supasorb 60–80# (2M).

9.7.5 **Experiment 20.** *Qualitative analysis by GC using retention data from two columns (polar and non-polar)*

Object

To determine the components in an unknown mixture by comparison of retention data with that of standard components obtained on polar (Carbowax 20M) and non-polar (SE30) stationary phases.

Keywords

Qualitative, two-column, homologous series.

Introduction

In addition to being a highly efficient separation technique, GC supplies much qualitative and quantitative information about sample constituents. With the use of general-type detectors the single piece of information which best describes the qualitative nature of the sample is the retention time, t_R. Within a homologous series it is expected that $\log_{10} t_R$ would be linearly related to the boiling point of the component. Such relationships are very useful in identification of particular sample components. For many types of compounds the boiling point is directly related to the length of the carbon chain (number of carbon atoms) and a plot of retention time versus the number of carbon atoms can be used for identification purposes. This approach, however, requires that one knows the basic type (i.e. alkane, alcohol, etc.) of compound whose retention time is being measured. In cases where this may be uncertain or of mixtures of solutes a two-column plot is

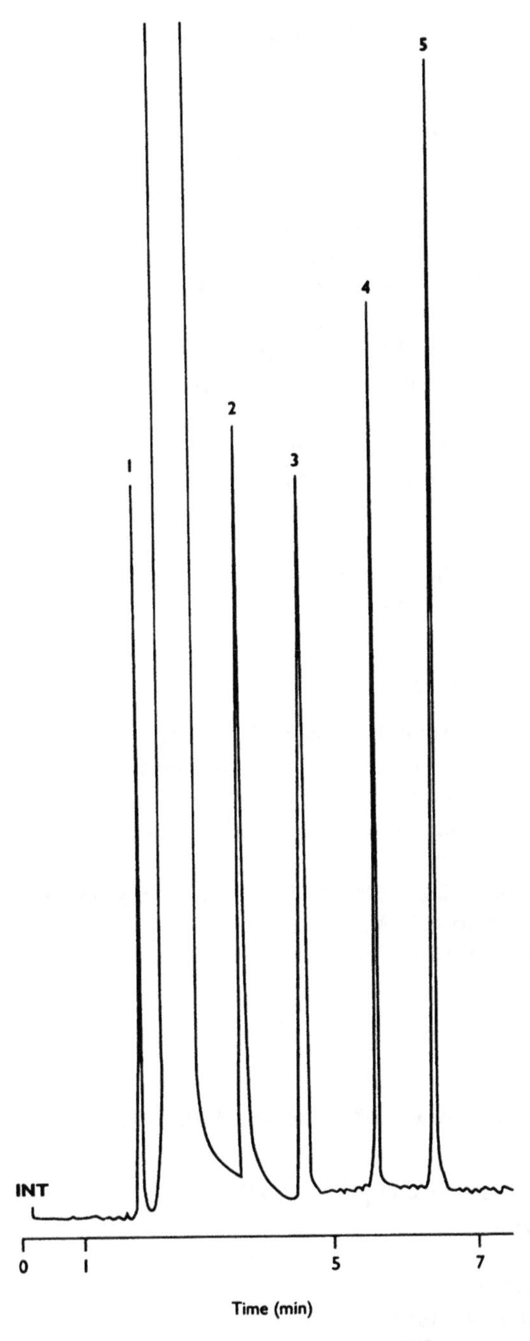

Figure 9.9 Gas chromatogram of standard mixture of whisky congeners, conditions and column as detailed in text. Peaks: (1) ethyl acetate; (2) *n*-propanol; (3) 2-methyl-1-propanol; (4) *n*-butanol; (5) 2-methyl-l-butanol.

necessary. Two components belonging to different homologous series may have identical retention times on a particular column. However, if the same two components are examined with a column of different polarity it is extraordinarily unlikely that the same retention times will again be encountered. Standard plots of retention times on column 1 versus retention times on column 2 for various homologous series are commonly used. An unknown, similarly chromatographed on the two columns, will have retention times which simultaneously fit only one of the lines, thus identifying the homologous series. Identification of the compound itself can also be made from the plots. The usual procedure is to use columns of substantially different properties in order to get a wide separation of slopes for the different homologous series. The range of compounds within a homologous series that can be identified from such a two-column plot is usually quite small (three or four compounds per series) because of the very large differences in retention times. Thus, pentanol may have a retention time of 0.5 min on an SE30 column while under the same conditions may have a retention time of several minutes on a polar column.

Materials and equipment

Alkanes: pentane, hexane, heptane and octane.
Esters: methyl, ethyl, *n*-propyl and *n*-butyl acetate.
Alcohols: methanol, ethanol, *n*-propanol and *n*-butanol.
Unknown(s): containing three of the above components, e.g. heptane, methyl acetate and *n*-propanol.
Columns: Carbowax 20M 10% on 60–80# Chromosorb W; Apiezon L 15% on 80–100# Chromosorb W.
GC fitted with dual FID.
Computing integrator.

Method

For the Hewlett Packard 5890 Series II the instrument settings are as follows.

Helium carrier gas: 15 psi (1 psi ≈ 6.9 kPa).
Hydrogen: 17 psi.
Air: 20 psi.
Column temperature: 95°C.
Injection port temperature: 150°C.
Detector temperature: 225°C.
Range/attenuation: 6/8.

Inject 1 μl of each member of the above homologous series on each column turn and record the retention time data. Inject 5 μl of the unknown(s) on each column and record the retention time data. Repeat the injections on the other column (Figure 9.10).

Calculations

1. Obtain the retention time for each compound on each column and tabulate.
2. On the same graph, plot retention time on the non-polar column versus retention time on the polar column for each homologous series.
3. Repeat 2, but plotting log retention times.
4. Identify the components of the mixture and note the elution order on each column.
5. Which graph, 2 or 3, gives the best plot and why?
6. What factors influence retention times.

9.7.6 ***Experiment 21.*** *A study of some important parameters in GC*

Object

To examine the influence of various parameters on gas chromatography performance. The parameters under study are

- column temperature;
- injection port temperature;
- sample size; and
- mobile phase flow-rate.

The effects of these parameters on the chromatographic system to be used are significant. These experiments demonstrate the effect on the recorded chromatogram and the need for optimising the above variables for better separations.

Keywords

Column and injection port temperature, sample size, mobile phase flow-rate, column efficiency.

(1) The effect of column temperature

Chromatograms obtained at constant temperature have early peaks which are sharp but closely spaced or even overlapping (peaks 1–4) and peaks eluted after a long time are broad but well separate (peaks 6, 7, 8) (Figure 9.11). To obtain well-resolved sharp peaks for each compound we can systematically vary the column temperature whilst the separation is progressing. This technique is called 'temperature programming'. The

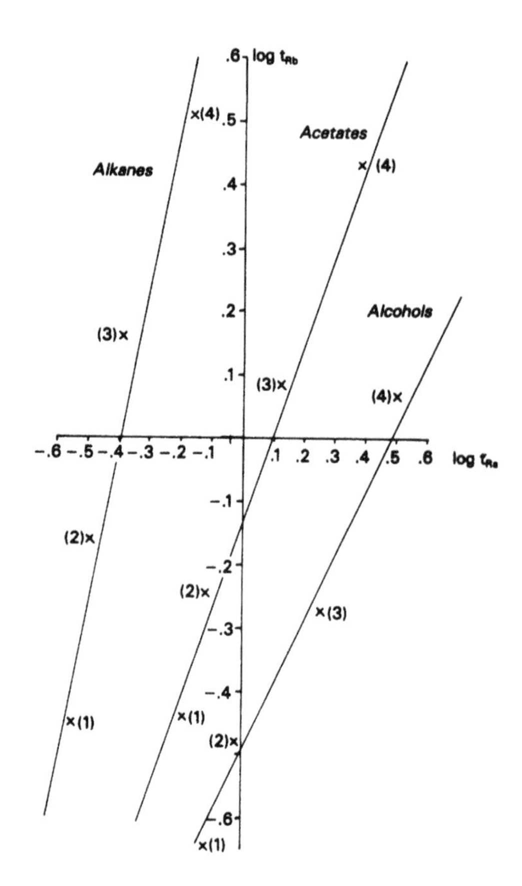

		Compound number in homologous series			
		1	2	3	4
n-Alkanes	t_{R_a}	0.29	0.33	0.44	0.66
	t_{R_b}	0.39	0.70	1.47	3.19
Alcohols	t_{R_a}	0.76	0.95	1.74	3.37
	t_{R_b}	0.23	0.32	0.55	1.10
Esters	t_{R_a}	0.61	0.83	1.36	2.42
	t_{R_b}	0.36	0.58	1.18	2.55

Figure 9.10 t_{Ra}, t_{Rb}: retention times on columns (a) and (b), respectively. *n*-alkanes: (1) pentane; (2) hexane; (3) heptane; (4) octane. *n*-alcohols: (1) methanol; (2) ethanol; (3) propanol; (4) butanol. Acetates: (1) methyl; (2) ethyl; (3) *n*-propyl; (4) *n*-butyl.

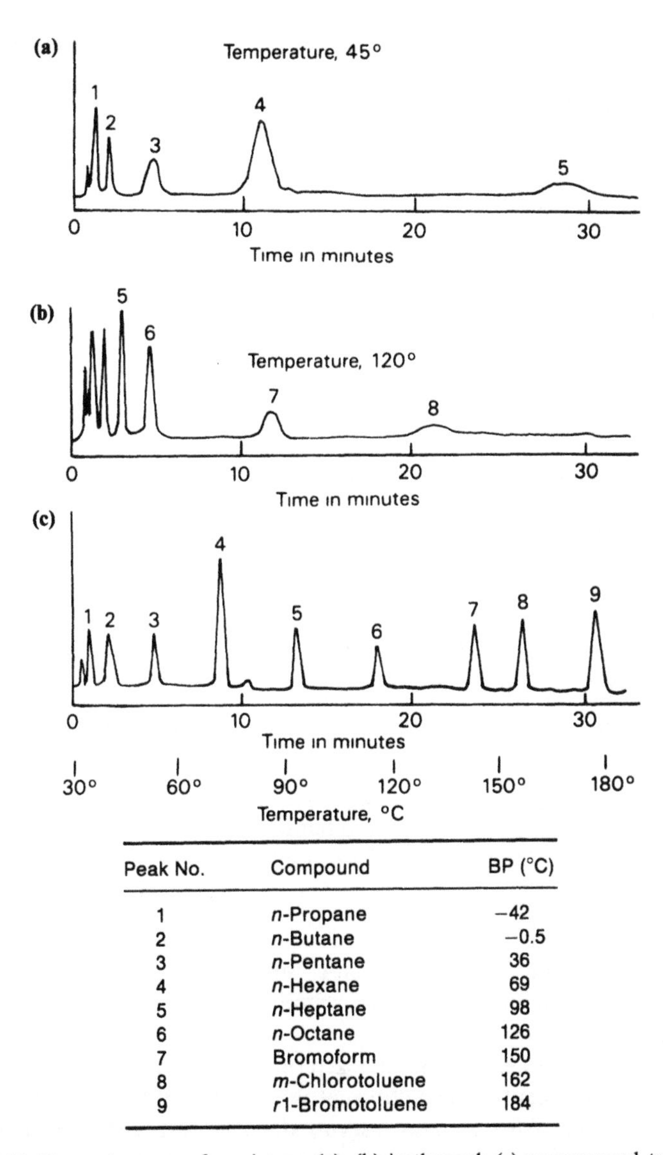

Peak No.	Compound	BP (°C)
1	*n*-Propane	−42
2	*n*-Butane	−0.5
3	*n*-Pentane	36
4	*n*-Hexane	69
5	*n*-Heptane	98
6	*n*-Octane	126
7	Bromoform	150
8	*m*-Chlorotoluene	162
9	*r*1-Bromotoluene	184

Figure 9.11 Chromatograms of a mixture: (a), (b) isothermal; (c) programmed temperature. (From Hapgood and Harris, *Anal Chem.*, **32** (1960) 450.)

column oven is set at a low temperature to enable the low boiling compounds to be resolved, once these are eluted the temperature is increased at a predetermined rate according to the boiling point spread of the remaining components of the mixture. This will progressively increase the rate at which the higher boiling components pass through the column thus, decreasing the retention time. Careful oven design and elaborate control electronics are required for accurate reproducible results. Careful design of the instrumentation and choice of suitable detectors usually overcomes the problem of base line drift associated with the increased volatility of the stationary phase.

Materials and equipment

n-Alkanes: pentane, hexane, heptane, octane, nonane, decane, dodecane.
Column: 3% SE30 on Chromosorb W (2 m).
GC fitted with temperature control unit and FID.
Computing integrator.

Method

For the PE8500 instrument the recommended settings are as follows.

Nitrogen carrier gas: $25\,ml\,min^{-1}$.
Hydrogen: $50\,ml\,min^{-1}$.
Air: $300\,ml\,min^{-1}$.
Injection temperature: 150°C.
Column temperature: isothermal, 135°C; programmed, initial 50°C—hold 1 min, $20°C\,min^{-1}$ increase to 250°C.

Ensure that the column has been 'cleaned' up at 250°C. Reduce the temperature to 130–140°C and then inject 5 µl of a mixture of equal amounts of the above alkanes. They will be eluted in order of boiling point. Next reduce the oven temperature to 50°C and set the temperature programme for an initial period of 1 minute, followed by a $20°C\,min^{-1}$ increase to 250°C. Repeat the injection.

1. Compare the two chromatograms, tabulate the retention data and the peak widths at half height for each component.
2. Plot the log retention time versus boiling point and versus molecular weight.
3. Obtain the area of each peak and determine the ratios relative to pentane for each component in each chromatogram. Would you expect the ratios to be the same? Also note the varying detector response to each alkane.

(2) Injection port temperature

Introduction

The injection port is a relatively simple device which must efficiently introduce the sample onto the column. The sample is injected through the septum quickly and vaporised very rapidly to produce a narrow 'plug' or band of sample on the column. A narrow band of sample will give the best resolution and efficiency. The latter is determined by calculating the number of theoretical plates, N, of the column for each injection.

Materials and equipment

As in part (1) above.

Method

Inject 1–5 μl samples of a mixture containing equimolar amounts of octane, nonane and decane with the injection port set to 50°C then 100°C, 150°C and 200°C. Ensure that the gas chromatograph has a column packed with a suitable stationary phase (e.g. SE30 silicone oil).

1. Calculate the number of theoretical plates (n) from one of the peaks as detailed below and plot n versus port temperature.
2. Comment on the effect of injection port temperature and decide on a general rule for choosing the optimum value. The boiling points of the components of the mixture are required.

(3) Sample size

Introduction

For optimum performance the sample size should be such that only the first theoretical plate in the column is saturated. Since a theoretical plate has a very small capacity large samples will overload the column. The optimum size is that which gives the sharpest symmetrical peaks with adequate sensitivity. The limit is thus a function of column capacity and detector performance.

Materials and equipment

As in part (1) above.

Method

Chromatograph settings as detailed in part (1) except column temperature 250°C and injection port temperature 275°C. Inject 5, 10, 20, 50, 100, 200 μl samples of an equimolar mixture or octane, nonane and decane on to the column.

1. Determine the number of theoretical plates (N) for each peak and plot N versus sample size.
2. Comment on the effect on the resolution of sample size.

(4) Selection of carrier gas and flow-rate in gas chromatography

Object

To determine the effect of carrier gas flow-rate on the efficiency of a given column and to construct a plot of H (HETP) mm versus u (flow-rate) mm min^{-1}.

Introduction

The choice of carrier gas is usually based on detector response and sensitivity, column efficiency and convenience in terms of availability cost and safety. Flow-rate markedly influences column efficiency. At low flow-rates the HETP is high, that is efficiency is low. As the flow-rate increases the HETP decreases, passes through a minimum and then slowly increases. Operation of the instrument at the flow-rate which corresponds to minimum HETP for a particular column would give maximum efficiency. The slope of the curve representing column efficiency is given by

$$H = L/N = H + A + B/u + C/u$$

where L is column length; N is the number of theoretical plates for the column; and u is the time averaged mean carrier gas velocity. The above equation is often called the van Deemter Equation (see Chapter 2). A, B and C are coefficients representing various parameters affecting gas flow through the column. The multiple path effect is a major contributor to A, molecular diffusion to B and resistance to mass transfer to C.

Materials and equipment

As in part (1) above. Standard mixture: n-hexane, n-heptane and n-octane (equimolar).

Method

For the PE8500 instrument the recommended settings are as follows.

Nitrogen carrier gas: 25 ml min^{-1}.
Hydrogen: 50 ml min^{-1}.
Air: 300 ml min^{-1}.
Injection temperature: 150°C.
Column temperature: 135°C.

Optimise the column and injector temperature to give good resolution and peak shape for the components of the standard mixture. Then inject the mixture (1–5 µl) at various carrier gas flow-rates. Determine the value of N

for various values of the flow-rate over the range 5–150 ml min^{-1}. Obtain at least eight readings and take extra readings in the region of the minimum value of H. Construct a plot of H versus u and hence determine the optimum flow-rate of carrier gas for the column. *Note*: The linear velocity of the carrier gas can be determined by dividing the retention time of an unretained peak (such as methane) by the column length.

$$u = \frac{t_{\mathrm{m}}}{L}$$

9.7.7 **Experiment 22.** *The determination of some chlorinated insecticides by capillary GC using an electron capture detector*

Object

The exercise is designed to investigate the separation and sensitivity limits of the analysis of a range of organochlorine pesticides using capillary GC with EC detection.

Keywords

Chlorinated pesticides, electron capture detection, capillary column.

Introduction

Pesticides are a group of substances which are used extensively, with new and varied formulations constantly being marketed. This expanding usage has resulted in the growing need and interest in the analysis for the presence of their residues in water courses and food and agricultural products. The analysis of these materials by GC is well documented and offers short analysis times and good sensitivity. The ECD is extremely sensitive to molecules containing electronegative substituents, such as chlorine, while being relatively insensitive to hydrocarbons. This selective sensitivity makes the detector especially valuable for the analysis of chlorinated pesticides. The utility of capillary columns has been improved with the advent of bonded phases which gives improved column lifetime and performance and also allows higher column temperatures to be employed. The complex mixtures commonly encountered in pesticide residue analysis frequently require the superior resolving power of these columns.

Materials and equipment

Standard: Suitable proprietary pesticide standard, e.g. Aldrich mixed standard 7—comprising α-HCH, γ-HCH, β-HCH, heptachlor, aldrin,

heptachlor epoxide, *pp*-DDE, dieldrin, *op*-TDE(DDD), endrin, *op*-DDT and *pp*-TDE(DDD) and *pp*-DDT at concentrations of 5 ng ml^{-1} in toluene.
Unknown(s): a river water or effluent sample spiked if required.
n-Hexane: fractionally distilled over potassium hydroxide pellets.
Aluminium oxide, neutral, Brockmann type 4, heated at 800°C for 5 h, cooled and deactivated by adding 10% (w/w) distilled water.
Anhydrous sodium sulphate (heated at 600°C for 5 h).
Column: 25 m, bonded phase fused silica HT5 (non-polar), film thickness 0.15 μm, diameter 0.53 mm.
Capillary GC fitted with ECD.
Computing integrator or data system.

Method

All glassware should be washed with detergent, rinsed in distilled water, soaked in chromic acid and rinsed with distilled water again then dried. For the PE8500 the recommended settings are as follows.

Helium: 7.0 psi (1 psi ≈ 6.9 kPa).
Column temperature: initial, 120°C hold for 1 min; ramp 1, 5°C min^{-1} to 225°C; ramp 2, 10°C min^{-1} to 300°C hold for 10 min.
Injection temperature: 225°C.
Detector temperature: 350°C.
ECD: saturation adjusted to 1%.
Injection: splitless; split relay opened after 1.5 min.

Sample preparation. The sample is extracted by shaking gently with 20 ml hexane (added by pipette) for 2 min. After marking the level of water sample in the bottle, distilled water is added to raise the hexane layer into the neck and 5 ml of hexane is removed. For sewage effluents, the hexane is 'cleaned up' by passing down a column consisting of approximately 1 g of alumina, topped by 0.5 g of anhydrous sodium sulphate. The column is washed with a further 15 ml hexane and the total eluate concentrated to 5 ml. A 2 μl aliquot of the hexane extract is injected into the chromatograph and the resulting chromatograms are compared with those obtained from standard solutions of known pesticides. The concentration of the pesticides in the unknown may be determined by comparison with peak areas of the standard (external standardisation) (Figure 9.12).

9.7.8 **Experiment 23.** *Analysis of paint shop vapours*

Object

To determine the time weighed averages (TWA) of toluene and methyl ethyl ketone (MEK) in a paint spraying booth.

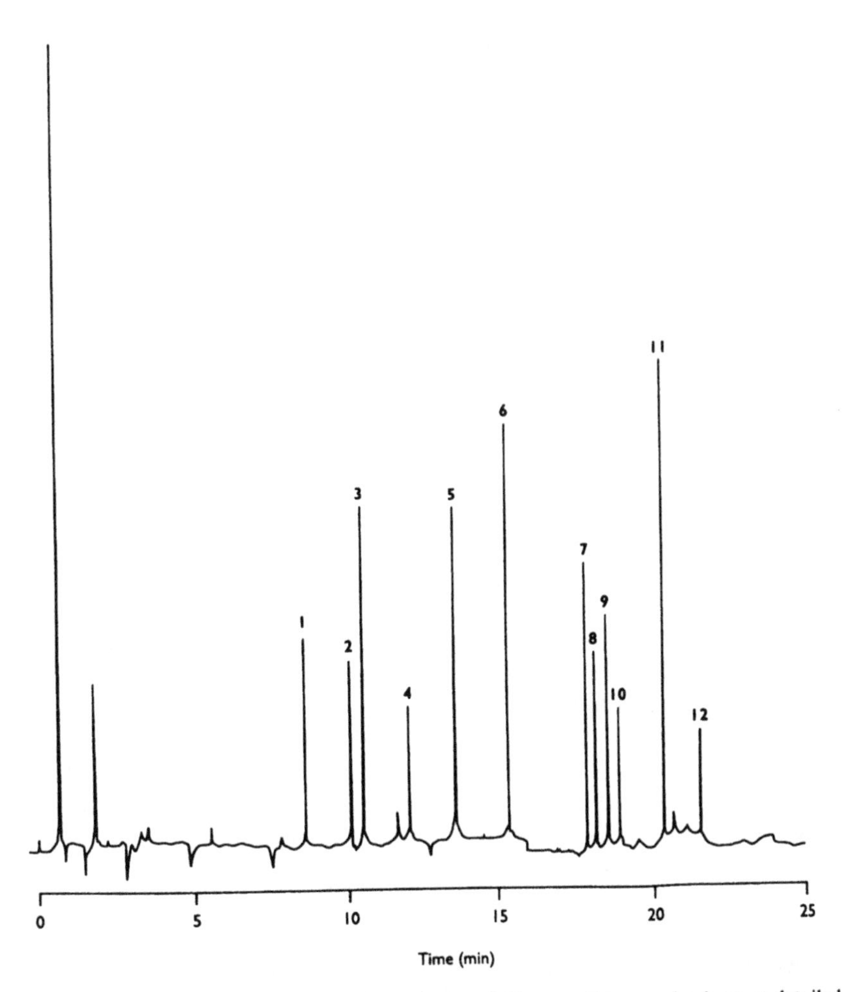

Figure 9.12 Chromatogram of standard mixture of pesticides, conditions and column as detailed in text. Peaks: (1) α-HCH; (2) γ-HCH; (3) β-HCH; (4) heptachlor; (5) aldrin; (6) heptachlor epoxide; (7) *p,p*-DDE; (8) dieldrin; (9) *o,p*-TDE (DDD); (10) endrin; (11) *o,p*-DDT and *p,p*-TDE (DDD); (12) *p,p*-DDT.

Keywords

Toluene, methyl ethyl ketone, TWA, STWA, HSE, quantitative, capillary, FID.

Introduction

The Health and Safety at Work Act places upon employers a duty to ensure, so far as is reasonably practicable, the health, safety and welfare at work of their employees. Exposure of personnel to hazardous substances should be kept as low as is reasonably practicable, primarily by the application of suitable material, plant and process control techniques. Exposure limits considered to represent good practice and realistic criteria for the control of exposure, plant design and use of protective equipment are set by the Health and Safety Executive. The exposure limits are used to assess compliance with the HSW Act. Both the long-term and short-term exposure limits are expressed as TWA concentrations which are simply airborne concentrations averaged over a specified period of time. The period for the long-term limit is normally 8 h; when a different period is used this is stated. The short-term exposure limits are normally expressed as 10 min TWAs. Concentration of gases and vapours are usually expressed as ppm, a measure of concentration by volume, as well as in $mg\,m^{-3}$, a measure of concentration by mass. In converting from ppm to $mg\,m^{-3}$, a temperature of 25°C and an atmospheric pressure of 1 bar were used. Concentrations of airborne particles (fumes, dust, etc.) are usually expressed in $mg\,m^{-3}$, with the exception of asbestos, which is expressed as fibres per millilitre of air (fibres ml^{-1}). The two principal areas in which workers may be exposed to solvent vapours in manufacturing plant are the metal degreasing shop and in paint spraying booths, and the common solvents encountered are MEK and toluene. Samples of these solvents can be analysed quantitatively by GC using *iso*-butylalcohol (IBA) as the internal standard. The response factors of MEK and toluene relative to IBA can be evaluated by running a sample containing the internal standard and each of the components, all in accurately known weights. The detector response factors (D_{RF}) for MEK and toluene (with IBA as internal standard) and amounts of these analytes in samples can be calculated as detailed in experiment 17.

Materials and equipment

Standard 1: MEK: Toluene: IBA (50:25:25%, w/w).
Standard 2: MEK: Toluene: IBA (33:33:33%, w/w).
Sample 1: sample A (470 mg) + IBA (150 mg).
Sample 2: sample B (400 mg) + IBA (150 mg).
Sample 3: sample C (57 mg) + IBA (20 mg).
Column: 10% Carbowax 20M on 60–80# Chromosorb W (2 m).
GC fitted with FID.
Computing integrator.

Method

For the Ai Cambridge GC94 the instrument settings are as follows.

Nitrogen carrier gas: 12 psi (note: 1 psi ≈ 6.9 kPa) (20–30 $ml\,min^{-1}$).
Hydrogen: 15 psi.

Table 9.4

Sample	Working period	Weight trapped (mg)	Sampling rates (litres min^{-1})
A	8.00–12.00	470	2
B	12.45–16.15	400	2
C	11.00–11.10	57	1.5

Air: 20 psi.
Column temperature: 80°C.
Injection port temperature: 200°C.
Detector temperature: 225°C.
Range/attenuation: 100.

Sampling details

The solvent vapours in the paint spraying shop were sampled using an appropriate solvent trapping system. The sampling detail is presented in Table 9.4. Inject 1 μl of each of the standards and from the resultant chromatograms determine the retention time values for MEK and toluene. Inject 1 μl of samples 1, 2 and 3. The analysis should be carried out in duplicate.

Results

From your chromatographic results determine the weights of MEK and toluene in samples *A*, *B* and *C*. By comparison with the HSE recommended limits does the atmosphere in the paint shop present a hazard? If so what changes in operating procedures need to be undertaken?

9.7.9 **Experiment 24.** *Determination of water content in solvents by capillary and packed GC using thermal conductivity detection*

Object

To compare and contrast the use of packed and capillary columns for the determination of water content in solvents using thermal conductivity detection (TCD).

Keywords

Water, packed and capillary columns, external standard, thermal conductivity detection.

Introduction

There is a requirement to determine the water content in a wide variety of non aqueous liquids, for example, solvents, gasoline and lubricating oils. A number of methods for the assay of water in such samples have been developed, the most important and popular of which involves the use of the Karl Fischer reagent. While the Karl Fischer methodology has been automated to a certain degree the hazards associated with the use of a reagent comprising iodine, sulphur dioxide, pyridine and methanol still persist and the method remains relatively time consuming and is subject to interferences. An alternative and rapid technique not subject to the above problems and capable of total automation is that of GC using either packed or capillary columns. Water does not, however, readily form ions in an FID and hence the detector of choice is the TCD. Although the latter is not as sensitive as the FID it has sufficient response to afford detection limits of 400 ppm or better.

Materials and equipment

Standard solutions: 500 ppm to 5% (v/v) range of standard water solutions prepared in absolute ethanol.
Unknown(s): proprietary solvents, e.g. methanol, acetone, methylated spirits or chloroform.
Column: (a) 25 m, bonded phase fused silica Carbowax, film thickness 1.25 μm, diameter 0.32 mm; (b) Porapak Q 80–100# (2 m).
Packed-capillary GC fitted with temperature programming unit and TCD.
Computing integrator.

Method

For the Perkin Elmer Series 8500 the recommended settings for column (a) are as follows.

Helium carrier: 12 psi.
Helium make-up: 20 ml min^{-1}.
Column temp: 55°C.
Injector temp: 200°C.
Injector: split/splitless.
Injection volume: 0.5 μl (using a 1 μl syringe).
Detector: filament current 240 mA, medium range sensitivity.

The recommended settings for column (b) are as follows.

Helium carrier: 20 ml min^{-1}.
Column temp: 160°C.
Injector temp: 250°C.

Detector: 200°C.
Injector: splitless.
Injection volume: 0.5 μl (using a 1 μl syringe).

Make a series of 0.5 μl injections of solvent to be analysed and optimise resolution for the components in the sample by adjusting column temperature and mobile phase flow-rate. Under the optimum conditions determined make at least three 0.5 μl injections of the standard solutions and of all sample solutions on both columns. Check the reproducibility of the area data. Repeat injections if greater than 5% deviation from mean of analyte peak. Check the linearity of detector response for both systems and by the method of external standardisation determine the concentration of water in the sample. Compare the order of elution of the analytes on the two columns and account for any differences (Figures 9.13 and 9.14). The experiment can be extended by using the technique of standard addition to determine the water content of methanol. A series of standard water solutions are prepared (e.g. 0.5, 1.0, 1.5 and 2.0% (v/v) water together with an accurate volume of methanol) these standards and the sample prepared in an analogous fashion are analysed as above. The water content in the solvent can be obtained by extrapolation of the calibration line; the intercept with the x-axis giving the concentration of water in the sample.

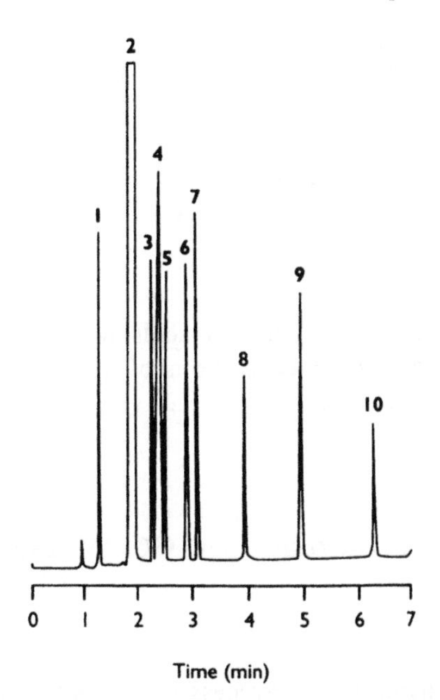

Figure 9.13 Chromatogram showing impurities in acetone. (From Barberio, *Laboratory Practice*, **40** (1989) 71. Peaks: (1) isopropanol; (2) acetone; (3) dichloromethane; (4) methanol; (5) ethylacetate; (6) ethanol; (7) tetrahydrofuran; (8) water; (9) dioxan; (10) unknown. Conditions as detailed above.

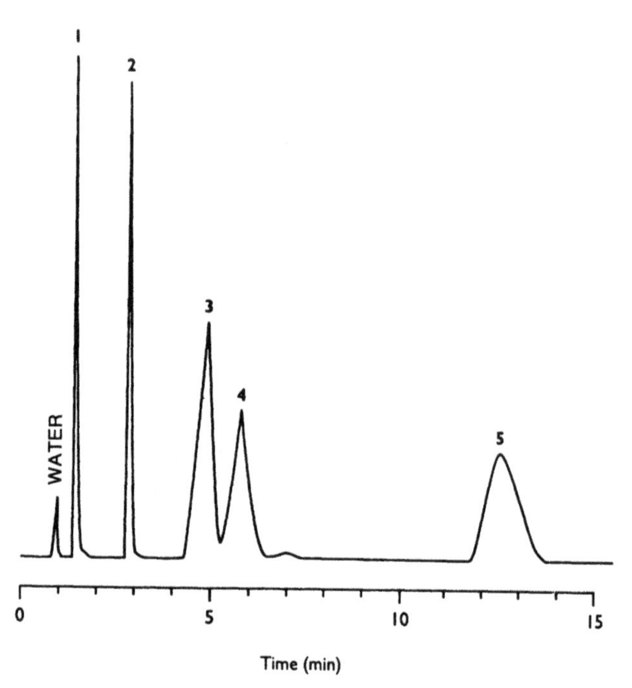

Figure 9.14 Chromatogram showing elution order of a range of common organic solvents with water present at 0.50% (v/v). Peaks: (1) methanol; (2) ethanol; (3) isopropanol; (4) diethylether; (5) dichloromethane. Conditions as detailed above.

9.7.10 ***Experiment 25.*** *Analysis of mineral acids as their cyclohexyl derivatives by capillary GC and GC–MS* [9]

Object

To investigate the reaction of mineral acids with cyclohexene oxides by GC–MS. To determine the concentrations of hydrochloric and nitric acid in a simulated stack emission.

Keywords

Mineral acids, acid oxides, derivatisation, GC–MS, capillary GC.

Introduction

There has been an increasing awareness of the deleterious environmental impacts associated with the presence of sulphur dioxide and the nitrogen oxides in power station and factory stack emissions. This resulting 'acid rain' has been responsible for the quite dramatic forest depletion and the acidification of rivers, streams and lakes in central and northern Europe and Scandinavia and the erosion of many historic buildings throughout Europe [10]. Standard methods of analysis of such emission are based typically on titrimetric or ion chromatographic methods. However, these methods are subject to a number of limitations (for example, the inability to distinguish the parent acid and the acid oxide) and interferences. An alternative approach to the analysis of not only mineral acids and their oxides but also short chain fatty acids and their anhydrides is based on their reaction with cyclohexene oxide:

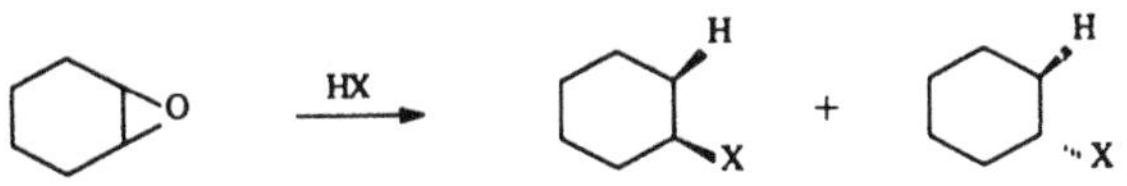

The derivatives formed are of good thermal stability and have adequate volatility for analysis by GC; though in some cases reaction of the parent analyte with cyclohexene oxide yields a mixture of *cis* and *trans* cyclohexyl derivatives.

Materials and equipment

Cyclohexene oxide supplied by Fluka Chemicals, 99% purity.
Concentrated hydrochloric and nitric acids ex BDH, Analar Grade.
Scrubber solution comprising cyclohexene oxide in dichloromethane (20% (v/v)).
Column: 30 m, bonded phase, BP20 (polyethylene glycol), film thickness 0.25 μm, diameter 0.22 mm supplied by SGE.
Capillary GC fitted with temperature programming unit and mass detector.

Method

Investigation of the reaction of hydrochloric and nitric acid with cyclohexene oxide. For the Hewlett Packard 5890 GC coupled to the 5971 Mass Sensitive Detector the instrument settings are as follows:

Helium: 9 psi.
Column temperature: initial, 150°C for 1 min; ramp, 10°C min^{-1} to 200°C.
Split: active 0.00–1.50 min; split ratio approximately 50 : 1.
Detector: total ion mode.

Prepare hydrochloric acid derivative(s) as follows. Place a small beaker containing the acid on a hot-plate stirrer inside a fume cupboard and heat gently. Sample the acid atmosphere by drawing vapour through a Dreshler bottle, containing scrubber solution (10 ml), connected to a suitable pump unit. Sample for approximately 30 min. Remove scrubber solution and dilute to 25 ml with dichloromethane. Use a similar procedure to prepare the nitric acid derivative(s). Make duplicate 0.5 μl injections of each solution. Print out the total ion chromatogram and the mass spectral detail for each chromatogram. Analyse data and by comparison with the spectral database assign structures to each peak.

Determination of the concentrations of a hydrochloric/nitric acid in a mixture. Prepare an external standard mixture for gas chromatographic analysis as follows. Add nitric acid (0.5 g) and hydrochloric acid (0.1 g) to dichloromethane (10 ml) followed by scrubber solution (10 ml). Stir for 15 min then dilute to 50 ml with dichloromethane. Generate a mixed acid vapour environment by placing a small beaker containing the acids in equal volume on a hot-plate stirrer inside a fume cupboard and heat gently. Sample the atmosphere as detailed in (a) but for 10 min. Note the volume sampled. Make 0.5 μl duplicate injections of standard and sample and hence determine the concentrations of the acids in the sample by direct comparison with peak areas of the standard (external standardisation). From this data and the volume of atmosphere sampled determine the short term exposure limit. By comparison with HSE recommended limits would this atmosphere pose a hazard? (See Figures 9.15–9.17.)

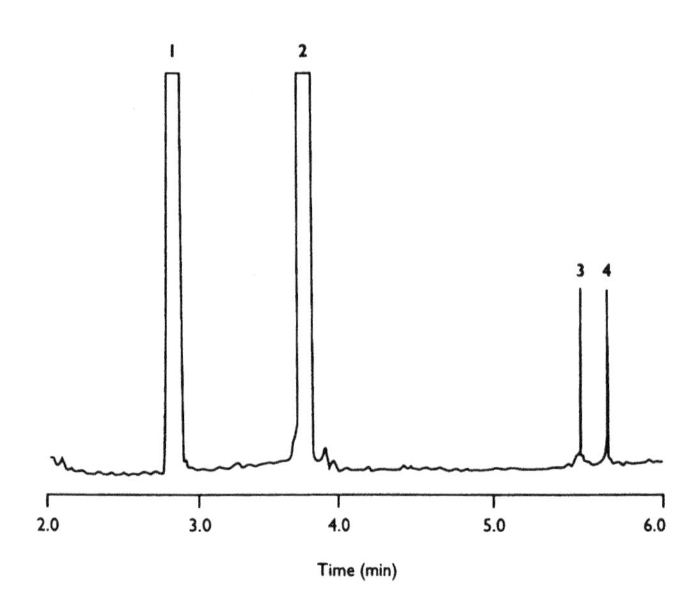

Figure 9.15 Total ion chromatograph of cyclohexene derivatives of N_2O_5, HNO_3 and HCl. Peaks: (1) 1,2-dinitratecyclohexane; (2) 1-hydroxy-2-nitrate cyclohexane; (3)/(4) *cis/trans*-1-chloro-2-hydroxycyclohexane.

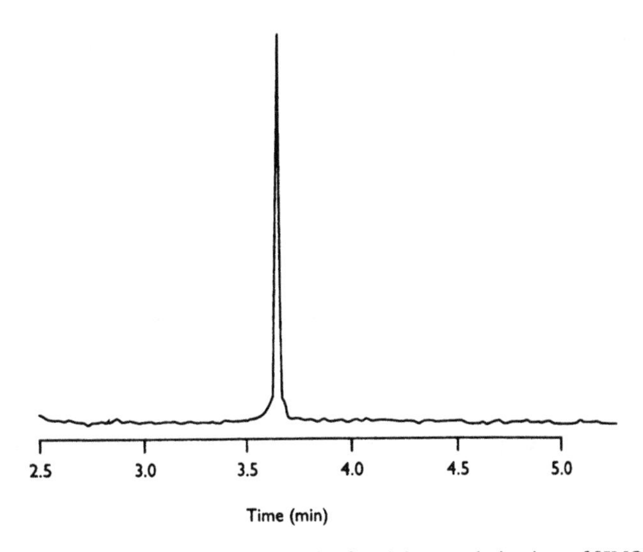

Figure 9.16 Total ion chromatograph of cyclohexene derivatives of HNO_3.

Addendum

The experiment can readily be extended to other systems, for example, a nitric acid/nitrogen oxide vapour produced by adding concentrated nitric acid to copper. The quantitative exercise could simply be carried out with flame ionisation detection; however, if both detection systems are available this forms the basis for a study and comparison of relative detector response using both total ion and selected ion monitoring.

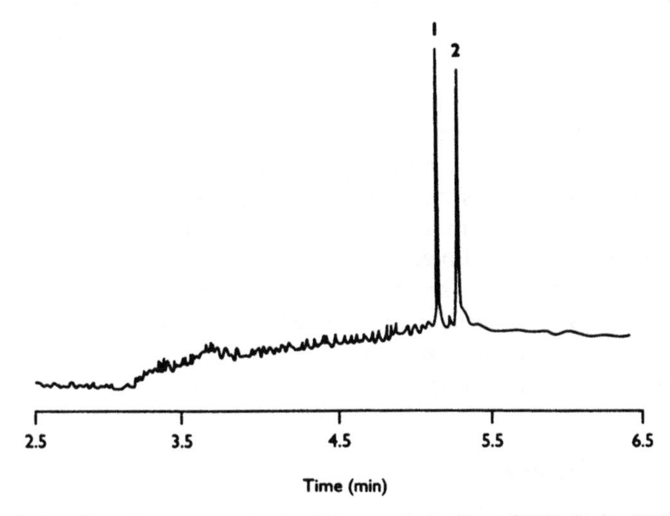

Figure 9.17 Total ion chromatograph of cyclohexene derivatives of HCl. Peaks: (1)/(2) *cis/trans* 1-chloro-2-hydroxycyclohexane.

9.8 SECTION F. High performance liquid chromatography

9.8.1 ***Experiment 26.*** *Analysis of barbiturates by reverse phase isocratic chromatography*

Object

(i) To investigate the influence of solvent strength on the degree of solution and analysis time for a range of barbiturates.
(ii) To determine the composition of a mixture of barbiturates.

Keywords

Reverse phase, solvent strength, barbiturates, ultraviolet detection.

Introduction

High efficiency column performance is achieved by the use of small diameter packing material (2–10 µm). The small size of the packing has the following consequences:

- the solvent must be pumped through under high pressure; and
- the stationary phase must be chemically bonded to the support material.

The most common stationary phase used is octadecylsilane (ODS) (see Chapter 6). Commonly a two-component solvent system is employed comprising a buffered aqueous solvent with methanol or acetonitrile added to enhance the solubility of non-polar organic sample components in the eluant. This is an example of reversed-phase chromatography. Using a solvent of constant composition for development of the chromatogram is called isocratic elution. If no single isocratic solvent mixture can be found that is effective for a mixture of components then solvent programming or gradient elution techniques may be used. Here elution is begun with a weak solvent and solvent strength is increased with time. The overall effect is to elute successively the more strongly retained substances and to achieve a reduction in the analysis time. The properties and uses of barbiturates have previously been discussed (see Experiment 18). The following exercise in quantitation uses the internal standard technique; it should be appreciated however that improved precision and accuracy may be obtained in HPLC using external standard calibrations (see Chapter 6).

Materials and equipment

Standard mixture: containing equal weights of barbitone, butobarbitone, phenobarbitone, amobarbitone and secobarbitone.

Unknown(s): containing a known weight (% w/w) of barbitone and two or more of the above barbiturates.
Column: Spherisorb—10 ODS, 250 mm × 4.6 mm i.d.
Mobile phase: solvent A—acetonitrile; solvent B—2 mM KH_2PO_4.
Equipment: gradient system with solvent programmer and variable wavelength detector.
Rheodyne valve fitted with 20 μl loop; computing integrator.

Method

The system should be set up as follows.

Mobile phase flow-rate: 2 ml min^{-1}
Detector wavelength: 225 nm.

Take 50 mg of the standard mixture and dilute with 20 ml of acetonitrile and make up to final volume of 25 ml with 2 mM KH_2PO_4. Find the optimum solvent composition for the separation of these components, i.e. that which requires the minimum analysis time but gives the necessary resolution of peaks (<10% overlap). Isocratic elution should be used with solvent mixtures of varying composition starting with 70% A, 30% B decreasing A down to about 25%. The response factors of the barbiturates should be calculated relative to barbitone. The order of elution of the components is

1. barbitone;
2. butobarbitone;
3. phenobarbitone;
4. amobarbitone; and
5. secobarbitone.

Hence, determine the composition of the unknown by taking 50 mg of mixture 2, which contains 20% by weight of barbitone and dilute to 25 ml as detailed for the standard solution. Two sets of solutions should be prepared and duplicate values for response factors and unknown composition obtained (Figure 9.18).

9.8.2 **Experiment 27.** *Ion pair chromatography of vitamins* [11, 12]

Object

To investigate the influence on retention times of vitamins due to varying the proportion of counter-ions in the eluant.

Keywords

Ion pair, pentane and heptane sulphonic acids, vitamins, reverse phase, ultraviolet detection.

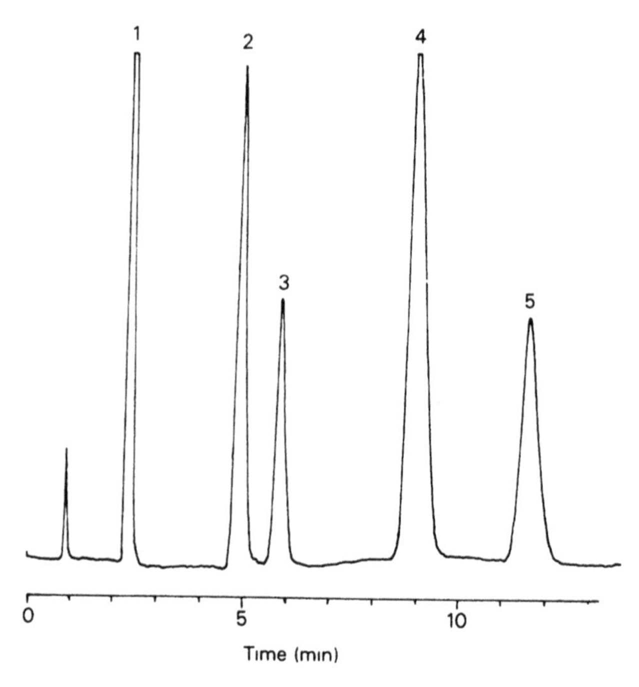

Figure 9.18 Chromatogram of a standard mixture of barbiturates, column and conditions as detailed in text. Eluant 25% solvent A/75% solvent B. Peaks: (1) barbitone; (2) butobarbitone; (3) phenobarbitone; (4) amobarbitone; (5) secobarbitone.

Introduction

A vitamin has been defined as a biologically active compound which acts as a controlling agent for an organism's health and growth. Vitamins are generally labile compounds and are susceptible to oxidation and breakdown when exposed to heat, oxygen and light; thus, HPLC in preference to GC has proved to be the method of analysis for such compounds. Vitamins may be classified as fat or water soluble. Chromatography of the former is generally achieved using reverse phase systems. Development of a general elution procedure for the latter class, however, has proved difficult due to the range of polarities encountered; for example, thiamine is strongly ionic while riboflavin has little tendency to ionise. Ion pair chromatography is a technique with which ionised compounds can be made to favour the organic stationary phase by using suitable counter-ions to form ion pairs according to the equation

$$\underset{}{A^+} + \underset{\text{counter-ion}}{B^-} \rightarrow \underset{\text{ion pair}}{[A^+B^-]}$$

The ion pairs behave as if they are non-ionic neutral species. For reverse phase ion pair chromatography, a non-polar surface (e.g. C_8 or C_{18}) is used as a stationary phase and an ionic alkyl compound is added to the aqueous mobile phase as a modifier. For the separation of acids, an organic base (e.g. tetrabutylammonium phosphate) is added to the eluant; for the separation of bases an organic acid (e.g. octane sulphonate) is used. The application of reversed phase ion pair chromatography to the separation of charged solutes has gained wide acceptance mainly because of the limitations of ion exchange to separate both neutral and ionic samples and because of the difficulty in separating ionic components by the reversed phase techniques of ion suppression. The combination of varying mobile phase strength and type of ion pairing reagent is often sufficient to enable the analyst to develop a separation. In addition to these parameters, it is also possible to mix two ion pairing reagents having different chain lengths to 'fine tune' or 'tailor' a separation. In some situations, when dealing with mixtures of ionic and non-ionisable compounds, mixing ion pair reagents to control retention is the only way to approach the development of a separation. For example, when a five-carbon alkyl chain is used to separate water-soluble vitamins, insufficient retention is observed. Changing the mobile phase counter-ion to a compound containing a seven-carbon alkyl chain increases the retention, particularly of the most strongly ionic compound. A mixture of counter-ions added to the mobile phase produces a retention proportional to the concentration of each counter-ion and is the approach by which the best separation can be achieved. The separation of water-soluble vitamins, niacin, pyridoxine, riboflavin, and thiamine, can be used to illustrate this approach.

Materials and equipment

Solvents: (a) methanol/1% acetic acid; (b) 2.5 mM pentanesulphonic acid and 2.5 mM heptanesulphonic acid in methanol: 1% acetic acid (25:75).
Solution 1: riboflavin (4 mg per 100 ml of methanol: 1% acetic acid (25:75)).
Solution 2: thiamine (20 mg) riboflavin (4 mg), pyridoxin (6 mg) and niacin (4 mg) in 100 ml of methanol: 1% acetic acid (25:75).
Column: μ-Bondapak column; 5 µm silica-ODS, 250 mm × 4.6 mm i.d.
Chromatograph, detector, integrator and injector system requirements as Experiment 26.

Method

The recommended instrument settings are as follows.

Mobile phase flow rate: 2 ml min^{-1}.
Detector wavelength: 270 nm.

Thiamine, pyridoxine and niacin are ionic, and riboflavin is non-ionic at the pH used to pair the ions of the ionic samples. The development of a

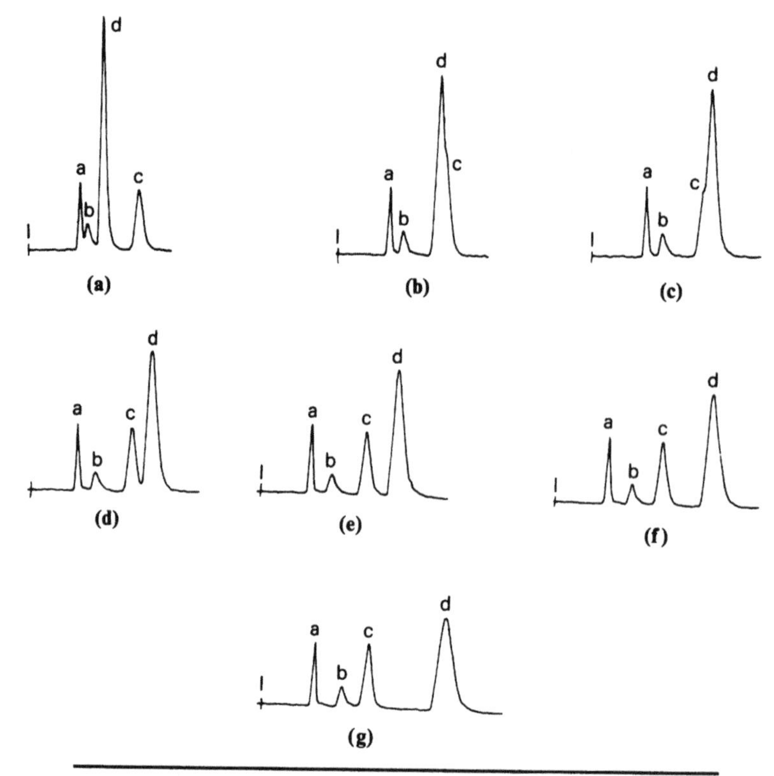

Chromatogram	Solvent %A:%B	Compound			
		a	b	c	d
1	100:0	2.34	2.70	4.99	3.33
2	75:25	2.38	3.00	Sh. > 4.53	4.53
3	60:40	2.40	3.14	Sh. < 5.23	5.23
4	50:50	2.40	3.20	4.80	5.64
5	40:60	2.42	3.37	4.85	6.20
6	25:75	2.42	3.50	4.82	6.98
7	0:100	2.43	3.66	4.75	8.10

Figure 9.19 Chromatograms and solvent data for standard mixture of vitamins. Solvent A: 2.5 mM pentane sulphonic acid in 25% methanol/75% 1% HOAc. Solvent B: 2.5 mM heptane sulphonic acid in 25% methanol/75% 1% HOAc. Peaks: (a) niacin; (b) pyridoxine; (c) riboflavin; (d) thiamine. See text for other details.

separation of these compounds requires a two-stop procedure. First adjust the methanol/water ratio (polarity of the mobile phase) to obtain good retention, approximately 6 min, of the non-ionic compound riboflavin (solution 1). Second, optimise the separation of the components in solution 2 by varying the relative proportions of pentane and heptanesulphonic acids while keeping the acid strength constant (5 mM). The base solvent for the acids is the methanol: 1% acetic acid composition established for the

6 min elution of riboflavin. Construct a graph of retention versus counter-ion composition for each component of the mixture (Figure 9.19).

9.8.3 *Experiment 28.* *Techniques in HPLC analysis of analgesics* [13, 14]

Object

This experiment demonstrates the techniques of solvent selection, gradient elution, pH control, ion pairing and wavelength selection in the analysis of an analgesic mixture using reversed phase HPLC on an octadecylsilane (ODS) column.

Keywords

Analgesics, reverse phase, isocratic and gradient elution, pH and ion pairing.

Introduction

Typical analgesic formulations contain components with a wide range of pK_a values. The HPLC analysis of such diverse samples can be aided by using techniques such as gradient elution, ion suppression and ion pairing. The relative sensitivity of the method can be adjusted by judicious choice of detector wavelength. The underlying principles of these techniques have been discussed in detail in Chapter 6.

Materials and equipment

Solvents: (a) phosphate buffers contain AR Na_2HPO_4 and NaH_2PO_4 (0.025 M), pH adjusted to requirements using NaOH or H_3PO_4; (b) tetrabutyl ammonium phosphate (0.005 M) buffered to pH 7.6; (c) 1% HOAc.
Mixture A: acetylsalicylic acid (86 mg); paracetamol (5.1 mg); salicylamide (87.8 mg); caffeine (28.1 mg); phenacetin (11.4 mg) in 50 ml of methanol; salicylic acid (6 mg) may be included because of the likelihood of its presence as an impurity in aspirin.
Mixture B: acetylsalicylic acid (86 mg) and salicylamide (87.8 mg) diluted as above.
Column: Spherisorb-10 ODS, 250 mm × 4.6 mm i.d.
Chromatograph: HPLC binary gradient system, variable wavelength UV detector.
Integrator: Hewlett-Packard 3390A.
Injector system: Rheodyne valve fitted with 10 μl loop.

Method

The recommended instrument settings are as follows.

Mobile phase flow-rate: 2 ml min^{-1}.
Detector wavelength: 254 nm; 235 nm.

Solvent selection and determination of an unknown. Isocratic elution of the analgesics may be investigated using a methanol/phosphate buffer (pH 7) and methanol/1% HOAc eluant. The methanol composition should be varied between 20 and 55%. For the optimum solvent composition the peaks can be identified by injecting the individual analgesics. Using the optimum solvent composition, a chromatogram with the detector set to 235 nm should be obtained and the relative responses compared with that recorded at 254 nm. The technique of gradient elution is demonstrated by injection of mixture A using the following conditions; initial solvent composition of approximately 15:85 methanol/pH 7 buffer, final solvent composition of approximately 60:40 methanol/pH 7 buffer, convex gradient (No. 3 of Waters Model 660 Solvent Programmer) using a run time of 12 min. When a suitable solvent composition has been determined the relative response factors of the individual components can be calculated as in experiment 25. An analgesic tablet may then be analysed quantitatively by crushing the tablet, extracting with methanol, making up to a known volume (e.g. 10 ml) and injecting the resulting solution into the HPLC sample valve via a luer filter.

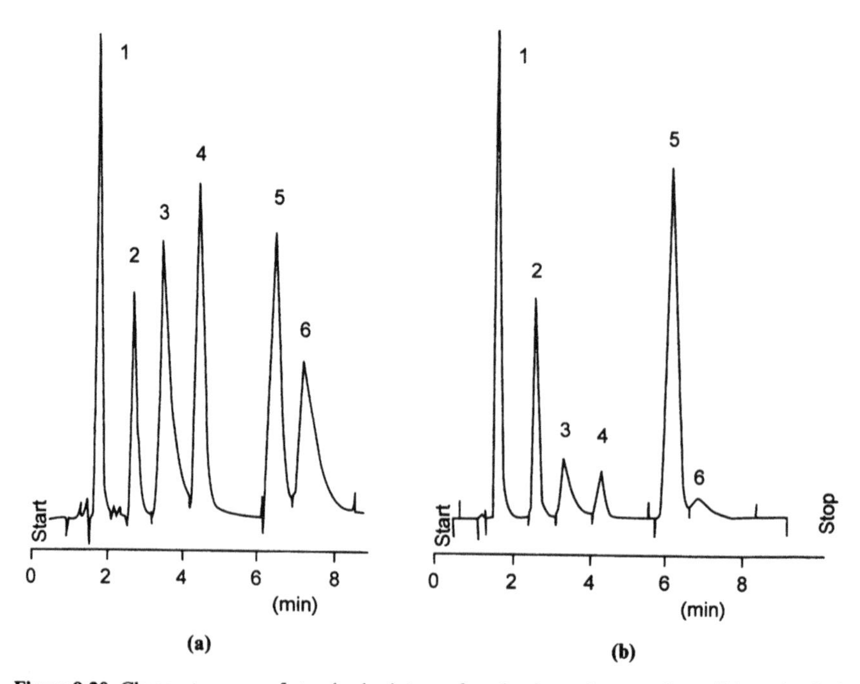

Figure 9.20 Chromatograms of standard mixture of analgesics, column and conditions detailed in text. Eluant 40% methanol/60% 1% HOAc. Detection (a) 235 nm (b) 254 nm. Peaks: (1) paracetamol; (2) caffeine; (3) salicylamide; (4) aspirin; (5) phenacetin; (6) salicylic acid.

pH effects. Solvents are prepared with the composition of approximately 45:55 methanol; buffer using phosphate buffers of pH 5.0, 7.0 and 9.0. After allowing the column to equilibrate to the solvent by flushing for 15 min, mixture A is injected and the peaks identified by injection of the individual solutes (Figure 9.20).

Ion pairing. The use of ion pairing in HPLC is investigated using tetrabutylammonium phosphate (TBAP) (solution ii), combined with methanol to give a solvent composition of approximately 45:55 methanol/TBAP. For comparison purposes, a solvent not containing TBAP is prepared using pH 7.6 phosphate buffer mixed with methanol in the same proportions as above. Mixture B is injected using both of the above solvents and the peaks are identified by comparison with chromatograms of the single analgesics obtained using the same solvents.

9.8.4 **Experiment 29.** *Analysis of amino acids as their DNP derivatives*

Object

To derivatise the amino acids of a protein hydrolysate with fluorodinitrobenzene (FDNB) and to identify and quantify by comparison with external standards containing the appropriate DNP amino acid derivatives.

Keywords

Amino acids, pre-column derivatisation, fluorodinitrobenzene.

Introduction

Derivatisation of samples in HPLC is undertaken principally for two reasons. First, there is no detector for HPLC that has universally high sensitivity for all solutes; hence a suitable chemical transformation of the solute can greatly extend the sensitivity and versatility of a selective detector. Second, sample derivatisation may be undertaken to enhance the detector response to sample bands relative to overlapping bands of no analytical interest. This experiment involves the pre-column derivatisation of sample with dinitrofluorobenzene so enhancing the spectrophotometric response of the sample, that is, the amino acids from a protein hydrolysate. The conversion of the amino acids to their apolar DNP derivatives allows them to be analysed using reverse phase chromatography. Thus, polar DNP-amino acids, such as DNP-aspartate, will elute early whereas apolar DNP-amino acids, such as DNP-alanine, will elute later. The DNP-amino acids are detected by an

Table 9.5

Mixture	DNP-alanine	DNP-aspartate	DNP-serine	DNP-threonine
I	10	50	10	50
II	20	40	20	40
III	30	30	30	30
IV	40	20	40	20
V	50	10	50	10

ultraviolet monitor set at 365 nm and the area under the peak determined by an integrating computer.

Materials and equipment

Standards: DNP-alanine, DNP-aspartate, DNP-serine and DNP-threonine in methanol (1 mg ml^{-1}).
Unknown: Protein sample.
Derivatising agent: fluorodinitrobenzene.
Column: Spherisorb-10 ODS, 250 MM × 4.6 i.d.
Solvents: A, methanol; B, water.
Chromatograph, detector, injection system and computing integrator requirements as for experiment 26.

Method

The recommended instrument settings are as follows.

Solvent flow: 2 ml min^{-1}.
Isocratic elution: 80% methanol, 20% H_2O.
Detector wavelength: 365 nm.

From the stock solutions of DNP-alanine, DNP-aspartate, DNP-serine and DNP-threonine provided make up working standards containing 0.05 mg ml^{-1} methanol of each of the DNP-amino acids. Inject 5 μl of each of these working standards and record the retention of each DNP-amino acid. Make up the calibration mixtures shown in Table 9.5, containing the volumes (μl) shown of each of the primary standards. To each calibration mixture add 1 ml methanol. Inject 5 μl of each calibration mixture. Identify the terminal amino acid of the protein sample as follows. React the protein with FDNB and then hydrolyse the derivative (6M HCl, 110°C, 20 h). Extract the hydrolysate with diethyl ether (3 × 5 ml) and evaporate to dryness; take up the residue in 1 ml of methanol. Chromatograph these solutions using the conditions established above.

1. Report the retention time of each DNP-amino acid and determine its reproducibility.
2. Construct calibration graphs.
3. Identify the *N*-terminal amino acid of protein X.

9.8.5 **Experiment 30.** *Analysis of paraben preservatives by HPLC with photodiode array detection*

Object

To illustrate the use of the photodiode array detector in HPLC method development.

Keywords

Parabens, lotions, photodiode array detection.

Introduction

Paraben is the generic name for a common group of preservatives used in various cosmetic and pharmaceutical preparations such as creams, emulsions and lotions. They are *p*-hydroxy benzoates and include the methyl, ethyl, propyl and butyl esters. The parabens are active against moulds, fungi and yeasts, but less active against bacteria. The higher esters are the most effective, but are limited in use by their lower solubility. Mixtures of two or more esters are generally more effective than the use of a single ester, since in this way a higher total concentration of preservative can be obtained in the solution and, in addition, the mixture may be active against a wider range of organisms—a mixture of methyl and propyl esters in the ratio of 2:1 is frequently encountered. In addition to microbiological contamination there is also the possibility of chemical contamination—for example, due to migration of phthalate plasticisers from the container or wrappings. The photodiode array (PDA) detector is perhaps one of the most versatile and useful HPLC detectors available today. Most PDAs monitor at least two different wavelengths to allow for the selective trace detection of compounds with different ultraviolet maxima. Complete ultraviolet spectra can be taken at any point during a chromatographic run to allow for comparison with a standard spectrum of a given analyte, which is an aid in qualitative analysis. Monitoring of 'peak purity' is also possible. In this technique, several spectra are collected at different times during the elution of a chromatographic peak. These spectra are then compared electronically for similarity, either by absorbance ratioing or full spectral overlay. Any differences between absorbance ratios or spectra indicate possible coelution of two or more components. The PDA is becoming increasingly important in both industrial and academic laboratories.

Materials and equipment

Solvents: A—methanol; B—water or 1% acetic acid.
Standards: methyl, ethyl, propyl and butyl *p*-hydroxy benzoates, benzophenone and diethylphthalate—1000 ppm in methanol.

Unknown: hand cream or lotion adulterated with diethylphthalate.
Column: spherisorb 5 μm, ODS 2, 250 mm × 4.6 mm.
Chromatograph: Perkin Elmer ternary gradient system comprising series 410 pump and controller with PDA detector.
Injector: Rheodyne valve fitted with 20 μl loop.
Integrator: HP 3390A or equivalent.

Method

The system should be set up as follows.

Mobile phase flow-rate 1.8 ml min^{-1}.
Detector wavelength 254 mm.

Pipette each of the standards (1 ml) into a 100 ml volumetric flask and dilute to volume with methanol/water (50:50) and store in the dark. The benzophenone serves as the internal standard. Weigh out approximately 0.5 g of sample, dissolve in 50 ml of methanol and dilute to 100 ml with water. Pass 2 ml of this solution through a pre-wet 'Sep-Pak' C18 cartridge. Discard the filtrate. Retrieve the 'analytes' by washing the cartridge with methanol (2 ml) and add 2 ml of 25 ppm benzophenone diluting to 5 ml with water. Obtain a chromatograph using a 80:20 methanol/water eluant. This initial operating condition should result in at least co-elution of two components in the standard mixture. The ultraviolet spectra should be obtained at each peak maxima, and in addition a purity index value determined for each peak. A peak index value >1 indicates peak inhomogeneity, i.e. co-eluting species. If this facility is not available on the PDA system then the spectral data should be obtained at each peak maxima and compared with reference spectra of each of the standard components to allow identification of the co-eluting peak(s). Subsequently, find the optimum solvent composition for separation of these components, i.e. that which requires the minimum analysis time but gives the necessary resolution of peaks (<10% overlap). Isocratic elution should be used with solvent mixtures of varying composition, starting with 80% A, 20% B and decreasing A in steps down to about 50%. Obtain spectra at each peak maxima and select appropriate wavelengths for optimum sensitivity. The response factors of the parabens and diethylphthalate should be determined relative to benzophenone. The order of elution of the components is

1. methyl paraben;
2. ethyl paraben;
3. propyl paraben;
4. diethylphthalate;
5. butyl paraben; and
6. benzophenone.

Hence, determine the composition of the unknown.

9.8.6 **Experiment 31.** *Determination of the amino acid composition of a peptide using pre-column derivatisation with* o*-phthalaldehyde and reverse phase HPLC and fluorescence detection*

Object

To determine the amino acid composition of a peptide using pre-column derivatisation with *o*-phthalaldehyde (OPA) and reverse phase HPLC and fluorescence detection.

Keywords

Pre-column derivatisation, *o*-phthalaldehyde, 2-mercaptoethanol, fluorescence detection.

Introduction

The separation and quantitation of amino acids have many applications in biochemistry and has traditionally been carried out using ion-exchange chromatography with post-column derivatisation using ninhydrin and detection with an ultraviolet–visible detector. This method has several drawbacks including requirements for specialised apparatus, long analysis times (1–8 h) and lack of sensitivity. Pre-column derivatisation with OPA and separation by reverse phase chromatography has become a popular method of analysis. Room temperature reaction of primary amines, OPA and a mercaptan produces a highly fluorescent thio substituted *iso*-indole derivative. The separation of the derivatives can be carried out on commercial HPLC instruments with no modification and limits of 50 fmols have been reported.

CHO, CHO + H_2N—CH(R)—COOH + HSH_2CH_2OH ⟶ $HOCH_2CH_2S$, N–R

Materials and equipment

Standard mixture of amino acids: aspartic acid, serine, histidine, alanine, methionine, phenylalanine, *iso*-leucine, leucine and lysine at 5 mM.
Internal standard: ornithine at 10 mM.
Sample column: 4 μm, C18, Nova-pak.
Mobile phase: solvent A, 20 mM sodium acetate buffered to pH 5.5; solvent B, methanol.
Chromatograph: Perkin Elmer 410 ternary gradient system coupled to Perkin Elmer 1000 M Fluorimeter with 7 μl flow cell; rheodyne valve with 10 μl loop.
Computing integrator.

Table 9.6

Standard	Amino acid	Internal standard
Standard 1	500 μM amino acids	1 mM ornithine
Standard 2	400 μM amino acids	1 mM ornithine
Standard 3	300 μM amino acids	1 mM ornithine
Standard 4	200 μM amino acids	1 mM ornithine
Standard 5	100 μM amino acids	1 mM ornithine

Method

The system should be set up as follows.

Mobile phase flow rate: 1.8 ml min^{-1}.
Gradient: 20% B to 100% B in 30 min; 100% B to 20% B in 10 min; 20% B to 20% B in 5 min; RETURN TO START.
Detection: excitation wavelength 337 nm; emission wavelength 455 nm.

From the supplied 5 mM mixed amino acid standard and 10 mM ornithine perform a series of dilutions to make the calibration mixtures shown in Table 9.6.

Derivatisation procedure

Using a finn-pipette add 250 μl OPA reagent (60 mg OPA/10 ml methanol) to 250 μl borate buffer pH 9.4; mix well. Add 25 μl of 2-mercaptoethanol (CARE—stench! the addition of mercaptoethanol must be carried out in the fume cupboard); mix well. To 250 μl of this mixture add 25 μl in the prepared sample mix, start a stopwatch and after 2 min inject 50 μl of this mixture onto the HPLC column. Keep mixing throughout the 2 min derivatisation. Start the gradient and the integrator. The same procedure is then performed on all the calibration mixtures and on the sample X. Do not begin the derivatisation until the HPLC is ready to begin the next analysis. Determine the retention time of the individual amino acids and comment on their reproducibility. Determine which amino acids are present in sample X. Construct calibration graphs for the amino acids by plotting peak area ratio against concentration. (Peak area of amino acid divided peak areas of the internal standard.) Determine the number of moles of amino acids in sample X and from the data given calculate the mole ratio of amino acids in sample X (Figure 9.21). [Also, using the following data, calculate the molarity of sample X. Sample X is a polypeptide whose molecular weight was calculated experimentally to be 2800. 20 mg of peptide was hydrolysed using vapour phase hydrolysis with 6 M HCl and heating at 105°C for 24 h. The hydrolysed peptide was then resuspended in 10 ml 0.1 M HCl. To a 5 ml aliquot of this suspension was added 5 ml of ornithine standard, the whole being diluted to a final volume of 50 ml.]

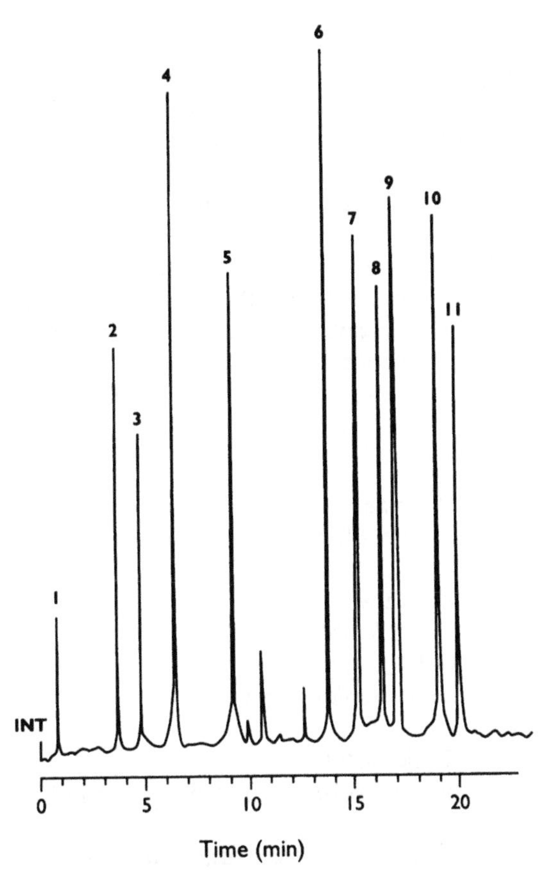

Figure 9.21 Chromatogram of standard mixture of derivatised amino acids; column and conditions as detailed in the text. Peaks: (1) aspartine; (2) seronine; (3) histidine; (4) glycine; (5) alanine; (6) methioine; (7) phenylalanine; (8) isoleucine; (9) ornithine; (10) lysine.

9.8.7 **Experiment 32.** *Analysis of inorganic anions in aqueous samples*

Object

To analyse qualitatively and quantitatively the inorganic anion contaminants in aqueous samples using an ion chromatograph system fitted with a suppressor column and conductivity cell.

Keywords

Ion chromatography, anions, Dionex.

Introduction

Initially, application of ion exchange to modern LC depended upon the analyte having a specific property, such as, ultraviolet absorbance, fluorescence or radioactivity. Conductivity detectors could not be used without modification, as the eluants used in ion exchange often contained complexing agents (citrate, EDTA), buffer solutions (carbonate, phosphate) and other various electrolytes required to achieve the desired resolution; thus the eluant itself produced an extremely high background conductivity making that due to the analyte species undetectable. This restriction was overcome by the work of Small, who developed a general technique for the removal of background electrolytes. The technique uses a secondary column, the scrubber column, which effectively removes ions arising from the background electrolyte leaving only the species of analytical interest, as the major conducting species in deionised water. Further detail on the principles and uses of such systems is presented in Chapter 6.

Materials and equipment

Standard solution: fluoride (2 ppm), chloride (3 ppm), nitrite (5 ppm), bromide (10 ppm), nitrate (30 ppm), phosphate (15 ppm), and sulphate (15 ppm).
Unknown: river or potable water sample.
Eluant: 1.8 mM Na_2CO_3/1.7 mM $NaHCO_3$.
Regenerant: 25 mM H_2SO_4.
Instrument: Dionex ion chromatograph.
Detector: Conductivity.
Integrator: Spectra Physics 4290.
Columns: Anion Separator (AS 4A/SC); Anion Guard (AG 4A/SC); Anion Suppressor (AFS).
Injector: multiport valve, air actuated, with 50 μl loop.

Method

The recommended instrument settings are as follows.

Eluant flow-rate: 2 ml min^{-1}.
Regenerant flow-rate: 3 ml min^{-1}.
Detector range: 10–30 μS.

All dilutions of standards and samples must be made with deionised water of conductivity <10 μS or better. A set of standards covering the following concentration range should be prepared (Figure 9.22).

F^-	2–0.2 ppm
Cl^-	10–1 ppm
NO_2^-	10–1 ppm
Br^-	5–1 ppm
PO_4^{3-}	10–4 ppm
NO_3^-	50–5 ppm
SO_4^{2-}	50–5 ppm

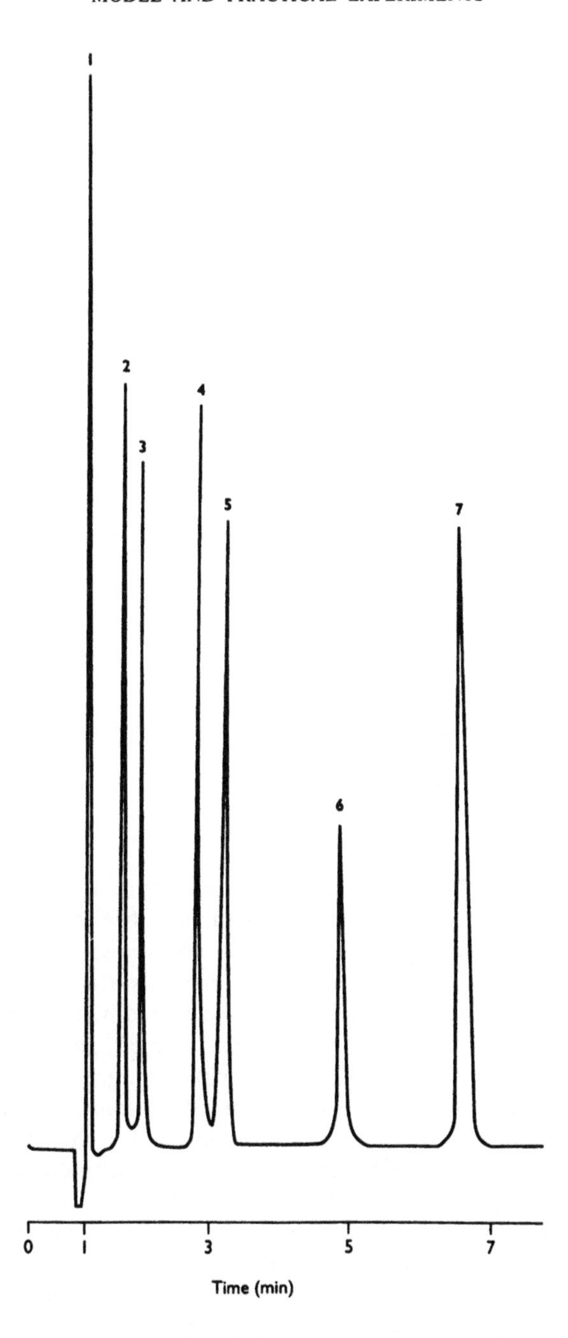

Figure 9.22 Chromatogram of a standard mixture of anions, column and conditions as described in text. Peaks: (1) fluoride; (2) chloride; (3) nitrite; (4) bromide; (5) nitrate; (6) phosphate; (7) sulphate.

Aqueous samples can be diluted such that the analyte levels lie within the calibration range.

1. Report the retention times of the analyte species.
2. Construct calibration graphs.
3. Analyse the sample provided, taking account of any diluted factors required. Alternative samples such as rain and river or tap water may be used diluted with ultra high water as required.

9.9 SECTION G. Capillary electrophoresis

9.9.1 ***Experiment 33.*** *Determination of inorganic cations by CE*

(Adapted with permission, from Dionex Application Note 91)

Object

To analyse a mixture of alkali and alkaline earth metals by capillary electrophoresis (CE) with indirect ultraviolet detection.

Keywords

Alkali and alkaline earth metals, capillary electrophoresis, indirect ultraviolet detection.

Introduction

Alkali and alkaline earth metals are routinely monitored in a variety of aqueous samples that are important to several industries such as the drinking and waste water industries, pharmaceutical companies, and metal plating industries. Although the most common methods for determining total concentrations of these metals have been spectroscopic techniques, such as atomic absorption (AA) and inductively coupled plasma and atomic emission spectroscopy (ICP–AES), ion chromatography has been the preferred method for the analysis of free metals in solution. More recently, CE has also been shown to be a viable option for the analysis of free metals in solution. CE is a simple, high resolution technique that separates charged species based on their relative mobilities under the influence of an applied electric field. In cation analysis by CE, complexing agents are added to the electrolyte to aid in the separation of the metal ions. The mobilities of the ions, and therefore their ease of separation, can be approximated from their equivalent ionic conductivities (see Figure 9.1). The equivalent ionic conductivities for alkali metals are somewhat different, but for the transition metals they are virtually identical and therefore more difficult to separate by CE. It is often possible to obtain separations between metal ions that would otherwise not be separable by adding complexing agents that complex the metals to different degrees. This application note describes the use of different complexing agents to modulate the selectivity of the separation, depending upon the ions of interest. The ions are detected by indirect photometric detection, using dimethyldiphenylphosphonium hydroxide (DDPOH) as the ultraviolet absorbing background visualising agent.

Materials and equipment

Electrolyte stock solutions: DDP iodide (25 mM) (dissolve 2.14 g in 250 ml of 18 MΩ deionised water. Since the salt does not instantly dissolve, sonication will speed up the process. DDP iodide is not soluble in water at concentrations above 25 mM); 18-crown-6 ether (40 mM) (dissolve 2.114 g in 200 ml of 18 MΩ deionised water); citric acid (500 mM) (dissolve 10.5 g citric acid monohydrate in 100 ml of 18 MΩ deionised water); α-hydroxyisobutyric acid (100 mM) (dissolve 2.08 g in 200 ml of 18 MΩ deionised water).
Column: fused silica capillary, 50 μm i.d. and 375 μm o.d. 45 cm to detector may be obtained from Dionex UK Ltd.
Capillary electrophoresis system.

To make 100 ml of the electrolyte, place 20 ml of 25 mM DDP hydroxide into a 100 ml plastic volumetric flask. Add 5 ml of 40 mM 18-crown-6 ether, 6 ml of 100 mM α-hydroxyisobutyric acid and bring to volume with 18 MΩ deionised water. The pH of this solution should be pH 4.5 ± 0.01.

Preparation of DDP hydroxide (25 mM). Convert the 25 mM DDP iodide solution into the hydroxide form by passing down an anion exchange column in the hydroxide state. Amberlite resin IRN 78 is a reasonably clean resin and is the resin of choice. Prepare the Amberlite resin for use by packing into a gravity fed column, then passing through it approximately 8 column volumes of 1 M NaOH, followed by sufficient 18 MΩ deionised water to bring the eluant pH down below pH 7. The first column volume of stock DDP iodide solution coming through the column should be discarded before beginning collection of the DDP hydroxide. Approximately 10 column volumes of DDP hydroxide may be collected before the column will require regeneration.

Stock standard solutions. Cation standards, including those of transition metals, should be made from salts wherever possible. Atomic absorption standards which are preserved in acid cause downward spikes in the electropherogram and should not be used. Prepare 1000 mg $litre^{-1}$ (1000 ppm) standards from pure (ACS grade) salts, by dissolving the appropriate weight of the salts.

Method

The recommended instrument settings for the Dionex CEST system are as follows.

Injection: hydrostatic, 100 mm for 30 s.
Polarity: (+), detector side cathodic.
Detection: indirect UV, 210 nm.
Temperature: ambient.
Control mode: constant voltage, 20 kV.

A set of multianalyte standards covering the following concentration ranges should be prepared by appropriate dilution of the stock standard solution.

NH_4^+	1.0–0.1 ppm
K^+	2.0–0.2 ppm
Ca^{2+}	2.0–0.2 ppm
Na^+	2.0–0.2 ppm
Mg^{2+}	1.0–0.1 ppm
Sr^{2+}	2.0–0.2 ppm
Ba^{2+}	4.0–0.5 ppm
Li^+	0.25–0.05 ppm

Aqueous samples can be diluted such that the analyte levels lie within the calibration range (Figure 9.23).

1. Report the retention times of the analyte species.
2. Construct calibration graphs.

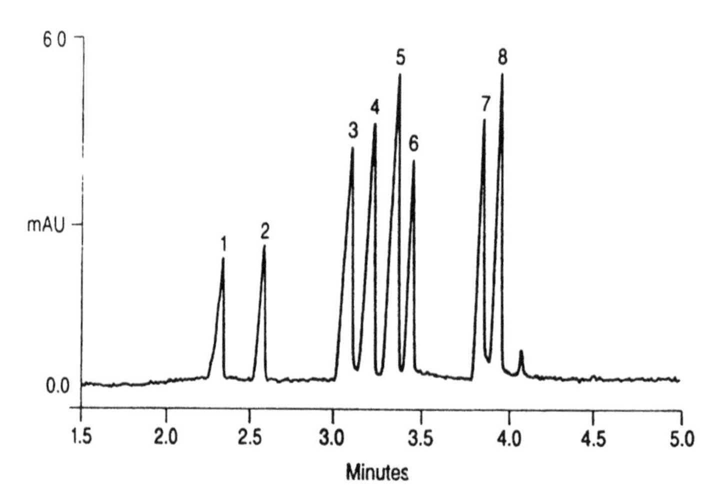

Figure 9.23 Electropherogram showing separation of alkali and alkaline earth metals and ammonium by capillary electrophoresis; column and conditions as detailed in text. Peaks (1) NH_4^+, (2) K^+, (3) Ca^{2+}, (4) Na^+, (5) Mg^{2+}, (6) Sr^{2+}, (7) Ba^{2+} and (8) Li^+. (Reproduced by permission of Dionex UK Ltd.)

3. Analyse the same provided, taking account of any dilution factors required.

9.9.2 ***Experiment 34.*** *Analysis of analgesics by CZE*
(Adapted with permission, from Dionex Application Note AU128)

Object

To analyse a proprietary analgesic formulation by capillary zone electrophoresis (CZE).

Keywords

Analgesics, capillary zone electrophoresis.

Introduction

CE is a fast and efficient analytical technique which can be employed for the analysis of pharmaceutical formulations, for example, analgesics. CE is capable of separating neutral and charged species in a single run with minimal sample and buffer consumption. Separation of impurities in the sample can easily be accomplished by CE. Automation in CE is important for reproducibility of measurements and provides quick screening of numerous samples. In CZE, the resolution of a mixture of compounds is strongly influenced by the pH of the buffer in the capillary. The pH affects the charge of the analytes and thus their electrophoretic mobilities. Additionally, the pH of the buffer affects the charge on the capillary surface, which contributes to the velocity of the electro-osmotic flow.

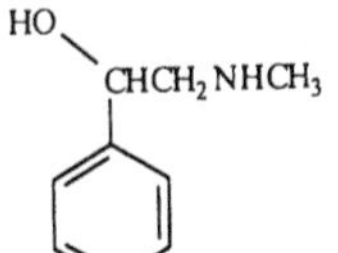

Phenylephrine

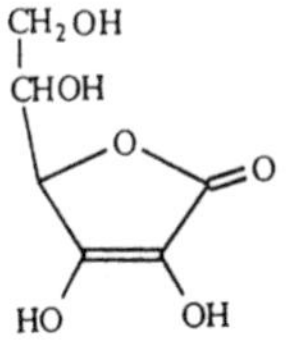

Ascorbic acid

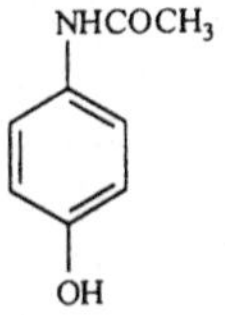

Acetaminophen

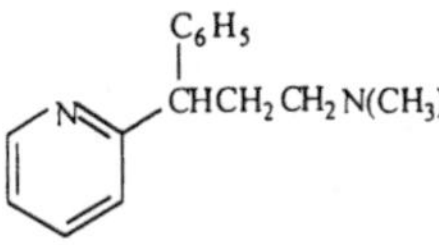

Pheniramine

Materials and equipment

Electrolyte: (a) 10 mM AR disodium hydrogen phosphate ($Na_2HPO_4.7H_2O$) adjusted to pH 6.0 with HCl; (b) 10 mM AR sodium dihydrogen phosphate ($NaH_2PO_4.H_2O$) (pH 6.9).
Stock solution: comprising phenylephrine (300 ppm), pheniramine (500 ppm), acetaminophen (400 ppm) and ascorbic acid (1000 ppm) in deionised water.
Unknown: pharmaceutical formulation, e.g. Neo Citran^{TM} or pseudo-unknown containing two or more of the above.
Capillary electrophoresis system.
Computing integrator.

Method

For the Dionex Capillary Electrophoresis System I (CES I) the recommended instrument settings are as follows.

Polarity: (+).
Capillary: 75 μm i.d. × 375 μm o.d. × 100 cm (95 cm to detector).
Control mode: constant voltage, 18 kV.
Detection: ultraviolet 215 nm.
Injection: gravity, 50 mm, 5 s.
Temperature: ambient, no air-cooling.

The electrolyte buffers should be filtered through a 0.45 μm membrane and degassed with sonication for at least 10 min before use. Analyte standard solutions should be prepared as follows:

1. 1 ml of stock solution diluted to 10 ml with deionised water (standard A).
2. 1 ml of stock solution diluted to 50 ml with deionised water (standard B).

Sample should be dissolved in deionised water to yield sample solution with analyte concentrations within limits of standards. Standards and sample should be analysed in duplicate. The concentration of analyte species in the unknown should be determined by comparison with peak area data of the standards.

Note

The separations should be performed using the (+) polarity power supply (detector side cathodic), thus ensuring that the electroosmotic flow will be towards the cathode. Maleic acid (from pheniramine maleate) will elute at retention times greater than ascorbic acid because it is a divalent anion, and its charge-to-mass ratio is greater than that of ascorbic acid. Similarly, the bromide anion (from dextromethorphan HBr), which is a highly mobile anionic species, will elute at much greater retention times than ascorbic acid (Figure 9.24).

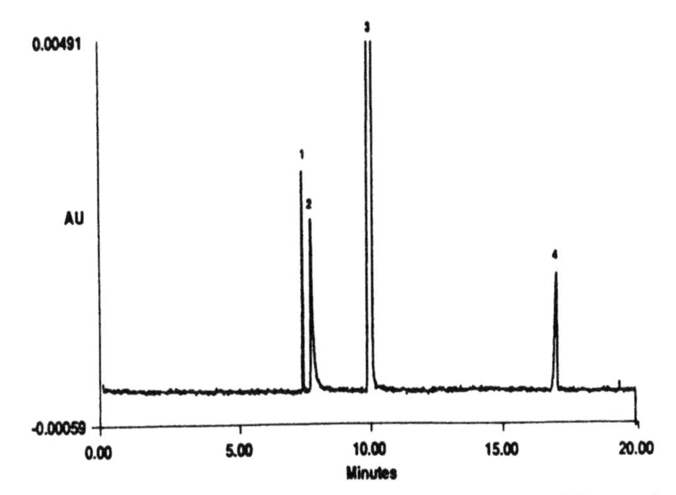

Figure 9.24 Electropherogram of standard mixture of analgesics; conditions as detailed in the text. Peaks (1) phenylephrine, (2) pheniramine, (3) acetaminophen and (4) ascorbic acid. (Reproduced with permission of Dionex UK Ltd.)

Neo Citran is a trademark of Sandoz Canada.

References

[1] Hunt, E.C., North, A.A. and Wells, R.A. *Analyst,* **80** (1955) 172.
[2] Jakubovic, A.O. and Knight, C.S. *Chromatographic and Electrophoretic Techniques*, Vol. 1 (Smith, I., ed.), p. 559. Heinemann, London, 1960.
[3] Still, W.C., Khan, M. and Mitra, A. *J. Org. Chem.*, **43** (1978) 2923.
[4] Sambrook, J., Fritsch, E.F. and Maniatis, T. *Molecular Cloning—A Laboratory Manual.* Cold Spring Harbor Press.
[5] Rice, G.W. *J. Chem. Educ.*, **64** (1987) 1055.
[6] Lehtonen, M. and Soumalainen, H. *Proc. Biochem.*, **14** (1979) 5.
[7] Dent, C.E. *J. Biochem.*, **43** (1948) 169.
[8] Anon. *Quart. J. Studies Alcohol*, **Suppl. 5** (1970).
[9] Smith, F.J. *et al. Analyst,* in preparation.
[10] Lean, G., Hinrickson, D. and Markham, A. *Atlas of the Environment.* Arrow Books, London, 1990.
[11] Wills, R.B.H., Shaw, C.G. and Day, W.R. *J. Chromatogr. Sci.*, **15** (1977) 262.
[12] Conrad, E.C. *Food Prod. Develop.*, **9** (1975) 97.
[13] Kagel, R.A. and Farwell, S.O. *J. Educ.*, **60** (1983) 163.
[14] Haddad, P., Hutchins, S. and Tuffy, M. *J. Educ.*, **60** (1983) 166.

10 Illustrative problems in chromatography

10.1 Introduction

The earlier chapters of the text have examined the qualitative and quantitative aspects of an extensive range of chromatographic techniques. The following is not intended to be an exhaustive set of problems, but is aimed at consolidating theoretical knowledge by application to a broad range of problem-types. The questions have been grouped under the following headings:

plate theory and resolution, van Deemter theory, volatility and temperature in GC, qualitative analysis, and quantitative analysis.

Before tackling these problems the student should be familiar with the principles and other relevant material presented in the main body of the text. For convenience the main relationships are presented before each of the above subsections.

10.2 Plate theory and resolution

One or more of the relationships detailed below may be used to solve the questions in this section. The equations are simply reproduced here for the reader's convenience. A detailed treatment of these equations and others is presented in Chapter 2.

$$k' = K\frac{V_S}{V_M} \qquad H = \frac{L}{N} \qquad N = 16\left(\frac{t'_R}{w_b}\right)^2$$

$$N_{eff} = 5.54\left(\frac{t_R}{w_h}\right)^2 \qquad t_R = t_M(1+k)$$

$$\alpha = \frac{t'_{R_B}}{t'_{R_A}} \qquad V_R = t_R F_C$$

$$R_S = \frac{2(t_{R_B} - t_{R_A})}{(w_{b,A} + w_{b,B})} \qquad R_S = \tfrac{1}{4}\sqrt{N_{eff}}\left[\left(\frac{\alpha - 1}{\alpha}\right)\right]\left[\frac{k}{(k+1)}\right]$$

10.2.1. Calculate (i) the effective theoretical plate number of a GC column for a compound retention time (t_R) of 3.64 min and with a peak width of 0.33 min. The void peak, t_0, is 0.21 min.

(Answer: $N = 1729$)

Table 10.1

Component	t_R (min)
Amobarbital	3.53
Pentobarbital	4.17
Phenobarbital	5.11
Alphenol	6.31
Methaqualone	10.74

10.2.2. Compare the peak width when $t'_R = 6.0$ min with the peak width when $t'_R = 3.64$ min, for a column where $N = 1600$.

(Answer: peak widths are 0.60 and 0.36 min)

10.2.3. The data shown in Table 10.1 were obtained on a reversed-phase HPLC column. All retention times were measured from the peak of a non retained component. Calculate the relative retention of each component with respect to alphenol. Determine the fraction of time that an average molecule of phenobarbital spends in the mobile phase if t_m is 43 s.

(Answer: $\alpha_{Amo} = 0.55; \alpha_{Pento} = 0.66; \alpha_{Pheno} = 0.81; \alpha_{Methaq} = 1.70$)

10.2.4. The retention times shown in Table 10.2 were recorded for the separation of a series of chlorinated insecticides on a reversed phase HPLC column. Assuming a flow-rate of 2.10 $cm^3\ min^{-1}$, calculate the retention volume for each insecticide.

(Answer: respective retention volumes are 7.85, 11.03, 14.51 and 22.55 cm^3)

10.2.5. Calculate H and N for a 25.0 cm column for which didodecylphthalate was found to have a retention time of 9.59 min and a peak width of 1.20 min.

(Answer: $N = 1022; H = 0.244$ mm)

10.2.6. Calculate H and N for a 30.0 cm column if 10-bromoanthracene has a retention time of 17.6 min and a half-peak width of 0.59 min.

(Answer: $N = 4930; H = 0.061$ mm)

Table 10.2

Pesticide	t_R (min)
DDT	3.73
Lindane	5.25
Dicofol	6.91
Hexachlorobenzene	10.74

10.2.7. The width at the base of a chromatographic peak is 31 s and the retention time of the peak is 8.32 min. Calculate the HETP for a column of 1.25 m length.

(Answer: $N = 4149$; $H = 0.301$ mm)

10.2.8. The retention time of an alkenylbenzene on a GC column 1.00 m in length was 11.0 min. The width of the chromatographic peak was 24 s. Calculate the HETP for the column.

(Answer: $N = 12\,100$; $H = 0.826$ mm)

10.2.9. A chromatographic peak has an asymmetry factor, measured at 10% of the peak height, of 1.58. The retention time of the peak is 5.57 min and the peak width at 10% of the peak height is 37 s. Determine N, the number of theoretical plates in the column.

(Answer: $N = 1305$)

10.2.10. Calculate the resolution between two chromatographic peaks which have retention times of 17.5 and 19.55 min and peak widths of 130 and 182 s, respectively.

(Answer: $R_S = 0.79$)

10.2.11. Calculate the resolution between two chromatographic peaks if the retention time corresponding to the first peak is 10.52 min, the retention time corresponding to the second peak is 11.36 min, and the widths of the two peaks are 0.38 and 0.48 min, respectively. Is the resolution adequate for analysis?

(Answer: $R_S = 1.95$; this corresponds to peak overlap of $<0.45\%$)

10.2.12. An asymmetrical chromatographic peak had a retention time of 12.75 min and a peak width after (to the right of) the retention time of 0.35 min. An adjacent peak had a retention time of 14.78 min and a peak width before at the leading edge of the peak of 0.55 min. Determine the resolution between the peaks.

(Answer: $R_S = 1.44$)

10.2.13. Calculate the asymmetry factor at 10% of the peak height for a chromatographic peak for which the peak width at 10% of the peak height is 0.38 min to the front of the peak and 0.75 min at the rear of the peak.

(Answer: asymmetry factor = 1.79)

10.2.14. Calculate the resolution for compounds A and B, where t_{Ra} and t_{Rb} are 5.73 and 6.36 min, respectively, and where the respective widths of peaks at their base are 0.56 min and 0.71 min.

(Answer: $R_S = 0.99$)

10.2.15. Determine the peak widths and possible overlap of compound A ($t_R = 4.4$ min) and compound B ($t_R = 5.0$ min), when chromatographed on (i) a column of length 900 mm and $H = 1.0$ mm, and (ii) on a less efficient column with $L = 900$ mm and $H = 3$ mm.

(Answer: (i) $w_{b,A} = 0.59$ min, $w_{b,B} = 0.44$ min, $R_S = 0.95$; (ii) $w_{b,A} = 1.02$ min, $w_{b,B} = 1.15$ min, $R_S = 0.55$)

10.2.16. A well packed column 25 cm long containing 10 mm porous ion exchange beads was used to separate roughly equal quantities of tri- and tetra-phosphate ions, $P_3O_{10}^{5-}$ and $P_3O_{13}^{6-}$, respectively. The $P_3O_{10}^{5-}$ ion eluted after 9.6 min the peak taking 0.5 min to pass. The $P_4O_{13}^{6-}$ ion is eluted after 10.0 min with a similar peak width. An unabsorbed material eluted after 2.0 min. Calculate: (i) the number of theoretical plates on the column (using adjusted retention times); (ii) the resolving power of the column; and (iii) the reduced plate height of the column.

(Answer: (i) $N = 3597$; (ii) $R_S = 0.80$; (iii) $H = 6.76 \times 10^{-3}$ cm)

10.2.17. On a capillary GC column 30 m long, the retention time of an unretained peak is 41.5 s and the retention time of *n*-dodecanol is 12.6 min. Determine the following: (i) the velocity of the mobile phase; (ii) the R_f value for the *n*-dodecanol zone; (iii) the velocity of the *n*-dodecanol zone.

(Answer: (i) $u = 72.3$ cm s^{-1}; (ii) $R_f = 0.063$; (iii) $u_{n\text{-dodecanol}} = 4.59$ cm s^{-1})

10.2.18. A particular chromatographic column has an efficiency corresponding to 3700 theoretical plates and has retention times for pyrene and 2-methylanthracene of 13.05 and 12.79 min, respectively. To what degree can these compounds be resolved on this column? How many theoretical plates would be required for unit resolution at those retention times?

(Answer: $R_S = 0.308$; $N = 39\,513$)

10.2.19. On a normal phase HPLC column 30 cm in length, operating with an efficiency of 3800 theoretical plates, the retention times of 5-aminoindole and 6-aminoindazole are 26.75 and 28.33 min, respectively. If these two compounds are to be separated with unit resolution, how many theoretical plates will be required? How long a column of this same type will be required in order to achieve this resolution if $H = 0.1$ mm?

(Answer: $N = 19\,380$; $L = 1.938$ m)

Table 10.3

Sample component	Retention time (min)
Air	0.15
Methyl acetate	2.33
Ethyl acetate	3.76
n-Propyl acetate	6.12
n-Butyl acetate	10.98

10.2.20. A sample containing two components was separated by TLC. After the chromatogram had developed for 15 min, the solvent front had moved 8.3 cm from the origin; component A had moved 7.5 cm; and component B had moved 2.3 cm. Calculate the retardation factors for components A and B.

(Answer: $R_{f,A} = 0.28$; $R_{f,B} = 0.90$)

10.2.21. The GLC data shown in Table 10.3 were obtained. Calculate the relative retention of components B, C and D with respect to component A.

(Answer: respective relative retentions are 1.66, 2.73 and 4.93)

10.2.22. The effective plate number N of an HPLC column is 1893. Calculate the resolution R_S achieved with this column for the two substances A and B, (a) when $k'_A = 1.15, k'_B = 1.27$ and (b) when $k'_A = 4.44$, $k'_B = 4.95$.

(Answer: (i) $R_S = 0.54$; (ii) $R_S = 0.90$)

10.3 Van Deemter

One or more of the relationships detailed below may be used to solve the questions in this section. The equations are simply reproduced here for the reader's convenience. A detailed treatment of these equations and others is presented in Chapter 2.

$$H = A + \frac{B}{u} + Cu$$

$$u_{OPT} = \sqrt{\frac{B}{C}} \qquad H_{MIN} = A + 2\sqrt{BC}$$

10.3.1. For a certain GC system, the parameters A, B and C have the numerical values 0.012, 0.25 and 0.0022, respectively, for the van Deemter equation, where the flow velocity, u, has the unit of cm s^{-1} and the HETP, H, is measured in centimetres.

(i) What is the optimum velocity of the carrier gas?
(ii) What is the HETP at optimum gas velocity?

Table 10.4

	A (cm)	B (cm^2 s^{-1})	C (s)
Column 1	0.15	0.47	0.005
Column 2	0.127	0.39	0.014

(iii) If the actual length of the column was 1 m, how many theoretical plates would be present?

(Answer: (i) $u_{OPT} = 10.66\,\text{cm s}^{-1}$; (ii) $H_{MIN} = 0.059\,\text{cm}$; (iii) $N = 1698$)

10.3.2. For a GC column operated under isothermal conditions the van Deemter constants were as follows: $A = 0.09\,\text{cm}$, $B = 0.37\,\text{cm}^2\,\text{s}^{-1}$, $C = 0.07\,\text{s}$. Calculate the values of H at several values of the linear velocity of the carrier gas (u), and from the graphical presentation of the results calculate:

(i) the optimum velocity of the carrier gas; and
(ii) the corresponding minimum value for the HETP.

(Answer: (i) $u_{OPT} = 2.30\,\text{cm s}^{-1}$; (ii) $H_{MIN} = 0.141\,\text{cm}$)

10.3.3. The constants A, B and C of the van Deemter Equation have the numerical values shown in Table 10.4 for two columns of equal length. Determine (i) which of the two columns gives the larger theoretical plate number, if the flow velocity of the carrier gas is $10.6\,\text{cm s}^{-1}$, and (ii) the optimum velocity (u_{OPT}) for each of the two columns?

(Answer: (i) column 1; (ii) $u_{OPT} = 1.77$ and $1.75\,\text{cm s}^{-1}$, respectively)

10.3.4. A capillary GLC column 50 m in length has an efficiency of 49 000 theoretical plates at a flow-rate of $15\,\text{cm s}^{-1}$ and an efficiency of 44 000 theoretical plates at a flow-rate of $40\,\text{cm s}^{-1}$. Determine (i) the optimum flow-rate, and (ii) the approximate efficiency obtained at that flow-rate?

(Answer: (i) $u_{OPT} = 21.66\,\text{cm s}^{-1}$; (ii) $N = 52\,410$)

10.3.5. The theoretical plate numbers for a GC system are found equal to 1000, 1540 and 1249, for linear carrier gas velocities of 1.05, 2.00 and $5.10\,\text{cm s}^{-1}$, respectively. Calculate the maximum possible theoretical plate number for this system, and the carrier gas velocity required to achieve it.

(Answer: (i) $u_{OPT} = 2.24\,\text{cm s}^{-1}$; (ii) $N = 1550$)

10.3.6. For a GC column the retention time t_R is 1.96 min for ethanol, 5.57 min for n-butanol, and 45 s for air. Calculate the expected retention times of n-propanol and n-pentanol.

(Answer: (i) $t_{R,PrOH} = 3.30\,\text{min}$, $t_{R,PrOH} = 9.40\,\text{min}$)

10.4 Volatility and temperature in gas chromatography

The questions in this section are concerned with the dependence of retention time on column temperature and carbon number. These relationships are expressed as follows:

$$\log t'_{\mathrm{R}} = \frac{A}{T} + B$$

$$\log t'_{\mathrm{R}} = An + B$$

10.4.1. A packed GC column gives a peak which has adjusted retention times of 25.27 and 18.34 min at 175 and 200°C, respectively. The column has an efficiency corresponding to 6200 theoretical plates. What is the highest column temperature which can be used such that the peak width will still be greater than 30 s? Assume that t_{R} does not differ greatly from t'_{R}.

(Answer: 225°C)

10.4.2. For a particular column, the adjusted retention time (t'_{R}) of n-decane is 20.0 min at a column temperature (T_{c}) of 220°C and 30.0 min at 210°C. If t'_{R} must not exceed 60.0 min, what is the lowest possible column temperature?

(Answer: 193°C)

10.4.3. The methyl ester of the 9-hydroxy linoleate has adjusted retention times of 65.5 and 55.3 min on a particular column at temperatures of 200 and 215°C, respectively. Will it be possible to elute this compound in less than 40 min, if the liquid phase in the column has a maximum operating temperature of 240°C?

(Answer: 245°C; therefore no!)

10.5 Qualitative analysis

The questions in this section are concerned in effect with the use of Kovats retention indices in GC. The Kovats retention index for a compound can be calculated from one of the following:

$$I = 100\left[\frac{\log t_{\mathrm{R,U}} - \log t_{\mathrm{R,Z}}}{\log t_{\mathrm{R,Z+1}} - \log t_{\mathrm{R,Z}}}\right] \qquad I = 100\left[n\left\{\frac{\log t_{\mathrm{R,U}} - \log t_{\mathrm{R,Z}}}{\log t_{\mathrm{R,Z+1}} - \log t_{\mathrm{R,Z}}}\right\} + z\right]$$

10.5.1. The data in Table 10.5 were obtained by using a GC column operated at constant temperature. Calculate the Kovats retention indices for sample

Table 10.5

Compound	Retention time (min)
Air	0.32
n-Pentane	4.49
A	6.23
n-Hexane	7.08
B	7.78
n-Heptane	11.52
C	16.24
n-Octane	18.11

components A, B and C. From a plot of I as a function of $\log(t_R - t_M)$ estimate the retention time of *n*-butane.

(Answer: $I_A = 572, I_B = 620, I_C = 777; t_{R,Butane} = 2.54\,\text{min}$)

10.5.2. The retention times, in seconds, in a GLC column are the following: air, 5.6; *n*-butane, 48.7; substance A, 85.4; *n*-hexane, 103.0; substance B, 152.0. Which one of substances A and B is a linear alkane?

(Answer: compound B, $I = 700$)

10.5.3. The unknown compounds U_1 and U_2 were co-injected into a GC with the homologous compounds A and B, having 5 and 8 carbon atoms in their molecules respectively. The retention times in minutes were the following: air, 0.5; compound A, 1.98; compound B, 11.7; compound U_1, 0.7; compound U_2, 3.4. What is the carbon number of compounds U_1 and U_2?

(Answer: U_1, 2; U_2, 6)

10.5.4. From the information given below identify the ester (A) which has the general formula

$$CH_3-\underset{}{\overset{CH_3}{\overset{|}{C}}}H.(CH_2)_n.COO(CH_2)_m.CH_3$$

Treatment of ester (A) with sodium methoxide in boiling methanol gave equimolar quantities of a methyl ester (B), and an alcohol (C). GC of the reaction product in the presence of *n*-octane and *n*-decane as internal standards gave the following adjusted retentions:

Component	C_8	B	C	C_{10}
Retention (s)	127	247	402	740

Under the same column conditions, authentic samples of methyl esters and alcohols gave the following retention data in Kovats' retention index units:

$$CH_3-\underset{}{\overset{CH_3}{\overset{|}{C}}}H.(CH_2)_n.COOCH_3 \quad CH_3.(CH_2)_m OH$$

$n = 4$	$I = 686$	$m = 4$	$I = 644$
$n = 5$	$I = 748$	$m = 5$	$I = 702$
$n = 10$	$I = 1064$	$m = 10$	$I = 989$
$n = 11$	$I = 1128$	$m = 11$	$I = 1048$

(Answer: $n = 7$, $m = 9$)

10.6 Quantitative analysis

Peak area data from chromatograms may be used to determine the concentrations or amounts of components in a mixture. The following and other equations are discussed more fully in Chapter 2:

$$D_{RF_X} = \frac{\text{Area}_{\text{Comp X}'} \times Amt \text{ Int St}}{\text{Area}_{\text{Int St}} \times Amt \text{ Comp X}}$$

$$Amt_X = \frac{\text{Area}_x}{(\text{Area}_{\text{Spike}} - \text{Area}_x)} \times \frac{Amt_{\text{Spike}} V_{\text{Spike}}}{V_X}$$

10.6.1. The GC data in Table 10.6 were obtained after injection of 2 μl portions of a sample component and standard solutions of the component. Calculate the concentration of the sample.

(Answer: 0.573 ml ml^{-1})

10.6.2. A mixture of the methyl ethanoate (*A*), methyl propanoate (*B*) and methyl butanoate (*C*) was analysed by GC. The area of each peak and the relative detector responses are given in Table 10.7. Use this information to determine the percentage of each compound in the mixture.

(Answer: *A*, 25.5%; *B*, 30.4%; *C*, 44.1%)

Table 10.6

Solution concentration (ml ml^{-1})	Peak area
0.200	1.43
0.400	2.86
0.600	4.29
0.800	5.73
1.000	7.16
Sample	4.10

Table 10.7

Compound	Area	Rel. det. response
A	14 171	1
B	21 573	1.28
C	34 755	1.42

Table 10.8

t_R	Area (%)	Area
1.29	14.98	9130
1.57	22.66	13 806
2.23	62.34	37 998

10.6.3. GC analysis of a mixture of ethanol, *n*-propanol and *n*-butanol on a carbowax-20M stationary phase gave the data shown in Table 10.8.

If the relative responses are ethanol 1.00, *n*-propanol 1.37 and *n*-butanol 1.63 calculate the % (v/v) composition of the mixture.

(Answer: ethanol 21.47% (v/v); *n*-propanol 23.71% (v/v); *n*-butanol 54.81% (v/v))

10.6.4. A mixture containing known amounts of methyl esters of stearic (*A*), palmitic (*B*) and lauric (*C*) acids in 10 cm^3 of diethyl ether, was analysed by GC. The peak area (Area 1) and the respective amount of ester present in this standard sample is given below. Calculate the response factors of palmitic and lauric methyl esters with respect to the methyl ester of stearic acid. A second sample (test) containing 75 mg of methyl stearate and unknown amounts of the other two esters in the same volume of diethyl ether as in the standard sample, was analysed. The peak area (Area 2) corresponding to each compound is given in Table 10.9. Determine the concentration (in mg cm^{-3}) of methyl palmitate and methyl laurate in the test sample.

(Answer: $D_{RF_{Palm}} = 1.12$; $D_{RF_{Lauric}} = 1.08$; conc. palmitic = 4.0 mg cm^{-3}; conc. lauric = 3.0 mg cm^{-3})

Table 10.9

Compound	Standard sample		Test sample	
	Amount (mg)	Area 1	Amount (mg)	Area 2
A	75	110 500	75	110 505
B	50	75 140	—	60 115
C	50	79 560	—	47 738

Table 10.10

	Standard		Test sample	
	t_R	Area	t_R	Area
Ethanol	1.29	9 130	1.30	9 157
Propanol	1.57	12 508	1.56	12 251
Butanol	2.23	14 881	2.25	19 682

10.6.5. Using the chromatographic data shown in Table 10.10, calculate the response factors of propanol and butanol with respect to ethanol and hence their concentrations in the test sample. The standard contains equal volumes of the alcohols, made up to 10 cm^3 with water. The test sample contains ethanol (1 cm^3) and analyte (2.5 cm^3), made up to 10 cm^3 with water.

(Answer: $D_{\mathrm{RF}_{\mathrm{Propanol}}} = 1.37$; $D_{\mathrm{RF}_{\mathrm{Butanol}}} = 1.67$; conc. propanol, 39.0% (v/v); conc. butanol, 52.8% (v/v))

10.6.6. Using the chromatographic data shown in Table 10.11 (A = barbitone, B = amobarbitone, C = secobarbitone), calculate the response factors of amobarbitone and secobarbitone with respect to barbitone and hence their amount in the test sample. The standard contains barbitone (100 mg), amobarbitone (50 mg) and secobarbitone (50 mg), made up to 100 cm^3 with water/ethanol. The test sample contains barbitone (100 mg) and analyte (125 g), made up to 100 cm^3 with water/ethanol.

(Answer: $D_{\mathrm{RF}_{\mathrm{Amo}}} = 1.047$; $D_{\mathrm{RF}_{\mathrm{Seco}}} = 0.808$; amount amo = 38.58 mg; amount seco = 49.86 mg)

10.6.7. A tablet weighing 1000 mg was analysed for aspirin, caffeine and paracetamol content. The tablet was crushed and extracted by a solvent mixture (methanol 50%, water 50%), 100 mg of phenaceten was added as the internal standard. The solution was then made up to 100 ml and subsequently analysed by HPLC. A solution of a mixture of standards was also analysed. This solution was obtained by taking 250 mg of each component and dissolving, making up to 100 ml using methanol/water (50:50). Using the

Table 10.11

RT	Area	Area (%)	RT	Area	Area (%)
0.26	7 335 600	97.315	0.24	9 512 000	98.186
2.41	105 020	1.393	2.41	97 273	1.004
4.91	54 984	0.729	4.94	39 295	0.406
6.88	42 408	0.563	6.82	39 187	0.405

RT = Retention time.

Table 10.12

RUN 1	Standard mixture 25% (w/w) of each component		
	Paracetamol	$t_R = 1.51$	38 801
	Caffeine	$t_R = 2.10$	14236
	Aspirin	$t_R = 2.65$	26 301
	Phenacetin	$t_R = 3.70$	39 594
RUN 2	Tablet solution		
	Paracetamol	$t_R = 1.54$	68 994
	Caffeine	$t_R = 2.11$	19050
	Aspirin	$t_R = 2.70$	53 480
	Phenacetin (internal standard)	$t_R = 3.80$	84922

results shown in Table 10.12, calculate the composition of the tablet. Note: the tablet contains the three active compounds and a binder.

(Answer: paracetamol, 8.29% (w/w); caffeine, 6.24% (w/w); aspirin, 9.48% (w/w)).

10.6.8. A sample of wine was analysed for ethanol content by GC using propan-1-ol (25 mg%) as the internal standard and 80 mg% standard ethanol solutions. A 2 ml aliquot of the wine was taken and diluted to 100 ml, replicate samples were taken. The samples for analysis were prepared by taking 400 μl of the standard solution or diluted wine and adding 1000 μl of propan-1-ol, the internal standard. 1 μl of this solution was then injected into the GC. Using the results shown in Table 10.13, calculate the ethanol content of the wine as % (v/v).

(Answer: $D_{RF_{EtOH}} = 0.335$; % ethanol (v/v) = 14.6)

10.6.9. The standard addition technique was used to assay a sample component by chromatography. From the results in Table 10.14, determine the concentration of the sample components.

Table 10.13

	Sample	Ethanol		Propanol	
		t_R (min)	Area	t_R (min)	Area
1	80 mg%	1.01	166408	1.62	156402
2	80 mg%	0.95	171848	1.59	157684
3	80 mg%	1.01	170208	1.59	157568
4	80 mg%	1.03	171066	1.61	161502
5	Wine	0.98	418628	1.60	158168
6	Wine	0.99	498642	1.64	157970
7	Wine	1.03	487992	1.63	157658
8	Wine	1.01	496630	1.61	157980

Notes: SG ethanol = 0.8.

Table 10.14

Added concentration (mg/ml)	Peak area
0 (sample)	3.72
1.23	7.00
3.47	12.70
4.89	16.30
6.24	20.01
7.15	22.25

Table 10.15

Pollutants	Concentration ($\mu g\, ml^{-1}$)
A	10
B	10
C	100
D	100

10.6.10. A sample of contaminated soil from an industrial waste tip has been analysed by ECD–GC for the pollutants 1,1,1-trichloroethane (*A*), 1,2-dichloroethane (*B*), 2,2,4-trimethylpentane (*i*-octane) (*C*), toluene (*D*). A sample of the soil (1.105 g) was extracted with 3 × 10 ml cyclohexane, the extracts combined and made up to 50 ml to form the analytical sample. Two analytical standards were also analysed. These were prepared from a stock solution by taking 1000 μl and 10 μl and diluting each to 10 ml with cyclohexane forming standard mixtures 1 and 2, respectively. Use the accompanying analytical and chromatographic data of Table 10.15 to calculate the amounts, in $\mu g\, g^{-1}$, of pollutants *A*, *B*, *C*, *D* in the soil.

Samples (1 μl) of standards 1 and 2 and the soil extract were injected onto a 25 m WCOT BP5 capillary column temperature programmed 50–260°C at $15°C\, min^{-1}$, helium flow-rate $1\, ml\, min^{-1}$, ECD, autoinjector with a precision of better than 99.4% (Table 10.16).

(Answer: $A = 31.5\, \mu g\, g^{-1}$; $B = 18.9\, \mu g\, g^{-1}$; $C = 1.68\, \mu g\, g^{-1}$; $D = 46.49\, \mu g\, g^{-1}$)

Table 10.16

Component	Retention time (min)	Peak areas		
		Standard 1	Standard 2	Soil extract
Cyclohexane	3.1	Solvent		
A	3.8	654 942	6571	456 651
B	4.1	385 531	3901	160 855
C	4.7	598 224	6011	2215
D	6.8	2 423 855	24 288	249 053

Table 10.17

Compound	t_R (min)	W (min)	Area	Det. res. factor
A	9.55	0.81	21 060	1.00
B	10.20	1.05	29 940	0.93
C	12.45	1.15	35 850	0.83

10.7 Composite questions

10.7.1. A mixture of three volatile compounds (A, B and C) was analysed by GC using a column of 50 cm length. The retention time (t_R), the peak width at base line (w_b), the area under each peak and the relative detector response factor, are given in Table 10.17. Under the same conditions a compound with no affinity for the stationary phase took 2.3 min to be eluted.

(i) Use data for compound A (see Table 10.17) to calculate: the effective plate number of the column and the effective HETP.
(ii) Use data for compounds B and C to determine the resolution power of the column.
(iii) Calculate the percentage of each compound in the mixture.

(Answer: (i) $N_{eff} = 1282$; (ii) $R_S = 2.05$; (iii) $A = 21.8\%$, $B = 33.4\%$, $C = 44.8\%$)

10.7.2. A mixture containing four hydroxy esters (A, B, C and D) in 10 cm^3 of diethyl ether, was analysed by HPLC using a 30 cm silica column. The peak area and the respective amount of each compound in a standard sample is given below. A second sample (test sample) of the same compound, but in different unknown proportions, was analysed. The peak area corresponding to each compound is given in Table 10.18.

(i) Use the data obtained from the analysis of the standard sample to determine the detector response factor for compounds B, C and D with respect to compound A.

Table 10.18

Compound	Amount (mg)	Standard sample	Test sample
		Peak area	Peak area
A	60	88 400	88 404
B	40	60 112	48 092
C	40	63 648	38 190
D	40	67 773	77 250

(ii) Given that the detector response factor for compound A is 1.00 and using the results obtained in (i) above, calculate the percentage (%, v/v) of each component of the test sample.

(iii) Given that the retention time of compound A is 20 min and its peak width at half height is 0.55 min, and also that under the same conditions a compound with no affinity for the stationary phase took 2.5 min to be eluted, determine the effective plate number on the column.

(Answer: (i) $D_{\mathrm{RF_B}} = 1.02$, $D_{\mathrm{RF_C}} = 1.08$, $D_{\mathrm{RF_D}} = 1.15$; (ii) $A = 37.1\%$, $B = 19.8\%$, $C = 14.9\%$, $D = 28.2\%$; (iii) $N_{\mathrm{eff}} = 5609$)

Appendix 1 Glossary of chromatographic terms

Definition of chromatography, IUPAC (1993) "Chromatography is a physical method of separation in which the components to be separated are distributed between two phases, one of which is stationary while the other moves in a definite direction."

Adjusted retention time t'_R, also known as corrected retention time, takes into account the dead time t_M of the column; see retention time and dead time.

$$t'_R = t_R - t_M$$

Adsorption chromatography mode of separation in which a solute or sample components are attracted to a solid surface, the stationary phase, by adsorption retention forces, the mobile phase may be a gas or liquid.

Adsorption retention forces attraction of a solute onto a solid stationary phase due to microporosity (pores 5–50 nm) and polar character (formation of van der Waal's forces and hydrogen bonding) of the surface, described by Langmuir isotherms (see isotherms).

Affinity chromatography separation effected by affinity of solute molecules for a bio-specific stationary phase consisting of complex organic molecules bonded to an inert support material, e.g. separation of proteins on a bonded antibody stationary phase. The technique is really selective filtration rather than chromatography.

Alumina Al_2O_3, slightly basic adsorbent used in liquid chromatography, particularly TLC, as a less acidic alternative to silica gel.

Anion exchange chromatography see ion exchange chromatography.

Asymmetry, A_s term used to describe non-symmetrical peaks measured by obtaining the ratio at 10% peak height h of the forward part, a and the rear part, b of a peak measured from the perpendicular line drawn from the peak maxima to the baseline.

$$A_S = \frac{b}{a} \quad \text{at } 10\% \ h$$

Back-flushing used to remove components in a sample held strongly on the stationary phase at the beginning of a column by reversing the flow of mobile phase after completion of an analysis.

Band (or zone) used to describe the distribution of a sample component (solute) as it travels through the chromatographic system.

Band or peak broadening the process whereby a solute band or spot spreads out and is diluted as it moves through the chromatographic system due to:

(i) variable pathways of solute through the stationary phase particles;
(ii) movement of solute molecules in the mobile phase;
(iii) mass transfer of solute molecules at the mobile phase–stationary phase boundary;

band broadening is measured as peak width, w_b; see van Deemter equation.

Band width see peak width.

Bonded stationary phase a stationary phase consisting of organic molecules bonded to a stationary phase support material (microparticulate silica gel), usually through siloxane bonds.

Bulk property detector an HPLC detector that responds to a bulk property of the mobile phase, such as refractive index, which is modified by the presence of solutes eluting from the column.

C_{18} see octadecylsilane.

Capacity factor or retention factor k, capacity of a stationary phase to attract a component, measured as the ratio of corrected retention time (t'_R) and column dead time (t_M)

$$k = \frac{(t_R - t_M)}{t_M} = \frac{t'_R}{t_M}$$

Capillary columns GC columns, e.g. WCOT, PLOT, having a bore of 0.1–0.7 mm with a 0.1–5 μm film of stationary phase coated onto the inner wall of a pure silica glass column 10–50 m long, high column efficiency, N_{eff}, and excellent resolution are achieved. Capacity of the stationary phase is very low so a sample injection splitter is used to introduce approximately 0.01 μl of sample onto the column.

Capillary column gas chromatography gas chromatography using columns 10–100 m long with an internal diameter of 0.1–0.7 mm, usually with the stationary phase bonded to the internal wall (WCOT). Capillary columns have high efficiencies (N_{eff}) and give rapid analysis times. The A term in the van Deemter equation is zero as the column does not contain stationary phase particles or packing.

Capillary electrophoresis, CE a high efficiency separation technique which is a cross between HPLC and gel electrophoresis. The separations are achieved by applying a high voltage (>1000 V cm^{-1}) across a narrow bore capillary column, 20–200 μm i.d. and 5–10 cm long, the ends of which dip into a buffer solution. The resulting electroosmotic and electrophoretic flow of buffer solution transports ionic species to a ultraviolet detector, separation occurring according to the electrophoretic properties of the sample components and pH of the buffer. Column efficiency is high with over 500 000 theoretical plates per column.

Carrier gas inert mobile phase gas for gas–liquid or gas–solid chromatography, usually nitrogen is used with packed columns, helium with capillary columns.

Cation exchange chromatography see ion exchange chromatography.

Chiral stationary phase used to separate optically active enantiomeric compounds, by bonding a stationary phase molecule that has enantiomeric properties to a solid support. The stationary phase therefore has specific optical retention characteristics; see bonded stationary phases.

Chromatogram display of separated components:

(i) column chromatography, chromatogram is obtained by plotting the detector signal against time to give a series of peaks each of which represents the distribution of an eluted component in the mobile phase;
(ii) planar chromatography, chromatogram is a visualised display of the separated spots from a TLC or paper chromatographic separation.

Coating efficiency CE, used as a measure of column quality for GC columns by comparing the theoretical minimum plate height (maximum column efficiency), H_{MIN}, with the plate height achieved in practice, H_{CALC}.

$$CE = \frac{H_{CALC}}{H_{MIN}} \times 100$$

Column chromatography mobile phase flows through a column containing the stationary phase material, either as a particulate packing or as a film on the inner wall (WCOT) of the column tubing.

Column efficiency N or N_{eff}; defined by the number of theoretical plates or equilibrium steps in a column and reflects the column's ability to separate a mixture of components with good resolution. Column efficiency depends on the number of equilibrium steps in a column, length L, or planar system, smallest equilibrium steps or plate heights (H) give highest efficiencies and values of N (see H, HETP):

N is the number of observed equilibrium steps calculated using t_R;
N_{eff} is the effective number of equilibrium steps calculated using t'_R

$$N = \frac{L}{H}\left(\frac{t_R}{4\sigma}\right)^2 = 16\left(\frac{t_R}{w_b}\right)^2 = 5.54\left(\frac{t_R}{w_h}\right)^2$$

$$N_{eff} = \left(\frac{t'_R}{4\sigma}\right)^2 = 16\left(\frac{t'_R}{w_b}\right)^2 = 5.54\left(\frac{t'_R}{w_h}\right)^2$$

Column oven thermostatically controlled enclosure for the chromatographic column, maintained to ±0.1°C, may be isothermal or temperature programmed.

Column packing stationary phase material in the form of small solid particles which may be coated with a liquid stationary phase. Particle size is 3–20 μm in HPLC, 50–500 μm (60–100 mesh) in GC.

Column switching use of two or more columns to effect a separation of a sample mixture. The mobile phase can be switched between the columns to direct specific fractions from the first column onto a second column, or switched to back-flush retained components off the first column.

Corrected retention time see adjusted retention time.

Counter-ion in ion pair chromatography, the ion of opposite charge to the component which forms a neutral ion pair that can be separated by reverse phase HPLC, also in ion chromatography the counter ion in solution that displaces the sample ion from the stationary phase.

Dead-time t_M, time taken for the mobile phase to pass through the column from injector to detector, also called the mobile phase hold-up time or void time.

Dead volume volume of mobile phase between the injection port and detector which does not contain stationary phase, causes excessive band broadening.

Degassing process to remove gases dissolved in an HPLC mobile phase by sparging with helium and/or using vacuum, to prevent microbubbles forming in the detector causing signal noise.

Detector response factors D_{RF}, the response of a detector varies for different molecules and is related to the property being measured, e.g. ionisation in

FIDs, UV absorption in HPLC ultraviolet detectors, and the molecular structure of the eluted components. D_{RF_x} is usually calculated with respect to a reference or internal standard:

$$D_{RF_x} = \frac{A_x}{A_{IS}} \times \frac{C_{IS}}{C_x}$$

where A is the peak area, x is the amount/concentration, x is the component, IS is the internal standard, D_{RF_x} is the detector response factor for component x.

Development separation procedure to obtain a chromatogram, also, visualisation of separated components in TLC, PC.

Diffusion coefficient D_M or D_S, movement of a molecule in solution, important in band broadening and is dependent on temperature, solvent viscosity and relative molar concentrations of solute and mobile phase (D_M) or stationary phase (D_S).

Diode array detector HPLC multichannel ultraviolet–visible detector consisting of a linear array of photodiodes onto which the full spectrum falls, i.e. 200–400 nm ultraviolet, 400–800 nm visible regions. Thus a complete spectrum can be obtained in less than 0.1 s, also several wavelengths can be monitored simultaneously to achieve maximum sensitivity for each component.

Displacement chromatography the separation process where a sample mixture is introduced at the start of the column and is then eluted with a mobile phase that is more strongly attracted to the stationary phase than the sample components. Individual components are therefore displaced by the mobile phase and by each other, in order of retention forces.

Distribution ratio, equilibrium constant K, the term describing the distribution of a solute between the mobile phase and stationary phase and reflects the retention forces. Also known as partition ratio K_P, for partition chromatography.

$$K = \frac{C_{SP}}{C_{MP}}$$

Effective theoretical plates N_{eff}, the effective or true number of separation steps in a column, takes into account the column dead time in the calculation of column efficiency (see column efficiency).

$$N_{eff} = \left(\frac{t'_R}{4\sigma}\right)^2 = 16\left(\frac{t'_R}{w_b}\right)^2 = 5.54\left(\frac{t'_R}{w_h}\right)^2$$

Efficiency see column efficiency.

Electron capture detector ECD; GC detector that uses a ^{63}Ni foil as a β-emitter, the output signal depends on the resulting electron flux between the foil and collector electrode, elution of compounds containing atoms with a high electron affinity causes a decrease in the flux and therefore in the signal, has high sensitivity (10^{-9} to 10^{-14} gs^{-1}) for halogens and poly-aromatic compounds.

Eluant an alternative term for the mobile phase.

Eluant strength refers to the polarity of the eluant or mobile phase and its ability to attract and retain the solutes or components being separated. An increase in eluant polarity reduces the distribution ratio K, and the elution time for a given component. The term is most frequently used in TLC and reverse phase HPLC, remember, 'like attracts like'.

Eluotropic series series of eluants or liquid mobile phases, arranged in order of their polar strengths used particularly in TLC, e.g. water > methanol > ethanol > butanol > chloroform > hexane.

Elution descriptive term for separated components passing through and eventually emerging from the column.

Elution chromatography components of a sample mixture in the mobile phase are separated by varying attractions for a stationary phase, the most common form of chromatography; see displacement and frontal chromatography.

Elution volume V_R, volume of mobile phase required to elute a component, that is, to transport a component through the chromatographic system.

$$V_R = t_R \times F_C$$

where F_C is the flow-rate of the mobile phase.

Equilibrium constant K, see distribution ratio.

Exclusion chromatography EC, separation of components in a mixture on the basis of molecular size. Usually a liquid mobile phase is used with a stationary phase consisting of a cross-linked polymer gel containing pores of a narrow size range. Components are retarded to varying degrees by the way the molecules partially fit into the pores, larger molecules are less retarded and elute first. EC is also known as size or steric exclusion chromatography.

Fast LC a form of HPLC using a low mobile phase pressure, short (5 cm) columns with 0.5 cm internal diameter and packed with 5–20 μm particles.

Flame ionisation detector FID, GC detector that employs a minute hydrogen air flame. Eluted components interact with the $OH^{\cdot}$ radicals in the flame by a series of steps to form positive ions, these ions move to a collector electrode to form the output signal, *S*. Response is proportional to the ability of the molecules to form ions in the flame, the detector response factor D_{RF}, and their concentration, C_{DET}, in the detector:

$$S \propto D_{RF} \times C_{DET}$$

Linear sensitivity range is 10^{-5} to 10^{-12} gs^{-1}, signal current is in picoamps, peak area is obtained by integration of the detector signal resulting from elution of a component band.

Flame photometric detector FPD, a selective GC detector for sulphur and phosphorus containing compounds. Separated components pass into a hydrogen-rich flame where they undergo a series of reactions to produce excited species HPO^* and S_2^*. The resulting atomic emission spectrum is monitored using narrow band pass filters (526 and 394 nm, respectively) and a photomultiplier detector, sensitivity is 10^{-5} to 10^{-11} gs^{-1}.

Flow-rate F_C, refers to the volumetric rate of flow of the mobile phase through the column, usually a constant flow-rate is maintained so retention times may be used to describe retention characteristics of each component.

Frontal chromatography frontal analysis, form of chromatography where the sample is continuously added to the column. Only the least retained component in a sample mixture is obtained in a pure form, the other components co-elute.

Fronting asymmetric peak shape where the front part of the peak tapers forward of the main peak often caused by overloading or poor injection techniques; see asymmetry.

Gas chromatography GC, employs a gaseous mobile phase, e.g. nitrogen or helium, and usually a liquid stationary phase, e.g. non-polar Apiezon (alkane grease), OV101 (polymethyl siloxane) or polar PEG 20M (polyethylene glycol). Separation is effected by competition between attraction of the components for the stationary phase and volatility at the column temperature being used, that is, retention forces versus partial vapour pressure in the mobile phase.

Gaussian peak a standard bell-shaped curve based on the mathematical description of a random distribution of, in the case of chromatography, component molecules in a band moving through the column.

Gel filtration or permeation chromatography GFC, uses an aqueous-based mobile phase and a dextran gel stationary phase with separation being effected by selective retention of solute molecules in the macropores of the cross-linked dextran; see exclusion chromatography.

Gel permeation chromatography GPC, see gel filtration.

Gradient elution refers to the progressive change of mobile phase composition (usually by increasing the proportion of organic solvent) in HPLC, that is, increasing the solvent strength and therefore progressively decreasing retention on the stationary phase and retention times.

Guard column short column containing the same stationary phase as the main column. It is located just before the main column to protect it by collecting strongly retained components in a sample mixture that would otherwise remain on the top of the main column.

H, HETP H is the height equivalent to a theoretical plate also called the equilibrium step height. It is a measure of column efficiency, H is approximately 0.5 mm in a GC capillary column and 0.01 mm in HPLC. $H = L/N$ where L is column length, N is number of theoretical plates in a column. H_{CALC} is the practical plate height of a column, H_{MIN} is the theoretical minimum plate height at optimum linear velocity and maximum column efficiency, and may be calculated in terms of the retention or capacity factor of a column; see N_{eff}, van Deemter equation, capacity factor and coating efficiency.

$$H_{MIN} = \frac{d_c}{2}\sqrt{\frac{(1 + 6k + 11k^2)}{3(1 + k)^2}}$$

High performance liquid chromatography HPLC, employs a liquid mobile phase and solid or bonded liquid stationary phase. High resolution and column efficiency is achieved by using 3–10 μm microporous or micropellicular column packing materials. In reverse phase chromatography an aqueous buffer containing varying amounts of a miscible organic solvent is used as the polar mobile phase, e.g. 1% acetic acid/methanol, 1% ammonium acetate/methanol and ODS as a non-polar stationary phase.

Hybrid techniques used to describe the interfacing of a secondary technique such as mass spectrometry or infrared spectrometry to GC or HPLC as an additional detector to help identify the separated components by using their mass or IR spectra, e.g. GC–MS, HPLC–MS.

Hydrophilic describes the 'water loving' or polar characteristics of molecules, used to describe the 'like attracts like' features of a mobile phase or stationary phase.

Hydrophobic describes the 'water hating' or non-polar characteristics of molecules, used to describe the 'like attracts like' properties of a mobile or stationary phase.

Injection port facility for introducing the sample mixture into the mobile phase. In GC the injector is maintained at least 50°C above the maximum column temperature to be used and the sample is introduced via a rubber septum using a microlitre syringe; for HPLC see multiport injection valves.

Integrator an electronic microprocessor based system for summing the detector signal arising from elution of a component. The detector signal is displayed as a Gaussian-like peak on the chromatogram, the area of which may be calculated by a trapezoidal integration technique, the peak maxima is also determined to give the retention time and peak height. Integrators are programmed to recognise a shoulder on the side of a peak and non-baseline valleys between peaks. Other parameters such as peak width and width at half height can be obtained.

Internal standard used for quantitative determinations with a known quantity of a reference standard added to *all* the solutions being analysed, the reason for this is to account for detector response factors and avoid the need for injection of accurate quantities of sample. The detector response factor is first calculated using a standard solution mixture, the amount of the analyte may then be calculated using the rearranged D_{RF_x} equation; see detector response factors.

$$C_x = \frac{A_x}{A_{\mathrm{IS}}} \times \frac{C_{\mathrm{IS}}}{D_{\mathrm{RF}_x}}$$

where A is the peak area, C is the amount/concentration, x is the component, IS is the internal standard, D_{RF_x} is the detector response factor for component x.

Ion exchange chromatography ions in a mM buffer solution (mobile phase) are carried through a column containing particles of a cation or anion exchange resin. The strength of the attraction of the ions for the resin counter-ions, e.g. SO_3^{2-} (cationic resin), N^+Me_3 (anionic resin), and the resulting exchange process retards the ions according to their ionic character. Conductivity detectors are usually employed.

Ion pair chromatography ions in a sample are paired to form a neutral 'ion pair' by addition of an ion pair reagent to the mobile phase, e.g. heptane sulphonic acid. The alkane chain gives non polar retention characteristics to the ion pair and therefore separation can be carried out using reverse phase HPLC.

Isocratic chromatography chromatographic separations, e.g. HPLC, using the same mobile phase solvent composition throughout the analysis.

Isothermal chromatography chromatographic separations using the same column temperature throughout the analysis, particularly in GC.

Isotherms plot of distribution ratios K, that is, a plot of concentration of the component in the stationary phase versus concentration in the mobile phase. Isotherms represent the relative attraction of a solute for the stationary and mobile phases, the plot is linear for a symmetrical Gaussian peak, non linear isotherms lead to unsymmetrical peaks, e.g. peak fronting and tailing. Langmuir isotherms describe the equilibrium process in adsorption chromatography, Nernst isotherms relate to partition chromatography.

Limit of detection see minimum detectable amount.

Limit of quantitation see method detection limit.

Linear elution chromatography used to describe the separation process in partition chromatography where a constant partition or distribution ratio is maintained for a given component throughout the separation. Thus the equilibrium that determines the partition of the solute between the mobile and stationary phases remains constant and the rate at which the solute moves through the column will be proportional to the time it spends in the mobile phase relative to the time spent retained on the stationary phase.

Linear range concentration range over which a detector has a constant and therefore linear response, D_{RF}, to a specific compound.

Linear velocity u, velocity of mobile phase through the column,

$$u = L/t_M$$

where L is the column length.

Liquid–liquid chromatography LLC, chromatography using liquid mobile and liquid stationary phases, e.g. reverse phase HPLC and TLC; see partition chromatography.

Liquid–solid chromatography LSC, chromatography using a liquid mobile phase and solid stationary phase, e.g. HPLC, TLC; see adsorption chromatography.

Loading the amount of stationary phase as coated onto the solid support material. In GC stationary phase loading is typically 1–10% (w/w). Loading also refers to the amount of sample mixture injected or loaded onto the column at the start of an analysis, overloading leads to 'fronting' of peaks and reduced column efficiency.

Macroporous resin stationary phase material formed from cross linked ion exchange resins that have pores of molecular size, 100–500 Å across, and large surface areas.

Method detection limit MDL, the minimum amount of an analyte that can be analysed by a given method with satisfactory precision, also referred to as limit of quantitation.

Microporous particles 3–10 μm particles used as a solid stationary phase or as the inert support for bonded stationary phases in HPLC, usually made from synthetic silica gel by a proprietary process involving the hydrolysis of $SiCl_4$ to form particles with a controlled pore size (*ca* 5–50 nm).

Minimum detectable amount MDA, minimum amount of an analyte that produces a measurable detector signal over and above the background noise, usually $>2 \times$ range of the signal noise, also referred to as limit of detection.

Mobile phase transports the solutes or components to be separated through a column or plate of stationary phase material. The solution properties of a liquid mobile phase compete with the retention forces of the stationary phase to determine the distribution ratio and hence elution time. In GC the gaseous mobile phase transports components in the vapour phase.

Multiport injection valve used in HPLC to introduce the sample onto the column, the valve has six ports, two connected to a sample loop and one each to the column, pump, syringe loading port and waste. The sample is first loaded into a sample loop, usually 5–20 μl capacity, with the valve in the bypass mode; rotation of the valve diverts the solvent through the sample loop whereupon the sample is swept onto the column. Sample size repeatability is better than 1%. The valves are also used for multicolumn switching.

Nitrogen phosphorus detector NPD, a selective GC detector in which nitrogen and phosphorous containing components are catalytically ionised on the surface of a hot rubidium or caesium bead in a reducing atmosphere. Sensitivity is similar to an FID, 10^{-5} to $10^{-12}\,gs^{-1}$.

Non-polar dispersion forces used to describe the retention of non-polar solutes on a non-polar liquid stationary phase. London's dispersion forces postulate an intermolecular induced dipole mechanism to account for attraction of a component onto a non-polar stationary phase.

Normal phase chromatography uses a polar stationary phase and a less polar or non-polar mobile phase, e.g. in TLC or HPLC with a silica gel stationary phase and hexane or methanol/chloroform/diethyl ether mobile phase.

Number of theoretical plates N, refers to the number of separation steps in a column, is used to describe the efficiency of a column for a given separation and is calculated using observed retention times; see column efficiency and effective theoretical plates.

$$N = \frac{L}{H} = \left(\frac{t_R}{4\sigma}\right)^2 = 16\left(\frac{t_R}{w_b}\right)^2 = 5.54\left(\frac{t_R}{w_h}\right)^2$$

Octadecylsilane, ODS a bonded stationary phase used in reverse phase chromatography (HPLC, TLC) consisting of a C_{18} alkane chain bonded to silica gel support particles through siloxane bonds to form a non polar stationary phase. C_2 and C_8 alkanes are also used; see bonded stationary phase.

On column injection the sample for analysis is introduced onto the end of the column packing to minimise dead volume and injector effects.

Optimum gas velocity, OGV average carrier gas velocity, u_{OPT}, in WCOT columns corresponding to H_{MIN} and maximum column efficiency.

Optimum practical gas velocity, OPGV average carrier gas velocity in WCOT columns that gives the best separation for a particular sample in the shortest analysis time whilst maintaining satisfactory resolution between peaks. OPGV $\approx$ 1.5–2.0 OGV.

Packed columns column chromatography where the stationary phase is in the form of uniform particles which are packed into a column, the particles may be a solid material, e.g. silica gel or alumina or particles coated with a liquid stationary phase.

Paper chromatography, PC a form of planar chromatography where the stationary phase consists of a sheet of special grade paper, the cellulose paper fibres form a matrix which act as the stationary phase and are able to retain a liquid stationary phase. The mobile phase solvent moves up the paper by capillary action carrying the solute molecules over the paper fibres, separation is affected by a mixture of partition and adsorption chromatography.

Partition chromatography chromatographic separations which involve a liquid stationary phase immobilised on an inert support material. The solutes are distributed between the mobile phase and stationary phases according to their respective partition coefficients.

Partition coefficient, K_P describes the distribution of a solute between a liquid or gaseous mobile phase and a liquid stationary phase on an inert support material. The solute molecules are partitioned between the two phases according to the retention forces of the stationary phase and the solvating properties of the mobile phase in LC or vapour pressure in GC. The equilibrium is temperature dependent; see distribution ratio:

$$K_P = \frac{C_{SP}}{C_{MP}}$$

where C is the concentration, SP is the stationary phase, MP is the mobile phase.

Partition retention forces intermolecular forces that result in attraction of solute molecules to the stationary phase. Polar retention forces include dipole–dipole attractions, van der Waal's forces and hydrogen bonding. Non-polar retention forces consist of London's dispersion forces arising from induced polarity in non-polar molecules, remember that 'like attracts like'.

Peak area area under a peak obtained from integration of the detector signal for a given component. Peak area is proportional to the number of molecules contained in an eluted band and hence may be used to calculate the proportion of each component in a mixture; see integration, detector response factors and internal standard.

Peak fronting an asymmetric peak with a gradual leading edge and a sharp trailing edge. Fronting is most frequently caused by overloading the column with sample when the stationary phase has insufficient capacity to maintain the equilibrium for the solute band, hence some of the solute is carried ahead of the band by the mobile phase.

Peak half height, $h_{0.5}$ used in assessing the width of a peak, $h_{0.5} = h/2$.

Peak height, h height of the peak at the peak maxima, used to obtain the retention time, the time from the start of the analysis, when the maximum concentration of solute is eluting from the column.

Peak tailing an asymmetric peak with a sharp leading edge and more gradual trailing edge resulting from a non-ideal isotherm. Tailing often occurs when some of the solute is retained to a higher degree than the majority of the solute and thus lags behind the rest of the peak.

Peak width, w_b width of a peak at the baseline, it is difficult to measure due to tailing of Gaussian peaks. Peak width at half height, w_h which corresponds to the inflexion point of the peak is a more accurate measurement, $w_b = 2w_h$. In terms of standard deviation $w_b = 4\sigma$, $w_h = 2.35\sigma$, where σ is the standard deviation of the Gaussian peak.

Peaks representation of the detector signal resulting from elution of component molecules obtained by continuously plotting the signal as the analysis proceeds. The distribution of molecules within an eluted band results in a Gaussian peak; see chromatogram.

Pellicular particles technical term for a synthetic HPLC column packing material consisting of microglass beads, 10–50 μm in diameter with a 2–3 μm surface film of active stationary phase material. These packings have only about 10% of the capacity of microporous materials but produce columns with greater separation efficiencies due to the very rapid equilibrium processes that occur with such regular particles and thin stationary phase films.

Phase ratio, β is used in GC to describe the thickness of the stationary phase, measured as the ratio of stationary phase volume, V_S, to mobile phase volume, V_M. Thicker stationary phase films give higher capacity factors and therefore longer retention times:

$$\beta = \frac{V_M}{V_S} = \frac{d_c}{4d_f}$$

where d_c is the column internal diameter, and d_f is the film thickness of the stationary phase.

Photoionisation detector, PID component molecules in the detector are ionised by photons from a high energy ultraviolet source, selectivity can be achieved by using a different source, e.g. 10.2 eV low energy krypton lamp for aromatics and alkenes, 11.7 eV argon lamp for alkanes, halogenated compounds as well as aromatics. Sensitivity is similar to a FID, 10^{-5} to 10^{-12} gs^{-1}.

Planar chromatography type of chromatography where the stationary phase is a flat film in the form of a thin layer of material coated onto a rectangular glass or inert plastic support plate (TLC) or as a paper sheet (PC). The mobile phase moves through the stationary phase by capillary action.

PLOT Porous layer open tubular column used in capillary GC, consists of a thin layer (1–5 μm) of solid stationary phase material, e.g. modified alumina, coated onto the inner wall of silica capillary tubing; see capillary GC.

Polar retention forces forces of attraction between a solute and the stationary phase due to dipole–dipole interactions, van der Waal's forces and hydrogen bonding.

Pre-columns short guard columns packed with the same stationary phase material as the analytical column but with a larger particle size to avoid reducing the overall efficiency of the system. Pre-columns are used to condition the mobile phase for the stationary phase and also to clean up the sample by removing any components that are strongly retained on the stationary phase.

Programmed chromatography analyses in which the factors which determine the partition or distribution equilibrium are varied, e.g. temperature programming in GC, solvent programming in HPLC.

Pyrolysis gas chromatography non-volatile materials, e.g. plastics, polymers, paint flakes, may be analysed by rapid heating to a high temperature (450–800°C in less than 5 s) so that pyrolysis and fragmentation of the molecular structure occurs. The resulting gaseous mixture of smaller molecular weight compounds is separated to produce a pyrogram that is a fingerprint for the sample under investigation.

Relative retention ratio see separation factor.

Relative retention time retention time corrected to allow for time the mobile phase takes to pass through the column; see retention volume.

$$t'_R = t_R - t_M$$

Relative retention volume V'_M retention volume corrected to take into account the volume of mobile phase in the system; see retention volume.

$$V'_R = V_R - V_M$$

Resolution a measure of the separation of two adjacent peaks (A and B), in the chromatogram. Resolution may be assessed by the valley between the peaks h_V or by peak width w_b; in quantitative analyses a baseline separation should be obtained. The separation of overlapping peaks may be measured as follows:

(i) 10% valley; if the peaks overlap then less than a 10% valley (<10% of the mean peak height) between the peaks should be specified, this corresponds to approximately a 2.5% peak overlap, 95% peak purity.

$$h_V \leq \frac{1}{10}\left(\frac{h_A + h_B}{2}\right)$$

(ii) resolution R_S calculated from a comparison of the separation as measured by the difference in retention time and peak width:

$$R_S = \frac{(t_{RB} - t_{RA})}{\frac{1}{2}(w_{bA} + w_{bB})} = \frac{\Delta t_R}{(w_{hA} + w_{hB})} \approx \frac{\Delta t_R}{2w_{hB}}$$

$R_S = 1.2$, peak overlap $< 1.0\%$
$R_S = 1.0$, peak overlap $\approx 2.3\%$, peak separation $= 4\sigma$
$R_S = 0.75$, peak overlap $\approx 6.5\%$, peak separation $= 3\sigma$
$R_S = 0.5$, peak overlap $\approx 16\%$, peak separation $= 2\sigma$
Resolution may also be defined in terms of efficiency, retention properties and separation capabilities of a column by the following equation derived from the theoretical work of J.H. Purnell.

$$R_S = \frac{\sqrt{N}}{4}\left(\frac{\alpha - 1}{\alpha}\right)\left(\frac{\bar{k}}{\bar{k} + 1}\right)$$

Retention factor, k IUPAC 1993 term for capacity factor. Capacity of a stationary phase to retain a component, measured as the ratio of corrected retention time (t'_R) and column dead time (t_M),

$$k = \frac{(t_R - t_M)}{t_M} = \frac{t'_R}{t_M}$$

Retention forces see polar, non polar and adsorption retention forces.

Retention indices, RI method of expressing relative retention characteristics of a wide range of components, the most well known being the Kovat's index. This compares log retention time of an unknown compound with log retention times of the *n*-alkanes eluted before and after.

$$\mathrm{RI} = 100\left(Z + \frac{\log t'_{RA} - \log t'_{RZ}}{\log t'_{RZ+1} - \log t'_{RZ}}\right)$$

where t'_{RA} is the corrected retention time of compound A, t'_{RZ}, t'_{RZ+1} is the corrected retention time of *n*-alkanes containing Z and $Z + 1$ carbon atoms.

Retention time, t_R time interval between introduction of a sample and elution. Corrected or relative retention time t'_R is a more accurate measure of delay due to retention forces and takes into account the system dead time, the time the mobile phase takes to pass through the system, t_M:

$$t'_R = t_R - t_M$$

Retention volume, V_R volume of mobile phase required to elute a component. Corrected or relative retention volume V'_R is a more accurate measure of the delay due to the retention forces by allowing for the volume of mobile phase in the system V_M. Usually a constant mobile phase flow-rate, F_C is maintained hence the more convenient term retention time is used.

$$V'_R = V_R - V_M \quad \text{and} \quad V_R = t_R \times F_C$$

Reverse phase chromatography RPC uses a mobile phase which is more polar than the stationary phase. Sample components are attracted to a hydrophobic stationary phase, e.g. ODS, according to the competing partition retention forces. The rate at which the components are carried through the chromatographic system depends on the solvent strength of the mobile phase for the individual components. The mobile phase consists of an aqueous buffer containing varying amounts of miscible organic solvents, RPC is used in HPLC and TLC.

R_f value retardation factor in TLC and PC as a measure of the retention of a component on the stationary phase as it is carried along in the mobile phase:

$$R_f = \frac{\text{distance moved by component, measured to most dense part of spot}}{\text{distance moved by solvent, measured to solvent front}}$$

A standard (R_{f_s}) is frequently analysed alongside a mixture to act as a reference, all R_f values are then calculated with reference to the standard to obtain R'_f:

$$R'_f = \frac{R_f}{R_{f_s}}$$

Sample injection splitter used in capillary GC to permit small samples to be introduced onto the column, e.g. for a split ratio of 1:100 and a 1 μl injection 0.01 μl would enter the column.

Selectivity see separation factor.

Separation factor, α used to describe the relative retention of two components, A and B on a given stationary phase, also known as the relative retention ratio;

$$\alpha = \frac{k_B}{k_A}$$

where k_A and k_B are the respective retention ratios (capacity factors) for A and B.

Separation number, SN used as a measure of efficiency in capillary columns. SN is obtained by comparing retention times with peak widths at half height for two *n*-alkanes containing Z and $Z + 1$ carbon atoms giving an indication of the number of peaks that could be contained between the alkanes with approximately baseline resolution ($R_S \approx 1.0$–1.2).

$$SN = \frac{(t_{RZ+1} - t_{RZ})}{(w_{hZ+1} + w_{hZ})}$$

Septum disc of silicone rubber or similar material that provides a barrier between the atmosphere and the carrier gas, allows syringe injections and is self-sealing. The under surface is purged by carrier gas to prevent a build-up of material and contamination of injections.

Sigma, σ standard deviation is a measure of the spread of the measurements or molecules that comprise a Gaussian distribution; see peak area and Gaussian peak.

Silica gel an amorphous material of varying pore size, consisting of siloxane –Si–O–Si– and silanol –Si–OH groups which confer the polar characteristics and chemical reactivity. It is most frequently used in TLC and HPLC as an adsorbent for normal phase chromatography or as the support particles for bonded phases in reverse phase chromatography; it is also used as a varying pore size packing in size exclusion chromatography.

Size exclusion, steric exclusion chromatography, SEC components in a sample are separated according to their molecular size by selective retention on a stationary phase having pores of varying sizes; see gel filtration.

Solid support for stationary phases robust inert material used as support for a liquid mobile phase, has a large surface area, common materials are silica gel for HPLC, processed 'brick' and celite particles for GC.

Solute an alternative term for a sample component that is present in the mobile phase or attracted on to the stationary phase as it moves through the chromatographic system.

Solvent programming changing the mobile phase composition as the chromatographic separation is proceeding. It is frequently used in reverse phase HPLC starting with a 'weak' solvent and progressively increasing the proportion of organic solvent to increase the solvent strength and therefore increase the rate at which components strongly attracted to the stationary phase move through the column.

Solvent strength describes the composition of a liquid mobile phase in terms of its ability to elute components in normal and reverse phase chromatography; see elutropic series.

Standard addition quantitative method of analysis where a sample is analysed for the analyte of interest (result A_1). A known or standard quantity (x mg litre^{-1}) of the same analyte is then added to the sample, calculated allowing for any dilution that may occur. The analysis is repeated (result A_2), the increase in peak area or calculated result is due to the added amount, hence the amount of the analyte in the original sample can be calculated:

peak area corresponding to added amount: $x = A_2 - A_1$

therefore amount, C of analyte in the mixture is: $C = \frac{(x \times A_1)}{(A_2 - A_1)}$

Stationary phase solid material or liquid immobilised on an inert support, which attracts components in the mobile phase according to characteristic retention forces and thus retards their progress through the column or plate; see mobile phase.

Stepwise elution change in composition of the mobile phase in HPLC at preset intervals, an alternative to solvent programming; see gradient elution.

Supercritical fluid chromatography, SFC a super critical fluid is produced when a gas, e.g. carbon dioxide is maintained above its critical pressure (73 atm) and temperature (31°C). The super critical fluid is used as the mobile phase with GC like capillary columns or HPLC reverse phase columns and a flame ionisation or flame photometric detector.

Suppressor column in ion chromatography; a short column located after the analytical column and before the detector. Its purpose is to convert the strong electrolyte ions of the eluant to a weak electrolyte thus reducing the background signal enabling sub-mg litre^{-1} concentrations of sample ions to be detected, for example, separation of anions using an anion ion exchange resin uses mM concentrations of sodium carbonate in the eluant. A suppressor in the form of a strong cation ion exchanger resin is used as a source of H^+ ions to exchange the Na^+ ions

$$2Na^+(\text{eluant}) + CO_3^{2-}(\text{eluant}) + 2H^+(\text{resin})$$
$$\rightarrow 2Na^+(\text{resin}) + CO_3^{2-}(\text{eluant}) + 2H^+(\text{eluant})$$

Temperature programming in GC, analysis of a mixture where the column temperature is progressively increased, thus increasing the volatility of the less volatile components and decreasing their retention on the stationary phase. Temperature programming enables samples containing components with a wide boiling point range to be separated and avoids unacceptably broad peaks due to long analysis times.

Thermal conductivity detector, TCD non-destructive GC detector consisting of a pair of heated platinum filaments (heated by a small low voltage current), one filament is in the sample carrier gas stream, the other in pure carrier gas acting as a reference. When a sample is eluted the resistance of the sample filament changes and the difference in current passing through the filaments is processed to form the detector signal. The preferred carrier gas is helium; linear sensitivity range is 10^{-2} to 10^{-6} gs^{-1}.

Thin layer chromatography, TLC a form of planar chromatography having a coating of a solid stationary phase, e.g. silica gel or alumina, on a plate of glass or inert plastic. The liquid mobile phase moves up the plate by capillary action, reverse phase TLC plates are available where a liquid stationary phase is immobilised on the solid support. The polarity of the mobile phase is carefully chosen to obtain the best separation; see elutropic series.

Trennzahl number, TZ see separation number.

Tswett, Mikhail Semenovich (1872–1919) Russian biochemist, 'father' of chromatography who in 1906 first used the term 'chromatography' (colour writing) to describe his work on the separation of plant pigments on a column of chalk using a petroleum ether eluant.

UV detector HPLC detector based on an ultraviolet–visible spectrophotometer using microsample flow cells (10 μl). Response to components depends on their absorption spectrum and absorptivity coefficient as defined by the Beer–Lambert law. Detector wavelength is set to give maximum sensitivity ideally at λ_{MAX}, however an optimum wavelength, λ_{OPT}, at which all the components have a satisfactory absorbance, may be used.

van Deemter equation accounts for the band broadening processes using a set of terms to explain the phenomena that contribute to bands of molecules spreading out as they pass through a column, expressed as H, the size of an equilibrium step or 'theoretical plate' (HETP);

$$H = A + \frac{B}{u} + Cu$$

where u is the mean flow-rate of the mobile phase. The A term is concerned with eddy diffusion, unequal pathways through the stationary phase particles, the B term accounts for the random movement or diffusion of component molecules in the mobile phase and to a lesser extent the stationary phase and the C term accounts for mass transfer processes occurring at the mobile–stationary phase boundary.

Void volume see dead volume.

WCOT Wall coated open tubular capillary columns for GC consist of a 0.1–5 μm film of liquid stationary phase coated or bonded on to the inner wall of a silica glass capillary column bore 0.1–0.7 mm.

Zone see band.

Appendix 2 Table of chromatography symbols

A	Peak area
A	Multipath term in van Deemter equation
A_S	Asymmetry
B	Longitudinal diffusion term in van Deemter equation
bp	Boiling point
C	Mass transfer term in van Deemter equation
d_c	Column internal diameter
d_f	Film thickness of stationary phase in WCOT
d_m	Density of mobile phase
d_p	Particle size of stationary phase
d_s	Density of stationary phase
D_{RF}	Detector response factor
D_M	Diffusion rate in the mobile phase
D_S	Diffusion rate in the stationary phase
F_C	Flow-rate of mobile phase through the column
F_S	Flow-rate through split vent of split–splitless GC injector
H	HETP, height equivalent to a theoretical plate, equilibrium step height
H_{CALC}	Observed value of H
H_{MIN}	Theoretical minimum value of H
h	Peak height
$h_{0.5}$	Peak half height
$h_{2.35\sigma}$	Peak half height when peak width is 2.35σ
h_R	Reduced plate height (IUPAC, h)
id	Internal diameter
IS	Internal standard
K	Distribution ratio equilibrium constant
K_{ADS}	Distribution ratio for adsorption chromatography
K_{IC}	Distribution ratio for ion chromatography
K_{IP}	Distribution ratio for ion pair chromatography
K_P	Distribution ratio for partition chromatography
k	Capacity factor, retention factor
L	Column length
MDA	Minimum detectable amount
MDL	Method detection limit
MP	Mobile phase
N	Number of theoretical plates, column efficiency
N_{eff}	Number of effective theoretical plates, effective column efficiency

od	Outside diameter
OPGV	Optimum practical gas velocity
P	Pressure
R	Gas constant
R_f	Retardation factor in TLC, PC
R'_f	Corrected retardation factor
RI	Retention index, Kovat's RI
R_S	Resolution between adjacent peaks
S	Signal output from detector
SN	Separation number
SP	Stationary phase
t	Time
t_M	Column dead time, time for unretained peak to pass through column
t_N	Net retention time
t_0	Replaced by t_M
t_R	Retention time
t'_R	Corrected retention time
T	Temperature
TZ	Trennzahl number (SN)
u	Linear velocity of mobile phase
u_{OPT}	Optimum linear velocity of mobile phase
u_{PRACT}	Practical linear velocity of mobile phase
v	Reduced velocity of the mobile phase
V_g	Specific retention volume
V_M	System dead volume, volume of mobile phase in the column
V_N	Net retention volume
V_R	Retention volume
V'_R	Corrected retention volume
V_S	Volume of stationary phase
w_b	Width of peak at base
w_h	Peak width at half peak height
Z	Number of carbon atoms in *n*-alkane
α	Separation factor
β	Phase ratio
ΔH	Enthalpy of adsorption
η	Viscosity
λ	Stationary phase packing factor
λ_{MAX}	Wavelength at which maximum absorbance occurs
λ_{OPT}	Optimum wavelength used in analysis (HPLC)
σ	Standard deviation
#	Sieve mesh size

Appendix 3 Abbreviations

AC	Affinity chromatography
AAS	Atomic absorption spectroscopy
AES	Atomic emission spectroscopy
ADC	Analogue to digital converter
ATD	Automated thermal desorption
CCGC	Capillary column gas chromatography
CE	Capillary electrophoresis
COSHH	Control of substances hazardous to health
FC	Flash chromatography
FTIR	Fourier transform infra red spectrometry
GC	Gas chromatography
GPC	Gel permeation chromatography
HPIC	High performance ion chromatography
HPLC	High performance liquid chromatography
IC	Ion chromatography
ICP	Inductively coupled plasma
IEC	Ion exchange chromatography
IR	Infra red spectroscopy
LC	Liquid chromatography
LIMS	Laboratory information management systems
MS	Mass spectrometry
PC	Paper chromatography
PC	Personal computer
SFC	Supercritical fluid chromatography
TLC	Thin layer chromatography
UV	Ultra violet spectroscopy
WCOT	Wall coated open tubular columns

Index